NUMERICAL SIMULATION OF FLUID MOTION

NUMERICAL SIMULATION OF FLUID MOTION

Proceedings of an International Conference
on the Numerical Simulation of Fluid Dynamic Systems
held at Monash University, Melbourne, 1976

Edited by

JOHN NOYE

Associate Dean,
Faculty of Mathematical Sciences,
The University of Adelaide,
South Australia

1978

NORTH-HOLLAND PUBLISHING COMPANY – AMSTERDAM • NEW YORK • OXFORD

ISBN: 0 444 85035 x

Publishers

NORTH-HOLLAND PUBLISHING COMPANY
AMSTERDAM • NEW YORK • OXFORD

Sole distributors for the U.S.A. and Canada
ELSEVIER NORTH-HOLLAND, INC.
52 VANDERBILT AVENUE,
NEW YORK, N.Y. 10017

PRINTED IN THE NETHERLANDS

PREFACE

This book is the outcome of the 1976 International Conference on Numerical Simulation of Fluid Dynamic Systems held at Monash University, Melbourne, Australia. The aim of the conference was first, to act as a medium by which mathematicians could describe the latest developments in the various fields of finite difference, finite element, spectral and other numerical methods, and second, to give an opportunity to engineers, physicists, oceanographers and meteorologists to report on their experiences in applying these methods to fluid flow problems.

This volume is more than a conference proceedings. Seventeen of the papers presented at the conference were chosen for inclusion and five others were produced afterwards in order to fulfil the aims of the conference. The additional articles filled gaps which would have occurred because some lecturers did not wish to contribute a paper to the proceedings. Articles, such as *An Introduction to Finite Difference Methods*, and *Galerkin Techniques: An Introduction*, were commissioned for this purpose.

Mention must be made of those who gave of their time to present lectures at the conference but who are not represented by a paper in the proceedings. They include Professor R. I. Tanner of Sydney University, who spoke on finite element techniques, Associate Professor G. de Vahl Davis of the University of New South Wales, who lectured on finite difference methods, and Dr. W. Bourke of the Australian National Meteorological Research Centre who gave a session on spectral methods. Others who presented papers not included in this book were Mr. O'Connor of the Monash University, Drs. Bharmal and Huilgol of the Flinders University of South Australia, and Dr. Antonia and Mr. Wood of the University of Newcastle.

A special mention must be made of the work done by Dr. R. A. Pearson who acted as secretary for the conference and who did much of the organisation. Thanks are also due to the two sponsors of the conference, the Commonwealth Scientific and Industrial Research Organisation of Australia and the Monash University of Melbourne.

Finally, my personal thanks go to Dr. E. Fredriksson of the North-Holland Publishing Company for arranging the printing of this book, thereby making its contents available to other scientists throughout the world.

John Noye,
The University of Adelaide,

CONTENTS

PART 2: APPLICATIONS

PART 1

NUMERICAL METHODS

Numerical Simulation of Fluid Motion, ed. John Noye
pp. 1-112

AN INTRODUCTION TO FINITE DIFFERENCE TECHNIQUES

B. J. Noye
University of Adelaide
South Australia

SUMMARY

Some of the more widely used finite difference methods available for solving the partial differential equations which govern fluid flow are described. In particular, both explicit and implicit methods of solving the transport equation are detailed. Ways of overcoming the more common problems encountered in this field of numerical simulation are also discussed. They include topics such as consistency of the finite difference method, stability of the finite difference equation and convergence of the finite difference solution.

TABLE OF CONTENTS

1. INTRODUCTION

1.1 General Considerations

In 1923 Richardson published the first application of finite difference methods to a problem in fluid dynamics, viz. the problem of weather prediction. The method he developed failed to satisfy what is known today as the stability condition, but the effect of the instability was not evident because his hand calculations progressed only through a small number of time steps. Today, more than one-half a century later, many scientists setting up a numerical model for the first time are perplexed by the development of oscillations which rapidly become unbounded in an otherwise smooth solution as their calculations progress step by step through time or space. The development of these oscillations are characteristic of an unstable finite difference method.

This article describes some of the finite difference methods available for solving partial differential equations, with particular application to those equations which model heat and mass transfer in fluids. It also discusses the more common difficulties encountered by the mathematician, physicist or engineer when he first enters this field of numerical simulation.

Section 1.2 introduces the transport equation which is the example used to illustrate most of the various finite difference techniques in the following sections. For particular initial and boundary conditions analytic solutions can be found for this particular equation. These solutions are very useful for comparison, in order to get an idea of the accuracy of any numerical method; that is, to determine whether the numerical solution converges to the solution of the original partial differential equation as the finite steps involved become smaller and smaller.

However, since numerical methods are employed only to solve problems to which an analytic solution is either very difficult or impossible to obtain, even with greatly simplified initial and boundary conditions, it is useful to know of methods which prove convergence and stability in these cases. Unfortunately, such methods have not been developed for most of the more complex (generally non-linear) problems and *ad hoc* approaches must be employed in which numerical methods become an experimental tool.

In Section 2, the various finite difference approximations to derivatives at a point in the solution space are described, and it is shown how to apply these approximations to the one-dimensional diffusion equation (which is a typical example of parabolic type partial differential equations) to obtain a very simple numerical solution of this equation.

In Section 3 the concepts of convergence, consistency and stability are considered in some detail, while in Section 4 some other finite difference methods of solving the one-dimensional diffusion equation are developed. Section 5 deals with methods of solving the transport equation in one or more space dimensions. In particular, ways of dealing with the boundary conditions on irregular boundaries in two and three space dimensions are described. Section 6 considers methods of improving the accuracy of solutions by varying the grid spacing, transforming the coordinate system used, and extrapolating the approximations found for various grid spacings to obtain the limiting value for a zero grid spacing. Section 7 summarises the various techniques considered.

It must be emphasised that the theoretical foundation of finite difference methods has not kept pace with their application. It is generally impossible to obtain a theoretical estimate of the accuracy of a numerical solution of a problem in which the non-linear terms in the equations governing the fluid flow are not small. Users of finite difference techniques must therefore apply every method available, such as grid refinement, comparison of the numerical solution with analytical solutions to simplified versions of the problem, and comparison with experimental results, in order to verify the quality of their results.

Readers who wish to use the methods described in this article should also study the original papers referenced and they should read more recent literature which discuss these methods. Publications such as Mathematics of Computation, Journal of Computational Physics, Computers and Fluids, Computer Methods in Applied Mechanics and Engineering, Numerical Methods in Engineering and the Journal of Fluid Mechanics, regularly include articles of relevance to finite difference techniques.

1.2 The Transport Equation

The behaviour of a fluid undergoing mass, vorticity or convective heat transfer, is described by a set of partial differential equations which are mathematical formulations of laws of conservation of mass, momentum and either concentration, or energy.

For instance, if the fluid is incompressible and has constant thermal conductivity k, the *heat equation*, a mathematical formulation of the law of conservation of thermal energy, is

$$\rho C_p \frac{DT}{Dt} = k\nabla^2 T, \tag{1.2.1}$$

where ρ is the density, C_p is the specific heat of the fluid at constant pressure, D/Dt is the operation of differentiation following the motion of the fluid, and T is its temperature. The viscous dissipation term, which is usually small, is omitted from Equation (1.2.1). In Cartesian coordinates this can be written

$$\frac{\partial T}{\partial t} + u\frac{\partial T}{\partial x} + v\frac{\partial T}{\partial y} + w\frac{\partial T}{\partial z} = \alpha\left(\frac{\partial^2 T}{\partial x^2} + \frac{\partial^2 T}{\partial y^2} + \frac{\partial^2 T}{\partial z^2}\right) . \tag{1.2.1a}$$

where $\alpha = k/\rho C_p$.

For fluids at rest, and for solids, $u=v=w=0$ and Equation (1.2.1) reduces to the *diffusion (conduction) equation*

$$\frac{\partial T}{\partial t} = \alpha\left(\frac{\partial^2 T}{\partial x^2} + \frac{\partial^2 T}{\partial y^2} + \frac{\partial^2 T}{\partial z^2}\right) . \tag{1.2.1b}$$

Equation (1.2.1a) is called the transport equation since it involves transfer of heat due to the motion of the fluid (convection) and due to diffusion (conduction). The three terms $u\partial T/\partial x$, $v\partial T/\partial y$ and $w\partial T/\partial x$ are therefore called convective terms, and the three terms $\alpha\partial^2 T/\partial x^2$, $\alpha\partial^2 T/\partial y^2$ and $\alpha\partial^2 T/\partial z^2$ are usually called diffusive terms.

Similar cylindrical and spherical forms of Equation (1.2.1) exist, but for the purpose of describing methods for their solution only the rectangular Cartesian forms will be considered.

In general, the transport equation

$$\frac{\partial \tau}{\partial t} + u\frac{\partial \tau}{\partial x} + v\frac{\partial \tau}{\partial y} + w\frac{\partial \tau}{\partial z} = \alpha\left(\frac{\partial^2 \tau}{\partial x^2} + \frac{\partial^2 \tau}{\partial y^2} + \frac{\partial^2 \tau}{\partial z^2}\right) \tag{1.2.2}$$

also describes the transfer of dissolved material or vorticity, as well as heat, to mention just three examples. In heat transfer problems τ is the temperature, in material transfer problems τ is the concentration and for vorticity transfer τ represents the vorticity.

Equation (1.2.2) is used in its various forms, such as:

the one-dimensional diffusion equation

$$\frac{\partial \tau}{\partial t} = \alpha \frac{\partial^2 \tau}{\partial x^2} , \tag{1.2.2a}$$

the one-dimensional convection equation

$$\frac{\partial \tau}{\partial t} + u\frac{\partial \tau}{\partial x} = 0 , \tag{1.2.2b}$$

the one-dimensional steady state transport equation

$$u\frac{\partial \tau}{\partial x} = \alpha \frac{\partial^2 \tau}{\partial x^2} , \tag{1.2.2c}$$

and so on. These forms will be used to illustrate the finite difference techniques described in later sections. For particular initial and boundary value conditions, analytic solutions (usually in terms of infinite series) can be found for these equations. The accuracy of the finite difference methods developed in this article can be determined by comparison with these exact solutions.

In the examples which follow, the coordinates will be non-dimensionalised so that the boundaries in the x-direction will be located at $x = 0$, $x = 1$, the time will be non-dimensionalised so that the dominant period in the solution τ will become unity, and the speed of the fluid flow will be non-dimensionalised accordingly.

1.3 Non-Dimensionalisation of the Transport Equation

The transport equation (1.2.2) can have the space and time variables non-dimensionalised in the following way.

Let L be a typical length, such as the distance between boundaries in the x-direction; for instance, the space region R over which (1.2.2) is to be solved may be defined by $0 \leqslant x \leqslant L$, $0 \leqslant y \leqslant B$, $0 \leqslant z \leqslant D$. Replace the x,y,z variables by the dimensionless space variables

$$\left.\begin{aligned} \bar{x} &= x/L, \\ \bar{y} &= y/L, \\ \bar{z} &= z/L, \end{aligned}\right\} \tag{1.3.1a}$$

so that the region R is now defined by $0 \leqslant \bar{x} \leqslant 1$, $0 \leqslant \bar{y} \leqslant \bar{B}$, $0 \leqslant \bar{z} \leqslant \bar{D}$ where $\bar{B} = B/L$, $\bar{D} = D/L$.

Let T_0 be a typical time, such as the period of an oscillation in the dependent variable τ . Replace the variable t by

$$\bar{t} = t/T_0 \tag{1.3.1b}$$

so that in terms of the dimensionless time variable $\bar{t}$ the period of the oscillation is 1.

Substitution of the new independent variables $\bar{x},\bar{y},\bar{z}$ and $\bar{t}$ defined by Equations (1.3.1) into the transport equation gives

$$\frac{\partial\tau}{\partial\bar{t}} + \frac{uT_0}{L}\frac{\partial\tau}{\partial\bar{x}} + \frac{vT_0}{L}\frac{\partial\tau}{\partial\bar{y}} + \frac{wT_0}{L}\frac{\partial\tau}{\partial\bar{z}} = \frac{\alpha T_0}{L^2}\left\{\frac{\partial^2\tau}{\partial\bar{x}^2} + \frac{\partial^2\tau}{\partial\bar{y}^2} + \frac{\partial^2\tau}{\partial\bar{z}^2}\right\} \qquad (1.3.2)$$

If τ_0 is some typical value of the dependent variable τ, this may also be non-dimensionalised by means of the substitution

$$\bar{\tau} = \tau/\tau_0 \quad . \qquad (1.3.3)$$

Substitution of (1.3.3) into Equation (1.3.2) yields

$$\frac{\partial\bar{\tau}}{\partial\bar{t}} + \bar{u}\frac{\partial\bar{\tau}}{\partial\bar{x}} + \bar{v}\frac{\partial\bar{\tau}}{\partial\bar{y}} + \bar{w}\frac{\partial\bar{\tau}}{\partial\bar{z}} = \alpha\left\{\frac{\partial^2\bar{\tau}}{\partial\bar{x}^2} + \frac{\partial^2\bar{\tau}}{\partial\bar{y}^2} + \frac{\partial^2\bar{\tau}}{\partial\bar{z}^2}\right\} \qquad (1.3.4)$$

where

$$\left.\begin{aligned} \bar{u} &= uT_0/L \\ \bar{v} &= vT_0/L \\ \bar{w} &= wT_0/L \\ \bar{\alpha} &= \alpha T_0/L \end{aligned}\right\} \qquad (1.3.5)$$

Dropping the over-bars with the exception of the dependent variable τ, we now have the non-dimensional form of the transport equation in the same form as Equation (1.2.2), now to be solved on the space region R defined by

$$0 \leqslant x \leqslant 1, \quad 0 \leqslant y \leqslant B, \quad 0 \leqslant z \leqslant D .$$

2. FINITE DIFFERENCE APPROXIMATIONS

2.1 The One-Dimensional Diffusion Equation

Many of the concepts associated with the numerical solution of partial differential equations by finite difference methods can be illustrated by considering the one-dimensional diffusion equation.

$$\frac{\partial\bar{\tau}}{\partial t} = \alpha\,\frac{\partial^2\bar{\tau}}{\partial x^2} \quad . \qquad (2.1.1)$$

assumed to have been suitably non-dimensionalised so that $\bar{\tau} \geqslant 0$ and $0 \leqslant x \leqslant 1$. For instance, the dependent variable $\bar{\tau}(x,t)$ may represent the temperature at time t at position x along a thin insulated rod of unit length with ends located at $x = 0$ and $x = 1$, as in Figure 2.1. If the rod is composed of homogeneous material then α, the coefficient of diffusion, is constant.

Certain initial and boundary conditions are required before Equation (2.1.1) can be solved, for example

$$\left.\begin{aligned} \bar{\tau}(x,0) &= f(x) \quad , & 0 \leqslant x \leqslant 1 \\ \bar{\tau}(0,t) &= g_0(t) \quad , & t \geqslant 0 \\ \bar{\tau}(1,t) &= g_1(t) \quad , & t \geqslant 0 \end{aligned}\right\} \quad , \qquad (2.1.2)$$

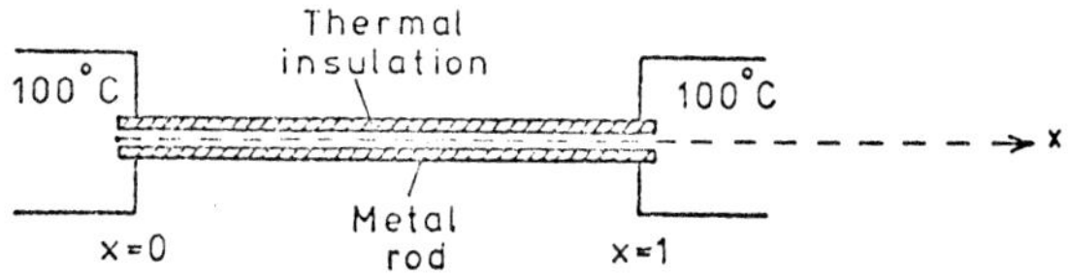

Figure 2.1: The insulated homogeneous metal rod.

Given the description of $f(x)$, $g_0(t)$ and $g_1(t)$, the solution $\bar{\tau}(x,t)$ is to be determined over the semi-infinite rectangle in x-t space defined by $0 \leqslant x \leqslant 1$, $t \geqslant 0$ (see Figure 2.2). This region, on which the solution of the given differential equation (2.1.1) is to be found subject to the initial and boundary conditions (2.1.2), is called the solution region.

For example, a thin insulated rod initially at a temperature of 0°C suddenly has the temperature at each end raised to 100°C and held at that value for all $t > 0$. Then the initial and boundary conditions (2.1.2) are

$$\left.\begin{aligned} \bar{\tau}(x,0) &= 0 \;, & 0 < x < 1 \\ \bar{\tau}(0,t) &= 100 \;, & t > 0 \\ \bar{\tau}(1,t) &= 100 \;, & t > 0 \end{aligned}\right\} . \qquad (2.1.3)$$

A difficulty occurs when the initial temperatures at $x = 0$ and $x = 1$, namely $\bar{\tau}(0,0)$ and $\bar{\tau}(1,0)$, are to be prescribed. In practice the temperature at $x = 0$ and $x = 1$ cannot be altered instantaneously, but we want

$$\bar{\tau}(0,0-) = 0 \text{ while } \bar{\tau}(0,0+) = 100$$
$$\bar{\tau}(1,0-) = 0 \text{ while } \bar{\tau}(1,0+) = 100.$$

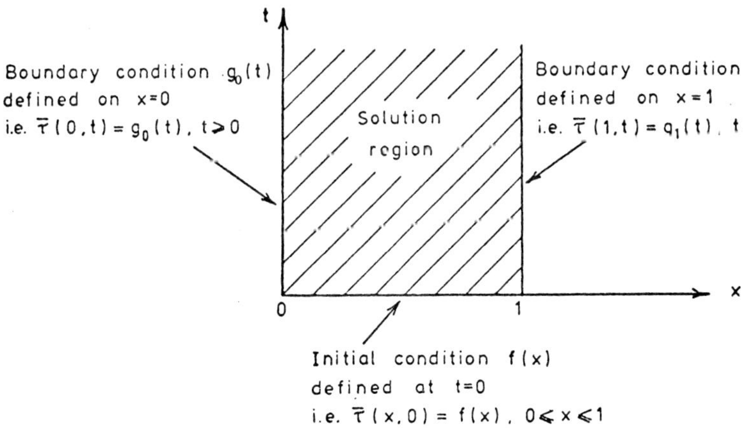

Figure 2.2: The solution region in x-t space for the one-dimensional diffusion equation.

The simplest way of defining $\bar{\tau}(0,0)$ and $\bar{\tau}(1,0)$ so the resulting problem is tractable mathematically is to give $\bar{\tau}(0,0)$ and $\bar{\tau}(1,0)$ the values of 100.

Equation (2.1.1) with initial and boundary conditions as now defined can be solved analytically by substituting $T = 100 - \bar{\tau}$ in (2.1.1) and (2.1.3), then applying the method of separation of variables (see Kreyszig, 1972, p.430) giving the solution

$$\bar{\tau}(x,t) = 100 - \sum_{m=1}^{\infty} \frac{400}{(2m-1)\pi} \sin\{(2m-1)\pi x\} \exp\{-\alpha(2m-1)^2\pi^2 t\} \quad , \quad (2.1.4)$$

which permits the value of the temperature τ at any point in the solution region $0 \leqslant x \leqslant 1$, $t \geqslant 0$ to be calculated. For a given x and t, this requires the summation of the series (2.1.4) on a computer. This is done by adding more and more terms of the infinite series until successive partial sums remain constant with the required accuracy.

It should be noted that the method of separation of variables is restricted to the solution of linear partial differential equations with constant coefficients, and boundary values which are equivalent to reflection and periodicity conditions.

An alternative method of solving the given partial differential equation is to use finite difference methods which are not restricted by such criteria.

2.2 The Finite Difference Grid

The solution region in x-t space is covered by a rectangular grid (sometimes called a mesh or net), with grid spacings of Δx, Δt in the x,t directions respectively, the values of Δx and Δt being assumed uniform for the present. The grid consists of the set of lines parallel to the t-axis given by

$$x = x_j \; , \; j = 0,1,\ldots,J, \qquad (2.2.1)$$

where $x_j = j\Delta x$ and $\Delta x = 1/J$, and the set of lines parallel to the x-axis

$$t = t_n , \; n = 0,1,2,\ldots, \qquad (2.2.2)$$

where $t_n = n\Delta t$, as shown in Figure 2.3.

Methods will be developed which determine approximately the values of $\bar{\tau}$ at interior points within the solution region where the lines defined by (2.2.1) and (2.2.2) intersect. The points with coordinates (x_j, t_n) are called gridpoints and are denoted (j,n). The terms mesh-point or node are sometimes used instead of grid-point. Grid-points within the solution region are termed interior; those for which $j = 0$ or J or $n = 0$ are called boundary points. In the diffusion problem under consideration the values of $\bar{\tau}$ on the boundaries of the solution region are defined by the initial and boundary conditions (2.1.3).

The following notation will be used for values of $\bar{\tau}$ and its derivatives at the grid-point (j,n) ,

$$\left.\begin{aligned} \bar{\tau}_j^n &= \bar{\tau}(j\Delta x, n\Delta t), \\ \left.\frac{\partial\bar{\tau}}{\partial t}\right|_j^n &= \left.\frac{\partial\bar{\tau}}{\partial t}\right|_{\substack{x = j\Delta x \\ t = n\Delta t}} , \end{aligned}\right\} \qquad (2.2.3)$$

and so on. Thus we denote the second spatial derivative at $x = (j-1)\Delta x$,

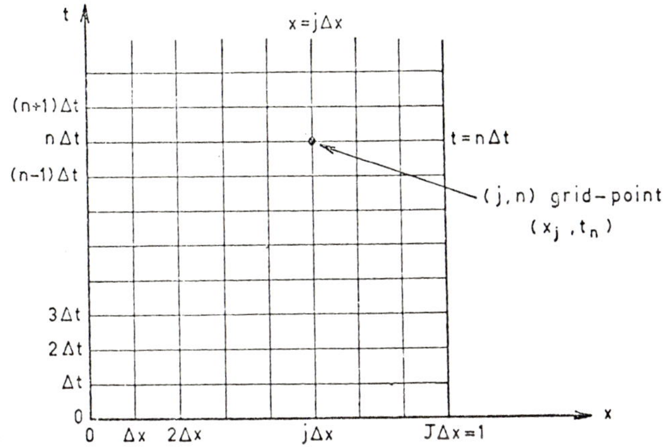

Figure 2.3: The finite-difference grid in the solution region for the one-dimensional diffusion equation.

$t = (n+1)\Delta t$, that is at the grid-point $(j-1,n+1)$, by

$$\left.\frac{\partial^2 \bar{\tau}}{\partial x^2}\right|_{j-1}^{n+1} .$$

2.3 Finite Difference Approximations to Derivatives

As a first step in developing a method of calculating the values of $\bar{\tau}$ at each interior grid-point, the space and time derivatives of $\bar{\tau}$ at the grid-point (j,n) must be expressed in terms of values of $\bar{\tau}$ at nearby grid-points. Taylor series expansions of $\bar{\tau}$ about the grid-point (j,n) with the remainder term expressed in Lagrange's form (see Thomas, 1968, p.639) will be used in this process, namely

$$\bar{\tau}(j\Delta x+\Delta x,n\Delta t) = \bar{\tau}(j\Delta x,n\Delta t)+\Delta x\left.\frac{\partial\bar{\tau}}{\partial x}\right|_{(j\Delta x,n\Delta t)} + \frac{(\Delta x)^2}{2!}\left.\frac{\partial^2\bar{\tau}}{\partial x^2}\right|_{(j\Delta x+\theta_1\Delta x,n\Delta t)}$$

where $0 < \theta_1 < 1$, or, in the notation defined by (2.2.3),

$$\bar{\tau}_{j+1}^{n} = \bar{\tau}_{j}^{n} + \Delta x \left.\frac{\partial\bar{\tau}}{\partial x}\right|_{j}^{n} + \frac{(\Delta x)^2}{2!}\left.\frac{\partial^2\bar{\tau}}{\partial x^2}\right|_{j+\theta_1}^{n} , \qquad (2.3.1)$$

and

$$\bar{\tau}(j\Delta x,n\Delta t+\Delta t) = \bar{\tau}(j\Delta x,n\Delta t)+\Delta t\left.\frac{\partial\bar{\tau}}{\partial t}\right|_{(j\Delta x,n\Delta t)} + \frac{(\Delta t)^2}{2!}\left.\frac{\partial^2\bar{\tau}}{\partial t^2}\right|_{(j\Delta x,n\Delta t+\theta_2\Delta t)}$$

where $0 < \theta_2 < 1$, which may be written

$$\bar{\tau}_{j}^{n+1} = \bar{\tau}_{j}^{n} + \Delta t \left.\frac{\partial\bar{\tau}}{\partial t}\right|_{j}^{n} + \frac{(\Delta t)^2}{2!}\left.\frac{\partial^2\bar{\tau}}{\partial t^2}\right|_{j}^{n+\theta_2} . \qquad (2.3.2)$$

These Taylor series expansions may be truncated after any number of terms using remainder terms of the form indicated.

The value of the continuous derivative $\partial\bar{\tau}/\partial t$ at the grid-point (j,n) may be given by one of a number of finite difference approximations. From Figure 2.4, which is a graph of $\bar{\tau}(j\Delta x,t)$ against t, it is clear that one approximation to the gradient $\partial\bar{\tau}/\partial t$ of the tangent to the curve at P is given by the gradient of the chord PQ which connects the points on the curve where $t = n\Delta t$ and $t = (n+1)\Delta t$. Thus we may write

$$\left.\frac{\partial\bar{\tau}}{\partial t}\right|_j^n \simeq \frac{\bar{\tau}_j^{n+1} - \bar{\tau}_j^n}{\Delta t} \quad . \qquad (2.3.3)$$

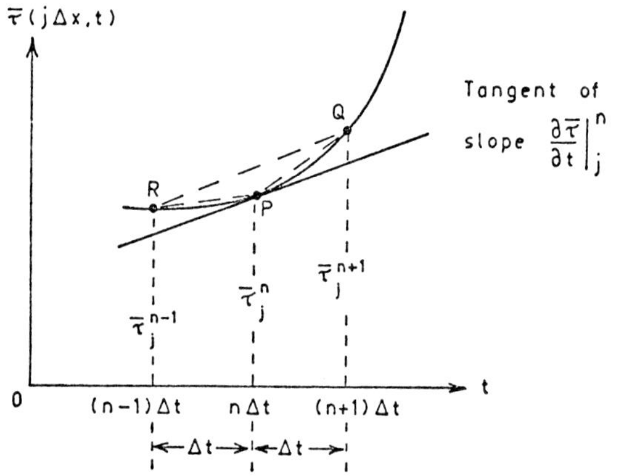

Figure 2.4: Approximations to the slope of the tangent to $\bar{\tau}(j\Delta x,t)$ at P.

Because the right hand side of this expression uses the value $\bar{\tau}_j^{n+1}$ which is forward of $\bar{\tau}_j^n$ in time, it is called the <u>forward difference approximation</u> to the time derivative evaluated at the grid-point (j,n).

By using the Taylor series (2.3.2) for $\bar{\tau}_j^{n+1}$ about the grid-point (j,n) we can estimate the order of the error involved in the forward difference approximation. Since

$$\frac{\bar{\tau}_j^{n+1} - \bar{\tau}_j^n}{\Delta t} = \frac{\left\{\bar{\tau}_j^n + \Delta t \left.\frac{\partial\bar{\tau}}{\partial t}\right|_j^n + \frac{(\Delta t)^2}{2!}\left.\frac{\partial^2\bar{\tau}}{\partial t^2}\right|_j^{n+\theta_1}\right\} - \bar{\tau}_j^n}{\Delta t}$$

$$= \left.\frac{\partial\bar{\tau}}{\partial t}\right|_j^n + \frac{\Delta t}{2}\left.\frac{\partial^2\bar{\tau}}{\partial t^2}\right|_j^{n+\theta_1}$$

where $0 < \theta_1 < 1$, we may write

$$\left.\frac{\partial\bar{\tau}}{\partial t}\right|_j^n = \frac{\bar{\tau}_j^{n+1} - \bar{\tau}_j^n}{\Delta t} + O(\Delta t), \text{ as } \Delta t \to 0 . \tag{2.3.4}$$

The term $O(\Delta t)$ is to be interpreted as meaning there exists a positive constant K, depending on $\bar{\tau}$, such that the difference between $\partial\bar{\tau}/\partial t$ at the grid-point (j,n) and its forward difference approximation is numerically less than $K\Delta t$ for all sufficiently small Δt. When there is no possibility of confusion, the qualifying statement "as $\Delta t \to 0$" is omitted. It is clear that the error involved in the forward difference approximation (2.3.3) is $O(\Delta t)$ which reduces in magnitude as the grid-spacing decreases. This improvement in accuracy is clear from Figure 2.4 because, as $\Delta t \to 0$ or $Q \to P$, the slope of PQ more closely approximates the slope of the tangent at P.

Also, from Figure 2.4 it is seen that an alternative way of approximating the value of $\partial\bar{\tau}/\partial t$ at the grid-point (j,n) is to use the slope of the chord RP as an approximation to the slope of the tangent at P, when

$$\left.\frac{\partial\bar{\tau}}{\partial t}\right|^n \simeq \frac{\bar{\tau}_j^n - \bar{\tau}_j^{n-1}}{\Delta t} . \tag{2.3.5}$$

Because the approximation on the right side of this equation uses the value $\bar{\tau}_j^{n-1}$ at a previous time level to $\bar{\tau}_j^n$ it is called the <u>backward difference approximation</u> to the time derivative. Like the forward difference approximation it has an error of $O(\Delta t)$.

From Figure 2.4 it is seen that a third way to approximate the value of $\partial\bar{\tau}/\partial t$ at the grid-point (j,n) is to use the slope of the chord RQ as the approximation to the slope of the tangent at P, when

$$\left.\frac{\partial\bar{\tau}}{\partial t}\right|_j^n \simeq \frac{\bar{\tau}_j^{n+1} - \bar{\tau}_j^{n-1}}{2\Delta t} \tag{2.3.6}$$

Intuitively, we would expect this to be a more accurate approximation than either the forward or backward difference approximations to $\partial\bar{\tau}/\partial t$. Because the time level $n\Delta t$ is centred between the time levels $(n+1)\Delta t$ and $(n-1)\Delta t$ at which the values $\bar{\tau}_j^{n+1}$ and $\bar{\tau}_j^{n-1}$ occur, the right hand side of (2.3.6) is called the <u>central difference approximation</u> to the first-order time derivative at the grid-point (j,n).

By considering the Taylor series expansions of $\bar{\tau}_j^{n+1}$ and $\bar{\tau}_j^{n-1}$ about the grid-point (j,n), truncated at the term in $(\Delta t)^3$, we find

$$\frac{\bar{\tau}_j^{n+1} - \bar{\tau}_j^{n-1}}{2\Delta t} = \frac{1}{2\Delta t}\left\{\begin{array}{l} \bar{\tau}_j^n + \Delta t \left.\dfrac{\partial\bar{\tau}}{\partial t}\right|_j^n + \dfrac{(\Delta t)^2}{2!}\left.\dfrac{\partial^2\bar{\tau}}{\partial t^2}\right|_j^n + \dfrac{(\Delta t)^3}{3!}\left.\dfrac{\partial^3\bar{\tau}}{\partial t^3}\right|_j^{n+\theta_1} \\ -\bar{\tau}_j^n + \Delta t \left.\dfrac{\partial\bar{\tau}}{\partial t}\right|_j^n - \dfrac{(\Delta t)^2}{2!}\left.\dfrac{\partial^2\bar{\tau}}{\partial t^2}\right|_j^n + \dfrac{(\Delta t)^3}{3!}\left.\dfrac{\partial^3\bar{\tau}}{\partial t^3}\right|_j^{n-\theta_2} \end{array}\right\}$$

where $0 < \theta_1 < 1$ and $0 < \theta_2 < 1$. Therefore

$$\left.\frac{\partial\bar{\tau}}{\partial t}\right|_j^n = \frac{\bar{\tau}_j^{n+1} - \bar{\tau}_j^{n-1}}{2\Delta t} + O\{(\Delta t)^2\} . \tag{2.3.7}$$

The error involved in using the central difference approximation for the first-order time derivative is therefore $O\{(\Delta t)^2\}$, which is much smaller than the error involved when either forward or backward difference approximations are used.

In a similar way, finite difference approximations to the second-order time derivative can be developed. For instance, the central difference approximation is given by the right hand side of

$$\left.\frac{\partial^2 \bar{\tau}}{\partial t^2}\right|_j^n \simeq \frac{\bar{\tau}_j^{n+1} - 2\bar{\tau}_j^n + \bar{\tau}_j^{n-1}}{(\Delta t)^2} \quad . \tag{2.3.7}$$

This can be verified by substituting the Taylor series expansions of $\bar{\tau}_j^{n+1}$ and $\bar{\tau}_j^{n-1}$ about the grid-point (j,n), truncated at the term in $(\Delta t)^4$, into the central difference approximation, giving

$$\frac{\bar{\tau}_j^{n+1} - 2\bar{\tau}_j^n + \bar{\tau}_j^{n-1}}{(\Delta t)^2}$$

$$= \frac{1}{(\Delta t)^2}\left\{\begin{array}{l} \bar{\tau}_j^n + \Delta t \left.\dfrac{\partial \bar{\tau}}{\partial t}\right|_j^n + \dfrac{(\Delta t)^2}{2!}\left.\dfrac{\partial^2 \bar{\tau}}{\partial t^2}\right|_j^n + \dfrac{(\Delta t)^3}{3!}\left.\dfrac{\partial^3 \bar{\tau}}{\partial t^3}\right|_j^n + \dfrac{(\Delta t)^4}{4!}\left.\dfrac{\partial^4 \bar{\tau}}{\partial t^4}\right|_j^{n+\theta_1} \\ -2\bar{\tau}_j^n \\ +\bar{\tau}_j^n - \Delta t \left.\dfrac{\partial \bar{\tau}}{\partial t}\right|_j^n + \dfrac{(\Delta t)^2}{2!}\left.\dfrac{\partial^2 \bar{\tau}}{\partial t^2}\right|_j^n - \dfrac{(\Delta t)^3}{3!}\left.\dfrac{\partial^3 \bar{\tau}}{\partial t^3}\right|_j^n + \dfrac{(\Delta t)^4}{4!}\left.\dfrac{\partial^4 \bar{\tau}}{\partial t^4}\right|_j^{n-\theta_2} \end{array}\right\}$$

where $0 < \theta_1 < 1$ and $0 < \theta_2 < 1$.

It follows that

$$\left.\frac{\partial^2 \bar{\tau}}{\partial t^2}\right|_j^n = \frac{\bar{\tau}_j^{n+1} - 2\bar{\tau}_j^n + \bar{\tau}_j^{n-1}}{(\Delta t)^2} + O\{(\Delta t)^2\} \ . \tag{2.3.8}$$

Thus the central difference approximation for the second-order time derivative has an error of $O\{(\Delta t)^2\}$, the same as the central difference approximation for the first-order derivative.

In a similar manner finite difference approximations to the spatial derivatives of $\bar{\tau}$ at the grid-point (j,n) may be obtained. For the first-order spatial derivative they are:

the forward difference approximation

$$\left.\frac{\partial \bar{\tau}}{\partial x}\right|_j^n = \frac{\bar{\tau}_{j+1}^n - \bar{\tau}_j^n}{\Delta x} + O(\Delta x) \ , \tag{2.3.9}$$

the backward difference approximation

$$\left.\frac{\partial \bar{\tau}}{\partial x}\right|_j^n = \frac{\bar{\tau}_j^n - \bar{\tau}_{j-1}^n}{\Delta x} + O(\Delta x) \ , \tag{2.3.10}$$

and the central difference approximation

$$\left.\frac{\partial \bar{\tau}}{\partial x}\right|_j^n = \frac{\bar{\tau}_{j+1}^n - \bar{\tau}_{j-1}^n}{2\Delta x} + O\{(\Delta x)^2\} \quad . \tag{2.3.11}$$

The central difference approximation to the second-order spatial derivative is

$$\left.\frac{\partial^2\bar{\tau}}{\partial x^2}\right|_j^n = \frac{\bar{\tau}_{j+1}^n - 2\bar{\tau}_j^n + \bar{\tau}_{j-1}^n}{(\Delta x)^2} + O\{(\Delta x)^2\} \quad . \tag{2.3.12}$$

2.4 An Approximate Solution to the Diffusion Equation

Approximate values τ_j^n to the true values $\bar{\tau}_j^n$ at the interior grid-point, that is at the grid-points (j,n) for $j = 1,2,\ldots,J-1$ and $n = 1,2,3,\ldots$, may be found for the diffusion equation (2.1.1) by substituting finite difference approximations to the derivatives involved. At the grid-point (j,n), the given equation is

$$\left.\frac{\partial\bar{\tau}}{\partial t}\right|_j^n = \alpha \left.\frac{\partial^2\bar{\tau}}{\partial x^2}\right|_j^n \quad , \tag{2.4.1}$$

and any of the various approximations for $\partial\bar{\tau}/\partial t$ and $\partial^2\bar{\tau}/\partial x^2$ described in Section 2.3 can be used. One of the simplest finite difference equations which approximates the partial differential equation (2.4.1) is obtained by using the forward difference approximation to the time derivative and the central difference approximation to the space derivative.

Substitution from Equations (2.3.4) and (2.3.12) into Equation (2.4.1) gives

$$\frac{\bar{\tau}_j^{n+1} - \bar{\tau}_j^n}{\Delta t} + O(\Delta t) = \alpha \frac{\bar{\tau}_{j+1}^n - 2\bar{\tau}_j^n + \bar{\tau}_{j-1}^n}{(\Delta x)^2} + O\{(\Delta x)^2\} \quad . \tag{2.4.2}$$

Solving the finite difference equation

$$\frac{\tau_j^{n+1} - \tau_j^n}{\Delta t} = \alpha \frac{\tau_{j+1}^n - 2\tau_j^n + \tau_{j-1}^n}{(\Delta x)^2} \tag{2.4.3}$$

therefore gives approximate values τ to the true solution $\bar{\tau}$ of the partial differential equation (2.4.1). Equation (2.4.3) is called the Forward-Time Centred-Space (FTCS) finite difference approximation to Equation (2.4.1). Clearly, the use of the finite difference approximation (2.4.3) involves a leading error of $O\{\Delta t + (\Delta x)^2\}$. The last statement is to be interpreted as meaning that there exist two positive constants K_1 and K_2, both depending on $\bar{\tau}$, such that the absolute value of the error is less than or equal to $K_1\Delta t + K_2(\Delta x)^2$ for all Δt and Δx which are sufficiently small.

Equation (2.4.3) may be rewritten in the form

$$\tau_j^{n+1} = s\tau_{j-1}^n + (1-2s)\tau_j^n + s\tau_{j+1}^n \tag{2.4.4}$$

where

$$s = \alpha\, \Delta t/(\Delta x)^2 \quad , \tag{2.4.5}$$

which is a formula giving the approximate value τ_j^{n+1} at the (n+1)th time level in terms of the values of τ_{j-1}^n, τ_j^n and τ_{j+1}^n at the n-th time level.

Application of Equation (2.4.4) with $n = 0$ gives the approximate values τ_j^1 at grid-points along the first time row $t = \Delta t$ in terms of the known boundary and initial values $\tau_j^0 = \bar{\tau}_j^0$, $j = 0,1,2,\ldots,J$. Application of Equation (2.4.4) with $n = 1$ then gives the approximate values τ_j^2 to the true solution $\bar{\tau}_j^2$ at grid-points along the second time level in terms of the computed values

τ_j^1 at the grid points on the first time level. An equation such as (2.4.4) which expresses one unknown value τ_j^{n+1} directly in terms of known values of τ_j^n, τ_j^{n-1}, and so on, is an <u>explicit</u> finite difference formula.

Returning to the diffusion equation (2.1.1) with boundary conditions given by (2.1.3), and letting $\alpha = 10^{-2}$, the explicit FTCS finite difference formula (2.4.4) can be used to obtain approximate solutions to the given problem at interior grid-points. Experience shows that at least ten intervals are required along the x-axis in order to obtain reasonably accurate answers, so $\Delta x = 0.1$ will be chosen to illustrate the method. Noting that Equation (2.4.4) takes the simple form

$$\tau_j^{n+1} = \tfrac{1}{2}(\tau_{j-1}^n + \tau_{j+1}^n), \tag{2.4.6}$$

when $s = \tfrac{1}{2}$, choose

$$\Delta t = \tfrac{1}{2}\frac{(\Delta x)^2}{\alpha} = 0.5 \quad .$$

In accordance with the initial conditions (2.1.3), the values of τ at grid-points along the time level $n = 0$ are $\tau_0^0 = 100$, $\tau_1^0 = \tau_2^0 = \tau_3^0 = \tau_4^0 = \tau_5^0 = \tau_6^0 = \tau_7^0 = \tau_8^0 = \tau_9^0 = 0$, $\tau_{10}^0 = 100$. Equation (2.4.4) is then applied with $n = 0$ and $j = 1,2,\ldots,9$ to calculate the approximate solutions τ_j^1 at each of the <u>internal</u> grid-points along the time level $n = 1$; that is,

$$\tau_j^1 = \tfrac{1}{2}(\tau_{j-1}^0 + \tau_{j+1}^0), \; j = 1,2,\ldots,9 \; .$$

is used. Values of τ_j^1 at grid-points on the boundaries $j = 0$ and J are specified by the boundary conditions of this problem and need not be calculated. The order in which the approximations τ_j^1 are calculated at the internal grid-points is immaterial since Equation (2.4.4) is explicit and is given entirely in terms of known values from previous time levels. The approximate values τ calculated are

$$\tau_0^1 = 100, \tau_1^1 = 50, \tau_2^1 = \tau_3^1 = \tau_4^1 = \tau_5^1 = \tau_6^1 = \tau_7^1 = \tau_8^1 = 0,$$

$$\tau_9^1 = 50, \; \tau_{10}^1 = 100.$$

Table 2.1 shows the results of these and calculations at further time-levels. As the time increases, heat diffuses inwards and the internal temperatures are seen to be increasing steadily and it would be found, if the calculations were continued, that the temperatures gradually approach the steady state value of 100 at all internal grid-points. Note that it takes five time steps for the boundary temperature of 100 to have any effect on the temperature at the centre of the bar using this finite difference scheme.

TABLE 2.1

Approximate solution of (2.1.1) using (2.4.4) with $s = \tfrac{1}{2}$

	$j\Delta x$	0.0	0.1	0.2	0.3	0.4	0.5	0.6	0.7	0.8	0.9	1.0
$n\Delta t$	n \ j	0	1	2	3	4	5	6	7	8	9	10
0.0	0	100	0	0	0	0	0	0	0	0	0	100
0.5	1	100	50	0	0	0	0	0	0	0	50	100
1.0	2	100	50	25	0	0	0	0	0	25	50	100
1.5	3	100	62.5	25	12.5	0	0	0	12.5	25	62.5	100
2.0	4	100	62.5	37.5	12.5	6.25	0	6.25	12.5	37.5	62.5	100
2.5	5	100	68.75	37.5	21.875	6.25	6.25	6.25	21.875	37.5	68.75	100

Clearly, a large number of time steps will be needed before the values calculated at internal grid-points in the solution region are close to 100. In order to reduce the number of arithmetic operations required a larger time step is preferable. Table 2.2 lists the results obtained when $\Delta t = 1$, for which $s = 1$ and Equation (2.4.4) becomes

$$\tau_j^{n+1} = \tau_{j-1}^n - \tau_j^n + \tau_{j+1}^n \qquad (2.4.7)$$

TABLE 2.2

Approximate solution of (2.1.1) using (2.4.4) with $s = 1$

	$j\Delta x$	0.0	0.1	0.2	0.3	0.4	0.5	0.6	0.7	0.8	0.9	1.0
$n\Delta t$	n \ j	0	1	2	3	4	5	6	7	8	9	10
0	0	100	0	0	0	0	0	0	0	0	0	100
1	1	100	100	0	0	0	0	0	0	0	100	100
2	2	100	0	100	0	0	0	0	0	100	0	100
3	3	100	200	-100	100	0	0	0	100	-100	200	100
4	4	100	-200	400	-200	100	0	100	-200	400	-200	100
5	5	100	700	-800	700	-300	200	-300	700	-800	700	100

Obviously, the values shown in Table 2.2 are physically impossible: the effect of changing from $s = \frac{1}{2}$ to $s = 1$ has been disastrous . The restrictions which apply to a particular finite difference method must therefore be found so that the user will obtain a reasonably accurate solution to the given partial differential equation. In order to establish these restrictions three important concepts will be studied in Section 3, those of convergence, consistency and stability.

2.5 The Finite Difference Method

The process of solving partial differential equations using finite difference techniques may be summarised as follows.

As a first step, a grid is developed to cover the solution region. At the grid-points, the approximate solution to the problem will be found.

As a second step, all dependent variables possible are initialised. These initial values may be exact, corresponding to a real initial situation for a transient problem, or they may be an approximation (perhaps a rough guess) for a steady state solution if the transient part is not of interest.

Once these two steps have been completed, the computational cycle begins. A finite difference equation, which is an analogue of the continuous partial differential equation to be solved, is constructed. This is used to calculate approximate values of the dependent variables at all grid-points necessary in the solution region. New values of the dependent variables are then calculated at the next time level and at new space positions incremented by some value Δx.

For instance, with the heat diffusion equation, the new temperature τ_j^{n+1} can be calculated approximately from the previously calculated temperatures τ_{j-1}^n, τ_j^n and τ_{j+1}^n using the FTCS formula (2.4.4). In this process the time is held constant at $t = n\Delta t$ and j is varied from $1,2,\ldots,(J-1)$ until the temperature at all points $j\Delta x$ in the grid are computed at a new time level $t = (n+1)\Delta t$. Once all these values have been found, the time is again increased by Δt and the process repeated to get the temperatures at the new time-level, $t = (n+2)\Delta t$, using the computed values at the time-level $t = (n+1)\Delta t$.

The process is continued until a predetermined time is reached or until the dependent variables at each space position $j\Delta x$ remain unchanged when computed at successive time levels. In the latter case a steady state is assumed to have been reached.

However, the finite difference analogue of the partial differential equation must satisfy certain conditions, or its solution will not approximate the required solution of the given partial differential equation.

3. CONVERGENCE, CONSISTENCY AND STABILITY

3.1 Convergence

A solution to a finite difference equation which approximates a given partial differential equation is said to be convergent if, at each point (x,t) in the solution region the finite difference solution approaches the solution of the partial differential equation as the grid-spacings Δx and Δt tend to zero. Formally, if we let $\bar{\tau}_j^n$ represent the exact solution of the partial differential equation and τ_j^n the exact solution of the approximating finite difference equation at the same point, then the finite difference solution is said to be convergent if

$$\tau_j^n \to \bar{\tau}_j^n \quad \text{as} \quad \Delta x \to 0, \quad \Delta t \to 0.$$

The difference between the true solution of the partial differential equation and the exact solution of the approximating finite difference equation is called the discretization error, e_j^n, where

$$e_j^n = \bar{\tau}_j^n - \tau_j^n \quad . \tag{3.1.1}$$

The exact solution of the finite difference equation is obtained when no numerical errors of any sort, such as round off errors, are introduced into the computations. The magnitude of the error e_j^n at the gridpoint (j,n) depends on the size of the grid spacings Δx and Δt, and on the magnitude of the higher order derivatives omitted in the finite difference approximations to the derivatives which occur in the given differential equation.

Guidance concerning the convergence of finite difference solutions calculated using a particular method, such as the FTCS approximation (2.4.4) to the diffusion equation, can be obtained by considering special cases in which corresponding exact solutions of the approximating finite difference equation and the original partial differential equation can be found. For instance, with initial condition $f(x) = a\sin(m\pi x)$ where m is an integer and boundary conditions $g_0(t) \equiv g_1(t) \equiv 0$ in (2.1.2), the diffusion equation (2.1.1) has the exact solution

$$\bar{\tau}(x,t) = a\sin(m\pi x).\exp(-\alpha m^2\pi^2 t), \tag{3.1.2}$$

while the approximating finite difference equation (2.4.4) has the exact solution

$$\tau_j^n = a\{1 - 4s\sin^2(m\pi/2J)\}^n \sin(m\pi j/J). \tag{3.1.3}$$

These solutions may be verified by substitution of (3.1.2), (3.1.3) into the appropriate equations (2.1.1), (2.4.4) respectively, and the given boundary and initial conditions. Similar results for other linear partial differential equations may be obtained using the variables separable method (see Kreyszig, 1972) and for other finite difference equations by using methods such as those described in Levy and Lessman (1959).

Whatever the value of m, Equation (3.1.2) shows that $\bar{\tau}(x,t)$ decreases in value as t increases. However, if $s > \frac{1}{2}$, then for some values of m

$$1 - 4s\sin^2(m\pi/2J) < -1 , \qquad (3.1.4)$$

so that for these m the magnitude of τ_j^n increases exponentially and its sign alternates as n (and therefore t) increases. Clearly, the values τ_j^n obtained at the grid-points (j,n) bear little similarity to the solutions of the diffusion equation at these points.

It should be noted that the exact solution (2.1.4) of the original heat diffusion problem (2.1.1) with initial and boundary conditions (2.1.3), contains terms of the form (3.1.2) in which m is an odd integer and $a = 400/m\pi$. Therefore for $s > \frac{1}{2}$, the relation (3.1.4) holds for some of these terms and a divergent oscillating solution to the approximating finite difference equation (2.4.3) is to be expected. The manner in which the exact solution of equation (2.4.3) with $s = 1$ diverges from the true solution of the given partial differential equation was shown in Table 2.2.

The discretization error for $s \leqslant \frac{1}{2}$ in this example, is the difference between

$$\bar{\tau}_j^n = a\sin(m\pi j/J).\exp(-\alpha m^2\pi^2 n\,\Delta t) \qquad (3.1.5a)$$

and

$$\begin{aligned}\tau_j^n &= a\sin(m\pi j/J).\exp n\{\ell n(1 - 4s\sin^2(m\pi/2J))\}\\ &= a\sin(m\pi j/J).\exp\{-\alpha m^2\pi^2 n\Delta t - \frac{\alpha}{2} m^4\pi^4(\Delta x)^2 n\Delta t(s-\tfrac{1}{6})\\ &\qquad + O(\Delta x)^4\} \ ,\end{aligned} \qquad (3.1.5b)$$

provided that J is large enough for $m\pi/2J$ to be much less than one. In that case the discretization error is

$$e_j^n = a\sin(m\pi j/J)\exp(-\alpha m^2\pi^2 n\Delta t)\ [1-\exp\{\frac{\alpha}{2}m^4\pi^4(\Delta x)^2 n\Delta t(s-\tfrac{1}{6}) + O(\Delta x)^4\}] \ . \qquad (3.1.6)$$

Therefore the approximation (3.1.3) will be very accurate until the number of time steps, n, becomes sufficiently large that

$$\exp\ \{\tfrac{1}{2}\alpha m^4\pi^4(\Delta x)^2 n\Delta t(s-\tfrac{1}{6}) + O(\Delta x)^4\} \qquad (3.1.7)$$

differs significantly from 1. Clearly, the choice $s = \frac{1}{6}$ should lead to particularly accurate results.

Exact solutions for suitable special initial and boundary conditions applied to the given partial differential equation and its finite difference analogue are often difficult to find. In such cases, an alternative procedure is to examine the behaviour of the discretization error e_j^n at the grid-point (j,n) and show that $|e_j^n|$ is bounded by some number K which tends to zero as $\Delta x \to 0$, $\Delta t \to 0$.

Again, considering the diffusion equation (2.1.1) with exact solution $\bar{\tau}_j^n$ and the explicit FTCS finite difference approximation (2.4.3) which has exact solution τ_j^n at the grid-point (j,n), Equation (3.1.1) gives

$$\tau_j^n = \bar{\tau}_j^n - e_j^n \ . \qquad (3.1.8)$$

Substitution of (3.1.8) into Equation (2.4.4) yields

$$e_j^{n+1} = s\, e_{j-1}^n + (1-2s)e_j^n + s\, e_{j+1}^n + \bar{\tau}_j^{n+1} - \bar{\tau}_j^n$$

$$- s\{\bar{\tau}_{j+1}^n + \bar{\tau}_{j-1}^n - 2\bar{\tau}_j^n\} \qquad (3.1.9)$$

Application of the truncated Taylor series expansion about the grid-point (j,n) gives

$$\bar{\tau}_{j+1}^n = \bar{\tau}_j^n + \Delta x \left.\frac{\partial\bar{\tau}}{\partial x}\right|_j^n + \frac{(\Delta x)^2}{2!}\left.\frac{\partial^2\bar{\tau}}{\partial x^2}\right|_{j+\theta_1}^n ,$$

$$\bar{\tau}_{j-1}^n = \bar{\tau}_j^n - \Delta x \left.\frac{\partial\bar{\tau}}{\partial x}\right|_j^n + \frac{(\Delta x)^2}{2!}\left.\frac{\partial^2\bar{\tau}}{\partial x^2}\right|_{j-\theta_2}^n ,$$

$$\bar{\tau}_j^{n+1} = \bar{\tau}_j^n + \Delta t \left.\frac{\partial\bar{\tau}}{\partial t}\right|_j^{n+\theta_3} ,$$

where $0 < \theta_1 < 1,\ 0 < \theta_2 < 1,\ 0 < \theta_3 < 1$.

Substitution of these expansions into Equation (3.1.9) gives

$$e_j^{n+1} = s\, e_{j-1}^n + (1-2s)e_j^n + s\, e_{j+1}^n$$

$$+ \Delta t \left.\frac{\partial\bar{\tau}}{\partial t}\right|_j^{n+\theta_3} - \frac{s(\Delta x)^2}{2}\left\{\left.\frac{\partial^2\bar{\tau}}{\partial x^2}\right|_{j+\theta_1}^n + \left.\frac{\partial^2\bar{\tau}}{\partial x^2}\right|_{j-\theta_2}^n\right\} . \qquad (3.1.10)$$

If $s \leqslant \frac{1}{2}$, so that $1-2s \geqslant 0$, and noting that $s = \alpha\Delta t/(\Delta x)^2 > 0$ since $\alpha > 0$ in any practical situation, Equation (3.1.10) may be written

$$|e_j^{n+1}| \leqslant s|e_{j-1}^n| + (1-2s)|e_j^n| + s|e_{j+1}^n|$$

$$+ \Delta t\left|\left.\frac{\partial\bar{\tau}}{\partial t}\right|_j^{n+\theta_3} - \frac{\alpha}{2}\left\{\left.\frac{\partial^2\bar{\tau}}{\partial x^2}\right|_{j+\theta_1}^n + \left.\frac{\partial^2\bar{\tau}}{\partial x^2}\right|_{j-\theta_2}^n\right\}\right| \qquad (3.1.11)$$

Let

$$M = \underset{\substack{j=1,2,3,\ldots,(J-1)\\ n=0,1,\ldots,(N-1)}}{\text{Maximum}} \left|\left.\frac{\partial\bar{\tau}}{\partial t}\right|_j^{n+\theta_3} - \frac{\alpha}{2}\left\{\left.\frac{\partial^2\bar{\tau}}{\partial x^2}\right|_{j+\theta_1}^n + \left.\frac{\partial^2\bar{\tau}}{\partial x^2}\right|_{j-\theta_2}^n\right\}\right| ,$$

where $t_N = N\Delta t$ is the last time step to be considered, and

$$E^n = \underset{j=1,2,\ldots,(J-1)}{\text{Maximum}} |e_j^n| .$$

Then Equation (3.1.11) becomes

$$|e_j^{n+1}| \leqslant s\, E^n + (1-2s)E^n + s\, E^n + \Delta t\, M ,$$

or

$$|e_j^{n+1}| \leqslant E^n + \Delta t\, M . \qquad (3.1.12)$$

Repeated application of (3.1.12) with $n = 0,1,2,\ldots,(N-1)$ yields

$$|e_j^N| \leqslant E^o + N\,\Delta t\, M \quad . \tag{3.1.13}$$

But there is no initial discretization error because the initial values are known exactly. Therefore $E^o = o$, giving

$$|e_j^N| \leqslant M\, t_N . \tag{3.1.14}$$

Now as $\Delta x \to 0$, $\Delta t = s(\Delta x)^2/\alpha \to 0$ for fixed s, and

$$M \to \underset{\substack{j = 1,2,\ldots,(J-1)\\ n = 0,1,\ldots,(N-1)}}{\text{Maximum}} \left| \left.\frac{\partial\bar{\tau}}{\partial t}\right|_j^n - \alpha \left.\frac{\partial^2\bar{\tau}}{\partial x^2}\right|_j^n \right|$$

$$\to 0 \; ,$$

because $\bar{\tau}_j^n$ is the exact solution to the diffusion equation (2.1.1), it follows that, as $\Delta x \to 0$ for fixed values of s in the range $0 < s \leqslant \frac{1}{2}$,

$$|e_j^N| \to 0 \; . \tag{3.1.15}$$

Therefore the solution to the finite difference approximation (2.4.4) converges on the exact solution to the given diffusion equation (2.1.1) as the grid becomes finer, provided $s \leqslant \frac{1}{2}$.

This method, unfortunately, is difficult to apply when the given partial differential equation is even reasonably complicated. In such cases the only indication of convergence may be the behaviour of the finite difference solutions as Δx and Δt are made smaller and smaller or by the application of Lax's equivalence theorem if consistency and stability can be determined (see Sections 3.2 - 3.4).

3.2 Consistency

Consider the FTCS finite difference approximation to the one-dimensional diffusion equation. The substitution of $\bar{\tau}_j^n$, the exact solution of the diffusion equation at the grid-point (j,n), for the approximation τ_j^n in the FTCS formula, gives

$$\bar{\tau}_j^{n+1} = s\,\bar{\tau}_{j-1}^n + (1-2s)\bar{\tau}_j^n + s\,\bar{\tau}_{j+1}^n \quad . \tag{3.2.1}$$

We wish to determine how closely Equation (3.2.1) corresponds to the diffusion equation (2.4.1) at the grid-point (j,n), that is, to

$$\left.\frac{\partial\bar{\tau}}{\partial t}\right|_j^n = \alpha \left.\frac{\partial^2\bar{\tau}}{\partial x^2}\right|_j^n \quad .$$

Substitution of the truncated Taylor series expansion about the grid-point (j,n) for each term of Equation (3.2.1) gives

$$\bar{\tau}_j^n + \Delta t \left.\frac{\partial\bar{\tau}}{\partial t}\right|_j^n + \frac{(\Delta t)^2}{2!}\left.\frac{\partial^2\bar{\tau}}{\partial t^2}\right|_j^n + \frac{(\Delta t)^3}{3!}\left.\frac{\partial^3\bar{\tau}}{\partial t^3}\right|_j^{n+\theta_1}$$

$$= s\left\{\bar{\tau}_j^n + \Delta x\left.\frac{\partial\bar{\tau}}{\partial x}\right|_j^n + \frac{(\Delta x)^2}{2!}\left.\frac{\partial^2\bar{\tau}}{\partial x^2}\right|_j^n + \frac{(\Delta x)^3}{3!}\left.\frac{\partial^3\bar{\tau}}{\partial x^3}\right|_j^n + \frac{(\Delta x)^4}{4!}\left.\frac{\partial^4\bar{\tau}}{\partial x^4}\right|_j^n \right.$$

$$\left. + \frac{(\Delta x)^5}{5!}\left.\frac{\partial^5\bar{\tau}}{\partial x^5}\right|_j^n + \frac{(\Delta x)^6}{6!}\left.\frac{\partial^6\bar{\tau}}{\partial x^6}\right|_{j+\theta_2}^n \right\}$$

$$+ s\left\{\bar{\tau}_j^n - \Delta x \left.\frac{\partial\bar{\tau}}{\partial x}\right|_j^n + \frac{(\Delta x)^2}{2!}\left.\frac{\partial^2\bar{\tau}}{\partial x^2}\right|_j^n - \frac{(\Delta x)^3}{3!}\left.\frac{\partial^3\bar{\tau}}{\partial x^3}\right|_j^n + \frac{(\Delta x)^4}{4!}\left.\frac{\partial^4\bar{\tau}}{\partial x^4}\right|_j^n\right.$$

$$\left. - \frac{(\Delta x)^5}{5!}\left.\frac{\partial^5\bar{\tau}}{\partial x^5}\right|_j^n + \frac{(\Delta x)^6}{6!}\left.\frac{\partial^6\bar{\tau}}{\partial x^6}\right|_{j-\theta_3}^n\right\}$$

$$+ (1-2s)\,\bar{\tau}_j^n , \qquad (3.2.2)$$

where $0 < \theta_1 < 1$, $0 < \theta_2 < 1$, $0 < \theta_3 < 1$.

Upon simplification, using $s = \alpha\Delta t/(\Delta x)^2$, Equation (3.2.2) becomes

$$\left.\frac{\partial\bar{\tau}}{\partial t}\right|_j^n = \alpha\left.\frac{\partial^2\bar{\tau}}{\partial x^2}\right|_j^n + E_T \qquad (3.2.3)$$

where

$$E_T = -\frac{\Delta t}{2}\left.\frac{\partial^2\bar{\tau}}{\partial t^2}\right|_j^n + \frac{\alpha(\Delta x)^2}{12}\left.\frac{\partial^4\bar{\tau}}{\partial x^4}\right|_j^n - \frac{(\Delta t)^2}{6}\left.\frac{\partial^3\bar{\tau}}{\partial t^3}\right|_j^{n+\theta_1}$$

$$+ \frac{s(\Delta x)^4}{6!}\left\{\left.\frac{\partial^6\bar{\tau}}{\partial x^6}\right|_{j+\theta_2}^n + \left.\frac{\partial^6\bar{\tau}}{\partial x^6}\right|_{j-\theta_3}^n\right\} . \qquad (3.2.4)$$

The term E_T is a measure of the error introduced by the use of the finite difference equation (2.4.4) to solve the diffusion equation (2.4.1) at the grid-point (j,n). This error is called the <u>truncation error</u>: it is the discrepancy between the approximating finite difference equation with the exact values $\bar{\tau}_j^n$ substituted for τ_j^n, and the partial differential equation which the difference equation replaces at the grid-point (j,n). Clearly, as the grid spacing gets smaller and smaller with the FTCS method, both Δx and Δt are reduced in size and the truncation error E_T gets smaller and smaller. In the limit as $\Delta x \to 0$, $\Delta t \to 0$, the finite difference equation (2.4.4) corresponds with the partial differential equation (2.4.1). This property is called <u>consistency</u> (or compatability). We say that a finite difference approximation is consistent with a partial differential equation if, as the grid-spacing tends to zero, the truncation error of the finite difference equation tends to zero.

For the FTCS finite difference approximation to the diffusion equation the truncation error (3.2.4) is seen to be generally $O\{\Delta t + (\Delta x)^2\}$. However, since $\bar{\tau}$ satisfies the diffusion equation (2.1.1), it also satisfies the equation

$$\frac{\partial^2\bar{\tau}}{\partial t^2} = \alpha^2\,\frac{\partial^4\bar{\tau}}{\partial x^4} , \qquad (3.2.5)$$

so the truncation error (3.2.4) may be written

$$E_T = -\frac{\alpha^2\Delta t}{2}\left(1 - \frac{1}{6s}\right)\left.\frac{\partial^4\bar{\tau}}{\partial x^4}\right|_j^n + O\{(\Delta t)^2+(\Delta x)^4\} . \qquad (3.2.6)$$

Therefore, if $s = \frac{1}{6}$, the first term in the above expression vanishes and the truncation error is now $O\{(\Delta t)^2+(\Delta x)^4\}$, or, what is the same thing, $O\{(\Delta x)^4\}$. In this case, the truncated error goes to zero faster than for any other value of s when the grid is made finer and finer. Therefore, the solution of the FTCS finite difference equation approaches the solution of the diffusion equation more rapidly when $s = \frac{1}{6}$ than for other values of $s \leqslant \frac{1}{2}$, as the grid-spacing is reduced.

Clearly, the property of consistency is necessary if the finite difference solution is to converge to the solution of the partial differential equation being approximated. However, this is not a sufficient condition, for even though the finite difference equation might approach a partial differential equation as the grid spacing tends to zero, it may not follow that the <u>solution</u> of the finite difference equation approaches the <u>solution</u> of the partial differential equation. For instance, if $\alpha = 10^{-2}$, choosing $\Delta x = 0.1$ and $\Delta t = 1$ so $s = 1$, the FTCS explicit finite difference equation simplifies to (2.4.7) which can be solved as in Table 2.2. Continuing to reduce Δx and Δt, so that s remains unchanged, the exact calculation of the finite difference solution at specific points in the solution region rapidly diverge (see Table 3.1) in contrast with solutions using $s = \frac{1}{2}$, $s = \frac{1}{6}$, with corresponding decreases in Δx.

TABLE 3.1

Finite difference approximations to $\bar{\tau}(0.4,8) = 45.03963$

J	Δx	$\Delta t/s$	$s = 1$	$s = 1/2$	$s = 1/6$
5	0.2	4	$\tau_2^2 = 100$	$\tau_2^4 = 50.00$	$\tau_2^{12} = 46.84500$
10	0.1	1	$\tau_4^8 = 12200$	$\tau_4^{16} = 47.54791$	$\tau_4^{48} = 45.49047$
15	0.066	0.44	$\tau_6^{18} = 197\times10^6$	$\tau_6^{36} = 45.61530$	$\tau_6^{108} = 45.23997$
20	0.05	0.25	∞	$\tau_8^{64} = 45.66436$	$\tau_8^{192} = 45.15231$
25	0.04	0.16	∞	$\tau_{10}^{100} = 45.24738$	$\tau_{10}^{300} = 45.11174$
40	0.025	0.0625	∞	$\tau_{16}^{256} = 45.19564$	$\tau_{16}^{768} = 45.06780$
50	0.02	0.04	∞	$\tau_{16}^{400} = 45.13947$	$\tau_{20}^{1200} = 45.05766$
100	0.01	0.01	∞	$\tau_{40}^{1600} = 45.06459$	$\tau_{40}^{4800} = 45.04414$

Note that the functional dependence of the truncation error E_T on Δt and Δx not only indicates how closely the solution of the differential equation satisfies the approximating finite difference equation (that is, it shows <u>consistency</u> if it exists), it also indicates how rapidly the solution of the finite difference equation approaches the solution of the given differential equation as $\Delta x \to 0$, $\Delta t \to 0$ <u>if</u> the solution is convergent. For instance, with $s = \frac{1}{6}$ the truncation error for the FTCS scheme was $O\{(\Delta t)^2+(\Delta x)^4\}$, whereas with $s = \frac{1}{2}$ it is $O\{\Delta t+(\Delta x)^2\}$. Since the discrepancy between the finite difference equation and the given differential equation tends to zero more rapidly with $s = \frac{1}{6}$ as the grid spacing decreases, a much greater rate of convergence of the solution of the finite difference equation to that of the diffusion equation occurs. (See Table 3.1.) However, with $s=1$, even though $E_T = O\{\Delta t+(\Delta x)^2\}$ and the finite difference equation <u>is</u> consistent with the diffusion equation, the solution of the difference equation does <u>not</u> converge to the solution of the diffusion equation.

3.3 Stability

A third important feature of a finite difference equation is its stability, a property concerned with the behaviour of errors produced in the finite difference solution due to errors introduced in previously calculated values. If it were possible to carry out all numerical operations to an infinite number of decimal places, the exact solution to the finite difference equation would be found. In practice, however, each calculation made by the computer is carried out to a finite number of significant figures which introduces a "round-off" error at every step, so that the finite difference solution actually found to Equation (2.4.4) is not τ_j^{n+1}, but $*\tau_j^{n+1}$ (say). $*\tau_j^n$ is called the numerical solution of the finite difference equation in contrast with its exact solution τ_j^n.

A set of finite difference equations is said to be stable if the cumulative effect of all the rounding off errors is negligible. More specifically, consider the errors $\xi_j^n = *\tau_j^n - \tau_j^n$ introduced at the grid-points (j,n) for $j = 0,1,2,\ldots,(J-1)$ and for time levels up to $N\Delta t$, all of which have absolute value $|\xi_j^n|$ less than some positive number δ (say). Then the finite difference equations are said to be stable if the maximum value of $|\xi_j^{n+1}|$ tends to zero as $\delta \to 0$. That is, the difference between the numerical solution $*\tau_j^{n+1}$ and the exact solution τ_j^{n+1} does not increase exponentially as the numbers of rows of calculations at successive time levels in the solution region is increased. If the errors do not increase exponentially as n increases, but persist as linear combinations of the initial errors, they are usually numerically tolerable provided their sum remains much smaller than the exact finite difference solution τ_j^n.

It is usually not possible to determine the exact value of the numerical error $(*\tau_j^n - \tau_j^n)$ at the grid-point (j,n) for an arbitrary distribution of errors at other grid-points. However, it can be estimated using certain standard methods, some of which will be discussed in this section. Note that the numerical solutions are invariably more accurate than these estimates indicate, because stability analysis always assume the worst possible combination of individual errors. For instance, it may be assumed that all errors have a distribution of signs so their total effect is additive, which is not always the case.

Note that stability does not refer to the difference between the solution of the finite difference equation and the given partial differential equation. This difference is the discretisation error, which is the province of convergence. Stability concerns only the calculation of the solution of the finite difference equation. The phenomenon shown in Table 2.2 is not due to instability, because the calculations were all performed exactly - no errors were introduced. Table 2.2 is an example of a non-convergent solution.

Each of the methods of stability analysis to be described give further insight into how a solution of a finite difference equation behaves when round-off errors effect the calculations. They are:

(1) Discrete perturbation stability analysis,
(2) The matrix method,
(3) von Neumann's stability analysis,
(4) Hirt's stability method.

These methods will be illustrated by application to the FTCS finite difference formula (2.4.4) which approximates the one-dimensional diffusion equation.

Note that in the calculation of τ_j^{n+1} using the FTCS formula we actually compute the numerical approximation

$$*\tau_j^{n+1} = \tau_j^{n+1} + \xi_j^{n+1} \tag{3.3.1a}$$

using numerical approximations from the previous time level $t = n\Delta t$, namely

$$*\tau^n_{j-1} = \tau^n_{j-1} + \xi^n_{j-1} \tag{3.3.1b}$$

$$*\tau^n_j = \tau^n_j + \xi^n_j \tag{3.3.1c}$$

$$*\tau^n_{j+1} = \tau^n_{j+1} + \xi^n_{j+1} \;. \tag{3.3.1d}$$

It can be shown that, for linear finite difference equations, the corresponding error terms satisfy the same homogeneous finite difference equation as the values of τ. For instance, application of Equation (2.4.4) means that we are actually calculating $*\tau^{n+1}_j$ using $*\tau^n_{j-1}$, $*\tau^n_j$ and $*\tau^n_{j+1}$, so that

$$*\tau^{n+1}_j = s(*\tau^n_{j-1}) + (1-2s)(*\tau^n_j) + s(*\tau^n_{j+1}). \tag{3.3.2}$$

Since the exact solutions τ satisfy the FTCS finite difference equation, subtraction of Equation (2.4.4) from Equation (3.3.2) followed by application of Equations (3.3.1) yields the homogeneous difference equation

$$\xi^{n+1}_j = s\,\xi^n_{j-1} + (1-2s)\xi^n_j + s\,\xi^n_{j+1} \;. \tag{3.3.3}$$

The boundary and initial values for this equation, assuming given boundary and initial values of the form (2.1.2), will all be zero and, unless some error is introduced in calculating the finite difference solution τ^n_j at some interior grid-point, the resulting errors throughout the solution region are all zero.

Discrete perturbation stability analysis

In this method a discrete perturbation, or error, ε is introduced into the finite difference scheme in the determination of τ^n_j at an arbitrary grid-point (j,n), and its effect on the computation of subsequent values of the solution of the finite difference equation is examined. Stability is indicated if the perturbation eventually dies out.

For a given value of s, application of (3.3.3) with zero errors up to some time level (n-1) and zero errors at grid-points everywhere along the n-th time level except one, where an error ε is introduced, will show whether the introduced perturbation increases without bound and the scheme is unstable. For example, Table 3.2 shows that the introduction of an error ε in the value of τ^1_4 in the FTCS finite difference scheme considered in Section 2.4, with $s=\frac{1}{2}$ (cf.Table 2.1), produces errors at later times which decrease in magnitude.

Clearly this scheme is stable with $s = \frac{1}{2}$.

TABLE 3.2

Error propagation due to introduction of error $\xi^1_4=\varepsilon$ in (3.3.3) with $s=\frac{1}{2}$.

	$j\Delta x$	0.0	0.1	0.2	0.3	0.4	0.5	0.6	0.7	0.8	0.9	1.0
$n\Delta t$	n \ j	0	1	2	3	4	5	6	7	8	9	10
0	0	0	0	0	0	0	0	0	0	0	0	0
0.5	1	0	0	0	0	ε	0	0	0	0	0	0
1.0	2	0	0	0	$\frac{1}{2}\varepsilon$	0	$\frac{1}{2}\varepsilon$	0	0	0	0	0
1.5	3	0	0	$\frac{1}{4}\varepsilon$	0	$\frac{1}{2}\varepsilon$	0	$\frac{1}{4}\varepsilon$	0	0	0	0
2.0	4	0	$\frac{1}{8}\varepsilon$	0	$\frac{3}{8}\varepsilon$	0	$\frac{3}{8}\varepsilon$	0	$\frac{1}{8}\varepsilon$	0	0	0
2.5	5	0	0	$\frac{1}{4}\varepsilon$	0	$\frac{3}{8}\varepsilon$	0	$\frac{1}{4}\varepsilon$	0	$\frac{1}{16}\varepsilon$	0	0
3.0	6	0	$\frac{1}{8}\varepsilon$	0	$\frac{5}{16}\varepsilon$	0	$\frac{5}{16}\varepsilon$	0	$\frac{5}{32}\varepsilon$	0	$\frac{1}{32}\varepsilon$	0

However, with $s = 1$, the introduction of the error $\xi_4^1 = \varepsilon$ into the FTCS finite difference equation produces a perturbation which rapidly increases in numerical value (see Table 3.3). Clearly, the finite difference equation (2.4.4) is unstable with $s = 1$.

TABLE 3.3

Error propagation due to $\xi_4^1=\varepsilon$ in (3.3.3) with $s=1$.

	$j\Delta x$	0.0	0.1	0.2	0.3	0.4	0.5	0.6	0.7	0.8	0.9	1.0
$n\Delta t$	$n \backslash j$	0	1	2	3	4	5	6	7	8	9	10
0	0	0	0	0	0	0	0	0	0	0	0	0
1	1	0	0	0	0	ε	0	0	0	0	0	0
2	2	0	0	0	ε	$-\varepsilon$	ε	0	0	0	0	0
3	3	0	0	ε	-2ε	3ε	-2ε	ε	0	0	0	0
4	4	0	ε	-3ε	6ε	-7ε	6ε	-3ε	ε	0	0	0
5	5	0	-4ε	10ε	-16ε	19ε	-16ε	10ε	-4ε	ε	0	0
6	6	0	14ε	-30ε	45ε	-51ε	45ε	-30ε	15ε	-5ε	ε	0

In general, if we wish to determine the values of s for which the finite difference equation (2.4.4) is stable, we must sweep through the grid as in Tables 3.2 and 3.3 using the general formula for error propagation, namely Equation (3.3.3) with arbitrary s. The only permissible values of s are those for which the resulting errors produced at higher time levels have moduli less than or equal to $|\varepsilon|$. With the only error at the n-th time level being $\xi_j^n = \varepsilon$, then the resulting errors at the $(n+1)$th time level are $\xi_{j-1}^{n+1} = s\varepsilon$ $\xi_j^{n+1} = (1-2s)\varepsilon$, $\xi_{j+1}^{n+1} = s\varepsilon$. These errors are not greater in magnitude than ε if

$$|s| \leqslant 1 \quad \text{and} \quad |1-2s| \leqslant 1,$$

both of which are satisfied if

$$0 \leqslant s \leqslant 1 . \tag{3.3.4}$$

Since $s = 0$ is impractical, as it implies $\Delta t = 0$, then the requirement for stability is

$$0 < s \leqslant 1 \quad .$$

The resulting errors at the $(n+2)$th time level are

$$\xi_{j-2}^{n+2} = s^2\varepsilon, \quad \xi_{j-1}^{n+2} = (2s-4s^2)\varepsilon, \quad \xi_j^{n+2} = (1-4s+6s^2)\varepsilon ,$$
$$\xi_{j+1}^{n+2} = (2s-4s^2)\varepsilon, \quad \xi_{j+2}^{n+2} = s^2\varepsilon .$$

These errors are equal to or less than ε, in magnitude, if

$$|s^2| \leqslant 1, \ |2s-4s^2| \leqslant 1 \ \text{and} \ |1-4s+6s^2| \leqslant 1;$$

that is, if

$$0 < s \leqslant \tfrac{2}{3}. \tag{3.3.5}$$

This process may be continued to higher time levels, but the algebra involved soon becomes prohibitively complicated. Consideration of the errors at time level (n+1) lead to the requirement $0 < s \leqslant 1$ for errors not to increase in magnitude; consideration of the errors at time level (n+2) leads to the requirement $0 < s \leqslant \frac{2}{3}$ for errors not to increase in magnitude. The asymptotic limit of this restriction as higher time levels are considered is required.

The initial set of round-off errors at the first time level will propagate through the grid as described by Equation (3.3.3) and at each step will have further round-off errors introduced. From an examination of Tables 3.2 and 3.3 it can be expected that after many time-steps the errors at any time level will be approximately the same in magnitude, all having the same sign or an alternating sign distribution.

If the errors at the n-th time level all have the same sign, then

$$\xi_{j-1}^{n} = \varepsilon, \quad \xi_{j}^{n} = \varepsilon, \quad \xi_{j+1}^{n} = \varepsilon$$

so that, by application of Equation (3.3.3),

$$\xi_{j}^{n+1} = \varepsilon \ ,$$

which yields no restriction on s for stability. If the distribution of errors at the n-th time level has alternate positive and negative signs, such as

$$\xi_{j-1}^{n} = -\varepsilon, \quad \xi_{j}^{n} = \varepsilon, \quad \xi_{j+1}^{n} = -\varepsilon \ ,$$

application of (3.3.3) yields

$$\xi_{j}^{n+1} = (1-4s)\varepsilon \ .$$

Stability therefore requires that

$$|1-4s| \leqslant 1$$

which yields the condition

$$0 < s \leqslant \tfrac{1}{2}. \qquad (3.3.6)$$

Examination of the error propagation at succeeding time levels does not change this restriction on s .

For a given α and a fixed space interval Δx, condition (3.3.6) places a limitation on the time step Δt, namely

$$\Delta t \leqslant (\Delta x)^2/2\alpha \ . \qquad (3.3.7)$$

The inequality (3.3.7) has important implications on the amount of computer time required to solve the FTCS finite difference approximation to the diffusion equation. Suppose the FTCS equation has been solved up to a certain time T with a mesh size Δx_1, using the maximum possible time step of $\Delta t_1 = (\Delta x_1)^2/2\alpha$. If the computation is repeated with a halved grid-spacing, say $\Delta x_2 = \Delta x_1/2$, in order to check convergence for example, then

$$\Delta t_2 = (\Delta x_2)^2/2\alpha = \Delta t_1/4 \ .$$

To find solutions until the same time T, the second example requires four times as many time levels in the solution region. Furthermore, at each time level the calculations take twice as long to perform, since there are twice as many interior grid-points at one time level. The required computer time is therefore increased eight-fold. Obviously, finite difference methods of solving the diffusion equation which do not have a stability restriction of the type (3.3.6) are highly desirable.

The matrix method

A more rigorous treatment of the manner in which errors propagate through the solution region is to express the governing equation (3.3.3) in matrix form and examine the eigenvalues of the associated matrix.

Substituting $j = 1,2,3,\ldots,(J-1)$ into (3.3.3) and noting that errors at the boundary are zero, so that $\xi_0^n = 0$ and $\xi_J^n = 0$, gives

$$\left.\begin{aligned}
\xi_1^{n+1} &= (1-2s)\xi_1^n + s\xi_2^n \\
\xi_2^{n+1} &= s\xi_1^n + (1-2s)\xi_2^n + s\xi_3^n \\
\xi_3^{n+1} &= s\xi_2^n + (1-2s)\xi_3^n + s\xi_4^n \\
&\;\;\vdots \\
\xi_{J-1}^{n+1} &= (1-2s)\xi_{J-2}^n + s\xi_{J-1}^n
\end{aligned}\right\} \qquad (3.3.8)$$

Equation (3.3.8) may be written in matrix form as

$$\underset{\sim}{\xi}^{n+1} = \underline{\underline{A}}\,\underset{\sim}{\xi}^n, \quad n = 1,2,\ldots, \qquad (3.3.9)$$

where $\underline{\underline{A}}$ is a $(J-1)$ square matrix and $\underset{\sim}{\xi}^n$ is a vector of length $(J-1)$, defined by

$$\underline{\underline{A}} = \begin{bmatrix} (1-2s) & s & & & 0 \\ s & (1-2s) & s & & \\ & s & (1-2s) & s & \\ & & \ddots & \ddots & \\ 0 & & & s & (1-2s) \end{bmatrix}, \quad \underset{\sim}{\xi}^n = \begin{bmatrix} \xi_1^n \\ \xi_2^n \\ \xi_3^n \\ \vdots \\ \xi_{J-1}^n \end{bmatrix}$$

Suppose the $(J-1)$ eigenvalues of $\underline{A}$ are λ_m, $m = 1,2,\ldots,(J-1)$ and that they are all different. Then the corresponding $(J-1)$ eigenvectors $\underset{\sim}{v}_m$ satisfy the relation

$$\underline{\underline{A}}\,\underset{\sim}{v}_m = \lambda_m \underset{\sim}{v}_m \quad , \qquad (3.3.10)$$

by definition of an eigenvector (see Wade, 1951). Also, because the λ_m are distinct, the eigenvectors $\underset{\sim}{v}_m$ form a linearly independent set, so that the initial error vector $\underset{\sim}{\xi}^o$ can be expressed uniquely in terms of the eigenvectors. That is, we can write

$$\underset{\sim}{\xi}^o = \sum_{m=1}^{J-1} c_m \underset{\sim}{v}_m \quad , \qquad (3.3.11)$$

where the C_m are constants which are the solutions of the (J-1) linear algebraic equations obtained by writing (3.3.11) in full, namely

$$\begin{bmatrix} \xi_1^o \\ \xi_2^o \\ \\ \xi_{J-1}^o \end{bmatrix} = C_1 \begin{bmatrix} v_{1,1} \\ v_{1,2} \\ \\ v_{1,J-1} \end{bmatrix} + C_2 \begin{bmatrix} v_{2,1} \\ v_{2,2} \\ \\ v_{2,J-1} \end{bmatrix} + \ldots + C_{J-1} \begin{bmatrix} v_{J-1,1} \\ v_{J-1,2} \\ \\ v_{J-1,J-1} \end{bmatrix}$$

Applying Equation (3.3.9) with $n = 0$ gives

$$\underset{\sim}{\xi}^1 = \underline{\underline{A}}\, \underset{\sim}{\xi}^0 = \sum_{m=1}^{J-1} C_m (\underline{\underline{A}}\, \underset{\sim}{v}_m) = \sum_{m=1}^{J-1} \lambda_m C_m \underset{\sim}{v}_m \ ;$$

then with $n = 1$ we find

$$\underset{\sim}{\xi}^2 = \underline{\underline{A}}\underset{\sim}{\xi}^1 = \sum_{m=1}^{J-1} \lambda_m C_m (\underline{\underline{A}}\, v_m) = \sum_{m=1}^{J-1} (\lambda_m)^2 C_m \underset{\sim}{v}_m \ ;$$

and so on, until

$$\underset{\sim}{\xi}^n = \sum_{m=1}^{J-1} (\lambda_m)^n C_m \underset{\sim}{v}_m \ . \qquad (3.3.12)$$

It follows that the errors are bounded as n increases if the eigenvalues λ_m all have absolute values less than or equal to one; that is, if

$$|\lambda_m| \leqslant 1, \ m = 1,2,\ldots,(J-1). \qquad (3.3.13)$$

Now, the eigenvalues of the tri-diagonal M-square matrix

$$\underline{\underline{B}} = \begin{bmatrix} b & c & & & 0 \\ a & b & c & & \\ & a & b & c & \\ & & & \ddots & \\ 0 & & & a & b \end{bmatrix}$$

are given by the formula

$$\lambda_m = b + 2 \ \mathrm{sgn}\ c \ \sqrt{ac} \ \cos\left\{\frac{m\pi}{M+1}\right\} \ , \ m=1,2,\ldots,M, \qquad (3.3.14)$$

where $\left.\begin{aligned} \mathrm{sgn}\ c &= +1 \ \text{ for } c > 0, \\ &= -1 \ \text{ for } c < 0. \end{aligned}\right\}$

This may be verified by substitution of the given eigen (characteristic) values λ_m into the characteristic equation $\underline{\underline{B}} - \lambda\underline{\underline{I}} = 0$. Therefore, the eigenvalues of matrix $\underline{\underline{A}}$ are

$$\lambda_m = (1-2s) + 2 \ s \ \cos(m\pi/J), \ m=1,2,\ldots,(J-1),$$

which simplifies to

$$\lambda_m = 1 - 4s \ \sin^2(m\pi/2J). \qquad (3.3.15)$$

Application of the stability condition (3.3.13) allows only values of s for which

$$|1 - 4s \ \sin^2(m\pi/2J)| \leqslant 1 \ ,$$

that is

$$-1 \leqslant 1 - 4s \sin^2(m\pi/2J) \leqslant 1. \qquad (3.3.16)$$

The right side of this inequality is satisfied for all m and s, while the left side requires

$$s \sin^2(m\pi/2J) \leqslant \tfrac{1}{2} \quad ,$$

which is true for all m if $s \leqslant \frac{1}{2}$. The FTCS finite difference equation is therefore stable for $s \leqslant \frac{1}{2}$.

It must be pointed out that if the eigenvalues λ_m are not distinct, the corresponding eigenvectors $\underset{\sim}{v}_m$ may not form a linearly independent set, in which case Equation (3.3.11) will not hold for an arbitrary initial error vector $\underset{\sim}{\xi}^0$. If the eigenvectors of $\underline{\underline{A}}$ are not distinct, the matrix method can be used only if a corresponding set of linearly independent eigenvectors $\underset{\sim}{v}_m$ can be found to satisfy (3.3.11).

<u>von Neumann stability analysis</u>.

This is the most commonly used method of stability analysis. However, it provides no information about the influence of boundary conditions on the numerical stability of the finite difference scheme. To obtain this information the matrix method must be used. The von Neumann (or Fourier series) method was developed in 1944 and it was rapidly revised and improved by von Neumann and his co-workers. The earliest complete description was given by O'Brien et. al. (1950).

In this method, the errors distributed along grid-points at one time level are expanded as a finite Fourier series. Then, the stability or instability of the finite difference equation is determined by considering whether separate Fourier components of the error distribution decay or amplify on progressing to the next time level. Thus the initial error vector $\underset{\sim}{\xi}^0$ is expressed as a finite complex Fourier series, so that at $x = j\Delta x$ it is

$$\xi_j^0 = \sum_{m=1}^{J-1} a_m e^{i(m\pi j \Delta x)} \quad ,$$

or

$$\xi_j^0 = \sum_{m=1}^{J-1} a_m e^{i\theta_m j} \qquad (3.3.17)$$

where $\theta_m = m\pi\Delta x$ and $i = \sqrt{-1}$. Because the finite difference equation being studied is linear, then so is the Equation (3.3.3) governing the error propagation, and it is sufficient to study the propagation of the error due to just the single term $e^{i\theta j}$ of the Fourier series representation (3.3.17). The coefficient a_m is a constant, so it has been omitted, and the subscript m has been dropped from θ_m .

A solution of the error equation (3.3.3) is therefore sought in the form

$$\xi_j^n = A^n \; e^{i\theta j} \qquad (3.3.18)$$

where the time dependence of this Fourier component of the error at $x = j\Delta x$ is contained in the coefficient $A^n = A(n\Delta t)$.

Substitution of the m-th Fourier components of the errors at the appropriate space locations at the n-th time level into Equation (3.3.3) gives the m-th Fourier component of the errors at the $(n+1)^{th}$ time level, namely

$$A^{n+1} e^{i\theta j} = s A^n e^{i\theta(j-1)} + (1-2s) A^n e^{i\theta j} + s A^n e^{i\theta(j+1)} .$$

This becomes, on dividing throughout by $e^{i\theta j}$,

$$A^{n+1} = A^n \{ s e^{-i\theta} + (1-2s) + s e^{i\theta} \} . \qquad (3.3.19)$$

Defining the amplification factor G for the m-th Fourier mode of the error, as it propagates one step forward in time, by

$$A^{n+1} = G A^n \qquad (3.3.20)$$

where G is a function of θ, that is $G = G(m,\Delta x)$, then the errors will remain bounded if the absolute value of G, called the gain, is never greater than one for all values of m (i.e. for all θ) and for all $s > 0$. The stability requirement is therefore that

$$|G| \leqslant 1 \text{ for all } \theta . \qquad (3.3.21)$$

Note that, in general, the amplification factor G is a complex number. Therefore the amplitude of the m-th Fourier component of the error is multiplied by $|G|$ and its phase is increased by $\arg G$.

Substitution of (3.3.20) into Equation (3.3.19) gives

$$G = 1 - 2s + s(e^{i\theta} + e^{-i\theta}) ,$$

which simplifies to

$$G = 1 - 4 s \sin^2(\theta/2) \qquad (3.3.22)$$

Clearly, this amplification factor has a different value for each θ, that is, for each Fourier component.

Our stability requirement for the FTCS finite difference equation is therefore, by (3.3.21),

$$|1 - 4s \sin^2(\theta/2)| \leqslant 1 ,$$

or

$$-1 \leqslant 1-4s \sin^2(\theta/2) \leqslant 1 .$$

The right-hand inequality is satisfied for all s and θ, whereas the left-hand inequality requires

$$s \sin^2(\theta/2) \leqslant \tfrac{1}{2} ,$$

which is true for all θ if

$$s \leqslant \tfrac{1}{2} .$$

This result is identical to that found by both discrete perturbation analysis and the matrix method.

The von Neumann method is the most commonly used way of determining stability criteria as it is generally the easiest to apply, the most straightforward and the most dependable. Unfortunately, it can only be used to establish necessary and sufficient conditions for stability of linear initial value problems with constant coefficients. Practical problems typically involve variable coefficients, non-linearities and various types of boundary conditions in which case this method can only be applied locally. If the complete non-linear difference equations are linearized in a small part of the solution region where its coefficients may be considered constant, then the conditions for the applicability of the von Neumann method are satisfied locally even though they are not satisfied over the whole solution region. Since the von Neumann method can provide useful stability information when applied locally at interior grid-points, similar local application of the method at the boundaries can provide useful boundary stability information (see Trapp and Ramshaw, 1976).

Hirt's stability analysis

Another method of stability analysis was developed by Hirt (1968). In this method, the terms of the finite difference equation are expanded in a Taylor series in order to develop a corresponding continuum partial differential equation. Stability is then determined from known stability properties of the resulting continuous differential equation.

Re-arranging the terms of the FTCS finite difference approximation of the diffusion equation (2.4.4) gives

$$\tau_j^{n+1} = s\{\tau_{j+1}^n + \tau_{j-1}^n\} + (1-2s)\tau_j^n \; .$$

Expanding each term of this equation in a truncated Taylor series about the grid-point (j,n) yields

$$\tau_j^n + \Delta t \left.\frac{\partial\tau}{\partial t}\right|_j^n + \frac{(\Delta t)^2}{2!}\left.\frac{\partial^2\tau}{\partial t^2}\right|_j^n + \frac{(\Delta t)^3}{3!}\left.\frac{\partial^3\tau}{\partial t^3}\right|_j^{n+\theta_1}$$

$$= s\left\{\begin{aligned} &\tau_j^n + \Delta x\left.\frac{\partial\tau}{\partial x}\right|_j^n + \frac{(\Delta x)^2}{2!}\left.\frac{\partial^2\tau}{\partial x^2}\right|_j^n + \frac{(\Delta x)^3}{3!}\left.\frac{\partial^3\tau}{\partial x^3}\right|_j^n + \frac{(\Delta x)^4}{4!}\left.\frac{\partial^4\tau}{\partial x^4}\right|_{j+\theta_2}^n \\ &+\tau_j^n - \Delta x\left.\frac{\partial\tau}{\partial x}\right|_j^n + \frac{(\Delta x)^2}{2!}\left.\frac{\partial^2\tau}{\partial x^2}\right|_j^n - \frac{(\Delta x)^3}{3!}\left.\frac{\partial^3\tau}{\partial x^3}\right|_j^n + \frac{(\Delta x)^4}{4!}\left.\frac{\partial^4\tau}{\partial x^4}\right|_{j-\theta_3}^n \end{aligned}\right\}$$
$$+ (1-2s)\tau_j^n \; .$$

Using the relation $s = \alpha\Delta t/(\Delta x)^2$, this becomes,

$$\left.\frac{\partial\tau}{\partial t}\right|_j^n = \alpha\left.\frac{\partial^2\tau}{\partial x^2}\right|_j^n - \frac{\Delta t}{2}\left.\frac{\partial^2\tau}{\partial t^2}\right|_j^n + O\{(\Delta t)^2 + (\Delta x)^2\} \; ,$$

or, on dropping the indices j and n,

$$\frac{\partial\tau}{\partial t} = \alpha\frac{\partial^2\tau}{\partial x^2} - \frac{\Delta t}{2}\frac{\partial^2\tau}{\partial t^2} + O\{(\Delta t)^2 + (\Delta x)^2\} \; . \qquad (3.3.23)$$

When $\Delta x \to 0$, $\Delta t \to 0$, this equation reverts to the one-dimensional diffusion equation. However, in a finite difference approximation, although Δt is small it is never zero. Therefore, while $(\Delta t)^2$ and $(\Delta x)^2$ may be considered negligible and will be discarded, the term containing Δt will be retained. Therefore Equation (3.3.23) may be written

$$\frac{\Delta t}{2}\frac{\partial^2\tau}{\partial t^2} - \alpha\frac{\partial^2\tau}{\partial x^2} + \frac{\partial\tau}{\partial t} = 0. \qquad (3.3.24)$$

This is a hyperbolic partial differential equation which has, for an arbitrary point (x,t) in the solution region, a domain of influence within the characteristic lines of slope $\pm\sqrt{\Delta t/2\alpha}$ through (x,t) as shown in Figure 3.1a. It is only within this domain that a perturbation, or an error, in the value of the dependent variable $\tau(x,t)$ is felt (see Ames, 1969).

The finite difference equation (2.4.4) also exhibits a domain of influence, as shown in Figure 3.1b. Every newly computed value τ_j^{n+1} depends on previously computed values $\tau_{j-1}^n, \tau_j^n, \tau_{j+1}^n$ at grid-points at the previous time level. Likewise, each computed value τ_j^n propagates its effect to $\tau_{j-1}^{n+1}, \tau_j^{n+1}, \tau_{j+1}^{n+1}$ which are neighbouring points at the next time level. These in turn propagate their effects to $\tau_{j\pm2}^{n+2}, \tau_{j\pm1}^{n+2}, \tau_j^{n+2}$, and so on. Thus the domain of influence of the finite difference approximation is within the characteristic lines of slope $\Delta t/\Delta x$, as shown in Figure 3.1b.

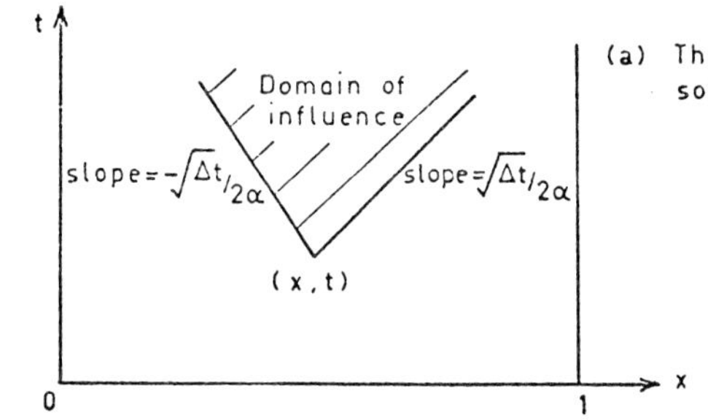

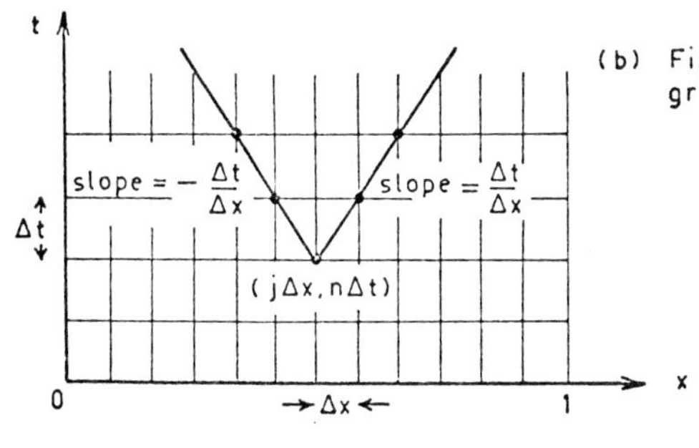

Figure 3.1: Domain of influence of the point $(x,t) = (x_j, t_n)$ for the hyperbolic equation (3.3.24).

For errors introduced into the finite difference approximation not to effect the solution to the differential equation (3.3.24), it is required that

$$\frac{\Delta t}{\Delta x} \leqslant \sqrt{\frac{\Delta t}{2\alpha}} \quad ,$$

or

$$\Delta t \leqslant \frac{(\Delta x)^2}{2\alpha} \qquad (3.3.25)$$

This is equivalent to

$$s = \frac{\alpha \Delta t}{(\Delta x)^2} \leqslant \tfrac{1}{2} \quad ,$$

so it is the same restriction as that obtained for the FTCS finite difference equation, obtained by the previous three methods.

Comparison of stability methods

There are many other methods of stability analysis in addition to the four described. Three others which deserve mention are those due to Friedrichs (see Hahn, 1958), Keller and Lax (see Richtmeyer and Morton, 1967) and Eddy (1949). The first two of these apply in practice only to the simplest finite difference equations. All three methods consider directly the amplification properties of the finite difference equations on introduced errors, rather than on their discrete Fourier components as in von Neumann's method.

The discrete perturbation method of stability analysis is a rather hit or miss affair compared with the more methodical matrix or von Neumann methods and for this reason it is not widely used. However, in certain cases (see Section 5) it easily produces stability criteria obtained only by much effort using other methods.

Hirt's method is not as formal as the matrix method or von Neumann's method and the full implication of some of the assumptions made are not clear. Although Hirt's method applied to simple problems produces similar results to both these methods, doubts exist about its usefulness with more complicated problems. Hence it is not widely used.

Like the von Neumann method, the matrix method is clearly formulated. It has the advantage that it can include the effect of boundary conditions which the von Neumann method neglects. However, the matrix method is more difficult to apply and requires considerable knowledge of matrix algebra (for example, as in Wade, 1951).

In summary, all four methods provide valuable information about the stability of a finite difference equation. At the present time the von Neumann method is favoured. However, none of them can be considered completely adequate, so when actually computing solutions to a finite difference equation they should be supplemented by numerical experimentation. Kusic and Lavi (1968) and Lavi (1969) present a non-iterative method of evaluating stability during the process of computing the solutions to a finite difference equation.

Each of the methods described has its own particular merits. The discrete perturbation method focusses attention on the individual calculations actually performed, rather than on related abstractions, so that the effect of the errors can be estimated. It also provides insight about the best way of incorporating boundary conditions. The matrix method includes the effect of the manner in which the boundary conditions are incorporated into the finite difference scheme, thereby giving the most comprehensive stability estimate. The von Neumann analysis provides information not only on attenuation (or damping) produced by the finite difference approximation, but also on any phase effects produced. It therefore gives an insight into resulting dispersion errors. Hirt's method also sheds light on dispersion errors.

It must be noted that the application of the discrete perturbation technique can be extended to non-linear problems in order to get an estimate of stability, but the basis of both the matrix method and the von Neumann stability analysis is that they apply only to linear systems with constant coefficients. They can however, give a rough estimate if applied to locally linearised forms of non-linear finite difference equations. John (1952) has shown that a mildly strengthened form of the von Neumann condition for boundaries is sufficient for stability of linear parabolic differential equations with variable coefficients. Lax (see Lax and Richtmeyer, 1956) has shown that a similar result holds for variable-coefficient hyperbolic differential equations.

3.4 Lax's Equivalence Theorem

There is an important connection between the consistency of a stable finite difference scheme and the convergence of its solution to that of the partial differential equation it approximates.

Lax's equivalence theorem (see Lax and Richtmeyer, 1956) states that if a finite difference approximation to a well-posed linear initial value problem is consistent, then stability is a necessary and sufficient condition for convergence. The two restrictions which apply to this theorem should be carefully noted. Firstly, the initial value problem must be well-posed; that is, the solution of the partial differential equation must depend continuously on the initial data. Secondly, the theorem only applies to linear problems. The important feature of linear equations required is that the sum of separate solutions is also a solution of the equation, which leads to the fact that the error terms themselves satisfy the homogeneous form of the finite difference equation which approximates the given differential equation. A corresponding theorem for the more important class of non-linear problems has yet to be derived.

This theorem is of considerable practical importance. For, while it is relatively easy to show that a finite difference equation is stable and that it is consistent with a partial differential equation, it is usually very difficult to show that the solution of a finite difference equation converges to the solution of the partial differential equation that it approximates. Lax's equivalence theorem bypasses the need to prove convergence, as carried out for the FTCS finite difference approximation to the diffusion equation in Section 3.1; because the given differential equation is linear and is well-posed it is sufficient to prove consistency and stability of the finite difference approximation.

4. SOLVING THE ONE-DIMENSIONAL DIFFUSION EQUATION

4.1 Explicit Methods

The FTCS method

By substituting the forward difference approximation for the time derivative and the central difference approximation for the space derivative in the equation

$$\left.\frac{\partial\bar{\tau}}{\partial t}\right|_j^n = \alpha \left.\frac{\partial^2\bar{\tau}}{\partial x^2}\right|_j^n \quad , \tag{2.4.1}$$

the classical FTCS explicit finite difference method for solving the one-dimensional diffusion equation was obtained, namely

$$\tau_j^{n+1} = s\,\tau_{j-1}^n + (1-2s)\tau_j^n + s\tau_{j+1}^n \tag{2.4.4}$$

where

$$s = \alpha\Delta t/(\Delta x)^2 \quad . \tag{2.4.5}$$

(See Section 2.1).

Then, in Section 3.1 it was shown that the solution τ_j^n of the finite difference equation (2.4.4) converged to the solution $\bar{\tau}_j^n$ of the diffusion equation (2.4.1) in the limit $\Delta x \to 0$, $\Delta t \to 0$, provided $s \leqslant \frac{1}{2}$. In Section 3.2 the finite difference equation (2.4.4), with $\bar{\tau}_j^n$ substituted for τ_j^n, was shown to be consistent with the one-dimensional diffusion equation; that is

$$\bar{\tau}_j^{n+1} = s\,\bar{\tau}_{j-1}^n + (1-2s)\bar{\tau}_j^n + s\,\bar{\tau}_{j+1}^n, \tag{3.2.1}$$

was shown to be the same as (2.4.1) in the limit as $\Delta x \to 0$, $\Delta t \to 0$. Finally, in Section 3.3, it was shown that the finite difference equation (2.4.4) was stable provided $s \leqslant \frac{1}{2}$; that is, round-off errors in the calculation of τ_j^n at any time level produced bounded errors in values of τ computed at later times.

Equation (2.4.4) simplifies to $\tau_j^{n+1} = \frac{1}{2}(\tau_{j-1}^n + \tau_{j+1}^n)$ when $s = \frac{1}{2}$. That is, the value of τ at $x = j\Delta x$ at the (n+1)th time level is just the arithmetic mean of the values of τ at $(j-1)\Delta x$ and $(j+1)\Delta x$ at the n-th time level. The process of taking the arithmetic mean can be carried out either numerically or graphically, and in the latter form has been familiar for many years in works on heat transfer under the name of Schmidt's method (see Carslaw & Jaeger, 1962).

Since the one-dimensional diffusion equation is linear and is well-posed, Lax's equivalence theorem indicates that consistency and stability of the FTCS finite difference approximation is necessary and sufficient for the finite difference solution to converge to the solution of the diffusion equation. This theorem is used to establish convergence of other finite difference solutions of the diffusion equation developed in this Section, since stability and consistency are much easier to prove than convergence.

Once convergence has been proved, the solution to the given partial differential equation can be obtained to any desired degree of accuracy, provided the spacing Δx and Δt of the finite difference grid is made small enough.

The FTCS method (2.4.4) is classified as explicit, because the value of τ_j^{n+1} at the (n+1)th time level may be calculated directly from known values of τ_j^n at previous time levels. Two other explicit methods will now be examined, those developed by Richardson (1910) and DuFort and Frankel (1953).

Richardson's method

This is a modification of the finite difference approximation (2.4.3) to Equation (2.4.1), in which the forward difference approximation for $\partial\tau/\partial t$ is replaced by a central difference form, giving

$$\frac{\tau_j^{n+1} - \tau_j^{n-1}}{2\Delta t} = \alpha\,\frac{\tau_{j+1}^n - 2\tau_j^n + \tau_{j-1}^n}{(\Delta x)^2}\ . \tag{4.1.1}$$

This can be written in the explicit form

$$\tau_j^{n+1} = \tau_j^{n-1} + 2s(\tau_{j-1}^n - 2\tau_j^n + \tau_{j+1}^n)\ . \tag{4.1.2}$$

Note that τ_j^{n+1} is calculated using values at the two previous time levels. Therefore an alternative procedure, such as the FTCS method with $n = 0$, is required to start the computations and calculate the approximations τ_j^1 at time level $n = 1$.

To determine the consistency of the finite difference equation (4.1.2), the values τ_j^n, which satisfy this equation, are replaced by values $\bar{\tau}_j^n$ which are solutions of the diffusion equation (2.4.1). Replacing these terms by their truncated Taylor series expanded about the grid-point (j,n) yields

$$\bar{\tau}_j^n + \Delta t\,\frac{\partial\bar{\tau}}{\partial t}\Big|_j^n + \frac{(\Delta t)^2}{2!}\,\frac{\partial^2\bar{\tau}}{\partial t^2}\Big|_j^n + \frac{(\Delta t)^3}{3!}\,\frac{\partial^3\bar{\tau}}{\partial t^3}\Big|_j^{n+\theta_1}$$

$$= \bar{\tau}_j^n - \Delta t\,\frac{\partial\bar{\tau}}{\partial t}\Big|_j^n + \frac{(\Delta t)^2}{2!}\,\frac{\partial^2\bar{\tau}}{\partial t^2}\Big|_j^n - \frac{(\Delta t)^3}{3!}\,\frac{\partial^3\bar{\tau}}{\partial t^3}\Big|_j^{n-\theta_2}$$

$$+ 2s \left\{ \begin{aligned} &\bar{\tau}_j^n - \Delta x \left.\frac{\partial \bar{\tau}}{\partial x}\right|_j^n + \frac{(\Delta x)^2}{2!} \left.\frac{\partial^2 \bar{\tau}}{\partial x^2}\right|_j^n - \frac{(\Delta x)^3}{3!} \left.\frac{\partial^3 \bar{\tau}}{\partial x^3}\right|_j^n + \frac{(\Delta x)^4}{4!} \left.\frac{\partial^4 \bar{\tau}}{\partial x^4}\right|_{j-\theta_3}^n \\ &-2\bar{\tau}_j^n \\ &+ \bar{\tau}_j^n + \Delta x \left.\frac{\partial \bar{\tau}}{\partial x}\right|_j^n + \frac{(\Delta x)^2}{2!} \left.\frac{\partial^2 \bar{\tau}}{\partial x^2}\right|_j^n + \frac{(\Delta x)^3}{3!} \left.\frac{\partial^3 \bar{\tau}}{\partial x^3}\right|_j^n + \frac{(\Delta x)^4}{4!} \left.\frac{\partial^4 \bar{\tau}}{\partial x^4}\right|_{j+\theta_4}^n \end{aligned} \right\},$$

where $\theta_1, \theta_2, \theta_3, \theta_4$ are constants, depending on j and n, each taking its value in the interval $(0,1)$. This equation simplifies to

$$\left.\frac{\partial \bar{\tau}}{\partial t}\right|_j^n = \alpha \left.\frac{\partial^2 \bar{\tau}}{\partial x^2}\right|_j^n + E_T \tag{4.1.3}$$

where

$$E_T = -\frac{(\Delta t)^2}{12} \left\{ \left.\frac{\partial^3 \bar{\tau}}{\partial t^3}\right|_j^{n+\theta_1} + \left.\frac{\partial^3 \bar{\tau}}{\partial t^3}\right|_j^{n-\theta_2} \right\} + \frac{\alpha(\Delta x)^2}{24} \left\{ \left.\frac{\partial^4 \bar{\tau}}{\partial x^4}\right|_{j-\theta_3}^n + \left.\frac{\partial^4 \bar{\tau}}{\partial x^4}\right|_{j+\theta_4}^n \right\}.$$

Richardson's method is therefore consistent, because the truncation error $E_T \to 0$ as $\Delta x \to 0, \Delta t \to 0$ and it is of a higher order of accuracy than the FTCS method because it has truncation error $O\{(\Delta t)^2 + (\Delta x)^2\}$ compared with $O\{\Delta t + (\Delta x)^2\}$.

To investigate the stability of Richardson's method the m-th Fourier component of the error distribution, $\xi_j^n = A^n e^{i\theta j}$, is substituted into the error equation corresponding to (4.1.3), namely

$$\xi_j^{n+1} = \xi_j^{n-1} + 2s\,(\xi_{j-1}^n - 2\xi_j^n + \xi_{j+1}^n), \tag{4.1.4}$$

giving

$$A^{n+1} e^{i\theta j} = A^{n-1} e^{i\theta j} + 2s\{A^n e^{i\theta(j-1)} - 2A^n e^{i\theta j} + A^n e^{i\theta(j+1)}\}. \tag{4.1.5}$$

If G is the amplification factor by which the m-th Fourier component of the error distribution at any time level is multiplied to get the corresponding Fourier component at the next time level, then

$$A^{n+1} = G\,A^n \quad \text{and} \quad A^{n-1} = \frac{1}{G} A^n \; . \tag{4.1.6}$$

Substitution of (4.1.6) into (4.1.5) yields, on simplification,

$$G - G^{-1} = 4s(\cos\theta - 1),$$

which gives the equation to be solved for G,

$$G^2 + 8s\gamma G - 1 = 0\;, \tag{4.1.7}$$

where $\gamma = \sin^2(\theta/2)$.

Equation (4.1.7) has two roots, G_1 and G_2, which satisfy the equations

$$G_1 G_2 = -1 \tag{4.1.8}$$

and

$$G_1 + G_2 = -8\gamma s\; . \tag{4.1.9}$$

It follows from (4.1.8) that either

$$|G_1| > 1 \quad \text{or} \quad |G_2| > 1,$$

in which case the errors at a given time level will contain some Fourier components which grow without bound as n increases; otherwise

$$G_1 = -G_2 = 1 \quad \text{(say)} ,$$

in which case (4.1.9) indicates that $s = 0$, when $\Delta t = 0$ and the method has no practical value. Richardson's method is therefore unconditionally unstable and has little practical value.

The DuFort-Frankel method

Another explicit method for solving the diffusion equation is that due to DuFort and Frankel (1953). This is a modification of Richardson's method in which the central grid-point value τ_j^n in the finite difference approximation for the diffusion term $\alpha\partial^2\bar{\tau}/\partial x^2$ in Equation (4.1.1) is replaced by its average at the $(n-1)$ and $(n+1)$ time-levels, giving

$$\frac{\tau_j^{n+1} - \tau_j^{n-1}}{2\Delta t} = \alpha \frac{\tau_{j+1}^n - (\tau_j^{n+1} + \tau_j^{n-1}) + \tau_{j-1}^n}{(\Delta x)^2} . \qquad (4.1.10)$$

Although a value of τ at the $(n+1)$th time level appears on the right-hand side of this equation, it is at the j-th space position, so Equation (4.1.10) can be solved explicitly as

$$\tau_j^{n+1} = \{2s(\tau_{j-1}^n + \tau_{j+1}^n) + (1-2s)\tau_j^{n-1}\}/(1+2s) . \qquad (4.1.11)$$

A stability analysis of Equation (4.1.11) by means of the von Neumann method leads to the amplification factor from one time level to the next for the m-th Fourier component of the error distribution, being

$$G = \frac{2s\cos\theta \pm (1-4s^2\sin^2\theta)^{\frac{1}{2}}}{1+2s} \qquad (4.1.12)$$

If s and θ are such that $1-4s^2\sin^2\theta \geqslant 0$, then both terms in the numerator of (4.1.12) are real. Also, $1+2s > 0$ so that

$$|G| \leqslant \frac{|2s\cos\theta| + |(1-4s^2\sin^2\theta)^{\frac{1}{2}}|}{1+2s}$$

and as

$$|2s\cos\theta| \leqslant 2s, \; 0 \leqslant 1-4s^2\sin^2\theta \leqslant 1$$

it follows that

$$|G| \leqslant \frac{2s+1}{1+2s} = 1.$$

Otherwise, $1-4s^2\sin^2\theta < 0$ so that

$$G = \frac{2s\cos\theta \pm i\,(4s^2\sin^2\theta-1)^{\frac{1}{2}}}{1+2s} ,$$

and

$$|G| = \left\{\frac{4s^2\cos^2\theta+(4s^2\sin^2\theta-1)}{(1+2s)^2}\right\}^{\frac{1}{2}}$$

$$= \left\{\frac{4s^2-1}{4s^2+4s+1}\right\}^{\frac{1}{2}}$$

$$\leqslant 1.$$

Therefore this finite difference method is stable for all positive s .

Like Richardson's formula, the DuFort-Frankel formula involves values of the dependent variable at grid-points along three time levels and it requires special starting procedures. These two methods are referred to as three level methods, in contrast with the FTCS explicit formula which is called a two level formula .

Alternatively, the stability of the DuFort-Frankel formula (4.1.11) may be investigated by matrix methods. Since this is a three-level formula, the matrix method for a two-level formula described in Section 3.3 requires modification.

With ξ_j^n designating the error in the computation τ_j^n , the error equation corresponding to (4.1.11) is

$$\xi_j^{n+1} = \frac{2s}{1+2s}\left(\xi_{j-1}^n+\xi_{j+1}^n\right) + \frac{1-2s}{1+2s}\,\xi_j^{n-1} \quad . \tag{4.1.13}$$

If $\underset{\sim}{\xi}^n$ represents the vector of error values at interior grid-points along the n-th time level, with zero error at the boundaries,

$$\underset{\sim}{\xi}^{n+1} = \left(\frac{2s}{1+2s}\right)\underline{\underline{A}}\underset{\sim}{\xi}^n + \left(\frac{1-2s}{1+2s}\right)\underset{\sim}{\xi}^{n-1} , \tag{4.1.14}$$

where

$$\underline{\underline{A}} = \begin{bmatrix} 0 & 1 & & & \\ 1 & 0 & 1 & & \\ & 1 & 0 & 1 & \\ & & \ddots & \ddots & \\ & & & 1 & 0 \end{bmatrix} , \quad \underset{\sim}{\xi}^n = \begin{bmatrix} \xi_1^n \\ \xi_2^n \\ \xi_3^n \\ \vdots \\ \xi_{J-1}^n \end{bmatrix}$$

Equation (4.1.14) can be written

$$\begin{bmatrix} \underset{\sim}{\xi}^{n+1} \\ \underset{\sim}{\xi}^n \end{bmatrix} = \begin{bmatrix} \frac{2s}{1+2s}\underline{\underline{A}} & \frac{1-2s}{1+2s}\underline{\underline{I}} \\ \underline{\underline{I}} & \underline{\underline{0}} \end{bmatrix} \begin{bmatrix} \underset{\sim}{\xi}^n \\ \underset{\sim}{\xi}^{n-1} \end{bmatrix} \tag{4.1.15}$$

where $\underline{\underline{I}}$ is the unit matrix and $\underline{\underline{0}}$ the zero matrix, both of order (J-1). Writing

$$\underset{\sim}{\zeta}^n = \begin{bmatrix} \underset{\sim}{\xi}^n \\ \underset{\sim}{\xi}^{n-1} \end{bmatrix} , \quad \underline{\underline{B}} = \begin{bmatrix} \frac{2s}{1+2s}\underline{\underline{A}} & \frac{1-2s}{1+2s}\underline{\underline{I}} \\ \underline{\underline{I}} & \underline{\underline{0}} \end{bmatrix} , \tag{4.1.16}$$

Equation (4.1.15) can be written

$$\underset{\sim}{\zeta}^{n+1} = \underline{\underline{B}}\,\underset{\sim}{\zeta}^{n} \quad , \; n = 1,2,3,\ldots \; . \qquad (4.1.17)$$

If the eigenvalues λ_m of $\underline{\underline{B}}$ are distinct then the corresponding eigen-vectors v_m form a linearly independent set and $\underset{\sim}{\zeta}^1$, any error distribution at the time levels 0 and 1, can always be written as a linear combination of the eigenvectors; that is

$$\underset{\sim}{\zeta}^1 = \sum_{m=1}^{2J-2} C_m \underset{\sim}{v}_m \quad . \qquad (4.1.18)$$

On application of Equation (4.1.17) with $n = 1,2,3,\ldots$, it follows that

$$\underset{\sim}{\zeta}^2 = \underline{\underline{B}}\,\underset{\sim}{\zeta}^1 = \sum_{m=1}^{2J-2} C_m(\underline{\underline{B}}\,\underset{\sim}{v}_m) = \sum_{m=1}^{2J-2} C_m \lambda_m \underset{\sim}{v}_m \; ,$$

$$\underset{\sim}{\zeta}^3 = \underline{\underline{B}}\,\underset{\sim}{\zeta}^2 \quad \sum_{m=1}^{2J-2} C_m(\lambda_m)^2 \underset{\sim}{v}_m \; ,$$

and so on, until finally

$$\underset{\sim}{\zeta}^n = \sum_{m=1}^{2J-2} C_m(\lambda_m)^{n-1} \underset{\sim}{v}_m \quad . \qquad (4.1.19)$$

Clearly, Equation (4.1.19) indicates that, for stability, the eigen-values of $\underline{\underline{B}}$ must satisfy the inequality $|\lambda_m| \leq 1$ for all m .

The eigenvalues of $\underline{\underline{B}}$ are, by definition, the roots of the characteristic equation $|\underline{\underline{B}}-\lambda\underline{\underline{I}}^*| = 0$, where $\underline{\underline{I}}^*$ is the unit matrix of order $(2J-2)$. Using (4.1.16) the characteristic equation can be rewritten

$$\begin{vmatrix} \frac{2s}{1+2s}\underline{\underline{A}} - \lambda\underline{\underline{I}} & \frac{1-2s}{1+2s}\underline{\underline{I}} \\ \underline{\underline{I}} & -\lambda\underline{\underline{I}} \end{vmatrix} = 0 \quad ,$$

so the eigen-values are the roots of

$$\left|-\lambda\underline{\underline{I}}\left(\frac{2s}{1+2s}\underline{\underline{A}}-\lambda\underline{\underline{I}}\right) - \left(\frac{1-2s}{1+2s}\right)\underline{\underline{I}}\right| = 0 \; ,$$

which simplifies to

$$|\lambda^2(1+2s)\underline{\underline{I}} - \lambda\, 2s\, \underline{\underline{A}} - (1-2s)\underline{\underline{I}}| = 0. \qquad (4.1.20)$$

The determinant on the left side of Equation (4.1.20) vanishes when λ takes the values λ_m given by

$$(1+2s)\lambda_m^2 - 4s\cos(m\pi/J).\,\lambda_m - (1-2s) = 0$$

for $m = 1,2,\ldots,(J-1)$. Solving this equation gives

$$\lambda_m = \frac{2s\cos(m\pi/J) \pm \{1-4s^2\sin^2(m\pi/J)\}^{\frac{1}{2}}}{1+2s} \quad . \qquad (4.1.21)$$

This equation is analogous to Equation (4.1.12), and using the same reasoning as then, it can be shown that $|\lambda_m| \leq 1$ for all $s > 0$ and all m . This shows that the DuFort-Frankel method is unconditionally stable.

On checking the consistency of the finite difference equation (4.1.11) by substitution of $\bar{\tau}$, the solution of the diffusion equation, for τ followed by expansion of each term in truncated Taylor series about the grid-point (j,n), the following relation is obtained:

$$\left.\frac{\partial\bar{\tau}}{\partial t}\right|_j^n = \alpha\left.\frac{\partial^2\bar{\tau}}{\partial x^2}\right|_j^n - \alpha\left(\frac{\Delta t}{\Delta x}\right)^2\left.\frac{\partial^2\bar{\tau}}{\partial t^2}\right|_j^n - \frac{(\Delta t)^2}{6}\left.\frac{\partial^3\bar{\tau}}{\partial t^3}\right|_j^{n+\theta_1} +$$

$$\frac{(\Delta x)^2}{24}\left\{\left.\frac{\partial^4\bar{\tau}}{\partial x^4}\right|_{j+\theta_2}^n + \left.\frac{\partial^4\bar{\tau}}{\partial x^4}\right|_{j-\theta_3}^n\right\} \tag{4.1.22}$$

where $\theta_1,\theta_2,\theta_3$ are constants in the range $(0,1)$, their values depending upon j and n. The DuFort-Frankel method is therefore consistent with the one-dimensional diffusion equation, with truncation error going to zero, only if $\Delta t/\Delta x \to 0$ as both Δx and $\Delta t \to 0$. Therefore, even though this method is unconditionally stable, $\Delta t \ll \Delta x$ is required for consistency. Note that, if Δt and $\Delta x \to 0$ in such a way that $\Delta t/\Delta x \to C$, a constant, then the finite difference equation (4.1.13) is consistent with the hyperbolic equation

$$\frac{\partial\bar{\tau}}{\partial t} = \alpha\left(\frac{\partial^2\bar{\tau}}{\partial x^2} - C^2\frac{\partial^2\bar{\tau}}{\partial t^2}\right), \tag{4.1.23}$$

and not with the parabolic one-dimensional diffusion equation (2.1.1).

In spite of the restriction $\Delta t \ll \Delta x$ required for consistency, the stabilizing effect of the substitution of the average value of τ_j^{n+1} and τ_j^{n-1} for τ_j^n in the finite difference approximation of the diffusion terms in various equations of fluid dynamics has been used successfully by many workers including Payne (1958), Fromm and Harlow (1963), Amsden and Harlow (1964), Hung and Macagno (1966), and Torrance (1968).

4.2 Implicit Methods

Each of the methods described previously is explicit. In them the finite difference equation contains at the new time level only the one unknown value τ_j^{n+1}, which is calculated explicitly from values of τ known at previous time levels. These methods are easy to program and require few computations at each time step. Unfortunately, each of the two usable methods is restricted to very small time steps Δt, the classical FTCS method because of stability requirements and the Dufort-Frankel method because of consistency requirements. For instance, if a numerical approximation at $t = 10$ is required and Δx must take the value 10^{-2}, then use of the FTCS method with $\alpha = 5$ implies that Δt must be chosen to satisfy

$$s = \frac{5\,\Delta t}{(10^{-2})^2} \leqslant \tfrac{1}{2} \quad ,$$

so

$$\Delta t \leqslant 10^{-5} \quad .$$

To generate a numerical solution at $t = 10$ would therefore require the computation of τ on at least one million rows of grid-points in the solution region.

The classical implicit method

By substituting the backward difference approximation for the time derivative and the central difference approximation for the space derivative into the diffusion equation (2.1.1) at the grid-point $(j,n+1)$ we get

$$\frac{\tau_j^{n+1} - \tau_j^n}{\Delta t} = \alpha\left\{\frac{\tau_{j+1}^{n+1} - 2\tau_j^{n+1} + \tau_{j-1}^{n+1}}{(\Delta x)^2}\right\} \tag{4.2.1}$$

Rearrangement yields the relation

$$-s\,\tau_{j-1}^{n+1} + (1+2s)\tau_j^{n+1} - s\tau_{j+1}^{n+1} = \tau_j^n \quad , \tag{4.2.2}$$

in which there are three unknown values of τ at the $(n+1)$th time level. With $n = 0$, substitution of $j = 1,2,\ldots,(J-1)$ in Equation (4.2.2) gives $(J-1)$ simultaneous equations for the $(J-1)$ unknown values of τ at internal grid-points along the first time level, in terms of known initial and boundary values. Similarly, putting $n = 1$ gives $(J-1)$ equations for the $(J-1)$ unknown values of τ at the second time-level in terms of boundary values and the computed values at the first time-level, and so on. Methods such as this, in which the unknown values of τ at any time-level are found by solving a set of algebraic equations are described as <u>implicit</u>. Equation (4.2.2) defines what is known as the classical implicit method (see Laasonen, 1949, and O'Brien, 1950).

Assuming values are known at the n-th time level we may substitute $j = 1,\ldots,(J-1)$ in Equation (4.2.2) to obtain the following set of simultaneous linear algebraic equations with unknowns τ_j^{n+1}, $j = 1,2,\ldots,(J-1)$:

$$\left.\begin{array}{llll}
(1+2s)\tau_1^{n+1} - s\,\tau_2^{n+1} & & & = d_1 \\
-s\,\tau_1^{n+1} + (1+2s)\tau_2^{n+2} - s\,\tau_3^{n+1} & & & = d_2 \\
\quad -s\,\tau_2^{n+1} + (1+2s)\tau_3^{n+1} - s\,\tau_4^{n+1} & & & = d_3 \\
\qquad \vdots & & & \ \ \vdots \\
\qquad -s\,\tau_{J-2}^{n+1} + (1+2s)\tau_{J-1}^{n+1} & & & = d_{J-1}
\end{array}\right\} \tag{4.2.3}$$

where
$$d_1 = \tau_1^n + s\tau_0^{n+1} \quad ,$$
$$d_j = \tau_j^n \quad , \ j = 1,2,\ldots,J-2,$$
$$d_{J-1} = \tau_{J-1}^n + s\tau_J^{n+1} \quad .$$

This tri-diagonal system may be solved by the elimination process first described by Thomas (1949) and outlined in Appendix 1. The Thomas algorithm makes full use of the fact that most of the coefficients of the unknown values of τ at the $(n+1)$th time level are zero's.

The process of solving the set of Equations (4.2.3) must be repeated at every time step. However, compensation for the extra arithmetic required at each time step occurs because the finite difference approximation (4.2.2) is stable for all values of s, so that large values of Δt may be used. In fact, application of the von Neumann stability analysis yields the amplification factor

$$G = \{1 + 2s(1 - \cos\theta)\}^{-1} \quad ,$$

so that $|G| \leqslant 1$ for all $s > 0$ and all θ. The classical implicit method is therefore unconditionally stable.

The consistency of this implicit finite difference method is established by substituting the solutions $\bar{\tau}$ of the given partial differential equation for the values of τ in the finite difference approximation (4.2.2), to obtain the relation

$$-s\,\bar{\tau}_{j-1}^{n+1} + (1+2s)\bar{\tau}_j^{n+1} - s\,\bar{\tau}_{j+1}^{n+1} = \bar{\tau}_j^n \quad . \tag{4.2.4}$$

The difference between Equation (4.2.4) and the one-dimensional diffusion equation is determined by expanding each term in a Taylor series about the grid-point (j,n). The Taylor series are

$$\bar{\tau}_{j+1}^{n+1} = \bar{\tau}_j^n + \Delta t \left.\frac{\partial\bar{\tau}}{\partial t}\right|_j^n + \Delta x \left.\frac{\partial\bar{\tau}}{\partial x}\right|_j^n$$

$$+ \frac{(\Delta t)^2}{2!} \left.\frac{\partial^2\bar{\tau}}{\partial t^2}\right|_j^n + \frac{\Delta t}{1!}\cdot\frac{\Delta x}{1!} \left.\frac{\partial^2\bar{\tau}}{\partial t \partial x}\right|_j^n + \frac{(\Delta x)^2}{2!} \left.\frac{\partial^2\bar{\tau}}{\partial x^2}\right|_j^n$$

$$+ \frac{(\Delta t)^3}{3!} \left.\frac{\partial^3\bar{\tau}}{\partial t^3}\right|_j^n + \frac{(\Delta t)^2}{2!}\cdot\frac{\Delta t}{1!} \left.\frac{\partial^3\bar{\tau}}{\partial t^2 \partial x}\right|_j^n + \frac{\Delta t}{1!}\cdot\frac{(\Delta x)^2}{2!} \left.\frac{\partial^3\bar{\tau}}{\partial t \partial x^2}\right|_j^n + \frac{(\Delta x)^3}{3!} \left.\frac{\partial^3\bar{\tau}}{\partial x^3}\right|_j^n$$

$$+ \ldots \quad ,$$

$$\bar{\tau}_{j-1}^{n+1} = \bar{\tau}_j^n + \Delta t \left.\frac{\partial\bar{\tau}}{\partial t}\right|_j^n - \Delta x \left.\frac{\partial\bar{\tau}}{\partial x}\right|_j^n$$

$$+ \frac{(\Delta t)^2}{2!} \left.\frac{\partial^2\bar{\tau}}{\partial t^2}\right|_j^n - \frac{\Delta t}{1!}\cdot\frac{\Delta x}{1!} \left.\frac{\partial^2\bar{\tau}}{\partial t \partial x}\right|_j^n + \frac{(\Delta x)^2}{2!} \left.\frac{\partial^2\bar{\tau}}{\partial x^2}\right|_j^n$$

$$+ \frac{(\Delta t)^3}{3!} \left.\frac{\partial^3\bar{\tau}}{\partial t^3}\right|_j^n - \frac{(\Delta t)^2}{2!}\,\frac{\Delta x}{1!} \left.\frac{\partial^3\bar{\tau}}{\partial t^2 \partial x}\right|_j^n + \frac{\Delta t}{1!}\,\frac{(\Delta x)^2}{2!} \left.\frac{\partial^3\bar{\tau}}{\partial t \partial x^2}\right|_j^n - \frac{(\Delta x)^3}{3!} \left.\frac{\partial^3\bar{\tau}}{\partial x^3}\right|_j^n$$

$$+ \ldots \quad ,$$

and

$$\bar{\tau}_j^{n+1} = \bar{\tau}_j^n + \Delta t \left.\frac{\partial\bar{\tau}}{\partial t}\right|_j^n + \frac{(\Delta t)^2}{2!} \left.\frac{\partial^2\bar{\tau}}{\partial t^2}\right|_j^n + \frac{(\Delta t)^3}{3!} \left.\frac{\partial^3\bar{\tau}}{\partial t^3}\right|_j^n \quad \ldots \quad .$$

Substitution of these expansions into Equation (4.2.4) yields the equivalent relation

$$\left.\frac{\partial\bar{\tau}}{\partial t}\right|_j^n = \alpha \left.\frac{\partial^2\bar{\tau}}{\partial x^2}\right|_j^n + E_T \tag{4.2.5a}$$

where E_T is given by

$$E_T = -\frac{\Delta t}{2} \left.\frac{\partial^2\bar{\tau}}{\partial t^2}\right|_j^n + \alpha\Delta t \left.\frac{\partial^3\bar{\tau}}{\partial t \partial x^2}\right|_j^n + \frac{\alpha(\Delta x)^2}{12} \left.\frac{\partial^4\bar{\tau}}{\partial x^4}\right|_j^n$$

$$+ O\{(\Delta t)^2 + \Delta t(\Delta x)^2\} \quad . \tag{4.2.5b}$$

Since $\bar{\tau}$ is a solution of the one-dimensional diffusion equation (2.1.1) it follows that $\partial^2\bar{\tau}/\partial t^2 = \alpha\partial^3\bar{\tau}/\partial t\partial x^2 = \alpha^2\, \partial^4\bar{\tau}/\partial x^4$, and $\Delta t = s(\Delta x)^2/\alpha$, so for a given s, E_T may be expressed by

$$E_T = \frac{\Delta t}{2}\left\{1 + \frac{1}{6s}\right\} \left.\frac{\partial^2\bar{\tau}}{\partial t^2}\right|_j^n + O\{(\Delta t)^2\} \quad . \tag{4.2.6}$$

Clearly, as $\Delta t \to 0$ for a given s, $E_T \to 0$ and formula (4.2.4) approaches Equation (4.1.1), so the finite difference equation is consistent with the one-dimensional diffusion equation. It is also clear that, for all $s > 0$, the truncation error E_T is $O\{\Delta t\}$, so that for a given Δx taking a large value of Δt unfortunately increases the error E_T. We must therefore compromise between taking large time steps, to save on the number of arithmetic operations needed, and increased inaccuracy produced by the larger truncation error which occurs. "Stability without accuracy has little to commend it." (Fox, 1962, p.243.)

The Crank-Nicolson method

Another implicit method used to solve the one-dimensional diffusion equation is due to Crank and Nicolson (1947). This method uses centred finite difference approximations for both time and space derivatives at the point $(j\Delta x, n\Delta t+\frac{1}{2}\Delta t)$ which is not a grid-point. The central difference approximation for the time derivative $\partial\bar{\tau}/\partial t$ at this point in the solution region is the same as the forward difference approximation for this derivative at the grid-point (j,n), and the spatial derivative $\partial^2\bar{\tau}/\partial x^2$ at $(j\Delta x, n\Delta t+\frac{1}{2}\Delta t)$ is approximated by the average of the central difference approximations of these spatial derivatives at the grid-points (j,n) and $(j,n+1)$.

The finite difference approximation to the diffusion equation at $(j\Delta x, n\Delta t+\frac{1}{2}\Delta t)$ is therefore

$$\frac{\tau_j^{n+1}-\tau_j^n}{\Delta t} = \tfrac{1}{2}\alpha\left\{\frac{\tau_{j+1}^{n+1}-2\tau_j^{n+1}+\tau_{j-1}^{n+1}}{(\Delta x)^2}+\frac{\tau_{j+1}^{n}-2\tau_j^{n}+\tau_{j-1}^{n}}{(\Delta x)^2}\right\}. \qquad (4.2.7)$$

This becomes, on writing the values of τ at the (n+1)th time level on the left side and the known values of τ at the n-th level on the right side,

$$-\tfrac{1}{2}s\tau_{j-1}^{n+1} + (1+s)\tau_j^{n+1} - \tfrac{1}{2}s\tau_{j+1}^{n+1} = \tfrac{1}{2}s\tau_{j-1}^{n} + (1-s)\tau_j^n + \tfrac{1}{2}s\tau_{j+1}^n. \qquad (4.2.8)$$

Because central differences have been used throughout, this finite-difference approximation has a truncation error of $O\{(\Delta t)^2 + (\Delta x)^2\}$.

Commencing with $n = 0$, then taking $n = 1,2,3,\ldots$ in turn, the right side of this equation is known for $j = 1,\ldots,(J-1)$. Application of the boundary conditions which define the values of $\tau_0^{n+1}, \tau_J^{n+1}$, then yields a set of simultaneous equations in the unknowns τ_j^{n+1}, $j = 1,\ldots,(J-1)$ similar to Equation (4.2.3), namely

$$\left.\begin{array}{rl}
(1+s)\tau_1^{n+1} - \tfrac{1}{2}s\tau_2^{n+1} & = d_1 \\
-\tfrac{1}{2}s\tau_1^{n+1} + (1+s)\tau_2^{n+1} - \tfrac{1}{2}s\tau_3^{n+1} & = d_2 \\
-\tfrac{1}{2}s\tau_2^{n+1} + (1+s)\tau_3^{n+1} - \tfrac{1}{2}s\tau_4^{n+1} & = d_3 \\
\vdots & \vdots \\
-\tfrac{1}{2}s\tau_{j-1}^{n+1} + (1+s)\tau_j^{n+1} - \tfrac{1}{2}s\tau_{j+1}^{n+1} & = d_j \\
\vdots & \vdots \\
-\tfrac{1}{2}s\tau_{J-2}^{n+1} + (1+s)\tau_{J-1}^{n+1} & = d_{J-1}
\end{array}\right\} \qquad (4.2.9)$$

where $d_1 = \tfrac{1}{2}s\tau_0^n + (1-s)\tau_1^n + \tfrac{1}{2}s\tau_2^n + \tfrac{1}{2}s\tau_0^{n+1}$,

$$d_j = \tfrac{1}{2}s\tau_{j-1}^n + (1-s)\tau_j^n + \tfrac{1}{2}s\tau_{j+1}^n, \quad j = 2,3,\ldots,J-2,$$

$$d_{J-1} = \tfrac{1}{2}s\tau_{J-2}^n + (1-s)\tau_{J-1}^n + \tfrac{1}{2}s\tau_J^n + \tfrac{1}{2}s\tau_J^{n+1},$$

are all known. This sytem can also be solved using the Thomas algorithm.

When the stability of the finite difference equation (4.2.8) is investigated, it is found that errors ξ_j^n in the computation of τ_j^n propagate according to the relation

$$-\tfrac{1}{2}s\xi_{j-1}^{n+1} + (1+s)\xi_j^{n+1} - \tfrac{1}{2}s\xi_{j+1}^{n+1} = \tfrac{1}{2}s\xi_{j-1}^{n} + (1-s)\xi_j^{n} + \tfrac{1}{2}s\xi_{j+1}^{n}, \quad (4.2.10)$$

for $j = 1,2,\ldots, (J-1)$. Application of the von Neumann method yields the amplification factor G for the m-th Fourier component of the error distribution at any time level,

$$G = \frac{1-2s\ \sin^2(\theta/2)}{1+2s\ \sin^2(\theta/2)} \quad ,$$

and as

$$|G| \leqslant 1$$

for all positive s and all θ, it is clear that the Crank-Nicolson method is unconditionally stable.

Alternatively, the stability criteria for this method may be found by the matrix method. Since no errors occur at the boundaries $j = 0,J$ then $\xi_0^n = \xi_0^{n+1} = 0$ and $\xi_J^n = \xi_J^{n+1} = 0$. Therefore, when stepping from the n-th time level to the (n+1)th time level, the error propagation may be represented by the matrix equation

$$\begin{bmatrix} (1+s) & -\tfrac{1}{2}s & & \\ -\tfrac{1}{2}s & (1+s) & -\tfrac{1}{2}s & \\ & \ddots & & \\ & & -\tfrac{1}{2}s & (1+s) \end{bmatrix} \begin{bmatrix} \xi_1^{n+1} \\ \xi_2^{n+1} \\ \vdots \\ \xi_{J-1}^{n+1} \end{bmatrix}$$

$$= \begin{bmatrix} (1-s) & \tfrac{1}{2}s & & \\ \tfrac{1}{2}s & (1-s) & \tfrac{1}{2}s & \\ & \ddots & & \\ & & \tfrac{1}{2}s & (1-s) \end{bmatrix} \begin{bmatrix} \xi_1^{n} \\ \xi_2^{n} \\ \vdots \\ \xi_{J-1}^{n} \end{bmatrix}$$

that is, by

$$(\underline{\underline{I}} + s\underline{\underline{B}})\underset{\sim}{\xi}^{n+1} = (\underline{\underline{I}} - s\underline{\underline{B}})\underset{\sim}{\xi}^{n} \quad ,$$

where

$$\underline{\underline{B}} = \begin{bmatrix} 1 & -\tfrac{1}{2} & & \\ -\tfrac{1}{2} & 1 & -\tfrac{1}{2} & \\ & \ddots & & \\ & & -\tfrac{1}{2} & 1 \end{bmatrix} , \quad \underset{\sim}{\xi}^{n} = \begin{bmatrix} \xi_1^{n} \\ \xi_2^{n} \\ \vdots \\ \xi_{J-1}^{n} \end{bmatrix} .$$

Hence

$$\underset{\sim}{\xi}^{n+1} = (\underline{\underline{I}} + s\underline{\underline{B}})^{-1}(\underline{\underline{I}} - s\underline{\underline{B}})\underset{\sim}{\xi}^{n} \quad . \quad (4.2.11)$$

Applying the same argument as that described in Section 3.3 it is seen that the finite difference scheme is stable when the moduli of the eigenvalues of

$$\underline{\underline{A}} = (\underline{\underline{I}} + s\underline{\underline{B}})^{-1}(\underline{\underline{I}} - s\underline{\underline{B}})$$

are each less than one. Equation (3.3.14) then gives

$$2\ \sin^2(m\pi/2J),$$

$m = 1,2,\ldots,(J-1)$, as the eigenvalues of $\underline{\underline{B}}$. But, when the eigenvalues of $\underline{\underline{B}}$ are λ , the eigenvalues of $f(\underline{\underline{B}})$ are $f(\lambda)$ if f is an elementary function (see Wade, 1951, p.109), so it follows that the eigenvalues of $\underline{\underline{A}}$ are

$$\lambda_m = \{1 + 2s \sin^2(m\pi/2J)\}^{-1}\{1 - 2s \sin^2(m\pi/2J)\} ,$$

$m = 1,2,\ldots,(J-1)$. Since the numerical values of λ_m are clearly less than one for all positive values of s, it follows that the Crank-Nicolson finite difference scheme has unrestricted stability.

The consistency of this finite difference scheme is investigated by substituting the exact solution $\bar{\tau}$ for the approximation τ in (4.2.8) and then expanding each term of the resulting equation in a Taylor series about the grid-point (j,n), yielding

$$\left.\frac{\partial\bar{\tau}}{\partial t}\right|_j^n = \alpha \left.\frac{\partial^2\bar{\tau}}{\partial \tau^2}\right|_j^n + \frac{\alpha(\Delta x)^2}{12} \left.\frac{\partial^4\bar{\tau}}{\partial t^4}\right|_j^n + O\{(\Delta t)^2 + (\Delta x)^4\} .$$

This gives Equation (4.1.1) in the limit as $\Delta x \to 0$, $\Delta t \to 0$, so the Crank-Nicolson equation is consistent with the original diffusion equation. It also has, as suggested previously, a truncation error which is $O\{(\Delta t)^2 + (\Delta x)^2\}$.

As mentioned earlier, the Crank-Nicolson equations (4.2.9) can readily be solved by the Thomas algorithm. Clearly, it requires more calculations per time-step than the classical implicit method to evaluate the d_j on the right side of the equations. Each d_j in Equation (4.2.9) includes two values of τ more than in the d_j of the classical implicit equations (4.2.3). What then is the advantage of the Crank-Nicolson scheme over the classical implicit method? Because the truncation error of Equation (4.2.9) is $O\{(\Delta t)^2 + (\Delta x)^2\}$ then the Crank-Nicolson equations are second order correct in both Δt and Δx; the classical implicit method was found to be second order correct in Δx but only first order correct in Δt . For the same degree of accuracy, a larger time increment can be used for the Crank-Nicolson method, so fewer time steps are necessary to compute values of τ for a given elapsed time.

The unconditionally stable Crank-Nicolson equation may be considered as a combination of the conditionally stable FTCS explicit method and the unconditionally stable classical implicit method. Consider the FTCS scheme which gives the values of τ_j^{n+1} in terms of values of τ_j^n by means of the equation

$$\tau_j^{n+1} = s\tau_{j-1}^n + (1-2s)\tau_j^n + s\tau_{j+1}^n . \qquad (2.4.4)$$

Application of the classical implicit equation gives the values of τ_j^{n+2} in terms of τ_j^{n+1} according to the relation

$$-s\tau_{j-1}^{n+2} + (1+2s)\tau_j^{n+2} - s\tau_{j+1}^{n+2} = \tau_j^{n+1} .$$

The result of equating these two equations is the finite difference equation, written between time levels n and $(n+2)$,

$$-s\tau_{j-1}^{n+2} + (1+2s)\tau_j^{n+2} - s\tau_{j+1}^{n+2} = s\tau_{j-1}^n + (1-2s)\tau_j^n + s\tau_{j+1}^n .$$

Allowing for the fact the time-increment is now $2\Delta t$, this is identical to the Crank-Nicolson scheme.

Therefore, every application of the Crank-Nicolson finite difference equation is, in effect, an application of the FTCS explicit equation followed by the classical implicit equation with the same time increment being used. This time-splitting of a stable finite difference equation into two equations, one of which may be unstable by itself, is used extensively to solve parabolic partial differential equations in two and three space dimensions. This is the basis of the Alternating Direction Implicit methods (see Section 5.3).

TABLE 4.1

Crank-Nicolson approximations to $\bar{\tau}(0.4,8) = 45.03963$

J	Δx	$\Delta t/s$	s = 1/2	s = 1	s = 2	s = 4	s = 8
10	0.1	1	$\tau_4^{16} = 45.15527$	$\tau_4^{8} = 45.17348$	$\tau_4^{4} = 45.18655$	$\tau_4^{2} = 45.68709$	$\tau_4^{1} = 37.61940$
20	0.05	0.25	$\tau_8^{64} = 45.06646$	$\tau_8^{32} = 45.06763$	$\tau_8^{16} = 45.07236$	$\tau_8^{8} = 45.08537$	$\tau_8^{4} = 45.09075$
40	0.025	0.0625	$\tau_{16}^{256} = 45.04620$	$\tau_{16}^{128} = 45.04628$	$\tau_{16}^{64} = 45.04658$	$\tau_{16}^{32} = 45.04777$	$\tau_{16}^{16} = 45.05257$
50	0.02	0.04	$\tau_{20}^{400} = 45.04383$	$\tau_{20}^{200} = 45.04386$	$\tau_{20}^{100} = 45.04129$	$\tau_{20}^{50} = 45.04447$	$\tau_{20}^{25} = 45.04644$
100	0.01	0.01	$\tau_{40}^{1600} = 45.04068$	$\tau_{40}^{800} = 45.04068$	$\tau_{40}^{400} = 45.04069$	$\tau_{40}^{200} = 45.04072$	$\tau_{40}^{100} = 45.04084$

The Crank-Nicolson method has been used to solve the problem presented in Section 2.1 and the results obtained are displayed in Table 4.1. Comparing these results with those in Table 3.1, obtained using the FTCS explicit method, it is clear that this implicit method gives much more accurate answers. For the same grid-spacing Δx, a relatively large value of Δt and therefore of s, say $s = 4.0$, still yields a more accurate answer than the FTCS method for any $s \leqslant 1/2$.

However, this increased accuracy is acquired only with an increase in the number of operations required to obtain the result. Since the time taken to perform a multiplication or division on a computer is generally much longer than that required to perform an addition or subtraction - for example, 2.9×10^{-6} secs. compared with 0.55×10^{-6} secs. on the CYBER 173 - only the number of multiplications and divisions required are considered when determining the computer time required in each of the above examples.

If the formula used with the FTCS method is written

$$\tau_j^{n+1} = s(\tau_{j-1}^n - 2\tau_j^n + \tau_{j+1}^n) + \tau_j^n ,$$

then only two multiplications are used for every application of the formula. Therefore, computing all the values of τ at each time level requires $2(J-1)$ multiplications. For example, the answer obtained in Table 3.1 when $J = 20$, $s = 1/2$ requires $64 \times 2 \times 19 = 2432$ such operations. When using the Crank-Nicolson method with the Thomas algorithm, to compute values of τ at each new time level requires $5(J-1)-4$ multiplications or divisions. In addition, at each time step the values d_j, for $j = 1,2,\ldots,(J-1)$, which appear on the right side of Equation (4.2.9) must be obtained and this requires approximately a further $2(J-1)$ multiplications. Hence, a total of about $7(J-1)-4$ multiplications or divisions are required to advance one time step. For example, when $J = 20$ and $s = 1/2$, 64 time steps had to be made to reach $t = 8$ and so $64 \times (7 \times 19 - 4) = 8256$ multiplications or divisions were required. Therefore, using the Crank-Nicolson method approximately quadrupled the time to reach $t = 8$ compared with the FTCS explicit method.

The method of weighted averages

Crandall (1955) used a weighted average instead of the Crank-Nicolson simple average of the central difference approximations to $\partial^2\bar{\tau}/\partial x^2$ at time-levels n and $(n+1)$ in Equation (4.2.7). The finite difference approximation to the diffusion equation then became

$$\frac{\tau_j^{n+1}-\tau_j^n}{\Delta t}=\alpha\left\{\beta\frac{\tau_{j+1}^{n+1}-2\tau_j^{n+1}+\tau_{j-1}^{n+1}}{(\Delta x)^2}+(1-\beta)\frac{\tau_{j+1}^{n}-2\tau_j^{n}+\tau_{j-1}^{n}}{(\Delta x)^2}\right\} \quad (4.2.12)$$

where $0 \leqslant \beta \leqslant 1$. As in the Crank-Nicolson method, three values of τ at the $(n+1)$th time level are expressed in terms of the known values of τ at the n-th time level, in the following way:

$$-\beta s\tau_{j-1}^{n+1}+(1+2\beta s)\tau_j^{n+1}-\beta s\tau_{j+1}^{n+1}$$
$$= s(1-\beta)\tau_{j-1}^n+[1-2s(1-\beta)]\tau_j^n+s(1-\beta)\tau_{j+1}^n\quad , \quad (4.2.13)$$

for $j = 1,2,\ldots,(J-1)$.

If the weighting factor $\beta = 1$, this method reduces to the classical implicit method considered at the start of this section. If $\beta = \frac{1}{2}$ the Crank-Nicolson method is obtained, and if $\beta = 0$ the classical FTCS explicit scheme is defined.

If $\beta \neq 0$, this scheme results in the following tri-diagonal set of $(J-1)$ simultaneous linear algebraic equations for the unknown values τ_j^{n+1}, $j = 1,2,\ldots,(J-1)$ in terms of the known values of τ_j^n, $j = 0,1,2,\ldots,J$, at the n-th time level and the known boundary values $\tau_0^{n+1}, \tau_J^{n+1}$ at the $(n+1)$th time level:

$$\left.\begin{array}{rl}
(1+2\beta s)\tau_1^{n+1}-\beta s\tau_2^{n+1} & = d_1\\
-\beta s\tau_1^{n+1}+(1+2\beta s)\tau_2^{n+1}-\beta s\tau_3^{n+1} & = d_2\\
-\beta s\tau_2^{n+1}+(1+2\beta s)\tau_3^{n+1}-\beta s\tau_4^{n+1} & = d_3\\
\vdots & \\
-\beta s\tau_{j-1}^{n+1}+(1+2\beta s)\tau_j^{n+1}-\beta s\tau_{j+1}^{n+1} & = d_j\\
\vdots & \\
-\beta s\tau_{J-2}^{n+1}+(1+2\beta s)\tau_{J-1}^{n+1} & = d_{J-1}
\end{array}\right\} \quad (4.2.14)$$

where

$$d_1 = s(1-\beta)\tau_0^n+[1-2s(1-\beta)]\tau_1^n+s(1-\beta)\tau_2^n+\beta s\tau_0^{n+1}\ ,$$

$$d_j = s(1-\beta)\tau_{j-1}^n+[1-2s(1-\beta)]\tau_j^n+s(1-\beta)\tau_{j+1}^n\ ,\ j = 2,3,\ldots,(J-2),$$

$$d_{J-1} = s(1-\beta)\tau_{J-2}^n+[1-2s(1-\beta)]\tau_{J-1}^n+s(1-\beta)\tau_J^n+\beta s\tau_J^{n+1}\ ,$$

are all known. With $\beta = 1$ this reduces to the system of Equations (4.2.3), and with $\beta = \frac{1}{2}$ this becomes the system (4.2.9). Equations (4.2.14) can be solved using the Thomas algorithm mentioned previously. With $\beta = 0$, they reduce to a statement of the FTCS explicit scheme.

Application of the von Neumann stability analysis to Equation (4.2.13) yields the amplification factor for the m-th Fourier component of the error distribution,

$$G = \frac{1 - 4s(1-\beta)\sin^2(\theta/2)}{1 + 4s\beta\sin^2(\theta/2)} \quad .$$

With $s > 0$ and $\beta \geqslant 0$, the denominator of G is always positive, so that the stability condition $|G| \leqslant 1$ requires

$$-1 - 4s\beta\sin^2(\theta/2) \leqslant 1 - 4s(1-\beta)\sin^2(\theta/2) \leqslant 1 + 4s\beta\sin^2(\theta/2). \tag{4.2.15}$$

Considering the right-hand inequality of (4.2.15) yields the condition

$$4s\sin^2(\theta/2) \geqslant 0,$$

which is always satisfied because $s > 0$. The left-hand inequality of (4.2.15) leads to the condition

$$2s(1-2\beta)\sin^2(\theta/2) \leqslant 1 \quad . \tag{4.2.16}$$

If $\beta < \frac{1}{2}$, this can be written

$$s \leqslant \frac{1}{2(1-2\beta)\sin^2(\theta/2)} \quad ,$$

which is satisfied for all θ if

$$s \leqslant \frac{1}{2(1-2\beta)} \quad . \tag{4.2.17}$$

If $\beta \geqslant \frac{1}{2}$, since $s > 0$ the left side of condition (4.2.16) is always negative or zero and the inequality is satisfied. The weighted average scheme defined by Equation (4.2.13) is therefore conditionally stable, requiring

$$\begin{aligned} 0 < s \leqslant \frac{1}{2(1-2\beta)} \quad & \text{for} \quad \beta < \tfrac{1}{2}, \\ s > 0 \quad & \text{for} \quad \beta \geqslant \tfrac{1}{2}. \end{aligned} \tag{4.2.18}$$

Substituting $\bar{\tau}$, the solution of the diffusion equation, for the approximate value τ in Equation (4.2.13), and expanding each term in a Taylor series about the grid-point (j,n) yields, on application of the relation $\partial\bar{\tau}/\partial t = \alpha\, \partial^2\bar{\tau}/\partial x^2$ and using $s = \alpha\,\Delta t/(\Delta x)^2$,

$$\left.\frac{\partial\bar{\tau}}{\partial t}\right|_j^n = \alpha \left.\frac{\partial^2\bar{\tau}}{\partial x^2}\right|_j^n + E_T \tag{4.2.19a}$$

where

$$E_T = -\Delta t \left.\frac{\partial^2 \bar{T}}{\partial t^2}\right|_j^n (\tfrac{1}{2}-\beta) + \frac{(\Delta x)^2}{12\alpha} \left.\frac{\partial^2 \bar{T}}{\partial t^2}\right|_j^n + O\{(\Delta t)^2 + (\Delta x)^4\}. \quad (4.2.19b)$$

Clearly, $E_T \to 0$ and (4.2.19a) yields the required diffusion equation in the limit as $\Delta t \to 0$, $\Delta x \to 0$. Thus the finite difference equation (4.2.13) is consistent with the one-dimensional diffusion equation and generally has truncation error of $O\{\Delta t + (\Delta x)^2\}$. The truncation error reduces to $O\{(\Delta t)^2 + (\Delta x)^2\}$ when $\beta = \tfrac{1}{2}$, as then the term in Δt vanishes. This is the result established previously, that the truncation error for the Crank-Nicolson method is second order in both Δt and Δx.

The accuracy of (4.2.13) can be further improved by suitably choosing β and s so both the terms in Δt and $(\Delta x)^2$ disappear from Equation (4.2.19b) The truncation error (4.2.19b) may be rewritten

$$E_T = \Delta t \left.\frac{\partial^2 \bar{T}}{\partial t^2}\right|_j^n \{-\tfrac{1}{2} + \beta + \frac{1}{12s}\} + O\{(\Delta t)^2 + (\Delta x)^4\} , \quad (4.2.20)$$

the first part of which vanishes if the factor

$$-\tfrac{1}{2} + \beta + \frac{1}{12s} = 0,$$

that is, if β and s are chosen to satisfy

$$\beta = \tfrac{1}{2} - \frac{1}{12s} \quad . \quad (4.2.21)$$

This finite difference approximation then has truncation error of $O\{(\Delta t)^2 + (\Delta x)^4\}$ or, since $\Delta t = s(\Delta x)^2/\alpha$, an error of $O\{(\Delta x)^4\}$. This is compatible with the stability relation

$$s < \frac{1}{2(1-2\beta)} \quad , \quad (4.2.22)$$

which is necessary for $\beta < \tfrac{1}{2}$, since the inequality (4.2.22) may be rewritten as

$$\beta > \tfrac{1}{2} - \frac{1}{4s} \quad ,$$

which is true for all β given by the relation (4.2.21).

The implicit methods are only advantageous if a value of s much greater than a half is chosen, otherwise the FTCS explicit method is more suitable. Equation (4.2.21) suggests that suitable values of s and β for a stable method with minimal truncation error might be $s = 1$ with $\beta = 5/12$, $s = 2$ with $\beta = 11/24$, $s = 3$ with $\beta = 17/36$, and so on.

These results for the stability and accuracy of the weighted average method are summarised in Figure 4.1. Each point in the (s,β) plane represents a different weighted-average finite difference scheme for solving the one-dimensional diffusion equation. All are implicit except those with $\beta = 0$. The range of stability is shown together with the family of formulae for which the truncation error is $O\{(\Delta t)^2 + (\Delta x)^2\}$ and $O\{(\Delta t)^2 + (\Delta x)^4\}$ instead of the usual $O\{\Delta t + (\Delta x)^2\}$. The actual stability limits depend on the size of the grid-spacing Δx and on the nature of the boundary conditions, but as Δx decreases there is a rapid approach to the limiting values shown. Note that, for formulae that have truncation error $O\{(\Delta x)^4\}$, better accuracy is obtained if the boundary conditions are incorporated with a truncation error for the spatial derivatives which also have truncation error $O\{(\Delta x)^4\}$.

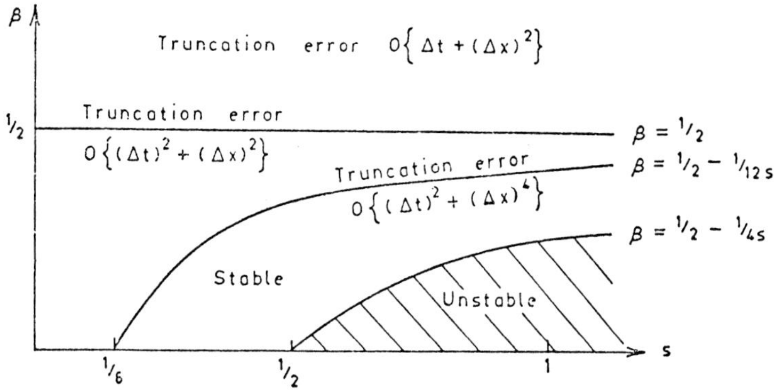

Figure 4.1: Properties of the weighted average implicit finite-difference approximations for the one dimensional diffusion equation.

The problem presented in Section 2 has been solved using the method of weighted averages with $J = 40$ and the results are presented in Table 4.2. As expected, the Crank Nicolson method ($\beta = 1/2$) gives more accurate results than either the explicit ($\beta = 0$) or fully implicit ($\beta = 1$) methods. However, the Crank-Nicolson results are also more accurate than those obtained using the Crandall method in which an optimal β is chosen for a given value of s according to Equation (4.2.21). This apparent disagreement with the theoretical result obtained above has also been noted by other workers in this field. For example, Moore, Kaplan and Mitchell (1975) found that the Crank-Nicolson method gave more accurate answers to the boundary value problem they were modelling than did the Crandall method.

TABLE 4.2

Numerical approximations to $\bar{\tau}(0.4,8) = 45.03963$ using the method of weighted averages with $J = 40$ and various s.

s	Explicit $\beta=0$	Fully Implicit $\beta=1$	Crank-Nicolson $\beta=1/2$	Crandall's Method $\beta=1/2-1/12s$
1/2	45.19564	44.98152	45.04620	45.06782
1	-	44.91719	45.04628	45.06790
2	-	44.78948	45.04658	45.06820
4	-	44.53794	45.04777	45.06941
8	-	44.05031	45.05257	45.07418
16	-	43.13556	45.06036	45.08018
32	-	41.52484	45.06501	45.08143

It is also interesting to note that the terms of order $(\Delta t)^2$ in Equation (4.2.20) can be made zero for a particular value of s and β as well as the term of order Δt. Crandall (1955) showed that choosing $s = \sqrt{5}/10$ and $\beta = (3-\sqrt{5})/6$ gives a finite difference approximation which has a truncation error of $O\{(\Delta x)^6\}$. These values of s and β have also been used to solve the problem presented in Section 2. For $J = 100$ the result $\tau(0.4,8) = 45.03557$ was obtained; for $J = 50$, 45.03700 and for $J = 40$ the answer obtained was 45.03505. A disadvantage of this method is that $\sqrt{5}$ is irrational and so it is impossible to find τ exactly at the time $t = 8.0$ secs. For example, the above results are those values calculated at times $t = 7.9984$, $t = 7.9962$ and $t = 7.9939$ respectively. Such higher order methods seem to be of little practical use since there are methods available such as Extrapolation (see Section 6) which give results just as accurate as the above but with a greatly reduced number of calculations.

4.3 Boundary conditions

In the diffusion problem considered so far the boundary conditions were taken to be known values of the dependent variable $\bar{\tau}$ at all time t. However, this need not be the case.

For example, in heat conduction problems, the most usual situation at the boundaries is not that they are held at a temperature which is known, but that heat is lost by conduction or radiation at a rate proportional to some power of the temperature difference between the surface of the body and its surroundings. This produces a relation involving the spatial derivative of temperature at the surface and the temperature of the surface. Note that in the analytical solution of such heat flow problems obtained using Fourier series, the use of non constant boundary conditions complicates determination of the coefficients of the series. However, the numerical techniques described previously require only minor modifications in order to obtain a solution to this problem.

In general, the two other ways of describing the boundary conditions are the use of a derivative boundary condition, or, a combination of a derivative boundary condition and a known value of the variable τ.

Derivative Boundary Conditions

A typical derivative boundary condition for the diffusion problem (2.1.1) is

$$\frac{\partial \bar{\tau}}{\partial x} = b(t) \tag{4.3.1}$$

where $b(t)$ is a known function of time defined at the boundary $x = 0$. Using the forward difference approximation in space at the $(n+1)$th time level, Equation (4.3.1) becomes

$$\frac{\bar{\tau}_1^{n+1} - \bar{\tau}_0^{n+1}}{\Delta x} = b^{n+1} + O(\Delta x), \tag{4.3.2}$$

where $b^{n+1} = b(t_{n+1})$. The corresponding finite difference approximation is

$$\tau_0^{n+1} = \tau_1^{n+1} - \Delta x b^{n+1} \quad . \tag{4.3.3}$$

Equation (4.3.3) can then be used with the FTCS or any other scheme being used in order to obtain the boundary value τ_0^{n+1} once τ_1^{n+1} is known. However, Equation (4.3.2) has a first order truncation error. Therefore no matter how accurate is the finite difference method used in the interior of the solution region, the overall accuracy is governed by the low order of accuracy of the boundary condition approximation.

To overcome this problem a method of higher order accuracy, such as one using the central difference approximation for the space derivative, could be used. Using this method with Equation (4.3.1) gives the equation

$$\frac{\bar{\tau}_1^{n+1} - \bar{\tau}_{-1}^{n+1}}{2\Delta x} = b^{n+1} + O\{(\Delta x)^2\} \; . \tag{4.3.4}$$

where $\bar{\tau}_{-1}^{n+1}$ is a fictitious value of τ prescribed at the exterior point (x_{-1}, t_{n+1}), where $x_{-1} = -\Delta x$. Equation (4.3.4) gives the finite difference equation

$$\tau_{-1}^{n+1} = \tau_1^{n+1} - 2\Delta x\, b^{n+1} . \tag{4.3.5}$$

Fictitious values such as τ_{-1}^{n+1} can be incorporated with the FTCS method in the following ways. Substitution of Equation (4.3.5) with n for $n+1$, viz.

$$\tau_{-1}^{n} = \tau_1^{n} - 2\Delta x\, b^{n} \; , \tag{4.3.6}$$

into Equation (2.4.4) with $j = 0$ gives

$$\tau_0^{n+1} = (1-2s)\tau_0^n + 2s\tau_1^n - 2s\Delta x\, b^n \quad . \tag{4.3.7}$$

The boundary value τ_0^{n+1} is therefore given in terms of the known values τ_0^n, τ_1^n found at the previous time-level, and the known derivative boundary value b^{n+1}.

As an example, consider the insulated rod depicted in Figure 2.1. The temperature, denoted by $\bar{\tau}(x,t)$ is initially $0°C$ throughout the rod. The right hand end of the rod is kept at $0°C$ for all time and at $t = 0$ a constant heat flux is suddenly imposed at the left hand end. The mathematical description of this problem is given by Equation (2.1.1) subject to the following boundary and initial conditions

$$\begin{aligned} \partial\bar{\tau}(0,t)/\partial x &= -1, \qquad && t > 0, \\ \bar{\tau}(1,t) &= 0 \; , && t > 0, \\ \bar{\tau}(x,0) &= 0 \; , && 0 < x \leqslant 1. \end{aligned}$$

The negative sign on the right side of the first equation indicates that there is a steady flow of heat into the rod at $x = 0$.

The analytical solution to this non-homogeneous problem may be obtained using the method of separation of variables (see Myers, 1971, p.103) and is

$$\bar{\tau}(x,t) = 1-x-\left(\frac{8}{\pi}\right)^2 \sum_{n=1}^{\infty} (2n-1)^{-2} \cos\{(2n-1)\pi x/2\}\exp\{-(2n-1)^2\pi^2\alpha t/4\} \quad .$$

This problem has also been solved using the FTCS explicit method, firstly with the forward difference approximation to the spatial derivative in the boundary condition at $x = 0$ as given by Equation (4.3.3) and then with the central difference approximation which is given by Equation (4.3.7). The value of $\bar{\tau}(0.4,8)$ obtained using the analytic result is 0.06665; the approximate value obtained using the FTCS finite difference method with $J = 20$, $s = \frac{1}{6}$, and incorporating Equation (4.3.3) is 0.07456, which has a relative error of 11.9%, and the value obtained using the FTCS method with Equation (4.3.7) is 0.06639 which is in error by 0.4%. The use of the central difference approximation for the derivative boundary condition at $x = 0$ gives a more accurate result than using the forward difference approximation.

The relative accuracy achieved using the two approximations to the derivative boundary condition considered, is more clearly shown in Figure 4.2, in which the distribution of percentage errors along the bar at time $t = 4.0$ is shown for both schemes. In each case the percentage error introduced at $x = 0$ increases from left to right along the bar. Also, for a fixed s the relative error decreases at a given position on the bar as time passes. For instance, using the central difference approximation to the boundary derivative at $x = 0$, the errors at $x = 0.4$ at the end of each of the first four seconds are respectively 10.85%, 3.27%, 1.69% and 1.07%.

However, this does not imply that, for a fixed Δx, more accurate answers can be obtained by reducing the size of Δt, since this has the effect of increasing the initial error. For instance, by using the central difference approximation to the boundary derivative with $J = 20$, $s = 1/12$, the relative error at $x = 0$, $t = 1$ becomes 1.33% compared with the figure of 1.05% with $s = \frac{1}{6}$.

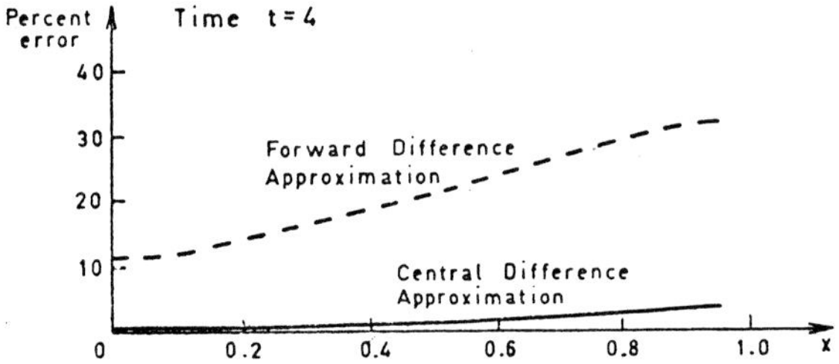

Figure 4.2: Percentage error on $0 \leq x < 1$ using different finite difference approximations to a derivative boundary condition.

When considering the method of weighted averages with the boundary condition at $x = 0$ specified by Equation (4.3.1), substitution of Equations (4.3.5), (4.3.6) into Equation (4.2.13) with $j = 0$, gives

$$(1+2s\beta)\tau_0^{n+1} - 2s\beta\tau_1^{n+1} = \{1-2s(1-\beta)\}\tau_0^n + 2s(1-\beta)\tau_1^n - 2s\Delta x\{(1-\beta)b^n + \beta b^{n+1}\} . \qquad (4.3.8)$$

This equation then becomes the first equation in the system of linear algebraic Equations (4.2.14) which can be solved using the Thomas algorithm.

Note that putting $\beta = 0$ into Equation (4.3.8) yields the corresponding FTCS formula (4.3.7).

Mixed Boundary Conditions

A general example of a mixed boundary condition is

$$\frac{\partial\bar{\tau}}{\partial x} = a\bar{\tau} + b \ , \tag{4.3.9}$$

on the boundary $x = 0$, where both a and b may vary with time, that is, $a = a(t)$, $b = b(t)$, and $a > 0$.

As for the derivative boundary condition, using the forward difference approximation for the spatial derivative at the $(n+1)$th time level with $j = 0$, Equation (4.3.9) becomes

$$\frac{\tau_1^{n+1} - \tau_0^{n+1}}{\Delta x} = a^{n+1}\tau_0^{n+1} + b^{n+1} + O(\Delta x) \tag{4.3.10}$$

where $a^{n+1} = a(t_{n+1})$. Rearranging Equation (4.3.10) yields the finite difference approximation

$$\tau_0^{n+1} = \frac{\tau_1^{n+1} - b^{n+1}\Delta x}{1 + a^{n+1}\Delta x} \ . \tag{4.3.11}$$

Equation (4.3.11) can now be incorporated into the iterative scheme being used so it covers the whole solution region including grid points on the boundary. However, since Equation (4.3.11) is only of first order accuracy, overall the scheme will be formally of first order accuracy even though in the interior of the solution region it is formally of higher accuracy.

To increase the order of accuracy when incorporating the boundary condition (4.3.9) at $x = 0$, the central difference approximation, which has a second order truncation error, may be used for the space derivative. Equation (4.3.9) then yields the fictitious value,

$$\tau_{-1}^{n+1} = \tau_1^{n+1} - 2a^{n+1}\Delta x \ \tau_0^{n+1} - 2b^{n+1}\Delta x \ . \tag{4.3.12}$$

Using this expression for τ_{-1}^{n+1} the values of τ at the boundary $x = 0$ can be calculated using any of the iterative schemes previously described, with $j = 0$. For example, substitution of Equation (4.3.12) with n for $n+1$, namely

$$\tau_{-1}^{n} = \tau_1^{n} - 2a^{n}\Delta x\tau_0^{n} - 2b^{n}\Delta x \ , \tag{4.3.13}$$

into the FTCS equation (2.4.4) with $j = 0$, yields

$$\tau_0^{n+1} = \{1-2s(1+a^n\Delta x)\}\tau_0^n + 2s\ \tau_1^n - 2s\ \Delta x\ b^n \ . \tag{4.3.14}$$

The boundary value τ_0^{n+1} is therefore presented in terms of values of τ known at the previous time level and the known coefficients a^n, b^n of Equation (4.3.9).

Using the mixed boundary condition (4.3.9) in the method of weighted averages, viz. Equation (4.2.13) with $j = 0$, gives

$$\{1+2s\beta(1+\Delta xa^{n+1})\}\tau_0^{n+1} - 2s\beta\tau_1^{n+1}$$
$$= \{1-2s(1-\beta)(1+\Delta xa^n)\}\tau_0^n + 2s(1-\beta)\tau_1^n - 2s\Delta x\{(1-\beta)b^n + \beta b^{n+1}\} \quad . \quad (4.3.15)$$

Equation (4.3.15) becomes the first equation in the system of linear algebraic equations (4.2.14). Again, this system of linear equations can be solved using the Thomas algorithm.

Note that, when $a(t) \equiv 0$, the mixed boundary condition (4.3.9) reduces to the derivative boundary condition (4.3.1) and the resulting equations involving τ_0^{n+1} reduce to the corresponding equations in the derivative boundary case; for example, Equation (4.3.14) reduces to Equation (4.3.7), and Equation (4.3.15) reduces to (4.3.8).

Effect on Stability

Consider the diffusion equation being solved with the mixed boundary condition (4.3.9) at $x = 0$ where a and b are constants, and the previous boundary condition still applying at $x = 1$, namely that $\bar{\tau}$ is known there. If Equation (4.3.12) is used to find τ_0^{n+1} in conjunction with the FTCS method, then the propagation of the errors ξ_j^n is given by the matrix equation (3.3.9), namely

$$\underset{\sim}{\xi}^{n+1} = \underline{\underline{A}}\,\underset{\sim}{\xi}^n \quad ,$$

where

$$\underline{\underline{A}} = \begin{bmatrix} \{1-2s(1+a\Delta x)\} & 2s & & 0 \\ s & (1-2s) & s & \\ & s & (1-2s) & \\ & & \ddots & \\ 0 & & s & (1-2s) \end{bmatrix} , \quad \underset{\sim}{\xi}^n = \begin{bmatrix} \xi_0^n \\ \xi_1^n \\ \xi_2^n \\ \vdots \\ \xi_{J-1}^n \end{bmatrix} .$$

As previously described in the matrix method of stability analysis (Section 3.3) the errors will not increase exponentially as n increases, provided the eigenvalue of $\underline{\underline{A}}$ with the largest modulus has a numerical value less than or equal to unity.

Application of Brauer's theorem assists with the determination of the conditions which result. This theorem states:

> "If R_j is the sum of the moduli of the terms along the j-th row of a matrix $\underline{\underline{A}}$, excluding the diagonal element a_{jj}, then every eigenvalue λ of $\underline{\underline{A}}$ lies inside or on the boundary of at least one of the circle $|\lambda - a_{jj}| = R_j$."

For $j = 2,3,4,\ldots$, $a_{jj} = 1 - 2s$ and $R_j = 2s$, and application of Brauer's theorem yields the same restriction as that found previously for non-derivative boundary conditions, viz.

$$0 \leqslant s \leqslant \tfrac{1}{2} \; . \tag{4.3.16}$$

For $j = 1$, $a_{jj} = 1 - 2s(1 + a\Delta x)$ and $R_j = 2s$, when Brauer's theorem indicates that λ, the eigenvalues of A, lie in the range

$$|\lambda - \{1 - 2s(1 + a\Delta x)\}| \leqslant 2s, \tag{4.3.17}$$

so that

$$1 - 2s(2 + a\Delta x) \leqslant \lambda \leqslant 1 - 2s\, a\Delta x \; .$$

For stability $|\lambda| \leqslant 1$ is required. Since $1 - 2s\, a\Delta x < 1$ for all s, Δx and a, the stability requirement is therefore

$$-1 \leqslant 1 - 2s(2 + a\Delta x)$$

or

$$s \leqslant 1/(2 + a\Delta x) \; . \tag{4.3.18}$$

This condition on s for the FTCS method incorporates the effect of the inclusion of the mixed boundary condition with constant coefficients when the central difference formulation for the spatial derivatives is used for the boundary condition.

4.4 Initial Conditions

Starting Procedures

Given the initial conditions at $t = 0$ there is no difficulty in starting the FTCS method, since going from the time level $n = 0$ to the level $n = 1$ is explicitly determined by Equation (2.4.4). Because the FTCS method approximates the diffusion equation (2.1.1.) with a truncation error of $O\{\Delta t\}$, the Richardson and Du Fort-Frankel methods, which have second order truncation errors in Δt, were developed (see Section 4.1). However, both these explicit methods calculate values of τ at one time level using values from the previous two time levels. Therefore there is a problem with starting these methods, because initial conditions are only given at the one time level, $t = 0$; there are no values of τ available for $t = -\Delta t$, the previous time level. To overcome this difficulty the FTCS method can be used to obtain values at the time level $n = 1$ from those at $n = 0$. Then, by using values at both these time levels, the Richardson or Du Fort-Frankel methods can be used to continue the process. However, because the FTCS method has a first order truncation error in Δt, while the Richardson and Du Fort-Frankel methods have second order truncation errors, the overall accuracy of this procedure can be considered as only of first order.

The classical implicit method, the Crank-Nicholson method and the method of weighted averages are all procedures by which values at one level are computed in terms of known values at only one previous time-level, so they require no special starting procedures. Therefore any second order accurate procedures, such as the Crank-Nicholson method, retain their second order accuracy after starting.

However, not all implicit methods are "two-level" methods such as these. Consider the following finite difference formulation (Richtmeyer, 1951) for the diffusion equation, in which a weighted average of finite difference approximations to the previous two time derivatives is used for the derivative $\partial\tau/\partial t$ at the (j,n) grid point;

$$(1+\beta)\left(\frac{\tau_j^{n+1}-\tau_j^n}{\Delta t}\right) - \beta\left(\frac{\tau_j^n-\tau_j^{n-1}}{\Delta t}\right)$$

$$= \alpha\left\{\frac{\tau_{j+1}^{n+1} - \tau_j^{n+1} + \tau_{j-1}^{n+1}}{(\Delta x)^2}\right\} \qquad (4.4.1)$$

where $0 \leqslant \beta \leqslant 1$. Equation (4.4.1) generally has a truncation error of $O\{\Delta t + (\Delta x)^2\}$, and is always stable. It can be rewritten

$$-s\tau_{j-1}^{n+1} + (1 + 2s + \beta)\tau_j^{n+1} - s\tau_{j+1}^{n+1} = (2\beta+1)\tau_j^n - \beta\tau_j^{n-1}, \qquad (4.4.2)$$

which indicates that it is an implicit "three-level" method.

Choosing $\beta = 0$ yields the classical implicit method, while choosing $\beta = \frac{1}{2}$ gives the formula

$$-s\tau_{j-1}^{n+1} + (3/2 + 2s)\tau_j^{n+1} - s\tau_{j+1}^{n+1} = 2\tau_j^n - 1/2\tau_j^{n-1} \qquad (4.4.3)$$

which has a second order truncation error, viz. $O\{\Delta t)^2 + (\Delta x)^2\}$, and is stable for all $s > 0$.

Richtmeyer has shown that Equation (4.4.2) is extremely useful when dealing with rapidly varying or discontinuous initial data because it damps the short wave-length components very rapidly.

Once again, there is a problem when starting the method. Equation (4.4.2) can be used only after applying the less accurate FTCS method at the first time step.

Initial-value, boundary-condition singularities

As noted in Section 2.1 singularities in the value of $\bar{\tau}$ at the intersection of the initial and boundary regions are often encountered. If the singularity is ignored in the application of a finite difference method, which was the case for the example involving heat diffusion considered in Section 2.4, then values of $\partial^4\bar{\tau}/\partial x^4$ and $\partial^2\bar{\tau}/\partial t^2$ are very large near $x = 0$ and $x = 1$, so that, for a particular pair of values of Δx and Δt the truncation error of the finite difference approximation is much larger near the singularity than near the centre of the rod. As a result, the method is less accurate near the ends of the rod, and much smaller values of Δx may be required there than in the centre of the rod. The use of variable size grids to permit this is described in Section 6.

A better approach, however, is to employ an asymptotic expansion near the singularity, which usually results in the application of a simple analytic expression for the solution which is valid near the singularity.

Consider the example of the insulated rod, at an initial temperature of zero with its ends suddenly heated to $\bar{\tau} = 100$ as described in Section 2.1. Due to symmetry, the region of interest is the interval $0 \leqslant x \leqslant \frac{1}{2}$. Initially only the region near $x = 0$ "feels" the sudden temperature disturbance so that, to start with, the problem is the same as that of a semi-infinite insulated rod initially at 0°C and suddenly heated to

$\bar{\tau} = 100$ at $x = 0$. Thus, for a short time the temperature may be considered to satisfy the diffusion equation (2.1.1) with initial condition

$$\bar{\tau}(x,0) = 0, \quad 0 < x < \infty \tag{4.4.4}$$

and boundary conditions

$$\tau(0,t) = 100, \ \tau(\infty,t) = 0, \ 0 \leqslant t < \infty. \tag{4.4.5}$$

With these conditions the diffusion equation can be solved by introducing new variables $T = 100 - \bar{\tau}$, $\eta = x/(\alpha t)^{\frac{1}{2}}$ which yields the solution

$$\bar{\tau}(x,t) = 100 \ \mathrm{erfc}\left(\frac{x}{2\sqrt{\alpha t}}\right) \ , \tag{4.4.6}$$

where $\mathrm{erfc}(x)$ is the complementary error function (see Kreyszig,1972). This solution is valid for the original diffusion problem until the centre of the rod, at $x = \frac{1}{2}$, begins to experience a significant increase in temperature. At this point in time, approximately $t = (100\alpha)^{-1}$, the finite difference method can be started using values for τ obtained by application of Equation (4.4.6).

5. THE TRANSPORT EQUATION

Finite difference methods for solving the transport equation in one, two and three dimensional space will now be considered.

5.1 The Transport Equation in One Dimension

The one dimensional transport equation

$$\frac{\partial\bar{\tau}}{\partial t} + u\frac{\partial\bar{\tau}}{\partial x} = \alpha\,\frac{\partial^2\bar{\tau}}{\partial x^2} \ , \tag{5.1.1}$$

describes the change in the scalar property $\bar{\tau}$ associated with diffusion and convection in a fluid moving with a speed u parallel to the x-axis. The term $u\partial\bar{\tau}/\partial x$ is the convection term and $\alpha\,\partial^2\bar{\tau}/\partial x^2$ is the diffusion term. Previously, in Sections 2 and 3, the pure diffusion problem was considered, viz.

$$\frac{\partial\bar{\tau}}{\partial t} = \alpha\,\frac{\partial^2\bar{\tau}}{\partial x^2} \ , \tag{2.1.1}$$

which is the same as (5.1.1) with $u = 0$, that is, the fluid is stationary.

In the following it will be assumed that the fluid moves with a constant speed u.

The Convection Equation

Pure convection is described by

$$\frac{\partial\bar{\tau}}{\partial t} + u\frac{\partial\bar{\tau}}{\partial x} = 0 \ , \tag{5.1.2}$$

the general solution of which is, for constant u, $\bar{\tau} = f(x - ut)$ where $f(x)$ is the initial distribution of $\bar{\tau}$. That is, the initial distribution of $\bar{\tau}$ is translated without change along the x-axis at the constant speed u.

Using a forward difference approximation to the time derivative and a central difference approximation in space (compare the FTCS method for the pure diffusion problem) Equation (5.1.2) becomes

$$\frac{\tau_j^{n+1} - \tau_j^n}{\Delta t} + u\left(\frac{\tau_{j+1}^n - \tau_{j-1}^n}{2\Delta x}\right) = 0. \qquad (5.1.3)$$

The resulting explicit finite difference equation is

$$\tau_j^{n+1} = \tau_j^n - \tfrac{1}{2}C(\tau_{j+1}^n - \tau_{j-1}^n) \qquad (5.1.4)$$

where $C = u\Delta t/\Delta x$ is called the Courant number.

If Equation (5.1.4) is analysed for stability using the von Neumann method, it is found that the amplification factor is

$$G = 1 - iC \sin \theta \, . \qquad (5.1.5)$$

Hence the gain $|G|$ is given by

$$|G|^2 = 1 + C^2 \sin \theta \geqslant 1, \qquad (5.1.6)$$

which implies Equation (5.1.4) is unstable for all values of Δx and Δt chosen.

The method known today as upwind differencing was first used by Courant, Isaacson and Rees (1952) in order to overcome the instability of Equation (5.1.3). Since then, the method has been used under various names with different rationales. For example, for some time meteorologists have known of the stabilizing effect of "upwind" (Forsythe and Wasow, 1960) or "weather" (Frankel, 1960) differencing and have applied it to incompressible fluid flow problems. Mathematicians generally refer to the finite difference equations as having "positive coefficients" (Forsythe and Wasow, 1960).

In the upwind differencing method a one sided, rather than space-centred, differencing is used for the space derivative $\partial\bar{\tau}/\partial x$ in the convection term, the direction of the difference being "upwind". That is, backward space differencing is used at the (j,n) grid point when $u > 0$ and forward space differencing is used when $u < 0$. This has the following physical basis. When determining the value of τ_j^{n+1} from known values of τ at the n-th time level, if $u > 0$ the fluid flow carries information from the position x_{j-1} to x_j, so a backward difference approximation is chosen for the spatial derivative in the convection term $u\,\partial\bar{\tau}/\partial x$. To use a forward difference approximation in such a case would be unrealistic, since information cannot be carried from the position x_{j+1} to x_j when $u > 0$. However, this choice would be appropriate if $u < 0$.

When $u < 0$ and u is constant, the forward time, forward space approximation to Equation (5.1.2) is

$$\frac{\tau_j^{n+1} - \tau_j^n}{\Delta t} + u\left(\frac{\tau_{j+1}^n - \tau_j^n}{\Delta x}\right) = 0 \qquad (5.1.7a)$$

Similarly for $u > 0$, the forward time, backward space approximation is

$$\frac{\tau_j^{n+1} - \tau_j^n}{\Delta t} + u\left(\frac{\tau_j^n - \tau_{j-1}^n}{\Delta x}\right) = 0 \ . \qquad (5.1.7b)$$

Both finite difference forms (5.1.7a,b) have a truncation error of $O\{\Delta t + \Delta x\}$. The resulting explicit finite difference equations are

$$\tau_j^{n+1} = (1 + C)\tau_j^n - C\,\tau_{j+1}^n, \quad u < 0 \ , \qquad (5.1.8a)$$

$$= (1 - C)\tau_j^n + C\,\tau_{j-1}^n, \quad u > 0. \qquad (5.1.8b)$$

When u is a positive constant, the errors ξ_j^n propagate according to the relation

$$\xi_j^{n+1} = (1 - C)\xi_j^n + C\,\xi_{j-1}^n \ . \qquad (5.1.9)$$

Application of the von Neumann method of stability analysis to this error equation yields the amplification factor

$$G = (1 - C + C\cos\theta) - iC\sin\theta \ ,$$

so the gain $|G|$ is given by

$$|G|^2 = 1 - 2C(1 - C)(1 - \cos\theta) \ . \qquad (5.1.10)$$

When $C \leqslant 1$, $|G|^2$ has a maximum value of one when $\cos\theta = 1$, since the second term on the right hand side of Equation (5.1.10) is always positive. The method is then stable.

When $C > 1$, $|G|^2$ has a maximum value of $(2C - 1)^2 > 1$ when $\cos\theta = -1$, and the method is unstable.

The required stability criterion is therefore

$$C \leqslant 1 \ . \qquad (5.1.11)$$

The relation (5.1.11) is called the Courant-Friedrichs-Lewy condition. This condition applies generally to explicit finite difference approximations to hyperbolic partial differential equations. It implies that $u\Delta t < \Delta x$, or the fluid should not travel more than one grid spacing in one time step.

If $C = 1$ when $u > 0$, Equation (5.1.8b) becomes

$$\tau_j^{n+1} = \tau_{j-1}^n \ , \qquad (5.1.12)$$

which is the exact solution to the convection equation. The initial distribution of τ has moved a distance Δx in the time Δt since $C = 1$ implies that $u\Delta t = \Delta x$. Condition (5.1.11) implies that $C = 1$ is also the stability limit.

Further for $C < 1$, the method introduces an artificial damping, the values of τ_j^n being reduced in magnitude at successive time levels in the same way as the errors ξ_j^n are reduced, since Equations (5.1.9) and (5.1.8b) are analogous. Checking the consistency of Equation (5.1.8b) shows, on substitution of the exact solution $\bar{\tau}$ for τ, that it is equivalent to

$$\frac{\partial\bar{\tau}}{\partial t} + u\frac{\partial\bar{\tau}}{\partial x} = \alpha' \frac{\partial^2\bar{\tau}}{\partial x^2} + O\{(\Delta x)^2 + (\Delta t)^2\} . \qquad (5.1.13)$$

where $\alpha' = \frac{1}{2}u\Delta x(1 - C)$. Clearly, as $\Delta x \to 0$, $\Delta t \to 0$ the coefficient $\alpha' \to 0$, as do the terms of $O\{(\Delta x)^2 + (\Delta t)^2\}$. Equation (5.1.13) is therefore consistent with the convection equation (5.1.2). Note that the upwind differencing method has introduced an artificial diffusion because of the non-physical coefficient α' of $\partial^2\bar{\tau}/\partial x^2$. For a given $C \neq 1$, large values of the speed u and large values of the space step Δx result in large values of α' and hence considerable artificial diffusion.

Similar results are obtained for $u < 0$.

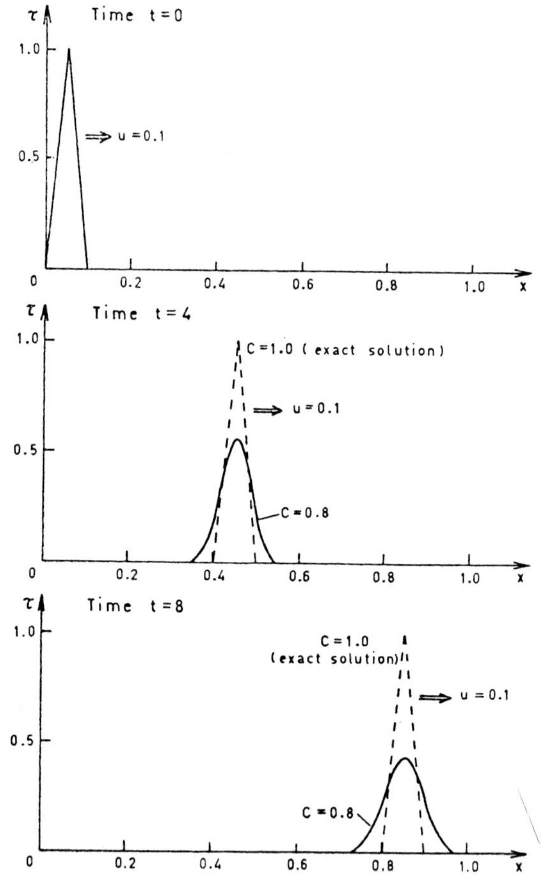

Figure 5.1: Solutions to the convection equation using upstream differencing with the Courant numbers C = 1 and 0.8.

An example which illustrates this artificial diffusion and damping is shown in Figure 5.1. The function $\bar{\tau}(x,t)$ which is governed by the convection equation (5.1.2) is initially given by

$$\bar{\tau}(x,0) = \begin{cases} 20x & , 0 \leqslant x \leqslant 0.05, \\ 20(0.1-x) & , 0.05 \leqslant x \leqslant 0.1 , \\ 0 & , 0.1 \leqslant x \leqslant 1.0, \end{cases}$$

and satisfies boundary conditions

$$\bar{\tau}(0,t) = \bar{\tau}(1,t) = 0 \qquad , t \geqslant 0 \quad .$$

On the interval $0 \leqslant x \leqslant 1$, the exact solution to this problem up to $t = 0.9u^{-1}$ is

$$\bar{\tau}(x,t) = \begin{cases} 0 & , 0 \leqslant x \leqslant ut, \\ 20(x-ut) & , ut \leqslant x \leqslant ut+0.05, \\ 20(0.1-x+ut) & , ut+0.05 \leqslant x \leqslant ut+0.1, \\ 0 & , ut+0.1 \leqslant x \leqslant 1.0 \quad . \end{cases}$$

That is, the peak shown at $t = 0$ in Figure 5.1 simply moves to the right with speed u. In the numerical solution, u was taken to be 0.1 units/second and clearly, when $C = 1$, the correct solution was obtained. However, for other values of C, the upstream finite-difference method had the peak moving to the right with the correct speed, but it underwent diffusion and damping. For $C < 1$ the peak spread out at its base and its height decreased as it moved downstream; this effect increased as C got smaller.

The Complete Transport Equation

Consider the complete one dimensional transport equation (5.1.1). If a central difference approximation is used for the spatial derivative in the convection term, which produces an unconditionally unstable method in the absence of diffusion, Equation (5.1.1) becomes

$$\frac{\tau_j^{n+1} - \tau_j^n}{\Delta t} + u\left(\frac{\tau_{j+1}^n - \tau_{j-1}^n}{2\Delta x}\right) = \alpha\left\{\frac{\tau_{j+1}^n - 2\tau_j^n + \tau_{j-1}^n}{(\Delta x)^2}\right\} \quad . \tag{5.1.14}$$

Written in explicit form this equation becomes

$$\tau_j^{n+1} = (\tfrac{1}{2}C + s)\tau_{j-1}^n + (1 - 2s)\tau_j^n + (-\tfrac{1}{2}C + s)\tau_{j+1}^n \quad . \tag{5.1.15}$$

The amplification factor in the von Neumann stability analysis of Equation (5.1.15) is found to be

$$G = 1 - 2s(1 - \cos\theta) - iC\sin\theta \quad . \tag{5.1.16}$$

The diffusion term in Equation (5.1.12) is responsible for the second term on the right of (5.1.16), and by reducing the real part of G below unity it exercises a stabilizing effect.

The value of the gain $|G|$ is given by the expression

$$|G|^2 = 1 + 4s^2 + 4s^2\cos^2\theta - 4s + 4s\cos\theta - 8s^2\cos\theta + C^2\sin^2\theta. \tag{5.1.17}$$

If $C \leqslant 2s$, then

$$|G|^2 \leqslant 1 - 4s(1 - 2s)(1 - \cos\theta)$$

so that, as $|\cos\theta| \leqslant 1$, $|G|^2 \leqslant 1$ provided $s \leqslant \frac{1}{2}$.

Therefore stablity is assured provided the two conditions

$$C \leqslant 2s \leqslant 1 \tag{5.1.18}$$

are satisfied. Substitution of $u\,\Delta t/\Delta x$ for C and $\alpha\Delta t/(\Delta x)^2$ for s gives

$$r = u\Delta x/\alpha \leqslant 2 \ , \tag{5.1.19}$$

where r can be thought of as a Reynolds number based on the grid size.

In Section 4.1 it was seen that the major disadvantage of applying the Richardson method of using central time differencing with central space differencing to solve the one dimensional diffusion equation, was that it gave a finite difference equation which was unconditionally unstable When this method is applied to the complete one dimensional transport equation (5.1.1), the following finite difference form with a truncation error of $O\{(\Delta t)^2 + (\Delta x)^2\}$ is obtained:

$$\frac{\tau_j^{n+1} - \tau_j^{n-1}}{2\Delta t} + u\left(\frac{\tau_{j+1}^n - \tau_{j-1}^n}{2\Delta x}\right) = \alpha\left(\frac{\tau_{j+1}^n - \tau_j^n + \tau_{j-1}^n}{(\Delta x)^2}\right) . \tag{5.1.20}$$

The resulting explicit finite difference equation, which involves three levels in time, is

$$\tau_j^{n+1} = \tau_j^{n-1} + 2s\ (\tau_{j+1}^n - 2\tau_j^n + \tau_{j-1}^n) - C(\tau_{j+1}^n - \tau_{j-1}^n) \ . \tag{5.1.21}$$

However, like the pure diffusion case, this method is also unconditionally unstable.

Application of the method of DuFort and Frankel (1953), which involves a small change to the Richardson method, produces a conditionally stable system. The centre node value τ_j^n in the approximation to the second-order space derivative $\partial^2\bar{\tau}/\partial x^2$ of the finite difference equation (5.1.20) is replaced by its average at time levels $(n-1)$ and $(n+1)$, giving

$$\frac{\tau_j^{n+1} - \tau_j^{n-1}}{2\Delta t} + u\left(\frac{\tau_{j+1}^n - \tau_{j-1}^n}{2\Delta x}\right) = \alpha\left\{\frac{\tau_{j+1}^n - (\tau_j^{n+1} + \tau_j^{n-1}) + \tau_{j-1}^n}{(\Delta x)^2}\right\} . \tag{5.1.22}$$

Since this equation has values from the $(n+1)$th time level on the right side which appear only at the space position j, it can be solved explicitly and

$$\tau_j^{n+1} = \left(\frac{1-2s}{1+2s}\right)\tau_j^{n-1} + \left(\frac{C+2s}{1+2s}\right)\tau_{j-1}^n - \left(\frac{C-2s}{1+2s}\right)\tau_{j+1}^n \ . \tag{5.1.23}$$

Application of the von Neumann stability analysis show that the only stability restriction on Equation (5.1.23) is the requirement that $C \leqslant 1$.

Using a Taylor series expansion about the grid point (j,n) with the exact value $\bar{\tau}$, replacing τ in Equation (5.1.23), produces the hyperbolic equation

$$\alpha\left(\frac{\Delta t}{\Delta x}\right)^2 \frac{\partial^2\bar{\tau}}{\partial t^2} + \frac{\partial\bar{\tau}}{\partial t} + u\,\frac{\partial\bar{\tau}}{\partial x} = \alpha\,\frac{\partial^2\bar{\tau}}{\partial x^2} + O\{\alpha\Delta t + (\Delta x)^2\}. \tag{5.1.24}$$

DuFort and Frankel (1953) suggest that the stability of the method is due to the introduction of the first term in equation (5.1.24), which makes the equation hyperbolic in nature. As seen with the diffusion equation in Section 4.1, the finite difference equation (5.1.12) is a consistent analogue of the Equation (5.1.1) only if the limits $\Delta x \to 0$, $\Delta t \to 0$ are taken such that $\Delta t/\Delta x \to 0$.

It is clear that the DuFort-Frankel finite difference scheme does have certain disadvantages, but it is a useful explicit conditionally stable method. In particular, although the formal order of the truncation error is $O\{\alpha\Delta t + (\Delta x)^2\}$ in a practical problem with fixed Δx and small α of $O\{\Delta t\}$ Equation (5.1.23) may have a second order truncation error of $O\{(\Delta t)^2 + (\Delta x)^2\}$ in the sense of small "size" of the actual error rather than true "order". For instance, Pearson (1965) has shown that the Du Fort-Frankel method is more accurate than the FTCS method when applied to certain problems.

Using the FTCS approximation for the time derivative and the diffusion terms of Equation (5.1.11), together with the upwind difference approximation to the convection term gives, for $u > 0$,

$$\frac{\tau_j^{n+1}-\tau_j^n}{\Delta t} + u\left(\frac{\tau_j^n-\tau_{j-1}^n}{\Delta x}\right) = \alpha\left\{\frac{\tau_{j+1}^n-2\tau_j^n+\tau_{j-1}^n}{(\Delta x)^2}\right\}, \qquad (5.1.25)$$

which can be rewritten

$$\tau_j^{n+1} = (C + s)\tau_{j-1}^n + (1 - C - 2s)\tau_j^n + s\tau_{j+1}^n . \qquad (5.1.26)$$

The amplification factor in the von Neumann stability analysis of Equation (5.1.26) is

$$G = 1 - (c + 2s)(1 - \cos\theta) - iC\sin\theta, \qquad (5.1.27)$$

so that the gain $|G|$ is given by

$$|G|^2 = 1 - 2C(1-C(1-\cos\theta) - 4s(1-2s)(1-\cos\theta) + 4Cs(1-\cos\theta)^2 - 4s^2\sin^2\theta \quad . \qquad (5.1.28)$$

The stability requirement that

$$|G|^2 \leqslant 1,$$

results in the condition

$$4Cs(1-\cos\theta) \leqslant 2C(1-C) + 4s(1-2s) + 4s^2\sin^2\theta/(1-\cos\theta). \qquad (5.1.29)$$

Since the maximum value of the left side of (5.1.29) is $8Cs$ which occurs when $\cos\theta = -1$, when the right side has the minimum value $2C(1-C)+4s(1-2s)$, the stability criterion becomes

$$8Cs \leqslant 2C(1-C) + 4s(1-2s),$$

or

$$C + 2s \leqslant 1. \qquad (5.1.30)$$

Note that for values of $\cos\theta$ other than -1, the left side of (5.1.29) is reduced by a factor between zero and unity, and the right side is increased by the term $4s^2\sin^2\theta/(1-\cos\theta)$ which is always positive. Therefore, if the inequality (5.1.30) is satisfied, then the condition (5.1.29) will be satisfied for all θ and stability is assured.

In order to check the consistency of the finite difference equation (5.1.26), the approximation τ is replaced by the exact solution $\bar{\tau}$, and a Taylor series expansion of each term is taken about $\bar{\tau}_j^n$. The resulting expression at the grid point (j,n) is

$$\frac{\partial\bar{\tau}}{\partial t} + u\frac{\partial\bar{\tau}}{\partial x} = (\alpha+\alpha')\frac{\partial^2\bar{\tau}}{\partial x^2} + O\{(\Delta t)^2 + (\Delta x)^2\} , \qquad (5.1.31)$$

where

$$\alpha' = \tfrac{1}{2}u\Delta x(1-C) \quad . \qquad (5.1.32)$$

Equation (5.1.32) indicates that the artificial diffusion effect will be minimized for C as close to unity as possible. However, in a practical problem with u variable this is difficult to achieve throughout the complete flow field and so some artificial diffusion will necessarily be present.

Clearly accurate solutions are not possible unless $\alpha' \ll \alpha$ which, in general, requires that $\tfrac{1}{2}u\Delta x \ll \alpha$.

Before leaving this section a caution must be given. By comparing Equations (5.1.30) and (5.1.11) it can be seen that the presence of a diffusion term when using the upwind finite difference approximation has a destabilizing effect. This is in direct contrast to the findings when the central difference approximation to the convection term was used and the presence of the diffusion term had a stabilizing effect.

If α is chosen to be equal to $-\alpha'$, then Equation (5.1.31) becomes the convection equation. This suggests that a possible method of modelling the convection equation using "upstream" differencing for the convective term, when the Courant number C is not equal to unity, is to model the transport equation (5.1.1) with $\alpha = -\alpha'$ at each grid point. Theoretically, this has the effect of cancelling the artificial diffusion term. This can be tested using the example considered previously, with the initial condition illustrated in the first diagram of Figure 5.1. Equation (5.1.31) is modelled using Equation (5.1.26) with

$$\alpha = -\alpha' = -u\Delta x(1-C)$$

and $u = 0.1$, $C = 0.8$. The results obtained are compared with the exact solution of the convection equation in Figure 5.2. Clearly the use of this correction procedure results in a better solution to the problem in that the shape of the peak more closely resembles the exact solution, that is, less diffusion and damping have occurred.

However, some negative values of τ appear behind the peak as it moves from left to right, reflecting the fact that using a negative value of α in the transport equation (5.1.26) represents a situation which is not physically meaningful - that is, having a negative coefficient of diffusion. The negative values of τ occur as part of an oscillation of increasing amplitude which appears behind the peak in the numerical solution. A further disadvantage of using this correction procedure is that the maximum values of the computed peaks now lag behind their true position.

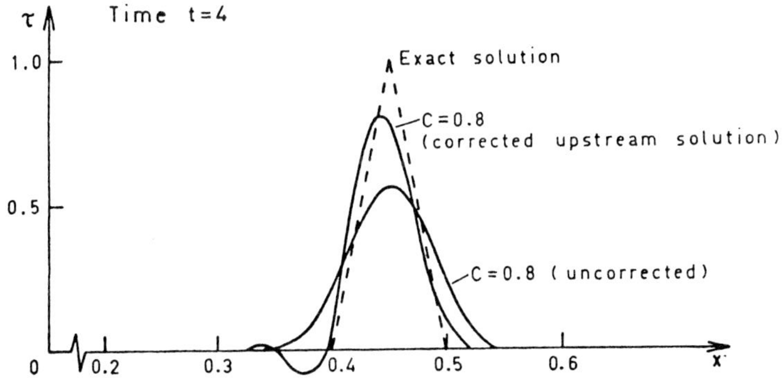

Figure 5.2: A comparison of the "corrected" upstream solution and the uncorrected solution , both for C = 0.8, and the exact solution, at time t = 4.

It is clear that usable solutions can be found using the upwind differencing method. However, the resulting introduction of artificial damping and diffusion must be considered when assessing the accuracy of the results obtained.

5.2 The Transport Equation in Multi-Dimensional Space - Explicit Methods

Finite difference methods of solving the two and three dimensional transport equations are now considered. The transport equation in two dimensional space is

$$\frac{\partial\bar{\tau}}{\partial t} + u\frac{\partial\bar{\tau}}{\partial x} + v\frac{\partial\bar{\tau}}{\partial y} = \alpha\left(\frac{\partial^2\bar{\tau}}{\partial x^2} + \frac{\partial^2\bar{\tau}}{\partial y^2}\right), \tag{5.2.1}$$

while in three dimensions, it is given by

$$\frac{\partial\bar{\tau}}{\partial t} + u\frac{\partial\bar{\tau}}{\partial x} + v\frac{\partial\bar{\tau}}{\partial y} + w\frac{\partial\bar{\tau}}{\partial z} = \alpha\left(\frac{\partial^2\bar{\tau}}{\partial x^2} + \frac{\partial^2\bar{\tau}}{\partial y^2} + \frac{\partial^2\bar{\tau}}{\partial z^2}\right). \tag{5.2.2}$$

Pure Diffusion

The equation for the pure diffusion problem in two space dimensions is

$$\frac{\partial\bar{\tau}}{\partial t} = \alpha\left(\frac{\partial^2\bar{\tau}}{\partial x^2} + \frac{\partial^2\bar{\tau}}{\partial y^2}\right) \tag{5.2.3}$$

The FTCS finite difference approximation for this equation is

$$\begin{aligned}\tau^{n+1}_{j,k} &= \tau^{n}_{j,k} + s_x(\tau^n_{j+1,k} - 2\tau^n_{j,k} + \tau^n_{j-1,k}) \\ &\quad + s_y(\tau^n_{j,k+1} - 2\tau^n_{j,k} + \tau^n_{j,k-1})\end{aligned} \tag{5.2.4}$$

where $\tau^n_{j,k} = \tau(j\Delta x,\, k\Delta y,\, n\Delta t)$ and $s_x = \alpha\Delta t/(\Delta x)^2$, $s_y = \alpha\Delta t/(\Delta y)^2$.

It can be shown, using von Neumann's stability analysis, that Equation (5.2.4) is stable if

$$s_x + s_y \leqslant \tfrac{1}{2} \;. \tag{5.2.5}$$

With equal grid spacings in the x and y coordinate directions, $\Delta x = \Delta y$ and $s_x = s_y = s$, so this relation becomes $s \leqslant \frac{1}{4}$, a condition which is twice as restrictive on the time step Δt as in the one dimensional case. Similarly, for the three dimensional diffusion equation

$$\frac{\partial\bar{\tau}}{\partial t} = \alpha\left(\frac{\partial^2\bar{\tau}}{\partial x^2} + \frac{\partial^2\bar{\tau}}{\partial y^2} + \frac{\partial^2\bar{\tau}}{\partial z^2}\right) \tag{5.2.6}$$

it can be shown that

$$s_x + s_y + s_z \leqslant \tfrac{1}{2} \tag{5.2.7}$$

or, if the grid-spacings are the same in the three spatial coordinate directions then $\Delta x = \Delta y = \Delta z$ and $s_x = s_y = s_z = s \leqslant \frac{1}{6}$.

The Pure Convection Equation

The equation which describes the pure convection problem in two-dimensions is

$$\frac{\partial\bar{\tau}}{\partial t} + u\frac{\partial\bar{\tau}}{\partial x} + v\frac{\partial\bar{\tau}}{\partial y} = 0 \;. \tag{5.2.8}$$

Using upwind differences for the spatial derivatives, the finite difference approximation for $u, v > 0$ is

$$\tau^{n+1}_{j,k} = \tau^n_{j,k} - C_x(\tau^n_{j,k} - \tau^n_{j-1,k}) - C_y(\tau^n_{j,k} - \tau^n_{j,k-1}), \tag{5.2.9}$$

where $C_x = u\Delta t/\Delta x$ and $C_y = v\Delta t/\Delta y$.

Application of the von Neumann stability analysis shows that Equation (5.2.9) is stable if

$$C_x + C_y \leqslant 1 \;. \tag{5.2.10}$$

With $C_x = C_y = C$, this becomes $C \leqslant \frac{1}{2}$, a condition which is twice as restrictive as the corresponding one dimensional case.

This method can also be used in the three dimensional problem and results in the stability condition

$$C_x + C_y + C_z \leqslant 1, \tag{5.2.11}$$

or, if $C_x = C_y = C_z = C$, that $C \leqslant \frac{1}{3}$.

The Transport Equation

Using central difference approximations for the spatial derivatives in Equation (5.2.1) with forward differencing in time, the resulting finite difference formula is

$$\tau^{n+1}_{j,k} = \tau^n_{j,k} - \tfrac{1}{2}C_x(\tau^n_{j+1,k} - \tau^n_{j-1,k}) - \tfrac{1}{2}C_y(\tau^n_{j,k+1} - \tau^n_{j,k-1})$$

$$+ s_x(\tau^n_{j+1,k} - 2\tau^n_{j,k} + \tau^n_{j-1,k}) + s_y(\tau^n_{j,k+1} - 2\tau^n_{j,k} + \tau^n_{j,k-1}). \tag{5.2.12}$$

The stability of Equation (5.2.12) may be determined by application of the von Neumann method in the following way. Each Fourier component of the error distribution $\xi^n_{j,k}$ at the n-th time level may be written

$$\xi^n_{j,k} = A^n e^{i\pi(m_x j\Delta x + m_y k\Delta y)}, \tag{5.2.13}$$

where A^n is the amplitude function at time level n of the particular component whose wave number in the x and y directions are m_x and m_y, and where $i = \sqrt{-1}$. Defining x and y phase angles as $\theta_x = \pi m_x \Delta x$ and $\theta_y = \pi m_y \Delta y$, Equation (5.2.13) becomes

$$\xi^n_{j,k} = A^n e^{i(j\theta_x + k\theta_y)} \tag{5.2.14}$$

Substituting into the error equation corresponding to the finite difference equation (5.2.11), gives the amplification G of an error passing from one time level to another as

$$G = 1 - 2\{(s_x + s_y) - (s_x\cos\theta_x + s_y\cos\theta_y)\} - i(C_x\sin\theta_x + C_y\sin\theta_y). \tag{5.2.15}$$

The value of the gain $|G|$ is therefore given by the equation

$$\left.\begin{aligned}|G|^2 = 1 &- 4\{(s_x + s_y) - (s_x\cos\theta_x + s_y\cos\theta_y)\} \\ &+ 4\{(s_x + s_y) - (s_x\cos\theta_x + s_y\cos\theta_y)\}^2 \\ &+ C_x^2\sin^2\theta_x + 2C_xC_y\sin\theta_x\sin\theta_y + C_y^2\sin\theta_y^2 .\end{aligned}\right\} \tag{5.2.16a}$$

In an analysis similar to that used to determine the stability conditions which apply to (5.1.15) for the one dimensional transport equation, let $C_x \leqslant 2s_x$ and $C_y \leqslant 2s_y$, so that (5.2.16a) becomes

$$|G|^2 \leqslant 1 - 4\{(s_x+s_y) - (s_x\cos\theta_x+s_y\cos\theta_y)\}$$
$$+ 4\{(s_x+s_y)^2-2(s_x+s_y)(s_x\cos\theta_x+s_y\cos\theta_y)+s_x^2\cos^2\theta_x$$
$$+ 2s_xs_y\cos\theta_x\cos\theta_y+s_y^2\cos^2\theta_y\}$$
$$+ 4\{s_x^2\sin^2\theta_x+2s_xs_y\sin\theta_x\sin\theta_y+s_y^2\sin^2\theta_y\}$$

or

$$|G|^2 \leqslant 1 - 4\{(s_x+s_y) - (s_x\cos\theta_x+s_y\cos\theta_y)\}$$
$$+ 4\{(s_x+s_y)^2-2(s_x+s_y)(s_x\cos\theta_x+s_y\cos\theta_y)+s_x^2+2s_xs_y\cos(\theta_x-\theta_y)+s_y^2\} .$$

Since the maximum value of $\cos(\theta_x-\theta_y)$ is 1, then

$$s_x^2 + 2s_xs_y\cos(\theta_x-\theta_y) + s_y^2 \leqslant (s_x+s_y)^2$$

and

$$|G|^2 \leqslant 1 - 4\{(s_x+s_y)-(s_x\cos\theta_x+s_y\cos\theta_y)\}\{1 - 2(s_x+s_y)\} \qquad (5.2.16b)$$

Because $|\cos\theta_x| \leqslant 1$ and $|\cos\theta_y| \leqslant 1$, it follows that $|G|^2 \leqslant 1$ so long as

$$s_x + s_y \leqslant \tfrac{1}{2} . \qquad (5.2.17)$$

If $s_x = s_y = s$, condition (5.2.17) becomes $s \leqslant \tfrac{1}{4}$ which is twice as restrictive as the one dimensional diffusion condition. Also, if $C_x = C_y = C$, then $C \leqslant 2s \leqslant \tfrac{1}{2}$, which is again twice as restrictive as the corresponding one dimensional condition.

Using the upwind difference approximation to the spatial derivatives in the convective terms, assuming $u,v > 0$, gives

$$\tau_{j,k}^{n+1} = \tau_{j,k}^n - C_x(\tau_{j,k}^n-\tau_{j-1,k}^n) - C_y(\tau_{j,k}^n-\tau_{j,k-1}^n)$$
$$+ s_x(\tau_{j+1,k}^n-2\tau_{j,k}^n+\tau_{j-1,k}^n) + s_y(\tau_{j,k+1}^n-2\tau_{j,k}^n+\tau_{j,k-1}^n), \qquad (5.2.18)$$

for which the stability requirement is

$$2s_x + 2s_y + C_x + C_y \leqslant 1 . \qquad (5.2.19)$$

In particular, if $C_x = C_y = C$ and $s_x = s_y = s$, it follows that the upwind difference method is stable if

$$C + 2s \leqslant \tfrac{1}{2} , \qquad (5.2.20)$$

which again is twice as restrictive as in the one dimensional case. In general, for any u and v the requirement on the time step is found to be

$$\Delta t \leqslant \left\{\frac{2\alpha}{(\Delta x)^2} + \frac{2\alpha}{(\Delta y)^2} + \frac{|u|}{\Delta x} + \frac{|v|}{\Delta y}\right\}^{-1} . \qquad (5.2.21)$$

Taylor series expansion of each term of Equation (5.2.18) about the (j,k,n) grid point, with $\bar{\tau}$ substituted for τ, yields

$$\frac{\partial\bar{\tau}}{\partial t} + u\frac{\partial\bar{\tau}}{\partial x} + v\frac{\partial\bar{\tau}}{\partial y} = (\alpha+\alpha_x')\frac{\partial^2\tau}{\partial x^2} + (\alpha+\alpha_y')\frac{\partial^2\tau}{\partial y^2}$$

$$+ O\{\Delta t + (\Delta x)^2\} \quad . \qquad (5.2.22)$$

where

$$\left.\begin{aligned} \alpha_x' &= \tfrac{1}{2}u\Delta x(1 - C_x), \\ \alpha_y' &= \tfrac{1}{2}v\Delta y(1 - C_y). \end{aligned}\right\} \qquad (5.2.23)$$

As $\Delta x \to 0$, $\Delta y \to 0$, $\Delta t \to 0$ then $\alpha_x' \to 0$, $\alpha_y' \to 0$ and the higher order terms on the right side of Equation (5.2.22) also tend to zero. Therefore, the finite difference equation (5.2.18) is consistent with the twc dimensional transport equation.

Even though the upwind differencing method leads to unconditional convergence of the approximation for the solution of the transport equation, the accuracy of this approximation is strongly influenced by the false diffusion introduced by the terms $\alpha_x'\partial^2\bar{\tau}/\partial x^2$ and $\alpha_y'\partial^2\bar{\tau}/\partial y^2$. Unless $\alpha_x' << \alpha$ and $\alpha_y' << \alpha$, the presence of this artificial diffusion may produce an approximate solution τ which differs greatly from the true solution. The central difference approximation does not have such an error, but it is only conditionally convergent.

The appearance of the coefficients α_x' and α_y' is important because at small values of α and large values of u the false diffusion introduced will dominate the true diffusion unless Δx is sufficiently small. de Vahl Davis and Mallinson (1974) give examples from their work on forced convection in two directions which illustrate this point. They showed that the effects of false diffusion were greater in regions where the fluid was moving at 45 degrees to the two coordinate directions. Also, for a nominal Reynolds number of 1000 it was found that the effective Reynolds number was as low as 240 in some regions.

Although the use of upwind differences has a beneficial effect on the stability of the solution, the method only achieves this at the cost of reduced accuracy, which may be dramatic at large Reynolds numbers and may make the method almost useless. Whenever the upwind differencing method is used, α must be large enough to ensure the method is reasonably accurate.

5.3 The Transport Equation in Multi Dimensional Space - Implicit Methods

Iterative Methods

The stability restrictions which applied to the explicit methods described in the previous section are overcome if implicit methods are used, thereby permitting the use of much larger time steps when using finite difference methods. For example, the Crank-Nicolson method for the two dimensional diffusion equation (5.2.3) is

$$\frac{\tau_{j,k}^{n+1}-\tau_{j,k}^{n}}{\Delta t} = \frac{\alpha}{2}\left\{\frac{\tau_{j+1,k}^{n+1}-2\tau_{j,k}^{n+1}+\tau_{j-1,k}^{n+1}}{(\Delta x)^2} + \frac{\tau_{j+1,k}^{n}-2\tau_{j,k}^{n}+\tau_{j-1,k}^{n}}{(\Delta x)^2}\right.$$

$$\left. + \frac{\tau_{j,k+1}^{n+1}-2\tau_{j,k}^{n+1}+\tau_{j-1,k}^{n+1}}{(\Delta y)^2} + \frac{\tau_{j,k+1}^{n}-2\tau_{j,k}^{n}+\tau_{j,k-1}^{n}}{(\Delta y)^2}\right\} \tag{5.3.1}$$

which can be rewritten

$$\tfrac{1}{2}s_x\ \tau_{j+1,k}^{n+1} + \tfrac{1}{2}s_x\ \tau_{j-1,k}^{n+1} - (1+s_x+s_y)\tau_{j,k}^{n+1} + \tfrac{1}{2}s_y\ \tau_{j,k+1}^{n+1} + \tfrac{1}{2}s_y\ \tau_{j,k-1}^{n+1} = d_{j,k}^{n}$$

where (5.3.2)

$$d_{j,k}^{n} = -\tfrac{1}{2}s_x\ \tau_{j+1,k}^{n} - \tfrac{1}{2}s_x\ \tau_{j-1,k}^{n} - (1-s_x-s_y)\tau_{j,k}^{n} - \tfrac{1}{2}s_y\ \tau_{j,k+1}^{n} - \tfrac{1}{2}s_y\ \tau_{j,k-1}^{n}$$

The values of τ on the left side are at the (n+1)th time level, and $d_{j,k}^{n}$ contains all the terms from (5.3.1) which are at the n-th time level. Knowing the values of τ at the n-th time level, those at the (n+1)th level can now be computed.

The system of Equations (5.3.2) is unconditionally stable. However, it does require the solution, at each time step, of a system of equations of order equal to the number of unknown values of τ, namely $(J-1)\times(K-1)$ in the case of a rectangular region in space with known boundary values. The major problem with this method is therefore the large amount of computational time needed to obtain a solution at each time level using conventional elimination methods of solution. Because this system is not tri-diagonal, the very rapid Thomas algorithm cannot be used. An iterative method is therefore used.

An iterative method for solving an equation is one in which a first approximation is used to determine a second approximation, which in turn is used to calculate a third, and so on. Any iterative method is said to be convergent when the difference between the exact solution and the successive approximations tend to zero as the number of iterations increase.

Iterative methods may be used to advantage when solutions are to be found for a set of linear algebraic equations which contain a large number of unknowns, but in which the individual equations only involve a few of these unknowns. The main advantage is speed. This is because iterative methods take advantage of the zeros which occur naturally in the matrix of coefficients for the system of equations, whereas many of the calculations of general elimination methods are concerned with introducing zero coefficients which may already exist.

Consider the two dimensional diffusion equation solved using the Crank-Nicolson method in which, for simplicity, the grid is equally spaced in both the x and y directions. Rewriting equation (5.3.2) using $\Delta x = \Delta y$ gives

$$\tfrac{1}{2}s\ \tau_{j,k-1}^{n+1} + \tfrac{1}{2}s\ \tau_{j-1,k}^{n+1} - (1+2s)\tau_{j,k}^{n+1} + \tfrac{1}{2}s\ \tau_{j+1,k}^{n+1} + \tfrac{1}{2}s\ \tau_{j,k+1}^{n+1} = d_{j,k}^{n}\ , \tag{5.3.3}$$

where $s = s_x = s_y$ and

$$d^n_{j,k} = -\tfrac{1}{2}s\ \tau^n_{j,k-1} - \tfrac{1}{2}s\ \tau^n_{j-1,k} - (1-2s)\tau^n_{j,k} - \tfrac{1}{2}s\ \tau^n_{j+1,k} - \tfrac{1}{2}s\ \tau^n_{j,k+1}\ .$$

The set of equations (5.3.3) can be written with the unknown values of $\tau^{n+1}_{j,k}$ occurring in the $\{k(J-1)+j\}$th place and the equations written in order such that the one with the right side $d^n_{j,k}$ is the $\{k(J-1)+j\}$th equation. Then the term $-(1+2s)\tau^{n+1}_{j,k}$ lies on the leading diagonal of the resulting set of $(J-1)\times(K-1)$ linear equations.

Written in this way the coefficient matrix formed from all equations of the form (5.3.3) is diagonally dominant for all s. Diagonal dominance of such a set of equations occurs when for each equation the absolute value of the coefficient of τ along the leading diagonal of the coefficient matrix is greater than, or equal to, the sum of the absolute values of the coefficients of the other unknown values of τ in that equation, and when strict inequality occurs in at least one equation. For example, for Equation (5.3.3) the requirement is that

$$(1+2s) \geqslant \tfrac{1}{2}s + \tfrac{1}{2}s + \tfrac{1}{2}s + \tfrac{1}{2}s \qquad (5.3.4)$$

which is true for all values of s. Since diagonal dominance of the coefficient matrix of a set of linear equations implies unconditional stability in the solution of these equations, it follows that the implicit method defined by Equation (5.3.3) is always convergent.

In order to solve (5.3.3) by iteration it is rewritten in the form

$$\tau^{n+1}_{j,k} = \frac{1}{2(1+2s)}\left\{s\ \tau^{n+1}_{j-1,k} + s\ \tau^{n+1}_{j+1,k} + s\ \tau^{n+1}_{j,k-1} + s\ \tau^{n+1}_{j,k+1} - 2d^n_{j,k}\right\} \qquad (5.3.5)$$

for all interior grid points. At boundary points the value of τ is either prescribed or must be found from some alternative scheme as described in Section 5.4. At any time level n, the value of $d^n_{j,k}$ is known, while the remaining terms on the right hand side of Equation (5.3.5) can be taken, at a first estimate, to be their values at the time level n. Equation (5.3.5) is therefore used for all j and k for which τ is unknown at the $(n+1)$th time level to produce, successively, a set of values for τ of improving accuracy The iterative form of Equation (5.3.5) is

$$\tau^{(p+1)}_{j,k} = \frac{1}{2(1+2s)}\left\{s\ \tau^{(p)}_{j-1,k} + s\ \tau^{(p)}_{j+1,k} + s\ \tau^{(p)}_{j,k-1} + s\ \tau^{(p)}_{j,k+1} - 2d^n_{j,k}\right\} \qquad (5.3.6)$$

with $\tau^{(o)}_{j,k} = \tau^n_{j,k}$. The bracketed superscript p denotes the iteration number. It is important to remember that the values of $\tau^{(p)}_{j,k}$ $p = 1,2,3,\ldots,$ are successive estimates of the value of $\tau^{n+1}_{j,k}$. A suitable solution is found when successive estimates of $\tau^{n+1}_{j,k}$ differ by less than a specified small quantity for all j and k - that is, when the process converges sufficiently.

As p increases, the iterates $\tau^{(p)}_{j,k}$ obtained will only converge if certain conditions on the coefficients of Equation (5.3.5) are satisfied. As mentioned previously, the implicit finite difference equations are convergent if the coefficient matrix is diagonally dominant. This is equivalent to requiring that the absolute value of τ on the left side of (5.3.5) is greater than the sum of the absolute values of the coefficients of τ on the right side. Clearly, since (5.3.4) applies here, the iteration defined by (5.3.5) is convergent.

This process is known as the Jacobi method. The order, in terms of j and k, in which the (p+1)th iterates for $\tau_{j,k}$ are calculated will not affect the values obtained or their rate of convergence, since they are calculated only in terms of p-th iterates.

However, this method is rather inefficient because it does not use the most up-to-date approximations for the values of τ in the right side of Equation (5.3.5). A better method is to use the most recent iterates as soon as they are available, that is to replace $\tau^{(p)}$ by $\tau^{(p+1)}$ as soon as it is calculated. When a computer is used, this replacement is made systematically. For instance, if the next iterates are computed in order of increasing j and k, with all values of j being used for a given k before the next value of k is taken, then $\tau^{(p+1)}_{j-1,k}$ and $\tau^{(p+1)}_{j,k-1}$ will have been found before $\tau^{(p+1)}_{j,k}$ is evaluated. Therefore it is more efficient to replace Equation (5.3.5) by

$$\tau^{(p+1)}_{j,k} = \frac{1}{2(1+2s)}\left\{s\,\tau^{(p+1)}_{j-1,k} + s\,\tau^{(p+1)}_{j,k-1} + s\,\tau^{(p)}_{j+1,k} + s\,\tau^{(p)}_{j,k+1} - 2d^{n}_{j,k}\right\}. \tag{5.3.7}$$

This procedure is called the Gauss-Seidel or unextrapolated Liebmann method and in this example it converges in less than half the number of iterations for the Jacobi method. Unlike the Jacobi method the order, in terms of j and k, of calculating the (p+1)th iterates $\tau^{(p+1)}_{j,k}$ is fixed.

There is an even better method available based on a modification of the Gauss-Seidel method. Equation (5.3.7) is rewritten as

$$\tau^{(p+1)}_{j,k} = \tau^{(p)}_{j,k} + D^{(p)}_{j,k} \tag{5.3.8}$$

where

$$D^{(p)}_{j,k} = \frac{1}{2(1+2s)}\left\{s\,\tau^{(p+1)}_{j-1,k} + s\,\tau^{(p+1)}_{j,k-1} + s\,\tau^{(p)}_{j+1,k} + s\,\tau^{(p)}_{j,k+1} - 2(1+2s)\tau^{(p)}_{j,k} - 2d^{n}_{j,k}\right\}$$

If we think of the term $D^{(p)}_{j,k}$ as a correction to be added to $\tau^{(p)}_{j,k}$ to make it more exact, then as the process converges $D^{(p)}_{j,k}$ tends to zero. The improvement to the Gauss-Seidel method is now achieved by "over relaxing"; that is, in Equation (5.3.8) the correction term is multiplied by a factor ω, called the relaxation factor, where usually $\omega > 1$, producing

$$\tau^{(p+1)}_{j,k} = \tau^{(p)}_{j,k} + \omega\, D^{(p)}_{j,k} \quad . \tag{5.3.9}$$

This improved Gauss-Seidel method is generally called the successive over-relaxation method (after Young, 1954), or, alternatively, the extrapolated Liebmann method method (Frankel, 1950). If the value of ω lies in the range $0 < \omega < 1$, then the process is called successive under-relaxation. This is not as widely used as successive over-relaxation.

With the successive over-relaxation method much computing time can be saved by determining ω_0 the optimum value of ω for convergence of the iterative scheme. A practical method for finding ω_0 has been given by de Vahl Davis and Isenberg (1975); they chose different values for ω, carried out five to ten test iterations, each time starting from the same initial set of values of τ, and selected the value of ω which gave the most rapid decrease in the difference between successive iterates.

It has been shown that using ω_0, the successive over-relaxation method is faster than the Gauss-Seidel method by a factor of the order of $2/\Delta x$. From this it is obvious that for small Δx it is important to be able to find ω_0.

Formulae for determining ω_0 can be found for certain finite difference methods. For example, solving the one dimensional diffusion problem using successive over-relaxation on the Crank-Nicolson formulae (4.2.8) it is found that (see Smith, 1969)

$$\omega_0 = 2/(1 + \sqrt{1 - \mu^2}\}$$

where

$$\mu = \frac{s}{1+s}\cos(\pi/J) \quad .$$

Here $(J-1)$ is the total number of internal mesh points along a time row, the boundary values being known.

For equations whose solutions require a large number of iterations, the rate of convergence of the Jacobi, Gauss-Seidel and successive over-relaxation method with optimal ω have been found theoretically to be proportional to

$$|\ln \mu|, \quad 2|\ln \mu| \quad \text{and} \quad |\ln(\omega_0 - 1)|$$

respectively. For example, with $J = 10$ and $s = 1$, then $\mu = 0.4756$ and $\omega_0 = 1.064$. Therefore $|\ln \mu| = 0.323$, $2|\ln \mu| = 0.646$, $|\ln(\omega_0-1)| = 1.19$, showing that, in this case, the successive over-relaxation method converges almost twice as fast as the Gauss-Seidel method. If possible, the successive over-relaxation method should be used with a value of ω as near as possible to the optimum value ω_0, because the rate of convergence can be very sensitive to small changes in ω. This is illustrated by Carré (1961) who considered a problem in which $\omega_0 = 1.9$. This value of the relaxation factor ω gave a rate of convergence forty times faster than that obtained using the Gauss-Seidel method ($\omega = 1$) and twice as fast as that given by reducing ω very slightly from 1.9 to 1.875.

Note that in each of these iterative methods, each approximation is found by a single calculation, and errors introduced do not prevent convergence of the process since the use of any incorrect values is equivalent to beginning with a new set of starting values. Hence the methods, if convergent, will converge to the correct values whatever the starting values.

The iterative methods already considered, such as the Gauss-Seidel and successive over-relaxation methods, are examples of point iterative methods. These methods involved explicit correction for the value of τ at each grid point; that is, the correction for a particular $\tau^n_{j,k}$ is the only unknown in the equation being used.

Faster convergence may be obtained using methods referred to as line (or row) iterative methods. These produce successive iterates in an implicit form, generally as a set of simple algebraic equations in tri-diagonal form, which permits the use of the Thomas algorithm described earlier.

For example, consider the solution of the penta-diagonal system given by Equation (5.3.3), namely

$$\tfrac{1}{2}s\ \tau_{j-1,k} + \tfrac{1}{2}s\ \tau_{j+1,k} - (1+2s)\tau_{j,k} + \tfrac{1}{2}s\ \tau_{j,k-1} + \tfrac{1}{2}s\ \tau_{j,k+1} = d^{n}_{j,k} \tag{5.3.10}$$

where the superscript (n+1) indicating the time level has been omitted for convenience. In the line iterative method, the next iterates for all values of τ along the k-th line of grid points $y = y_k$ is determined successively for increasing k, viz. $k = 1,2,3,\ldots,(K-1)$. For a given value of k, the unknown (p+1)th iterates for $\tau_{j-1,k}, \tau_{j,k}, \tau_{j+1,k}$ can be written in terms of the (p+1)th iterate for $\tau_{j,k-1}$, which is known from the calculation for the row (k-1), and the p-th iterate for $\tau_{j,k+1}$, which is known from the previous iteration. The set of equations to be solved is therefore given by

$$\tfrac{1}{2}s\ \tau^{(p+1)}_{j-1,k} - (1+2s)\tau^{(p+1)}_{j,k} + \tfrac{1}{2}s\ \tau^{(p+1)}_{j+1,k} = D^{(p)}_{j,k}. \tag{5.3.11}$$

for $j = 1,2,\ldots,(J-1)$, where the right side is the known value

$$D^{(p)}_{j,k} = d^{n}_{j,k} - \tfrac{1}{2}s\ \tau^{(p+1)}_{j,k-1} - \tfrac{1}{2}s\ \tau^{(p)}_{j,k+1} \quad .$$

Once the (p+1)th iterates for τ^{n+1} along the k-th row of grid points have been computed using the Thomas algorithm, those along the (k+1)th row can be found in the same way. Taking $k = 1,2,\ldots,(K-1)$ in order, the (p+1)th iterates for τ^{n+1} can be found at all grid points on the (n+1)th time level.

The most important advantage of line iterative methods such as the one described is that, for a given value of $\tau^{n}_{j,k}$, less iterations are required for convergence than by using any of the point iterative methods. However, it must also be remembered that the use of implicit methods in place of explicit methods means that more arithmetic operations are needed to produce each new iterate.

When the full transport equation is considered the form of the implicit finite difference equations change. Using the central difference approximation to the convection terms the finite difference approximation to Equation (5.2.1) evaluated at the point $(j\Delta x, k\Delta y, n\Delta t+\tfrac{1}{2}\Delta t)$, is

$$\frac{\tau^{n+1}_{j,k} - \tau^{n}_{j,k}}{\Delta t} = P^{n+1}_{j,k} + P^{n}_{j,k} \tag{5.3.12}$$

where

$$P^{n}_{j,k} = -\tfrac{1}{2}u\frac{(\tau^{n}_{j+1,k} - \tau^{n}_{j-1,k})}{2\Delta x} - \tfrac{1}{2}v\frac{(\tau^{n}_{j,k+1} - \tau^{n}_{j,k-1})}{2\Delta y}$$

$$+ \tfrac{1}{2}\alpha\frac{(\tau^{n}_{j+1,k} - 2\tau^{n}_{j,k} + \tau^{n}_{j-1,k})}{(\Delta x)^2} + \tfrac{1}{2}\alpha\frac{(\tau^{n}_{j,k+1} - 2\tau^{n}_{j,k} + \tau^{n}_{j,k-1})}{(\Delta y)^2}$$

and $P^{n+1}_{j,k}$ is a similar expression in terms of τ^{n+1}. Equation (5.3.12) has a truncation error of $O\{(\Delta t)^2 + (\Delta x)^2 + (\Delta y)^2\}$. Rearranging it gives the implicit finite difference equation

$$(s_y+\tfrac{1}{2}C_y)\tau^{n+1}_{j,k-1} + (s_x+\tfrac{1}{2}C_x)\tau^{n+1}_{j-1,k} - 2(1+s_x+s_y)\tau^{n+1}_{j,k}$$

$$+ (s_x-\tfrac{1}{2}C_x)\tau^{n+1}_{j+1,k} + (s_y-\tfrac{1}{2}C_y)\tau^{n+1}_{j,k+1} = Q^n_{j,k} , \qquad (5.3.13)$$

where $Q^n_{j,k}$ contains all the terms at time level n .

Diagonal dominance and hence stability is assured if

$$2(1+s_x+s_y) > (s_y+\tfrac{1}{2}C_y) + (s_x+\tfrac{1}{2}C_x) + |s_x-\tfrac{1}{2}C_x| + |s_y-\tfrac{1}{2}C_y| \quad .$$

A number of cases need to be considered, depending on the sign of $s_x-\tfrac{1}{2}C_x$ and $s_y-\tfrac{1}{2}C_y$. For simplicity, assume that $s_x = s_y = s$ and $C_x = C_y = C$. Then the stability requirement is

$$1+2s \geqslant s+\tfrac{1}{2}C + |s-\tfrac{1}{2}C| \quad .$$

For $C < 2s$, stability of Equation (5.3.13) is assured. However, if $C > 2s$, the requirement becomes

$$C \leqslant 2s + 1 \quad . \qquad (5.3.14)$$

Therefore, C may exceed $2s$, but only by unity. Clearly, if the cell Reynolds number $r = u\Delta x/\alpha$ is less than 2, then the above condition (5.3.14) is met.

If the convection terms in the transport equation evaluated at the point $(j\Delta x, k\Delta y, n\Delta t+\tfrac{1}{2}\Delta t)$ are approximated by upwind differences, the corresponding finite difference equation for $u > 0$, $v > 0$ is

$$(s_y-C_y)\tau^{n+1}_{j,k-1} + (s_x-C_x)\tau^{n+1}_{j-1,k} - 2(1+s_x+s_y-\tfrac{1}{2}C-\tfrac{1}{2}C_y)\tau^{n+1}_{j,k}$$

$$+ s_x\tau^{n+1}_{j+1,k} + s_y\tau^{n+1}_{j,k+1} = Q^n_{j,k} , \qquad (5.3.15)$$

where $Q^n_{j,k}$ contains all the terms in τ^n. Stability of (5.3.15) is assured if there is diagonal dominance of the coefficient matrix; that is, if

$$2(1+s_x+s_y-\tfrac{1}{2}C_x-\tfrac{1}{2}C_y) \geqslant |s_y-C_y| + |s_x-C_x| + s_x + s_y \quad .$$

Again, for simplicity, let $s_x = s_y = s$ and $C_x = C_y = C$. The above stability requirement then becomes

$$1+2s-C \geqslant |s-C| + s \ .$$

If $C < s$, stability is assured. However, if $C > s$, the requirement becomes

$$C \lesssim s + \tfrac{1}{2} \quad ,$$

so that C may exceed s but only by $\tfrac{1}{2}$. Clearly, if the cell Reynolds number r is less than unity, then the above condition is satisfied.

Alternating Direction Implicit Methods

Using the Crank-Nicolson method to solve the two dimensional diffusion equation (5.2.3) gave the unconditionally stable finite difference equation (5.3.2). The resulting system of equations is not tri-diagonal so the Thomas algorithm cannot be used in their solution.

The alternating direction implicit method, sometimes called the method of variable direction, was introduced by Peaceman and Rachford (1955) and Douglas (1955) in order to develop an implicit finite difference method which only required the inversion of a tri-diagonal system of equations. This method requires two sets of calculations in order to take one step forward in time; a set of intermediate values τ^* with no physical significance are computed at the midpoint of the normal time interval Δt.

In the first step, values of $\tau^*_{j,k}$ are calculated for the time level $(n+\frac{1}{2})\Delta t$ using central difference approximations for both $\partial^2\bar{\tau}/\partial x^2$ and $\partial^2\bar{\tau}/\partial y^2$, the former being written at the time level $(n+\frac{1}{2})\Delta t$ and the latter at the time-level $n\Delta t$. The resulting finite difference approximation is

$$\frac{\tau^*_{j,k}-\tau^n_{j,k}}{\frac{1}{2}\Delta t} = \alpha\left\{\frac{\tau^*_{j+1,k}-2\tau^*_{j,k}+\tau^*_{j-1,k}}{(\Delta x)^2} + \frac{\tau^n_{j,k+1}-2\tau^n_{j,k}+\tau^n_{j,k-1}}{(\Delta y)^2}\right\} . \quad (5.3.16)$$

Using this equation, the unknown values of τ^* at the time level $(n+\frac{1}{2})\Delta t$ are computed along a line in the x-direction using known values of τ which lie along a line in the y-direction at the time level $n\Delta t$. This is seen from the rewritten form of Equation (5.3.16),

$$-\tfrac{1}{2}s_x\tau^*_{j-1,k} + (1+s_x)\tau^*_{j,k} - \tfrac{1}{2}s_x\tau^*_{j+1,k} = d^n_{j,k} \quad (5.3.17)$$

where

$$d^n_{j,k} = \tfrac{1}{2}s\,\tau^n_{j,k-1} + (1-s_y)\tau^n_{j,k} + \tfrac{1}{2}s_y\tau^n_{j,k+1} .$$

The system of equations obtained for a given k by substituting $j = 1,2,3,\dots$ in Equation (5.3.16) is tri-diagonal, so it can be solved, using the Thomas algorithm. On a rectangular grid where $j = 0,1,2,\dots,J$ and $k = 0,1,2,\dots,K$, then $(K-1)$ applications of the Thomas algorithm gives the intermediate values $\tau^*_{j,k}$ over the whole grid.

Checking the stability of the finite difference equation (5.3.16) by the von Neumann method, the error distribution at the time-level $n\Delta t$ is considered to have spatial Fourier components of the form $\xi^n_{j,k} = A^n e^{i(j\theta_x+k\theta_y)}$ while it has the component $\xi^*_{j,k} = A^*e^{i(j\theta_x+k\theta_y)}$ at the time-level $(n+\frac{1}{2})\Delta t$. The amplification factor of this error when computing values at time level $(n+\frac{1}{2})\Delta t$ from those at time level $n\Delta t$ is

$$G' = \frac{1-2s_y\sin^2(\theta_y/2)}{1+2s_x\sin^2(\theta_x/2)} . \quad (5.3.18)$$

For $\theta_x \to 0$ and $\theta_y \to \pi$, $|G'| > 1$ for sufficiently large values of s_y. This step is therefore only conditionally stable.

At the second step, values of $\tau^{n+1}_{j,k}$ are computed from the values of $\tau^*_{j,k}$ found at the first step, by writing the central difference approximation for $\partial^2\bar{\tau}/\partial x^2$ at the intermediate time level $(n+\frac{1}{2})\Delta t$ and the central difference approximation for $\partial^2\bar{\tau}/\partial y^2$ at the new time level $(n+1)\Delta t$. The resulting finite difference approximation is

$$\frac{\tau^{n+1}_{j,k}-\tau^*_{j,k}}{\frac{1}{2}\Delta t} = \alpha\left\{\frac{\tau^*_{j+1,k}-2\tau^*_{j,k}+\tau^*_{j-1,k}}{(\Delta x)^2} + \frac{\tau^{n+1}_{j,k+1}-2\tau^{n+1}_{j,k}+\tau^{n+1}_{j,k-1}}{(\Delta y)^2}\right\} \quad (5.3.19)$$

which can be rewritten in the implicit form

$$-\tfrac{1}{2}s_y\,\tau^{n+1}_{j,k-1} + (1+s_y)\,\tau^{n+1}_{j,k} - \tfrac{1}{2}s_y\,\tau^{n+1}_{j,k+1} = d^*_{j,k} \quad (5.3.20)$$

where

$$d^*_{j,k} = \tfrac{1}{2}s_x\,\tau^*_{j-1,k} + (1-s_x)\,\tau^*_{j,k} + \tfrac{1}{2}s_x\,\tau^*_{j+1,k} \quad .$$

In contrast to Equation (5.3.17), Equation (5.3.20) is implicit in the y-direction. Along each grid line $x = j\Delta x$ a tri-diagonal system of equations is obtained which can be solved by means of the Thomas algorithm. By letting j take all its possible values, the values of $\tau^{n+1}_{j,k}$ can be found over the whole grid, that is, (J-1) applications of the Thomas algorithm are needed. Note that values of the intermediate variable $\tau^*_{j,k}$ have no physical significance; they should be discarded as soon as the values at the next time level have been calculated.

Application of the von Neumann stability analysis shows that the second step (5.3.20) is also only conditionally stable, the amplification factor of the error propagation being

$$G'' = \frac{1-2s_x\,\sin^2(\theta_x/2)}{1+2s_y\,\sin^2(\theta_y/2)} \quad . \quad (5.3.21)$$

However, the double process in which Equation (5.3.17) is followed by Equation (5.3.20) at each time step has an amplification factor of

$$G = G'G'' = \frac{1-2s_y\,\sin^2(\theta_y/2)}{1+2s_x\,\sin^2(\theta_x/2)} \cdot \frac{1-2s_x\,\sin^2(\theta_x/2)}{1+2s_y\,\sin^2(\theta_y/2)} \quad (5.3.22)$$

and $|G| \leq 1$ for all values of $s_x, s_y, \theta_x, \theta_y$. The two conditionally stable steps have combined to produce a method which is always stable.

Although it may not seem plausible that intermediate values which have no significance, computed half a time step forward, can be used to obtain correct values another half time step later, the Crank-Nicolson equation for the one-dimensional diffusion equation is the result of such a process. This was shown in Section 4.2.

This method also has truncation error of $O\{(\Delta t)^2 + (\Delta x)^2 + (\Delta y)^2\}$. The second order accuracy is obvious when Equations (5.3.15) and (5.3.19) are added to give

$$\frac{\tau^{n+1}_{j,k}-\tau^{n}_{j,k}}{\Delta t} = \alpha\left\{\frac{\tau^{*}_{j+1,k}-2\tau^{*}_{j,k}+\tau^{*}_{j-1,k}}{(\Delta x)^2} + \tfrac{1}{2}\left(\frac{\tau^{n}_{j,k+1}-2\tau^{n}_{j,k}+\tau^{n}_{j,k-1}}{(\Delta y)^2} + \frac{\tau^{n+1}_{j,k+1}-2\tau^{n+1}_{j,k}+\tau^{n+1}_{j,k-1}}{(\Delta y)^2}\right)\right\}. \quad (5.3.23)$$

The truncation error is of order $(\Delta t)^2$ because the central difference approximation was used about the time level $(n+\frac{1}{2})\Delta t$ for the derivative $\partial\tau/\partial t$. The error is also second order in Δx and Δy because central difference approximations were used for the spatial derivatives $\partial^2\bar{\tau}/\partial x^2$ and $\partial^2\bar{\tau}/\partial y^2$.

The alternating direction implicit method can also be applied to solve the three dimensional diffusion problem. The most obvious method, described by Brian (1961), uses the same process as the two dimensional case, that is, the time interval Δt is divided up into three equal parts. Hence, three calculations are performed with two sets of artificial intermediate values being calculated at $(n+\frac{1}{3})\Delta t$ and $(n+\frac{2}{3})\Delta t$. This method, however, has only first order accuracy in time and becomes unstable for $s > 3/2$.

Another method for the advancement of the diffusion equation (5.2.6) over the time interval Δt, which is second order accurate in time and is unconditionally stable, is described by Douglas (1962). This method is once again a three-step process but differs from Brian's method in that the two intermediate fictitious values are not calculated at the times $(n+\frac{1}{3})\Delta t$ and $(n+\frac{2}{3})\Delta t$ but rather they are used to find successive approximations for τ at the (n+1)th time level. That is, this process is similar to a three-step iterative procedure for calculating τ^{n+1} in which the formula for the iteration process changes at each step.

This method gives the following finite difference equations where $\tau^{(n+1)*}_{j,k,\ell}$ and $\tau^{(n+1)**}_{j,k,\ell}$ refer to the first and second approximations respectively of the value of $\tau^{(n+1)}_{j,k,\ell}$ at the grid-point $(x_j, y_k, z_\ell, t_{n+1})$:

$$\frac{\tau^{(n+1)*}_{j,k,\ell}-\tau^{n}_{j,k,\ell}}{\Delta t} = \alpha\left\{\tfrac{1}{2}\left(\frac{\tau^{(n+1)*}_{j+1,k,\ell}-2\tau^{(n+1)*}_{j,k,\ell}+\tau^{(n+1)*}_{j-1,k,\ell}}{(\Delta x)^2} + \frac{\tau^{n}_{j+1,k,\ell}-2\tau^{n}_{j,k,\ell}+\tau^{n}_{j-1,k,\ell}}{(\Delta x)^2}\right) + \frac{\tau^{n}_{j,k+1,\ell}-2\tau^{n}_{j,k,\ell}+\tau^{n}_{j,k-1,\ell}}{(\Delta y)^2} + \frac{\tau^{n}_{j,k,\ell+1}-2\tau^{n}_{j,k,\ell}+\tau^{n}_{j,k,\ell-1}}{(\Delta z)^2}\right\} \quad (5.3.24a)$$

which is implicit in $\tau^{(n+1)*}_{j-1,k,\ell}, \tau^{(n+1)*}_{j,k,\ell}$ and $\tau^{(n+1)*}_{j+1,k,\ell}$;

$$\frac{\tau^{(n+1)**}_{j,k,\ell}-\tau_{j,k,\ell}}{\Delta t} = \alpha\left\{\tfrac{1}{2}\left(\frac{\tau^{(n+1)*}_{j+1,k,\ell}-2\tau^{(n+1)*}_{j,k,\ell}+\tau^{(n+1)*}_{j-1,k,\ell}}{(\Delta x)^2}+\frac{\tau^{n}_{j+1,k,\ell}-2\tau^{n}_{j,k,\ell}+\tau^{n}_{j-1,k,\ell}}{(\Delta x)^2}\right)\right.$$
$$+\tfrac{1}{2}\left(\frac{\tau^{(n+1)**}_{j,k+1,\ell}-2\tau^{(n+1)**}_{j,k,\ell}+\tau^{(n+1)**}_{j,k-1,\ell}}{(\Delta y)^2}+\frac{\tau^{n}_{j,k+1,\ell}-2\tau^{n}_{j,k,\ell}+\tau^{n}_{j,k-1,\ell}}{(\Delta y)^2}\right)$$
$$\left.+\frac{\tau^{n}_{j,k,\ell+1}-2\tau^{n}_{j,k,\ell}+\tau^{n}_{j,k,\ell-1}}{(\Delta z)^2}\right\} \tag{5.3.24b}$$

which is implicit in $\tau^{(n+1)**}_{j,k-1,\ell}$, $\tau^{(n+1)**}_{j,k,\ell}$ and $\tau^{(n+1)**}_{j,k+1,\ell}$;

$$\frac{\tau^{n+1}_{j,k,\ell}-\tau^{n}_{j,k,\ell}}{\Delta t} = \alpha\left\{\tfrac{1}{2}\left(\frac{\tau^{(n+1)*}_{j+1,k,\ell}-2\tau^{(n+1)*}_{j,k,\ell}+\tau^{(n+1)*}_{j-1,k,\ell}}{(\Delta x)^2}+\frac{\tau^{n}_{j+1,k,\ell}-2\tau^{n}_{j,k,\ell}+\tau^{n}_{j-1,k,\ell}}{(\Delta x)^2}\right)\right.$$
$$+\tfrac{1}{2}\left(\frac{\tau^{(n+1)**}_{j,k+1,\ell}-2\tau^{(n+1)**}_{j,k,\ell}+\tau^{(n+1)**}_{j,k-1,\ell}}{(\Delta y)^2}+\frac{\tau^{n}_{j,k+1,\ell}-2\tau^{n}_{j,k,\ell}+\tau^{n}_{j,k-1,\ell}}{(\Delta y)^2}\right)$$
$$\left.+\tfrac{1}{2}\left(\frac{\tau^{n+1}_{j,k,\ell+1}-2\tau^{n+1}_{j,k,\ell}+\tau^{n+1}_{j,k,\ell-1}}{(\Delta z)^2}+\frac{\tau^{n}_{j,k,\ell+1}-2\tau^{n}_{j,k,\ell}+\tau^{n}_{j,k,\ell-1}}{(\Delta z)^2}\right)\right\} \tag{5.3.24c}$$

which is implicit in $\tau^{n+1}_{j,k,\ell-1}$, $\tau^{n+1}_{j,k,\ell}$ and $\tau^{n+1}_{j,k,\ell+1}$. This method becomes only conditionally stable if in Equation (5.3.24c) the most recent approximation $\tau^{(n+1)**}$ is used instead of $\tau^{(n+1)*}$ in the finite difference approximation to the derivative $\partial^2\overline{\tau}/\partial x^2$ (see Richtmeyer and Morton, 1967).

Alternating direction implicit methods are used widely for the solution of parabolic and hyperbolic partial differential equations in both two and three dimensions. However, Birkoff and Varga (1959) point out that the unconditional stability of the method applied to the diffusion equation (see Douglas 1955, 1957) does not necessarily apply to differential equations with variable coefficients nor to non-rectangular regions. Nevertheless, numerical experimentation indicates that they are applicable to a wider range of partial differential equations than those with constant coefficients prescribed on a rectangular domain.

When alternating direction implicit methods are applied to the complete transport equation, it is found (see Pearson, 1964) that the inclusion of the advection terms does not change the unconditional stability as it does with the fully implicit method, which is only conditionally stable when applied to the transport equation.

Locally One Dimensional Methods

Recently, Russian numerical analysts such as D'Yakonov (1963), Samarskii (1964), Yanenko (1971) and Marchuk (1975) have developed a splitting of time dependent partial differential equations in two or more space variables, so that the resulting set of partial differential equations are one dimensional in space. These methods are called <u>locally one dimensional</u>.

By this method the two dimensional diffusion equation (5.2.3) is split into the pair of equations

$$\tfrac{1}{2}\frac{\partial \tau}{\partial t} = \alpha \frac{\partial^2 \tau}{\partial x^2} \qquad (5.3.25a)$$

and

$$\tfrac{1}{2}\frac{\partial \tau}{\partial t} = \alpha \frac{\partial^2 \tau}{\partial y^2} \quad . \qquad (5.3.25b)$$

In advancing a calculation from the time level n to time level $(n+1)$ it is assumed that Equation (5.3.25a) holds from $t = n\Delta t$ to $t = (n+\tfrac{1}{2})\Delta t$ giving a set of values $\tau^{(n+\frac{1}{2})*}$ at each point in space, which have no physical significance and must be discarded, from which the set of values of τ at $t = (n+1)\Delta t$ are computed using Equation (5.3.25b).

The split equations (5.3.24) can be solved by either explicit or implicit methods. Application of the FTCS method to these two equations leads to the explicit forms

$$\tau^{(n+\frac{1}{2})*}_{j,k} = s_x\, \tau^n_{j-1,k} + (1-2s_x)\tau^n_{j,k} + s_x\, \tau^n_{j+1,k} \qquad (5.3.26a)$$

and

$$\tau^{n+1}_{j,k} = s_y\, \tau^{(n+\frac{1}{2})*}_{j,k-1} + (1-2s_y)\tau^{(n+\frac{1}{2})*}_{j,k} + s_y\, \tau^{(n+\frac{1}{2})*}_{j,k+1} \quad . \qquad (5.3.26b)$$

The stability condition for the pair of equations (5.3.26) can be determined by the von Neumann method. Inserting

$$\xi^n_{j,k} = A^n e^{i(j\theta_x + k\theta_y)} \quad ,$$

the Fourier component of the error distribution at the n-th time level, into the corresponding error equations to (5.3.25a,b) gives the amplification factor

$$G = \{1-4s_x \sin^2(\theta_x/2)\}\{1-4s_y \sin^2(\theta_y/2)\} \qquad (5.3.27)$$

for the propagation from one time level to another. Therefore the stability requirement $|G| \leqslant 1$ holds if

$$|1-4s_x \sin^2(\theta_x/2)|\,|1-4s_y \sin^2(\theta_y/2)| \leqslant 1 \ . \qquad (5.3.28)$$

This condition is satisfied if both $|1-4s_x \sin^2(\theta_x/2)| \leqslant 1$ and $|1-4s_y \sin^2(\theta_y/2)| \leqslant 1$. Using the results of Section 3.3 this is true if both $s_x \leqslant \tfrac{1}{2}$ and $s_y \leqslant \tfrac{1}{2}$.

When grid spacings in the x and y coordinate directions are the same, i.e. $s_x = s_y = s$, the stability requirement becomes $s \leqslant \tfrac{1}{2}$. This is an improvement on the usual explicit restriction of $s \leqslant \tfrac{1}{4}$ which holds for the FTCS method of solving the two dimensional diffusion equation. This is because Equations (5.3.26) relate the value of τ at one point at the $(n+1)$ time level with values at nine points at the n-th time level, since elimination of $\tau^{(n+\frac{1}{2})*}$ from these finite difference equations gives

$$\left.\begin{aligned}\tau^{n+1}_{j,k} &= s_x s_y \tau^n_{j-1,k-1} + (1-2s_x)s_y\tau^n_{j,k-1} + s_x s_y \tau^n_{j+1,k-1}\\ &+ s_x(1-s_y)\tau^n_{j-1,k} + (1-2s_x)(1-2s_y)\tau^n_{j,k} + s_x(1-s_y)\tau^n_{j+1,k}\\ &+ s_x s_y \tau^n_{j-1,k+1} + (1-2s_x)s_y \tau^n_{j,k+1} + s_x s_y \tau^n_{j+1,k+1}\end{aligned}\right\} \quad (5.3.29)$$

This contrasts with the FTCS explicit method (5.2.4) which relates a value of τ at any time level with values of τ at only five points at the previous time level.

Two other features of this locally one dimensional method are worth noting. Firstly, (5.3.29) approximates the two dimensional diffusion equation (5.2.3) with a truncation error of $O\{\Delta t + (\Delta x)^2\}$, the same as the FTCS method of solution. Secondly, the use of Equations (5.3.26a,b) is more economical than a direct application of (5.3.29) since, for τ calculated at every gridpoint at the $(n+1)$th level, fewer arithmetic operations are required using the pair of Equations (5.3.26) than the number required when (5.3.29) is used.

Applying the Crank-Nicolson method gives the approximation to the Equation (5.3.25a) at the point $(j\Delta x, k\Delta y, n\Delta t+\frac{1}{2}\Delta t)$ in the solution space,

$$\frac{1}{2}\left\{\frac{\tau^{(n+\frac{1}{2})*}_{j,k} - \tau^n_{j,k}}{\frac{1}{2}\Delta t}\right\} = \frac{\alpha}{2}\left\{\frac{\tau^{(n+\frac{1}{2})*}_{j+1,k} - 2\tau^{(n+\frac{1}{2})*}_{j,k} + \tau^{(n+\frac{1}{2})*}_{j-1,k}}{(\Delta x)^2} + \frac{\tau^n_{j+1,k} - 2\tau^n_{j,k} + \tau^n_{j-1,k}}{(\Delta x)^2}\right\} .$$

Rearrangement yields the implicit finite difference equation

$$-\frac{1}{2}s_x\,\tau^{(n+\frac{1}{2})*}_{j-1,k} + (1+s_x)\tau^{(n+\frac{1}{2})*}_{j,k} - \frac{1}{2}s_x\,\tau^{(n+\frac{1}{2})*}_{j+1,k} = d^n_{j,k} \quad (5.3.30a)$$

where $d^n_{j,k}$ contains only values of τ from the n-th time level, viz.

$$d^n_{j,k} = \frac{1}{2}s_x\,\tau^n_{j-1,k} + (1-s_x)\tau^n_{j,k} + \frac{1}{2}s_x\,\tau^n_{j+1,k} .$$

Similarly, the Crank-Nicolson approximation to Equation (5.3.25b) yields the implicit formula

$$-\frac{1}{2}s_y\,\tau^{n+1}_{j,k-1} + (1+s_y)\tau^{n+1}_{j,k} - \frac{1}{2}s_y\,\tau^{n+1}_{j,k+1} = d^{(n+\frac{1}{2})*}_{j,k} \quad (5.3.30b)$$

where $d^{(n+\frac{1}{2})*}_{j,k}$ contains only the artificial values of $\tau^{(n+\frac{1}{2})*}$ which are discarded at each time step, viz.

$$d^{(n+\frac{1}{2})*}_{j,k} = \frac{1}{2}s_y\,\tau^{(n+\frac{1}{2})*}_{j,k-1} + (1-s_y)\tau^{(n+\frac{1}{2})*}_{j,k} + \frac{1}{2}s_y\,\tau^{(n+\frac{1}{2})*}_{j,k+1} .$$

The von Neumann stability analysis readily shows that the system of Equations (5.3.30a,b) is unconditionally stable.

In a similar way, the time-dependent diffusion equation (5.2.6) in three dimensions can be split into

$$\frac{1}{3}\frac{\partial\bar{\tau}}{\partial t} = \alpha\frac{\partial^2\bar{\tau}}{\partial x^2}, \tag{5.3.31a}$$

$$\frac{1}{3}\frac{\partial\bar{\tau}}{\partial t} = \alpha\frac{\partial^2\bar{\tau}}{\partial y^2}, \tag{5.3.31b}$$

and

$$\frac{1}{3}\frac{\partial\bar{\tau}}{\partial t} = \alpha\frac{\partial^2\bar{\tau}}{\partial z^2} \tag{5.3.31c}$$

Again this gives a set of one dimensional equations to be solved by either explicit or implicit methods for each time step of length $\Delta t/3$ to give values of τ^{n+1}. Thus Equation (5.3.31a) is solved over the interval $t = n\Delta t$ to $t = (n+\frac{1}{3})\Delta t$ to give a fictitious set of values $\tau^{(n+\frac{1}{3})^*}$ at the time $t = (n+\frac{1}{3})\Delta t$; Equation (5.3.31b) is then used with the values $\tau^{(n+\frac{1}{3})^*}$ at the beginning of the time interval $(n+\frac{1}{3})\Delta t$ to $(n+\frac{2}{3})\Delta t$ to obtain a fictitious set of values $\tau^{(n+\frac{2}{3})^*}$; these are used at the start of a further time step of $\Delta t/3$ with Equation (5.3.31c) to give an approximate value for $\bar{\tau}$ at the time $t = (n+1)\Delta t$

Gourlay and Mitchell (1969) have pointed out an interesting connection between the alternating direction implicit method given by Equations (5.3.17) and (5.3.19) for the two dimensional time dependent diffusion equation, and the locally one dimensional method (5.3.30).

Because they are relatively new, not many applications of locally one dimensional methods can be found in the literature, but their use is increasing, particularly because of their rather simple form when applied to three dimensional problems.

An Explicit-Implicit Method

Consider the time dependent diffusion equation (5.2.3) to be solved on the two-dimensional rectangular region $R: 0 \leq x \leq 1,\ 0 \leq y \leq K\Delta y$ with $\bar{\tau}$, the dependent variable, defined initially on R and for all time t on the boundary.

The following pseudo-explicit method (see Noye, 1977, 1978) which is based on the linearity of equation (5.2.3) and its corresponding finite

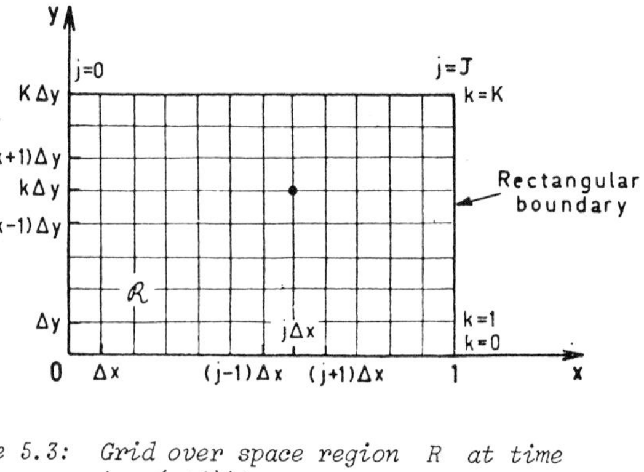

Figure 5.3: Grid over space region R at time t = (n+1)Δt.

difference approximation, can be applied to obtain values of τ at the interior points $(j\Delta x, k\Delta y)$, $j = 1,2,\ldots,J-1$, $k = 1,2,\ldots,K-1$, in R at the time $t = n\Delta t$, $n = 1,2,3,\ldots$. (See Figure 5.1).

Application of the Crank-Nicolson method to Equation (5.2.3) yields the unconditionally stable implicit scheme (5.3.2). This equation may be rewritten in the form

$$\tau_{j,k+1} = 2(1+s_x s_y^{-1}+s_y^{-1})\tau_{j,k} - s_x s_y^{-1}(\tau_{j+1,k}+\tau_{j-1,k}) - \tau_{j,k-1} + 2s_y^{-1}d^n_{j,k}, \tag{5.3.32}$$

where, for convenience, the superscripts $(n+1)$ indicating that the values of τ are those at the new time level have been omitted and the term $d^n_{j,k}$ contains the known terms from the previous time level n. This is a quasi-explicit form for the computation of τ at time level $(n+1)$ since, for $j = 1,2,\ldots,(J-1)$, Equation (5.3.32) defines values of τ explicitly at grid points along the line $y = (k+1)\Delta y$ in terms of values of τ along the two rows of grid points immediately below, namely along $y = k\Delta y$ and $y = (k-1)\Delta y$, as shown in Figure 5.1. Values of $\tau_{j,2}$ for $j = 1,2,3,\ldots,(J-1)$, can be calculated using Equation (5.3.32) with $k = 1$, provided the values of $\tau_{j,1}$, $j = 1,2,\ldots,(J-1)$, are known in addition to the boundary values $\bar{\tau}_{j,0}$ for $j = 0,1\ldots,J$, and $\bar{\tau}_{0,1}$, $\bar{\tau}_{J,1}$. Then taking $k = 2,3,4,\ldots,(K-2)$, in order, the grid can be marched through to give $\tau_{j,k}$ at all interior grid-points. Finally, with $k = K-1$, in Equation (5.3.32), the values of $\tau_{j,K}$, for $j = 1,2,\ldots,(J-1)$, can be computed.

The method just described does not utilise the boundary conditions along the line $y = K\Delta y$. In fact, for each different set of starting values $\tau_{j,1}$ for $j = 1,2,\ldots,J-1$, different values of τ will be obtained along $y = K\Delta y$. Our problem then is to determine the starting values $\tau_{j,1}$ so that the given boundary values are obtained along $y = K\Delta y$. If the vector $\underset{\sim}{e} = (e_1\ e_2\ \ldots\ e_{J-1})^T$ denotes the given boundary values along $y = K\Delta y$, then, with the correct starting values, the computed values

$$\underset{\sim}{\tau}_K = (\tau_{1,K}\ \tau_{2,K}\ \ldots.\ \tau_{J-1,K})^T$$

must be such that

$$\underset{\sim}{\tau}_K = \underset{\sim}{e} .$$

Since the values of $\tau_{j,K}$ are linearly related to the values of $\tau_{j,1}$ used in the first application of (5.3.32), then

$$\underset{\sim}{\tau}_K = A\underset{\sim}{\tau}_1 + \underset{\sim}{b} \tag{5.3.33}$$

where

$$\underset{\sim}{\tau}_1 = (\tau_{1,1}\quad \tau_{2,1}\quad \ldots.\quad \tau_{J-1,1})^T ,$$

$\underset{\sim}{b}$ is the constant vector,

$$\underset{\sim}{b} = (b_1\quad b_2\quad \ldots.\ b_{J-1})^T ,$$

and $\underline{\underline{A}}$ is the $(J-1)$ square matrix

$$\underline{\underline{A}} = [a_{i,j}],\ i,j = 1,2,\ldots,(J-1).$$

If $\underline{\underline{A}}$ and b can be found, the required starting vector s to be substituted for $\tilde{\tau}_1$ in order to obtain the correct boundary values along $y = K\Delta y$ is found by putting $\tau_K = e$ in Equation (5.3.33), giving

$$\underset{\sim}{s} = \underline{\underline{A}}^{-1}(\underset{\sim}{e} - \underset{\sim}{b}). \qquad (5.3.34)$$

The vector b and the columns a_j of the matrix $\underline{\underline{A}}$ may be found in the following manner. Firstly, start the marching process with $\underset{\sim}{\tau}_1 = \underset{\sim}{0}$ to obtain the end vector $\underset{\sim}{\tau}_K = \underset{\sim}{\varepsilon}^{(0)}$, where

$$\underset{\sim}{\varepsilon}^{(0)} = (\varepsilon_{1,0} \quad \varepsilon_{2,0} \quad \ldots \quad \varepsilon_{J-1,0})^T \quad .$$

From Equation (5.3.33) it is seen that $\underset{\sim}{b}$ is known, since

$$\underset{\sim}{\varepsilon}^{(0)} = \underset{\sim}{b} \quad .$$

Then progressively start the marching process with the vector $\tilde{\tau}_1$ consisting of elements of the Kronecker delta form; that is, start with $\underset{\sim}{\tau}_1 = \underset{\sim}{s}^{(\ell)}$, $\ell = 1,2,\ldots,(J-1)$, where

$$\underset{\sim}{s}^{(\ell)} = (\delta_{1,\ell} \quad \delta_{2,\ell} \quad \ldots \quad \delta_{\ell,\ell} \quad \ldots \quad \delta_{J-1,\ell})^T$$

$$= (0 \quad 0 \quad \ldots \quad 1 \quad \ldots \quad 0 \quad)^T \quad . \qquad (5.3.35)$$

Let the corresponding end vectors $\underset{\sim}{\tau}_K = \underset{\sim}{\varepsilon}^{(\ell)}$, obtained on marching through the mesh with $\underset{\sim}{\tau}_1 = \underset{\sim}{s}^{(\ell)}$, be

$$\underset{\sim}{\varepsilon}^{(\ell)} = (\varepsilon_{1,\ell} \quad \varepsilon_{2,\ell} \quad \ldots \quad \varepsilon_{J-1,\ell})^T \quad .$$

Substitution of the values $\underset{\sim}{\tau}_1 = \underset{\sim}{s}^{(\ell)}$ and $\underset{\sim}{\tau}_K = \underset{\sim}{\varepsilon}^{(\ell)}$ into Equation (5.3.33) gives

$$\underline{\underline{A}}\underset{\sim}{s}^{(\ell)} = \underset{\sim}{\varepsilon}^{(\ell)} - \underset{\sim}{\varepsilon}^{(0)}. \qquad (5.3.36)$$

Since

$$\underline{\underline{A}}\underset{\sim}{s}^{(\ell)} = \begin{bmatrix} a_{1,1} & a_{1,2} & \cdots & a_{1,\ell} & \cdots & a_{1,J-1} \\ a_{2,1} & a_{2,2} & \cdots & a_{2,\ell} & \cdots & a_{2,J-1} \\ \cdot & \cdot & & \cdot & & \cdot \\ \cdot & \cdot & & \cdot & & \cdot \\ \cdot & \cdot & & \cdot & & \cdot \\ \cdot & \cdot & & \cdot & & \cdot \\ \cdot & \cdot & & \cdot & & \cdot \\ \cdot & \cdot & & \cdot & & \cdot \\ a_{J-1,1} & a_{J-1,2} & \cdots & a_{J-1,\ell} & \cdots & a_{J-1,J-1} \end{bmatrix} \begin{bmatrix} 0 \\ 0 \\ \cdot \\ \cdot \\ 1 \\ \cdot \\ \cdot \\ \cdot \\ 0 \end{bmatrix} = \begin{bmatrix} a_{1,\ell} \\ a_{2,\ell} \\ \cdot \\ \cdot \\ \cdot \\ \cdot \\ \cdot \\ \cdot \\ a_{J-1,\ell} \end{bmatrix} = \underset{\sim}{a}^{(\ell)}$$

where $a^{(\ell)}$ is the ℓ-th column of the matrix $\underline{\underline{A}}$, it is seen from Equation (5.3.36) that

$$\underset{\sim}{a}^{(\ell)} = \underset{\sim}{\varepsilon}^{(\ell)} - \underset{\sim}{\varepsilon}^{(0)} \qquad (5.3.37)$$

Thus, from the end vectors $\varepsilon^{(\ell)}$ determined by marching through the mesh with starting vectors $\tilde{s}^{(\ell)}$, $\ell = 0,1,\dots,(J-1)$, the vector b and the matrix $\underline{\underline{A}}$ can be constructed. Application of Equation (5.3.34) then gives the set of starting values $\underset{\sim}{\tau_1} = \underset{\sim}{s}$ required to give the set of boundary values $\underset{\sim}{\tau_K} = \underset{\sim}{e}$.

One more sweep through the grid, starting with $\tau_1 = \underset{\sim}{s}$ then gives the values of $\tau_{j,k}$, the solutions of the diffusion equation (5.2.3) in the rectangular region R at time level (n+1) which satisfy the given boundary conditions.

Examination of the error propagation in the quasi-explicit finite difference scheme (5.3.22) by means of discrete perturbation analysis indicates that an error $\xi_{j,k}$ in the computed value of $\tau_{j,k}$ produces an error of $2(1+s_x s_y^{-1}+s_y^{-1})\,\xi_{j,k}$ in the value of $\tau_{j,k+1}$. Therefore, in marching from one row of grid points to the next, the error is magnified by

$$|G| = 2(1 + s_x s_y^{-1} + s_y^{-1}) > 1,$$

so the scheme is unconditionally unstable.

However, the method is of use so long as the amplification of round off errors using (5.3.32) does not affect the first four or five figures in the values of τ calculated along the row of grid points at $y = K\Delta y$. Progressing one grid spacing in the y direction produces a loss of one significant figure if the magnification of the round off error is 10, there is a loss of two figures if the magnification is 10^2, and so on. It follows that the loss of significant figures is $\log_{10} 2(1 + s_x s_y^{-1} + s_y^{-1})$ in marching one grid spacing in the y-direction using this method. Since Equation (5.3.32) must be applied (K-1) times before the top boundary is reached, there will be at most a loss of

$$L = (K-1)\log_{10} 2\{1 + s_x s_y^{-1} + s_y^{-1}\} \tag{5.3.38}$$

significant figures on marching through the grid using this method.

Unfortunately, the final march through the grid with the correct starting values will produce a further loss of accuracy of the same amount, L. In order to avoid this, the linear nature of the problem can be exploited in the following way. The complete array of values $\tau_{j,k}^{(\ell)}$, $j = 1,2,\dots,(J-1)$, $k = 1,2,\dots,K$, is stored on a scratchtape or permanent file for each set of starting values of the form $\tilde{s}^{(\ell)}$ given by Equation (5.3.35). Then, for the set of correct starting values

$$\underset{\sim}{s} = (s_1\ s_2\ \dots\ s_{J-1})^T = s_1\,\underset{\sim}{s}^{(1)} + s_2\,\underset{\sim}{s}^{(2)} + \dots + s_{J-1}\,\underset{\sim}{s}^{(J-1)}$$

computed from Equation (5.3.34), the required array $\tau_{j,k}$ is given by

$$\tau_{j,k} = s_1\,\tau_{j,k}^{(1)} + s_2\,\tau_{j,k}^{(2)} + \dots + S_{J-1}\,\tau_{j,k}^{(J-1)} + \left(1 - \sum_{m=1}^{J-1} s_m\right)\tau_{j,k}^{(0)}$$

for $j = 1,2,\dots,(J-1)$, $k = 1,2,\dots,K$.

An estimate of the loss of significant figures when using this method can be found from Equation (5.3.38). If $K = 16$ and $s_x = \frac{1}{4}s_y = 2$, then the possible loss of significant figures would be of the order $15 \log_{10} 2.75 \simeq 6.6$. Clearly, the larger the ratio of s_y/s_x and the greater the value of s_y, the smaller is the loss of accuracy in this method. On a computer which stores 12 significant figures of each number, five figure accuracy would be retained. With $K = 40$, the number of significant figures lost would be approximately $39 \log_{10} 2.75 \simeq 17.1$. Using double precision with storage of 24 significant figures of each number, at least six figure accuracy is achieved.

In practice, the accuracy of the results is indicated from a comparison of the computed values $\tau_{j,K}$ with the given boundary values $\bar{\tau}_{j,K} = e_j$. The effect of a discrete perturbation at the point $(\ell\Delta x, \Delta y)$ along the line $k = 1$ is indicated by the values of $\varepsilon^{(\ell)}_{j,K} - \varepsilon^{(0)}_{j,K}$ for $j = 1,2,\ldots,(J-1)$, because the starting vector $\tilde{s}^{(\ell)j}$ contains a perturbation of unity in the value of $\tau_{\ell,1}$.

This method has been used to solve certain linearised problems involving wind induced circulation in lakes (for example, see Noye, 1977 and 1978).

5.4 Boundary Conditions at Irregular Boundaries

When the physical boundary is irregular in two or three dimensions, that is, when segments of the boundary are not parallel to one of the spatial co-ordinate axes, finite differences can still be employed, but care must be taken if the same accuracy is desired for the finite difference approximations to the boundary conditions as the accuracy for the finite difference method applied to the interior of the space region. For instance, suppose the two dimensional diffusion equation (5.2.3) is to be solved in some given region R such as that shown in Figure 5.4.

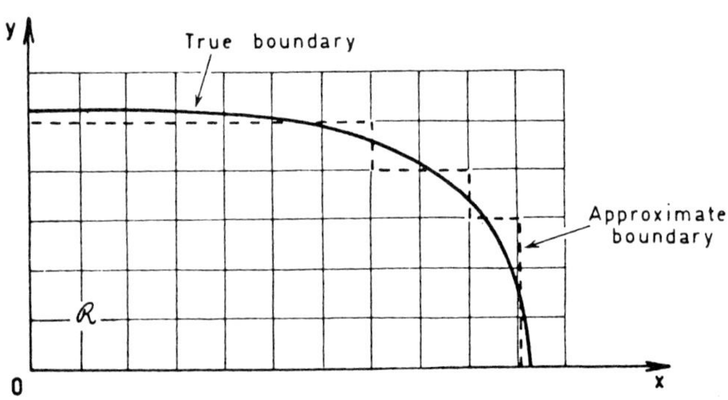

Figure 5.4: Replacing true boundary with approximate boundary through grid points.

At different parts of the curved boundary either the value of the dependent variable $\bar{\tau}$ is specified, or its derivative $\partial\bar{\tau}/\partial n$ along the normal n to the boundary is specified. In the process of trying to find a numerical solution for Equation (5.2.3) these two possible boundary conditions are treated in quite different ways, and hence they will be discussed separately.

Dependent Variable Specified on the Boundary (The Dirichlet Problem)

The most obvious way to handle this type of boundary condition is to choose the grid point which lies closest to the boundary, and at that grid point let τ take the boundary value. This has the effect of reshaping the boundary so that it conforms to the discrete points of the grid.

The dashed line in Figure 5.4 shows how the boundary would be drawn through the grid points if this method is used. The values of the dependent variable allotted to these points are those which apply to the original boundary. In this method the finite difference equation is not written about these new boundary points; they are written only for the interior grid points. Once the boundary is altered in this way the problem can be solved by any of the methods discussed previously.

The obvious disadvantage of treating the boundary condition in the above way is the low order of accuracy which results when the finite difference equations are applied at points near the boundary. Also, a small grid spacing must be used to ensure that grid points lie close to the actual boundary.

There is another method of dealing with this sort of boundary condition which results in a higher order of accuracy. The extra work which is required is generally worthwhile since a coarser grid can be used with the same truncation error. In this method the interior points are divided into two disjoint sets. Regular points are defined as those for which no neighbouring grid point in any direction lies outside the boundary. In Figure 5.5 these points are denoted by open circles. Interior points having neighbouring points on the grid which lie outside the boundary are termed irregular points and are denoted by the closed circles. For regular points the usual finite difference equation applies. For instance, the second order derivative $\partial^2\bar{\tau}/\partial y$ at the regular gridpoint (x_j, y_k) can be represented by the central difference approximation similar to Equation (2.3.12), namely

$$\left.\frac{\partial^2\bar{\tau}}{\partial y^2}\right|_{j,k}^{n} = \frac{\bar{\tau}^n_{j,k+1} - 2\bar{\tau}^n_{j,k} + \bar{\tau}^n_{j,k-1}}{(\Delta y)^2} + O\{(\Delta y)^2\} \quad , \tag{5.4.1}$$

while the first order derivative $\partial\bar{\tau}/\partial y$ can be approximated by an equation similar to (2.3.11), namely

$$\left.\frac{\partial\bar{\tau}}{\partial y}\right|_{j,k}^{n} = \frac{\bar{\tau}_{j,k+1} - \bar{\tau}_{j,k-1}}{2\Delta y} + O\{(\Delta y)^2\} \ . \tag{5.4.2}$$

For irregular points this finite difference approximation is altered in the following way. Discrete points are placed on the boundary at the intersection of the boundary with each grid line. These are shown as closed squares in Figure 5.5. Consider the grid point $P(x_j, y_k)$ in Figure 5.5. Its neighbour in the positive y direction is less than a whole increment Δy from it. Suppose the distance is $\rho\Delta y$ where $0 < \rho < 1$. Thus the

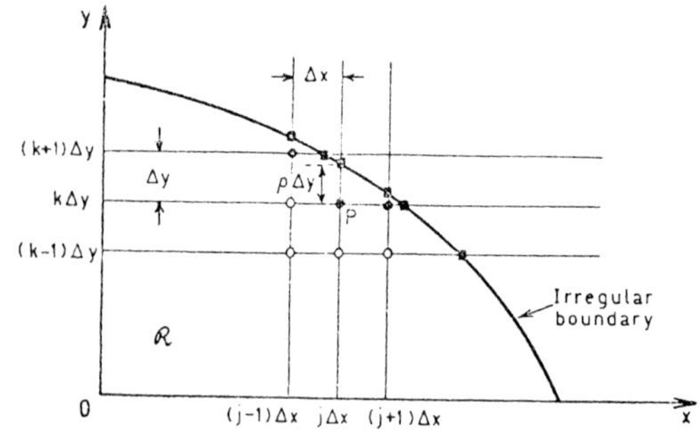

Figure 5.5: Incorporating boundary values into finite difference equations.

central difference approximation for $\partial^2\bar{\tau}/\partial x^2$ can be written in the normal manner, but this cannot be done for $\partial^2\bar{\tau}/\partial y^2$ since the two appropriate neighbouring points at which τ is defined along the $x = j\Delta x$ grid line are spaced unequally.

Application of the Taylor series expansion about the point (x_j, y_k) at time $t = n\Delta t$ gives the approximation for the second derivative to be

$$\left.\frac{\partial^2\bar{\tau}}{\partial y^2}\right|_{j,k}^{n} = \frac{2\{B^n_{j,k+1} - (\rho+1)\bar{\tau}^n_{j,k} + \rho\,\bar{\tau}^n_{j,k-1}\}}{\rho(\rho+1)(\Delta y)^2} + O\{(1-\rho)\Delta y\} \qquad (5.4.3)$$

where, at time $t = n\Delta t$, $B^n_{j,k+1}$ is the known value of $\bar{\tau}$ on the boundary where it cuts the grid line $x = j\Delta x$, between $y = k\Delta y$ and $y = (k+1)\Delta y$. This finite difference approximation is only first order correct although the truncation error becomes second order as ρ tends to 1. A second order correct approximation to the first order derivative $\partial\bar{\tau}/\partial y$ written for unequally spaced points is

$$\left.\frac{\partial\bar{\tau}}{\partial y}\right|_{j,k}^{n} = \frac{B^n_{j,k+1} - (1-\rho^2)\bar{\tau}^n_{j,k} - \rho^2\bar{\tau}^n_{j,k-1}}{\rho(\rho+1)\Delta y} + O\{\rho(\Delta y)^2\}\ . \qquad (5.4.4)$$

This approximation is required when convective terms occur in the differential equation to be solved. For each irregular point either the x-derivatives, the y- derivatives, or perhaps both, will need to be altered. With the altered finite difference equations for the irregular points, the set of finite difference equations for all interior points is complete. The set of equations thus obtained can be solved using any appropriate method.

<u>Derivative Normal to the Boundary Prescribed</u> (The Neumann Problem)

Of the two methods available for handling this type of boundary condition, the simpler was proposed by Fox (1944). The interior gridpoints are again divided into regular and irregular points. In both cases the centred finite difference approximations to the spatial derivatives are based on Equation (5.4.1). However, in the case of the irregular points at least one of the x or y spatial derivatives will include a fictitious value of τ from an exterior gridpoint. Thus, in Figure 5.6 the gridpoint $P(j,k)$ is an irregular point and the finite difference approximations to the first and second order derivatives $\partial\bar{\tau}/\partial y$ and $\partial^2\bar{\tau}/\partial y^2$ at that point, which are based on Equations (5.4.1), (5.4.2) include a value of $\bar{\tau}$ prescribed at the exterior point $E(j,k+1)$. An estimate for this value of $\bar{\tau}$ can be found in terms of the geometry of the boundary, the value of the normal derivative on the boundary and two interior gridpoints, in the following way.

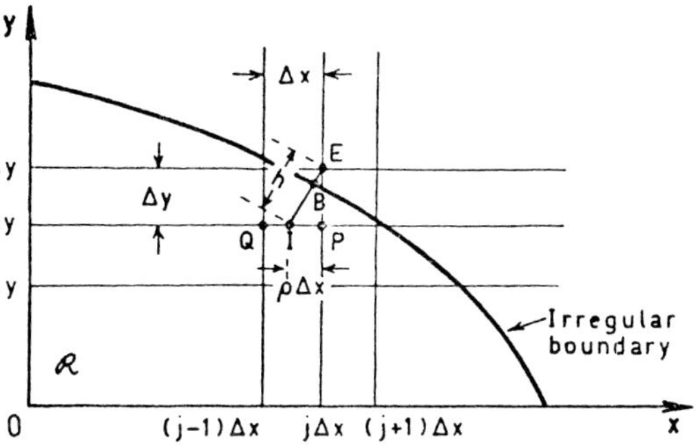

Figure 5.6: Approximating derivative boundary values.

From the exterior point E a line is drawn normal to the boundary and extended until it intersects a grid line joining two adjacent interior grid points, such as at I in Figure 5.6. An estimate of the value of τ at I can be obtained by taking a weighted average of the values of τ at the two grid points P and Q. Thus

$$\tau_I^n = \rho\,\tau_{j-1,k} + (1-\rho)\tau_{j,k}^n \quad , \tag{5.4.5}$$

where the length $PI = \rho\Delta x$. A value $\tau_{j,k+1}^n$ is then allotted to the exterior point E by relating the values of τ at E and I through the derivative boundary condition. If $\partial\tau/\partial n$ takes the value b^n at the point B at time $t = n\Delta t$, where EI, of length h, intersects the boundary, then the approximation

$$\frac{\tau_{j,k+1}^n - \tau_I^n}{h} = b^n \quad , \tag{5.4.6}$$

combined with (5.4.5), leads to the relation

$$\tau_{j,k+1}^n = \rho\,\tau_{j-1,k}^n + (1-\rho)\tau_{j,k}^n + b^n h \ . \tag{5.4.7}$$

This value of $\tau_{j,k+1}^n$ is then used in Equations (5.4.1) and (5.4.2) so finite difference approximations for the derivatives $\partial^2\bar{\tau}/\partial y^2$ and $\partial\bar{\tau}/\partial y$ can be written entirely in terms of values of τ at interior grid points. Thus (5.4.1) leads to the approximation

$$\left.\frac{\partial^2\bar{\tau}}{\partial y^2}\right|_{j,k}^{n} = \frac{\rho\,\tau_{j-1,k}^n - (1+\rho)\tau_{j,k}^n + \tau_{j,k-1}^n + b^n h}{(\Delta y)^2} \quad , \tag{5.4.8}$$

and (5.4.2) leads to

$$\left.\frac{\partial\bar{\tau}}{\partial y}\right|_{j,k}^{n} = \frac{\rho\,\tau_{j-1,k}^n + (1-\rho)\tau_{j,k}^n - \tau_{j,k-1}^n + b^n h}{\Delta y} \quad . \tag{5.4.9}$$

The set of finite difference equations which result on substituting these approximations to the derivatives in the transport equation, can be solved by any of the methods described in Sections 5.2 and 5.3.

The alternative method of dealing with the derivative boundary condition in multi-dimensional space is more complicated, and is described by Viswanathan (1957), Fox (1962) and Ames (1969).

6. DIFFERENT GRID AND COORDINATE SYSTEMS

6.1 General Considerations

In the previous sections the transport equation was differenced on a uniformly spaced grid in the rectangular cartesian coordinate system. However, this may not be the most appropriate system to use.

The grid and coordinate system should be chosen so that numerous grid points occur in regions of primary concern, and as far as possible, the boundary should pass through gridpoints. While a finite difference approximation developed on a uniform grid may be the simplest and formally the most accurate scheme to use, it can prove deficient in some respects. For example, when computing the flow around solid bodies or solving other boundary layer problems, there are quite large velocity gradients in certain regions of the flow near the body. If there are not enough grid points across the boundary layer to resolve the velocity, the numerical solution may be a very poor approximation to the true solution in this layer. Ideally, to increase the resolution in this region we could still use a uniform grid but with a much smaller grid spacing over the complete area of flow. However, this would lead to unacceptably large computational times.

To avoid excessive numbers of calculations, while still obtaining a higher resolution in particular regions, a non-uniform grid system may be used with a fine grid in the area of interest and a coarse grid over the rest of the flow field. Different types of non-uniform grids are discussed in Section 6.2. For instance, it is possible to have a coarse grid which changes abruptly to a fine grid in the region of interest or, alternatively, to have a grid in which there is a gradual reduction from a large to a small grid size. Grids need not be rectangular or even regular; specially shaped grids of variable size may be used in problems involving irregular boundaries (for example, see Thoman and Szewczyk, 1969).

While there are considerable benefits to be gained by having a non-uniform grid, these may be offset by other problems introduced by the change in grid size. There may be wave distortion due to a phase change at the interface of two different grids or a wave reflection off the interface. It may also be difficult to match the finite difference approximations at the interface of different grids. The introduction of a change in grid spacing also may adversely affect the formal truncation error and the stability of the system.

Because of these difficulties, a coordinate transformation is often preferred to a change in grid shape or size in order to obtain higher resolution in a particular area of flow. It is often desirable to make a change of coordinate system if a problem can be handled more easily in another system. For example, by aligning a coordinate surface along a boundary, it is easier to apply boundary conditions. However, changing the coordinate system or making a coordinate transformation can also lead to complications which are outlined later.

6.2 Abrupt Changes in Grid Spacing

The simplest non-uniform grid scheme has a large constant grid spacing over most of the flow field with an abrupt change to a smaller constant grid spacing to cover the region of particular interest. In making this change of grid size the second order spatial accuracy which occurs with central differences on a uniform grid should be retained; a decrease in computational accuracy due to the change in grid size may more than offset the gain in accuracy due to the use of a finer grid.

Crowder and Dalton (1971) considered the case where the grid spacing is increased from a constant value Δx_1 to the larger value Δx_2 at a point J (see Figure 6.1), in order to see how the order of the truncation error of the approximations to the spatial derivatives $\partial\bar{\tau}/\partial x$ and $\partial^2\bar{\tau}/\partial x^2$ of a function $\bar{\tau}$ was affected by this change. Expanding the function $\bar{\tau}$ in a Taylor series forward and backward from the J-th grid point, at a particular time level, the superscript for which is omitted in the following, gives

$$\bar{\tau}_{J+1} = \bar{\tau}_J + \Delta x_2 \left.\frac{\partial\bar{\tau}}{\partial x}\right|_J + \frac{(\Delta x_2)^2}{2!}\left.\frac{\partial^2\bar{\tau}}{\partial x^2}\right|_J + \frac{(\Delta x_2)^3}{3!}\left.\frac{\partial^3\bar{\tau}}{\partial x^3}\right|_J + O\{(\Delta x_2)^4\} , \quad (6.2.1)$$

$$\bar{\tau}_{J-1} = \bar{\tau}_J - \Delta x_1 \left.\frac{\partial\bar{\tau}}{\partial x}\right|_J + \frac{(\Delta x_1)^2}{2!}\left.\frac{\partial^2\bar{\tau}}{\partial x^2}\right|_J - \frac{(\Delta x_1)^3}{3!}\left.\frac{\partial^3\bar{\tau}}{\partial x^3}\right|_J + O\{(\Delta x_1)^4\} . \quad (6.2.2)$$

Eliminating the terms with $\partial^2\bar{\tau}/\partial x^2|_J$, by multiplying Equation (6.2.1) by $(\Delta x_1)^2$ and substracting Equation (6.2.2) multiplied by $(\Delta x_2)^2$, yields

$$\left.\frac{\partial\bar{\tau}}{\partial x}\right|_J = \frac{\bar{\tau}_{J+1}-(\nu^2-1)\bar{\tau}_J-\nu^2\bar{\tau}_{J-1}}{\Delta x_2(\nu+1)} - \frac{\Delta x_1\Delta x_2}{6}\frac{\partial^3\bar{\tau}}{\partial x^3} + O\{(\Delta x_1)^3\} , \quad (6.2.3)$$

where $\nu = \Delta x_2/\Delta x_1$. The quasi-central difference approximation to the first derivative at the point where the grid spacing changes is therefore

$$\left.\frac{\partial\bar{\tau}}{\partial x}\right|_J = \frac{\tau_{J+1}-(\nu^2-1)\ \tau_J-\nu^2\tau_{J-1}}{\Delta x_2(\nu+1)} , \quad (6.2.4)$$

which has a truncation error of $O\{\Delta x_1\Delta x_2,(\Delta x_1)^3\}$.

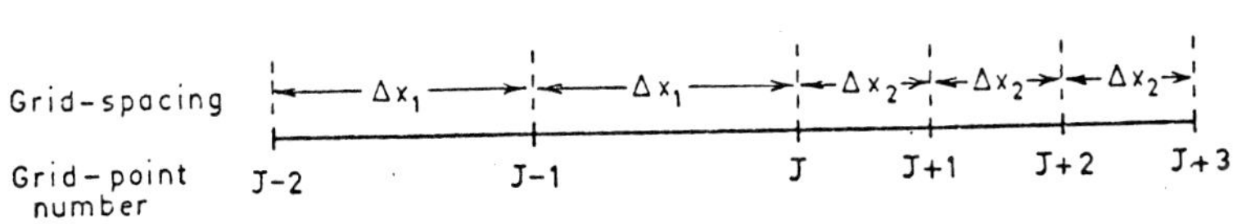

Figure 6.1: Sudden change in grid-spacing at J-th grid-point.

Similarly, by eliminating the terms with $\partial\bar{\tau}/\partial x|_J$ from Equations (6.2.1) and (6.2.2.) gives

$$\left.\frac{\partial^2\bar{\tau}}{\partial x^2}\right|_J = \frac{\bar{\tau}_{J+1}-(\nu+1)\bar{\tau}_J + \nu\bar{\tau}_{J-1}}{\frac{1}{2}\Delta x_1\Delta x_2(\nu+1)} - \frac{(\Delta x_2-\Delta x_1)}{3}\left.\frac{\partial^3\tau}{\partial x^3}\right|_J + O\{(\Delta x_2)^2\} \quad . \qquad (6.2.5)$$

Therefore the quasi-central difference approximation to the second derivative at x_J is

$$\left.\frac{\partial^2\bar{\tau}}{\partial x^2}\right|_J = \frac{\tau_{J+1}-(\nu+1)\tau_J+\nu\tau_{J-1}}{\frac{1}{2}\Delta x_1\Delta x_2(\nu+1)} \quad . \qquad (6.2.6)$$

Note that the choice $\Delta x_1 = \Delta x_2$ in (6.2.4) and (6.2.6) yield the usual central difference approximations (2.3.11) and (2.3.12) for a uniform grid spacing.

Because the finite difference approximation (6.2.6) has a truncation error of $O\{\Delta x_2-\Delta x_1,(\Delta x_2)^2\}$, any finite difference methods for second order partial differential equations based upon this approximation will involve a second order error $O\{(\Delta x_1)^2\}$ at the interface where the grid size changes abruptly, only if

$$\Delta x_2 = \gamma\Delta x_1, \quad \gamma = 1 + k\Delta x_1 , \qquad (6.2.7)$$

and k is of order one.

Therefore, the truncation error is worse than that with a uniform grid, rather than an improvement, unless the grid spacing is changed gradually. Crowder and Dalton (1971) compared results obtained using six different grids and concluded that a uniform square mesh was the best grid to use for the Poiseuille flow problem. This was contrary to their original hypothesis that using non-uniform, non-square grids would give a minimum overall error with their finite difference solutions. Commenting on these findings, Blottner and Roache (1971) pointed out that such results are not generally true and that they probably occurred because change of grid spacing from 0.05 to 0.10 by Crowder and Dalton did not satisfy condition (6.2.7). Some applications where variable grids give better results than a constant grid are described by Moretti and Salas (1970) and Anthes (1970). MacCormack (1971) and Chavez and Richards (1970) also indicate that the loss of accuracy of the solution in a finite difference approximation in which grid size varies is not usually as bad as may be indicated by the formal truncation error, particularly if an isolated change in mesh size is used.

When a variable grid spacing is used one of the main problems is to obtain consistent difference equations at interfaces in the grid without introducing instability. For explicit finite difference approximations, stability conditions such as

$$s = \alpha\Delta t/(\Delta x)^2 \leqslant 1/2$$

for the diffusion equation, and the Courant-Friedrichs-Lewy condition

$$C = u\Delta t/\Delta x \leqslant 1$$

for the convection equation, are needed. The first of these requires

$$\Delta t \leqslant (\Delta x)^2/2\alpha \quad , \tag{6.2.8}$$

while the second needs

$$\Delta t \leqslant \Delta x/u \quad . \tag{6.2.9}$$

From Equations (6.2.8) and (6.2.9) we can see that if the grid size Δx is reduced, then the time step Δt may also have to be reduced if the method is to remain stable, so that the number of computations required may increase significantly. Since such restrictive conditions do not apply to implicit methods, these are more useful in such cases, although it must be remembered that truncation errors do increase as Δt increases, thereby reducing accuracy even with implicit methods.

Osher (1970) considered difference equations appropriate for the solution of the diffusion equation (2.1.1) on a grid with an abrupt change in grid size and showed that they were stable over the entire spatial domain, especially at the interface. Ciment (1971) studied the solution of the convection equation (5.1.2) across a change in grid size and showed that certain differencing schemes, such as the Lax-Wendroff scheme (Lax and Wendroff, 1960, 1964) are stable approximations to this equation over the whole domain. Similar stability considerations are discussed by Venit (1973) and Ciment and Sweet (1973).

A further complication arising from the use of a non-uniform grid is the possibility of wave reflection off the interface between the regions of different grid size. Browning et al. (1973) state this need not be a problem but they point out another phenomenon, causing wave distortion, which cannot be avoided. If a wave is already well represented in the coarse grid it should propagate through the interface without any difficulty. But any wave which is poorly represented in the coarser grid will change phase speed when it passes through the interface into a finer grid region. If this wave passes back into the coarse grid at some other place there can be a serious interaction of the part which propagated through the fine grid with the part of the wave which remained in the coarse grid.

6.3 Variable Grid Spacing

One way of avoiding some of the problems caused by a sudden change in grid size is to use a graded grid spacing in which the intervals are varied continuously from a coarse grid covering most of the flow field to a fine grid in the region of special interest. One advantage of this gradual reduction in grid size is the decrease in computer memory necessary and computational time required to obtain improved solutions over the same area as an abrupt change to the smallest grid scale.

Sundquist and Veronis (1970) have constructed a simple finite difference grid with non-constant intervals that gives the same accuracy as the uniform grid when derivatives are represented by centred differences. Consider the interval (0,1) divided into J intervals of increasing length as shown in Figure 6.2. The grid points are numbered $j = 0,1,2,3,\ldots,J$, the value of x at the j-th grid point being denoted x_j and the interval between x_j and $x_{j\pm1}$ denoted Δx_j. Now consider $\bar{\tau}(x)$ on the range $(0,1)$ with $\bar{\tau}_j = \bar{\tau}(x_j)$. From the Taylor series expansion of $\bar{\tau}_{j-1}$ and $\bar{\tau}_{j+1}$ about the point x_j the following finite difference forms for the first and second order space derivatives are obtained by eliminating the terms in $\partial^2\bar{\tau}/\partial x^2|_j$ and $\partial\bar{\tau}/\partial x|_j$, respectively :

$$\left.\frac{\partial\bar{\tau}}{\partial x}\right|_j = \frac{\bar{\tau}_{j+1} + (\nu_j^2 - 1)\bar{\tau}_j - \nu_j^2\bar{\tau}_{j-1}}{\Delta x_j(\nu_j + 1)} - \frac{\Delta x_j\,\Delta x_{j-1}}{6}\left.\frac{\partial^3\bar{\tau}}{\partial x^3}\right|_j + O\{(\Delta x_j)^3\} \tag{6.3.1}$$

$$\left.\frac{\partial^2\bar{\tau}}{\partial x^2}\right|_j = \frac{\bar{\tau}_{j+1} - (\nu_j + 1)\bar{\tau}_j + \nu_j\bar{\tau}_{j-1}}{\tfrac{1}{2}\Delta x_j\,\Delta x_{j-1}(\nu_j + 1)} - \frac{(\Delta x_j - \Delta x_{j-1})}{3}\left.\frac{\partial^3\bar{\tau}}{\partial x^3}\right|_j + O\{(\Delta x_j)^2\} \tag{6.3.2}$$

where $\nu_j = \Delta x_j/\Delta x_{j-1}$. By choosing $\Delta x_j - \Delta x_{j-1} = O\{(\Delta x_j)^2\}$ the first terms on the right hand side of Equations (6.3.1) and (6.3.2) approximate the derivatives on the left hand side with a truncation error of order $O\{(\Delta x_j)^2\}$

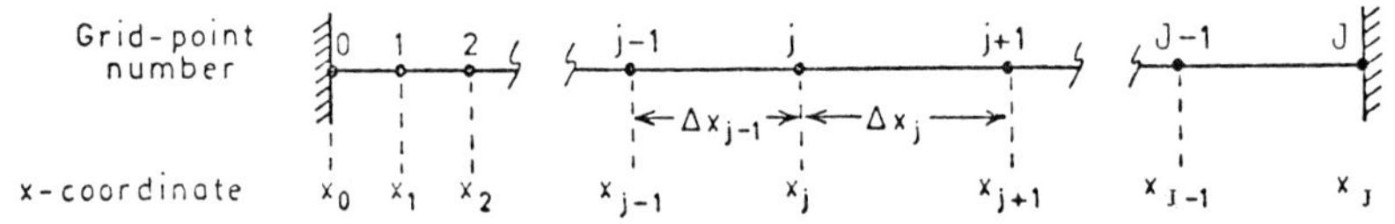

Figure 6.2: Nonuniform grid-spacing used in a finite-difference scheme by Sundquist and Veronis (1970).

In this way they solved the boundary layer problem proposed by Stommel (1948), using a non-uniform grid with spacings

$$\Delta x_0 = h, \ \Delta x_j = \Delta x_{j-1}(1+\gamma\Delta x_{j-1}),$$

where γ is a constant of $O(1)$, and compared the results with those obtained using a uniform grid. With the uniform grid the solution exhibited large variations from the exact solution with an oscillatory behaviour near the boundary, while the non-uniform grid solution was much more accurate near the boundary and no oscillations were evident. While this method of choosing the grid allows some improvement in resolution near the boundary, overall it still requires a large number of grid points to significantly reduce the grid spacing there.

Kalnay de Rivas (1972) investigated a different kind of grid variation. Consider a function $\bar{\tau}(x)$ defined on a non-uniform grid as in Figure 6.3. Suppose grid intervals are varied by defining a stretched coordinate μ where $x = x(\mu)$ in such a way that grid intervals $\Delta\mu$ are constant. If the function being studied is defined on the region $0 \leqslant x \leqslant 1$ with a boundary layer at $x = 0$, then $x(\mu)$ should have the properties:

(i) $dx/d\mu$ should be finite over the whole interval - if it becomes infinite at some point, then there is poor resolution near that point;

(ii) $dx/d\mu = 0$ at $x = 0$, $dx/d\mu \neq 0$ if $x \neq 0$, which ensures high resolution near the point $x = 0$.

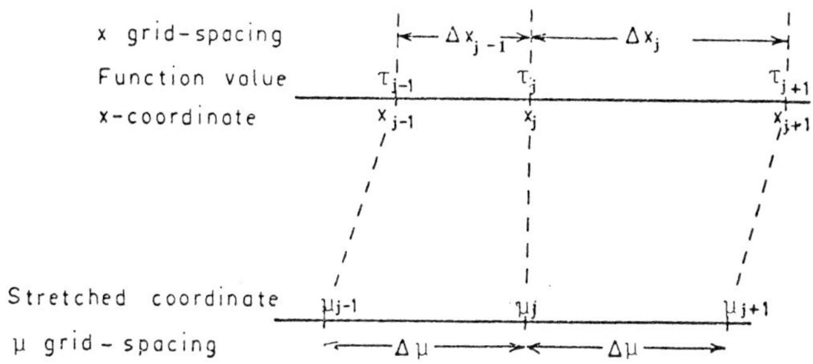

Figure 6.3: Nonuniform grid defined through the use of a stretched coordinate.

By taking Taylor series expansions about the point x_j to obtain the approximations to the first and second derivatives, it is found that the truncation error is second order in $\Delta\mu$ even near the boundary layer where the errors may be first order with respect to Δx_j. However, since the intervals Δx_j are very small here, the overall error remains small.

The Taylor series analysis suggests that a convenient form for the function x is $x = \mu^2$. If there are boundary layers expected at $x = 0$ and $x = 1$ a suitable choice of stretched coordinate is $x = \sin^2(\pi\mu/2)$.

Both types of stretched coordinates have been successfully used in two dimensional numerical models of the atmosphere of Venus (Kalnay de Rivas, 1971). Comparing the percentage errors introduced by the Sundqvist-Veronis grid for Stommel's problem with those introduced by using the stretched coordinate $x = \mu^2$, Kalnay de Rivas found that the latter is more accurate and that there is no tendency for relative errors to grow as $x \to 0$.

In order for a fine grid to follow regions in which large velocity gradients occur, it is necessary to change the grid scheme as a solution changes with time. This changing of the grid as the solution changes is called "rezoning" and it may affect the results obtained.

A two dimensional grid in which Δx and Δy varies is often used to describe irregular boundaries in a rectangular grid. While this approach usually gives acceptable results the complications previously described which arise from the change in grid size can cause difficulties.

Geometrical shapes other than the rectangular grid may have advantages for matching irregular boundaries. In particular triangular meshes have been successfully used in problems by Winslow (1966) and Williamson (1969). A hexagonal grid was used by Sadourny and Morel (1969) for a spherical surface.

In their analysis of the formation of a shock in a viscous flow Moretti and Salas (1970) used both constant and variable size grid spacings. With the constant grid size their solutions exhibited oscillations while with a stretched mesh these oscillations were no longer present and the solution was more accurate.

There are many geophysical situations in which variable grids are used in order to get a higher resolution in certain areas. For example, ocean models may require higher resolution over continental boundary regions than over other areas of the ocean because of the rapidly changing shore line and associated shallow waters. Variable grids are also useful in modelling atmospheric phenomena. Numerical experiments by Anthes (1970) and Harrison (1973) indicate that a variable grid may give superior results to a constant grid when forecasting the movement and development of tropical storms.

6.4 Coordinate Transformations

While grid refinement has proved useful for improving the solution of a flow in special regions such as boundary layers, a coordinate transformation is generally preferred for increasing resolution in a particular area. Similarly the preferred method for treating non-rectangular boundaries is to express the governing partial differential equation in a non-rectangular coordinate system in which boundaries are coordinate surfaces. For example, flow over a cylinder of elliptic cross-section is naturally calculated in elliptic cylinder coordinates. Roache (1974) describes a number of examples in which different coordinate systems are used for specific problems.

Coordinate transformations may also be used to align coordinate surfaces along physical boundaries and so simplify the modelling of such boundaries. Thus, problems involving circular boundaries are usually solved more easily in polar coordinates because they avoid the use of complicated finite difference formulae near the curved boundary. Consider the two dimensional diffusion equation (5.2.3); in polar coordinates in two dimensional space this becomes

$$\frac{\partial \bar{\tau}}{\partial t} = \alpha\left\{\frac{\partial^2 \bar{\tau}}{\partial r^2} + \frac{1}{r}\frac{\partial \bar{\tau}}{\partial r} + \frac{1}{r^2}\frac{\partial^2 \bar{\tau}}{\partial \theta^2}\right\} . \qquad (6.4.1)$$

If Equation (6.4.1) is differenced on a uniform rectangular grid in the $r - \theta$ plane this corresponds to differencing (5.2.3) on a non-uniform grid consisting of radial and circular lines in the x-y plane.

Increasing resolution in certain areas may be achieved with a coordinate stretching transformation, such as an exponential stretch. For example, Pao and Dougherty (1969) transformed the cartesian coordinates (x,y) to the coordinates (X,Y) by means of the relations

$$\left.\begin{aligned} x &= X \\ y &= b\{\exp(aY)-1\} \quad , \end{aligned}\right\} \qquad (6.4.2)$$

where a and b are arbitrary constants.

While such stretching transformations have the same purpose as that of an expanding grid, in that both increase the number of grid points in special regions, the two methods are fundamentally different. When the original equations are differenced on an expanding grid, there is a deterioration in the formal truncation error, whereas when the transformed equations are differenced on a regular grid there is no such loss in formal accuracy. Consequently, the stretching transformation method is preferred to a change in grid spacing.

Blottner and Roache (1971) noted that, for the problem considered by Crowder and Dalton (1971), a large improvement in accuracy of solution is achieved by transforming the independent variables, rather than by changing the grid spacing. In fact, the exact solution to this problem can be obtained by means of a parabolic transformation.

While the formal truncation error is not degraded by such coordinate transformations, other problems do arise. Stability and convergence may be affected by the introduction of new terms in the transformed equations which can lead to the computation time per step being increased. Also the boundary conditions for the system must be rederived in the new system. There is also the possibility of wave distortion due to phase changes and damping.

The numerical solution of equations, such as the Navier-Stokes, is often hindered by the fact that with the cartesian coordinate system and the standard finite difference approximation, we have to deal with complicated boundaries and so it is difficult to use the correct boundary conditions. Gal-Chen (1975) used a special coordinate transformation to change a domain with an irregular boundary into a regular domain; the transformed system had flat boundaries and homogeneous boundary conditions. Again, while there are considerable advantages to be obtained by this transformation, difficulties may arise due to extra terms being obtained in the transformed governing equations. A more serious problem can occur if the transformation causes new singularities to appear in the transformed equations. Since these singularities are not part of the true solution they may considerably affect numerical stability. Nevertheless,the transformation approach to flow over irregular bodies, such as atmospheric flows over the earth's surface, seems preferable to that of retaining the rectangular cartesian system and using special techniques to cope with the lower flow boundary.

6.5 Improving the Accuracy of Solutions

Reducing the grid-spacing

No useful results which connect the magnitude of the discretization error to the size of the grid spacing have yet been found. However, it seems reasonable to assume that the error will decrease as the grid-spacing is reduced (for example, see Table 3.1). One therefore expects that using smaller and smaller grid spacings will eventually produce successive finite difference solutions which differ by less than some required amount.

However, this approach can be very uneconomical. For instance, when solving the one dimensional diffusion equation using the FTCS method with a fixed value of the parameter s, a reduction of the grid spacing from $\Delta x = 0.1$ (the second row in Table 3.1) to $\Delta x = 0.01$ (the last row) increased the required number of calculations 1000-fold in order to compute τ at a given position and time. In general, for a given s, the number of arithmetic operations in this case is proportional to $1/(\Delta x)^3$. Solving the time dependent diffusion equation in two and three space dimensions with the same decrease in size of grid spacing is even more uneconomical. If the FTCS method with fixed s is used in the three dimensional case, then the number of arithmetic operations is multiplied by a factor of q^5 when the grid-spacing is divided by q .

Higher order finite difference approximations

More accurate solutions to a problem can be found by representing the derivatives in the given partial differential equation by finite difference approximations of a higher order of accuracy. In this way, the truncation error of the finite difference approximation to the partial differential equation is reduced. For example, in Section 1.4 it was seen that a central difference approximation to a spatial derivative on a grid with spacing h has a truncation error of $O\{h^2\}$, whereas the forward and backward difference approximations have a truncation error of $O\{h\}$. From the point of view of accuracy it is preferable to use central difference approximations for spatial derivatives.

Finite difference equations of higher order of accuracy may contain a larger number of different values of the dependent variable τ, but for many operators such as the Laplacian the increase is not large.

There are many finite difference representations for the Laplacian in two dimensions,

$$\nabla^2\bar{\tau} = \partial^2\bar{\tau}/\partial x^2 + \partial^2\bar{\tau}/\partial y^2 \ , \tag{6.5.1}$$

a number of these being listed in Bickley (1948) and Thom and Apelt (1961). The Laplacian occurs in the two dimensional diffusion equation (5.2.3) which may be written in terms of the Laplacian operator ∇^2 as

$$\partial\bar{\tau}/\partial t = \alpha\nabla^2\bar{\tau} \quad . \tag{6.5.2}$$

Consider a square grid, that is, one with equal spacings in each of the coordinate directions so $\Delta x = \Delta y = h$. Using the central difference approximation to both space derivatives in (6.5.1) at the grid point (j,k,n) gives

$$\nabla^2\bar{\tau}\Big|_{j,k} = \frac{\bar{\tau}_{j+1,k}-2\bar{\tau}_{j,k}+\bar{\tau}_{j-1,k}}{h^2} + \frac{\bar{\tau}_{j,k+1}-2\bar{\tau}_{j,k}+\bar{\tau}_{j,k-1}}{h^2} + O(h^2)$$

$$= 1/h^2\left\{\begin{array}{c} +\ \bar{\tau}_{j,k+1} \\ \\ +\ \bar{\tau}_{j-1,k} - 4\bar{\tau}_{j,k} + \bar{\tau}_{j+1,k} \\ \\ +\ \bar{\tau}_{j,k-1} \end{array}\right\} + O(h^2) \quad . \tag{6.5.3}$$

The five point finite difference approximation to the Laplacian, with a second order truncation error, is often written in the stencil form

$$\nabla^2\tau = \frac{1}{h^2}\begin{bmatrix} & 1 & \\ 1 & -4 & 1 \\ & 1 & \end{bmatrix}\tau \quad . \tag{6.5.4}$$

A similar nine point finite difference approximation with a fourth order truncation error under certain conditions is given by

$$\nabla^2\tau = \frac{1}{6h^2}\begin{bmatrix} 1 & 4 & 1 \\ 4 & -20 & 4 \\ 1 & 4 & 1 \end{bmatrix}\tau \tag{6.5.5}$$

This may be verified by expanding each term of the right side of Equation (6.5.5) in a Taylor series about the point (x_j, y_k) to obtain the result

$$\frac{1}{6h^2}\left\{\begin{matrix} \bar{\tau}_{j-1,k+1} + 4\bar{\tau}_{j,k+1} + \bar{\tau}_{j+1,k+1} \\ +4\bar{\tau}_{j-1,k} - 20\bar{\tau}_{j,k} + 4\bar{\tau}_{j+1,k} \\ +\bar{\tau}_{j-1,k-1} + 4\bar{\tau}_{j,k-1} + \bar{\tau}_{j+1,k-1} \end{matrix}\right\} = \left.\frac{\partial^2\bar{\tau}}{\partial x^2}\right|_{j,k} + \left.\frac{\partial^2\bar{\tau}}{\partial y^2}\right|_{j,k} + O(h^4)\ , \tag{6.5.6}$$

provided $\partial^4\bar{\tau}/\partial x^4$ and $\partial^4\bar{\tau}/\partial y^4$ are all $O(h^2)$. Unfortunately this is not true in many cases, in particular in most fluid dynamics problems. As a result, this form may be even less accurate than the five-point formula (6.5.4) (see, for example, Cyrus and Fulton, 1967, and Jenssen and Straede, 1968).

Proper higher order local accuracy for the Laplacian can be obtained with difference approximations which use τ values two grid-points either side of the point (j,k). For example, the spatial derivative $\partial^2\bar{\tau}/\partial x^2$ can be written in the form (see Jenssen and Straede, 1968)

$$\left.\frac{\partial^2\bar{\tau}}{\partial x^2}\right|_{j,k} = \frac{-\bar{\tau}_{j+2,k} + 16\bar{\tau}_{j+1,k} - 30\bar{\tau}_{j,k} + 16\bar{\tau}_{j-1,k} - \bar{\tau}_{j-2,k}}{12(\Delta x)^2} + O\{(\Delta x)^4\}\ ,$$

and there is a similar expression for $\partial^2\bar{\tau}/\partial y^2$. For the Laplacian the stencil with fourth order truncation error is therefore

$$\nabla^2\tau = \frac{1}{12h^2}\begin{bmatrix} & & -1 & & \\ & & 16 & & \\ -1 & 16 & -60 & 16 & -1 \\ & & 16 & & \\ & & -1 & & \end{bmatrix}. \tag{6.5.7}$$

When methods of improved accuracy are developed for grid points in the interior of the solution region, it should be kept in mind that the formulation of the boundary conditions should be of comparable accuracy. There is little point in using finite difference approximations to the space derivatives in a partial differential equation with truncation error of sixth order, if the finite difference approximations to the space derivatives in the boundary conditions are of second order.

However, the use of finite difference equations based on derivative approximations of higher order accuracy does not necessarily mean that the finite difference solution obtained will be more accurate. Cheng (1970), for example, indicates that generally second "order" methods are optimal.

Examples given by Hamming (1962), Cyrus and Fulton (1967) and Cheng (1968) clearly show that some lower "order" methods produce results at least as accurate as corresponding methods of higher "order". The lack of success with the latter may have been due to poor treatment of the boundary conditions; often the accuracy of the solution is restricted by the "order" of the boundary value approximation. The order of the finite difference approximations to the boundary conditions and to the partial differential equation should be consistent.

One of the reasons for the lack of improvement when higher order methods are used to solve partial differential equations is that the order of the method is significant only as $\Delta x \to 0$, $\Delta y \to 0$, $\Delta z \to 0$, $\Delta t \to 0$, because the "order" of the truncation error is only that of the leading term which has been omitted. This "order" is more significant to ordinary differential equations where much smaller grid spacings, Δx say, can be used than the Δx, Δy, Δz, Δt for partial differential equations. This is because finite difference solutions in the former case only involve storage of arrays of dimension J, whereas difficulty is often encountered when arrays of dimension $J \times K \times L \times N$, the product of the number of grid points in the x-direction with the numbers in the y,z and t-directions, are required to be stored in the solution of time dependent three dimensional partial differential equations.

However, the possibility of improved accuracy is not the only factor to be considered when higher order methods are evaluated. In practice, stability, rate of convergence of the solution and storage requirements must be considered. For instance, stability is important, since conditional stability requirements restrict the size of the time step which may be used with time dependent problems and this can greatly increase the number of calculations required. Even with unconditionally stable methods, such as most of the implicit methods, the rate of convergence of the solutions found (say) by iterative methods and the amount of storage required on the computer, become important.

Reduction of the Discretization Error by Extrapolation

This method, introduced by Richardson (1910) and called by him the deferred approach to the limit, may lead to considerable improvements of numerical results when solving partial differential equations by finite difference methods. Applications of this method to equations to be solved on rectangular regions in space are described in Richardson and Gaunt (1910) and Salvadori (1951).

The reduction of the error depends upon there being a reliable estimate of the discretization error as a function of the grid spacing. Let $\bar{\tau}$ be the exact solution of the differential equation and τ_1, τ_2 the finite difference solutions at a particular point in the solution region computed using grid spacings h_1 and h_2 respectively. If the leading term in the discretization error is of $O\{h^m\}$, then

$$\bar{\tau} - \tau_1 = Kh_1^m \qquad (6.5.8a)$$

and

$$\bar{\tau} - \tau_2 = Kh_2^m \quad . \qquad (6.5.8b)$$

Eliminating K and solving for $\bar{\tau}$ gives a more accurate estimate of the solution in terms of the two approximate solutions τ_1 and τ_2, namely

$$\bar{\tau} = \frac{h_2^m \tau_1 - h_1^m \tau_2}{h_2^m - h_1^m} \quad . \tag{6.5.9}$$

Thus $\bar{\tau}$ is found provided a reliable estimate of m is available.

When solving the steady state diffusion problem, namely

$$\nabla^2 \bar{\tau} = 0 \tag{6.5.10}$$

using the five point finite difference approximation (6.5.3) to the Laplacian, the resulting finite difference equation is

$$\tau_{j,k} = \tfrac{1}{4}\{\tau_{j,k-1} + \tau_{j-1,k} + \tau_{j+1,k} + \tau_{j,k+1}\}$$

which has a truncation error of $O\{h^4\}$. If the region R on which this is to be solved is rectangular and smooth values of τ are defined on the boundary, the solution $\tau_{j,k}$ over the complete region R is of the form

$$\bar{\tau}_{j,k} = \tau_{j,k} + O\{h^4\} \ .$$

It follows that $m = 4$, and application of Equation (6.5.9) with $h_2 = \tfrac{1}{2}h_1$ gives

$$\bar{\tau} = \frac{16}{15}\tau_2 - \frac{1}{15}\tau_1 \ .$$

When the value of m is not known, it can be determined from three finite difference solutions obtained for the same point using grid spacings which have been successively reduced by multiplying by some fraction $f < 1$ so that

$$h_3 = fh_2 = f^2 h_1 \quad .$$

Since

$$\tau_2 - \tau_1 = K(h_1^m - h_2^m) = Kh_1^m\{1-f^m\} \quad ,$$

and

$$\tau_3 - \tau_2 = K(h_2^m - h_3^m) = f^m K h_1^m \{1-f^m\} \ ,$$

then

$$\frac{\tau_3 - \tau_2}{\tau_2 - \tau_1} = f^m \quad , \tag{6.5.11}$$

so

$$m = \log_f \left\{\frac{\tau_3 - \tau_2}{\tau_2 - \tau_1}\right\} \quad . \tag{6.5.12}$$

Application of Equation (6.5.9) with $h_2 = fh_1$ gives

$$\bar{\tau} = \frac{\tau_2 - f^m \tau_1}{1 - f^m} . \qquad (6.5.13)$$

Substitution of (6.5.11) for f^m in this equation yields

$$\bar{\tau} = \frac{\tau_1 \tau_3 - \tau_2^2}{\tau_1 - 2\tau_2 + \tau_3} , \qquad (6.5.14)$$

which gives a better estimate of the exact solution at any grid point.

This method of reducing the discretization error has been applied to the diffusion problem presented in Section 2 which was solved using the FTCS explicit method in Section 3. From the results presented in Table 3.1 it is seen that, when $s = \frac{1}{2}$ and $J = 10, 20, 40$ the respective values of $\tau(0.4,8)$ obtained were $\tau_1 = 47.54791, \tau_2 = 45.66436$, $\tau_3 = 45.19564$. The first two approximations to the exact value $\bar{\tau} = 45.03963$ have relative errors of 5.6% and 1.4% respectively, and the third has a relative error of 0.35%. Substitution of these values into Equation (6.5.14) gives the value $\tau = 45.04036$ which has a relative error of 0.0016%.

When this method was applied using values of τ computed for $s = \frac{1}{6}$ and $J = 10, 20, 40$ respectively, the answer obtained was accurate to four decimal places. Clearly, this is a useful way of increasing the accuracy of the result at a given grid point without having to excessively increase the number of calculations required. To obtain the result $\tau = 45.04036$ with $s = \frac{1}{2}$ required $16 + 64 + 256 = 336$ applications of the explicit finite difference equation (2.4.4). This result is more accurate than that obtained in Table 3.1 using $J = 100$ and $s = \frac{1}{6}$, which required 4800 applications of the equation (2.4.4). Note that the more accurate values are only obtained on the coarsest grid used, such as for $J = 10$ in the above case.

This method may be of uncertain value near boundaries which are not straight (since boundary value interpolation is then required), near corners with interior angles greater than 180° and near boundaries on which the given values of the dependent variable or its normal derivative are not spatially smooth. For details about these restrictions see Wasow (1955) and Forsythe and Wasow (1960).

7. SUMMARY

Analytical methods of solving partial differential equations are usually restricted to linear cases with simple geometries and boundary conditions. The increased availability of more and more powerful digital computers has made more common the use of numerical methods for solving such equations, in addition to non-linear equations with more complicated boundaries and boundary conditions.

In the previous sections, one particular numerical method, the method of finite differences, has been described. This method is based on the representation of the continuously defined function $\bar{\tau}(x,y,z,t)$ and its derivatives in terms of values of τ defined at particular, discrete points called gridpoints. These are generally the points of intersection of a grid of straight lines,

$x=x_j$, $y=y_k$, $z=z_\ell$, $t=t_n$, $j,k,\ell,n = 1,2,3,\ldots,$ that cover the domain of $\bar{\tau}$. From the appropriate Taylor's series expansions of $\bar{\tau}$ about such gridpoints, forward, backward and central difference approximations to derivatives of $\bar{\tau}$ can be developed for use in the given partial differential equation and its initial and boundary conditions.

The resulting numerical model is composed of the finite difference analogues (the difference equations) of the mathematical model (the partial differential equation and the initial and boundary conditions) that describes the original fluid flow problem. While the mathematical model represents the fluid flow in terms of the dependent variable $\bar{\tau}$ defined continuously in the domain of the independent variables z,y,z and t, the finite difference numerical model defines the dependent variable pointwise at the discrete gridpoints (x_j, y_k, z_ℓ, t_n) within the domain of the mathematical model. The resulting numerical solution of the finite difference equation is a table of numbers, each one being associated with one point in the grid. Distinct finite difference numerical models of the same mathematical model are obtained by using different grids and various finite difference approximations to the continuous derivatives involved in the mathematical model.

The finite difference method can be used to solve even the most complicated mathematical models of fluid flows where the models consist of partial differential equations and their associated auxiliary conditions. The basic techniques used have been described in this article and are illustrated by considering a number of simple models based on the transport equation, which is a mathematical model of the variation in space and time of fluid properties such as temperature and vorticity. More information about these techniques may be found in books by Ames (1969), Forsythe and Wasow (1960), Fox (1962), Richtmeyer and Morton (1967) and Roache (1974), as well as in scientific and engineering journals such as the Journal of Computational Physics.

The formulation of a well posed mathematical model of a fluid system requires both a reduction of the system to a mathematical model and a demonstration of the accuracy of the model. The formulation of a well posed numerical model requires both the derivation of a set of difference equations based on the mathematical model and a demonstration of the adequacy of these difference equations. The latter involves an analysis of the consistency, convergence and stability of the numerical model.

The difference equations and auxiliary conditions of the numerical model must be <u>consistent</u> with the differential equations and initial and boundary conditions of the mathematical model. The numerical model is consistent if the truncation error , that is, the discrepancy between the finite difference approximations and the continuous derivatives, tends to zero as the grid spacings get smaller and smaller. Second, the solution to the finite difference equation must <u>converge</u> to the solution of the differential equation as the grid spacings get smaller and smaller. That is, the difference between the exact solutions of the numerical and mathematical models should vanish as the grid spacings tend to zero. Finally, it must be possible to solve the finite difference equation on a computer. As computers can store only a finite number of digits to represent each number (between 7 and 24 decimal digits) at each step of any calculation, round-off errors may occur. The computation is <u>stable</u> if the growth of these errors is reasonable or controlled.

A numerical model with consistent equations, convergent solutions and stable error propagation forms a computationally stable scheme which gives results which closely approximate the exact solution of the mathematical model. This was illustrated in Sections 2, 4 and 5, where finite difference solutions obtained from various numerical models were compared with analytical solution of the original mathematical models.

Convergence of the numerical solution of particular finite difference schemes is often difficult to prove. However, for mathematical models which are well posed, linear, initial-value boundary-value problems, Lax's Equivalence Theorem can be used. This theorem states that the solution of a numerical model which is consistent with such a mathematical model, converges if and only if the finite difference equation is stable.

In principle, the method of finite differences can be used on non-linear mathematical models, but consistency, stability and convergence are more difficult to prove. For instance, the matrix method and von Neumann's method of stability analysis no longer apply, since they only apply to linear finite difference equations. In many cases numerical experimentation, such as solving the finite difference equation using progressively smaller grid spacings and examining the behaviour of the sequence of values of τ obtained at given points (x,y,z,t), is the only method available.

The different methods of obtaining a finite difference numerical model corresponding to a particular mathematical model may result in either explicit or implicit finite difference equations. The former may be solved directly with a minimum of programming effort and computation time in order to advance from one time level to another in an initial-value problem, but are generally only conditionally stable. It was found that the FTCS explicit method for solving the one dimensional diffusion equation can be assumed to be stable only if $s = \alpha\Delta t/(\Delta x)^2 \leqslant \frac{1}{2}$. This restricts the size of the time increments possible with an initial-value problem of fixed grid size in space.

However, implicit finite difference schemes are often unconditionally stable. Thus, the Crank-Nicolson method for solving the one dimensional diffusion equation was found to be stable for all values of s. This implies computational stability for any size of the time increment. However, the size of Δt is still limited by the accuracy required in the answers; very large values result in poor answers because of unacceptably large truncation errors produced.

Implicit finite difference schemes depend on the solution of a large number of linear algebraic equations. These may be solved by elimination processes, such as the Thomas algorithm used for the tri-diagonal system of linear algebraic equations obtained when the Crank-Nicolson finite difference method is used to solve the one dimensional diffusion equation. Alternatively, iterative methods such as successive over relaxation may be used. Iterative methods may be used at each time level when solving numerical models of multi-dimensional equations in space. Recently, alternating direction implicit methods and locally one dimensional methods have been increasingly used to solve initial-value, boundary-value problems involving more than one space dimension.

Another recently developed numerical technique which shows great promise for solving equilibrium type problems is the finite element method (see Zienkiewicz, 1967). This method is not classified as a finite difference technique since it differs from finite difference methods in two ways. First, instead of involving a set of grid points at the intersection of a set of parallel lines, it employs a series of "finite elements" connected to discrete points on the boundary of the space region and covering the region. Second, the left side of the partial differential equation, written in the form $\phi=0$, is replaced over the region by its equivalent integral form, the value of which is minimized using the calculus of variations. The resulting set of linear simultaneous equations connecting values of the dependent variable at various points over the space region involved, usually has a strongly banded, sparse coefficient matrix. Methods such as the Thomas algorithm can then be used to solve these equations.

APPENDIX : The Thomas Algorithm

The system

$$
\begin{aligned}
b_1\tau_1 + c_1\tau_2 \quad &= d_1 \\
a_2\tau_1 + b_2\tau_2 + c_2\tau_3 \quad &= d_2 \\
a_3\tau_2 + b_3\tau_3 + c_3\tau_4 \quad &= d_3 \\
&\vdots \\
a_j\tau_{j-1} + b_j\tau_j + c_j\tau_{j+1} \quad &= d_j \\
&\vdots \\
a_M\tau_{M-1} + b_M\tau_M \quad &= d_M
\end{aligned}
\tag{A1.1}
$$

in which the a's, b's, c's and d's are known, is usually solved for the unknown values τ_j, $j = 1,2,\ldots,M$, by an elimination procedure attributed to Thomas (1949). The first equation listed in (A1.1) is used to eliminate τ_1 from the second equation, the new second equation is used to eliminate τ_2 from the third equation and so on until finally, the new second-last equation is used to eliminate τ_{M-1} from the last equation giving one equation with only one unknown, τ_M. The unknowns $\tau_{M-1}, \tau_{M-2}, \ldots, \tau_2, \tau_1$ are then found in turn by back-substitution. This method is therefore a particular example of the Gauss elimination method with back substitution.

Suppose τ_{j-2} has been eliminated from the (j-1)th equation in (A1.1), which is now written

$$\beta_{j-1}\,\tau_{j-1} + \gamma_{j-1}\,\tau_j = \delta_{j-1}\ . \tag{A1.2}$$

This is now used to eliminate τ_{j-1} from the j-th equation in (A1.1) giving the result

$$\left(b_j - \frac{a_j\gamma_{j-1}}{\beta_{j-1}}\right)\tau_j + c_j\,\tau_{j+1} = d_j - \frac{a_j\,\delta_{j-1}}{\beta_{j-1}} \tag{A1.3}$$

A comparison of Equations (1.2) and (A1.3) shows that the coefficients $\beta_j, \gamma_j, \delta_j$, are given in terms of $\beta_{j-1}, \gamma_{j-1}, \delta_{j-1}$ and a_j, b_j, c_j, d_j by the relations

$$\beta_j = b_j - a_j\,\gamma_{j-1}/\beta_{j-1}\ , \tag{A1.4a}$$

$$\gamma_j = c_j\ , \tag{A1.4b}$$

$$\delta_j = d_j - a_j\,\delta_{j-1}/\beta_{j-1}\ , \tag{A1.4c}$$

for $j = 2,3,\ldots,M$ with $\beta_1 = b_1$, $\gamma_1 = c_1$ and $\delta_1 = d_1$.

When $j = M$, we obtain

$$\beta_M\,\tau_M = \delta_M$$

since $c_M = 0$. Thus

$$\tau_M = \delta_M/\beta_M$$

and the remaining values of τ are found by substitution into a rearranged form of Equation (A1.2), namely

$$\tau_{j-1} = (\delta_{j-1} - \gamma_{j-1}\tau_j)/\beta_{j-1}\ , \ j = M, M-1, \ldots, 2. \qquad \text{(A1.5)}$$

The complete algorithm, with c_j substituted for γ_j in the previous results, is:

$$\text{Set}\ \ \beta_1 = b_1,\quad \delta_1 = d_1\ .$$

For $j = 2,3,\ldots,M$ compute

$$K = a_j/\beta_{j-1} \quad (K \text{ is a temporary storage}),$$

$$\beta_j = b_j - K\,c_{j-1},$$

$$\delta_j = d_j - K\,\delta_{j-1}.$$

Then

$$\tau_M = \delta_M/\beta_M\ .$$

For $j = M-1, M-2, \ldots, 1$ compute

$$\tau_j = (\delta_j - c_j\,\tau_{j+1})/\beta_j\ .$$

This method is stable with regard to propagation of round-off error as long as $|b_j| \geqslant |a_j| + |c_j|$ for all j.

The normal Gauss elimination methods require about $M^3/3$ multiplications and divisions (about 3,000 if $M = 30$) to solve a system of M linear algebraic equations in M unknowns. The Thomas method which takes advantage of the tri-diagonal nature of the system requires only $5M - 4$ multiplications and divisions (146 if $M = 30$) since the algorithm eliminates all multiplications by zero coefficients of τ which take place when the general Gauss elimination method is applied.

REFERENCES

Ames, W.F. (1969), *Numerical Methods for Partial Differential Equations*, Barnes and Noble, Inc., New York, New York.

Amsden, A.A., and Harlow, F.H. (1964), "Slip Instability", *Physics of Fluids*, Vol. 7, pp. 327-334.

Anthes, R.A. (1970), "Numerical Experiments with a Two Dimensional Horizontal Variable Grid", *Monthly Weather Review*, Vol. 98, No. 11, pp. 810-822.

Bickley, W.G. (1948), "Finite-Difference Formulae for the Square Lattice", *Quarterly Journal of Mechanics and Applied Mathematics*, Vol. 1, pp. 35-42.

Birkhoff, G., and Varga, R.S. (1959), "Implicit Alternating Direction Methods", *Transactions American Mathematical Society*, Vol. 92, pp. 13-24.

Blottner, F.G., and Roache, P.J. (1971), "Nonuniform Mesh Systems", *Journal Computational Physics*, Vol. 8, pp. 498-499.

Brian, P.L.T. (1961), "A Finite-Difference Method of High-Order Accuracy for the Solution of Three-Dimensional Transient Heat Conduction Problems", *American Institute of Chemical Engineering Journal*, Vol. 7, No. 3, pp. 367-370.

Browning, G., Kreiss, H.O., and Oliger, J. (1973), "Mesh Refinement", *Mathematics of Computation*, Vol. 27, No. 121, pp. 29-39.

Carre, B.A. (1961), "The Determination of the Optimum Accelerating Factor for Successive Over-Relaxation", *Computer Journal*, Vol. 4, No. 1, pp. 73-78.

Carslaw, H.S., and Jaeger, J.C. (1962), *Conduction of Heat in Solids*, Oxford University Press, London.

Chavez, S.P., and Richards, C.G. (1970), *A Numerical Study of the Coanda Effect*, 70-FIcs-12, The American Society of Mechanical Engineers, United Engineering Center, New York, New York.

Cheng, S.I. (1968), *Accuracy of Difference Formulation of Navier-Stokes Equations*, A.M.S. Department, Princeton University, Princeton, New Jersey.

Cheng, S.I. (1970), "Numerical Integration of Navier-Stokes Equations", *A.I.A.A. Journal*, Vol. 8, No. 12, pp. 2115-2122.

Ciment, M. (1971), "Stable Difference Schemes with Uneven Mesh Spacings", *Mathematics of Computation*, Vol. 25, No. 114, pp. 219-227.

Ciment, M., and Sweet, R.A. (1973), "Mesh Refinements for Parabolic Equations", *Journal of Computational Physics*, Vol. 12, pp. 513-525.

Courant, R., Isaacson, E., and Rees, M. (1952), "On the Solution of Nonlinear Hyperbolic Differential Equations by Finite Differences", *Communications on Pure and Applied Mathematics*, Vol. 5, pp. 243-255.

Crandall, S.H. (1955), "An Optimum Implicit Recurrence Formula for the Heat Conduction Equation", *Quarterly of Applied Mathematics*, Vol. 13, No. 3, pp. 318-320.

Crank, J., and Nicolson, P. (1947), "A Practical Method for Numerical Evaluation of Solutions of Partial Differential Equations of the Heat-Conduction Type", *Proceedings of the Cambridge Philosophical Society*, Vol. 43, No. 50, pp. 50-67.

Crowder, H.J., and Dalton, C. (1971), "Errors in the Use of Nonuniform Mesh Systems", *Journal of Computational Physics*, Vol. 7, pp. 32-45.

Cyrus, N.J., and Fulton, R.E. (1967), *Accuracy Study of Finite Difference Methods*, NASA TN D-4372, National Aeronautics and Space Administration, Langley Research Center, Langley Station, Hampton, Virginia.

de Vahl Davies, G., and Mallinson, G.D. (1976), "An Evaluation of Upwind and Central Difference Approximations by a Study of Recirculating Flow", *Computers and Fluids*, Vol. 4, pp. 29-43.

Douglas, J., Jr. (1955), "On the Numerical Integration of $\partial^2u/\partial x^2 + \partial^2u/\partial y^2 = \partial u/\partial t$ by Implicit Methods", *Journal Society of Industrial Applied Mathematics*, Vol. 3, No. 1, pp. 42-65.

Douglas, J., Jr. (1957), "A note on the Alternating Direction Implicit Method for the Numerical Solution of Heat Flow Problems", *Proceedings of the American Mathematical Society*, Vol. 8, pp. 409-412.

Douglas, J., Jr. (1962), "Alternating Direction Methods for Three Space Variables, *Numerische Mathematik*, Vol. 4, pp. 41-63.

DuFort, E.C., and Frankel, S.P. (1953), "Stability Conditions in the Numerical Treatment of Parabolic Differential Equations", *Mathematical Tables and Other Aids to Computation*, Vol. 7, pp. 135-152.

D'Yakonov, E.G. (1963), "Difference Schemes with Split Operators for Multi-Dimensional Unsteady Problems", *U.S.S.R. Comp. Math.*, Vol. 4, No. 2, pp. 92-110.

Eddy, E.P. (1949), *Stability in the Numerical Solution of Initial Value Problems in Partial Differential Equations*, NOLM 10232, Naval Ordnance Laboratory, White Oak, Silver Spring, Maryland.

Forsythe, G.E., and Wasow, W. (1960), *Finite Difference Methods for Partial Differential Equations*, Wiley, New York.

Fox, L. (1944), "Solution by Relaxation Methods of Plane Potential Problems with Mixed Boundary Conditions", *Quarterly of Applied Mathematics*, Vol. 2, pp. 251-257.

Fox, L. (1962), *Numerical Solution of Ordinary and Partial Differential Equations*, Addison-Wesley Publishing Company, Inc., Reading, Massachusetts.

Frankel, S.P. (1950), "Convergence Rates of Iterative Treatments of Partial Differential Equations", *Mathematical Tables and Other Aids to Computation*, Vol. 4, pp. 65-75.

Fromm, J.E., and Harlow, F.H. (1963), "Numerical Solution of the Problem of Vortex Street Development", *Physics of Fluids*, Vol. 6, No. 7, pp. 975-982.

Gal-Chen, T. (1975), "On the Use of a Coordinate Transformation for the Solution of the Navier Stokes Equations", *Journal of Computational Physics*, Vol. 17, pp. 209-228.

Gourlay, A.R., and Mitchell, A.R. (1969), "The Equivalence of Certain Alternating Direction and Locally-One Dimensional Difference Methods", *SIAM Journal on Numerical Analysis*, Vol. 6, No. 1, pp. 37-46.

Hahn, S.G. (1958), "Stability Criteria for Difference Schemes", *Communications on Pure and Applied Mathematics*, Vol. 11, pp. 243-255.

Hamming, R.W. (1962), *Numerical Methods for Scientists and Engineers*, McGraw-Hill Book Company, Inc., New York.

Harrison, E.J. (1973), "Three-Dimensional Numerical Simulations of Tropical Systems Utilizing Nested Finite Grids", *Journal of the Atmospheric Sciences*, Vol. 30, p. 1528.

Hirt, C.W. (1968), "Heuristic Stability Theory for Finite-Difference Equations", *Journal of Computational Physics*, Vol. 2, pp. 339-355.

Hung, T.K., and Macagno, E.O. (1966), "Laminar Eddies in a Two-Dimensional Conduit Expansion", *La Houille Blanche*, Vol. 21, No. 4, pp. 391-400.

Isenberg, J., and de Vahl Davis, G. (1975), "Finite Difference Methods in Heat and Mass Transfer", in *Topics in Transport Phenomena*, ed. C. Gutfinger, Hemisphere Publishing Company, Washington, D.C.

Jenssen, D., and Straede, J. (1968), "The Accuracy of Finite Difference Analogues of Simple Differential Operators", *Proceedings of the WMO/IUGG Symposium on Numerical Weather Prediction*, Tokyo, November 26 - December 4, 1968.

John, F. (1952), "On the Integration of Parabolic Equations by Difference Methods", *Communications on Pure and Applied Mathematics*, Vol. 5, pp. 155-211.

Kalnay de Rivas, E. (1971), *The Circulation of the Atmosphere of Venus*, Ph.D. Thesis, Department of Meteorology, M.I.T., Massachusetts, U.S.A.

Kalnay de Rivas, E. (1972), "On the Use of Non-Uniform Grids in Finite-Difference Equations", *Journal of Computational Physics*, Vol. 10, pp. 202-210.

Kreyszig, E. (1972), *Advanced Engineering Mathematics*, 3rd edition, John Wiley & Sons, Inc., New York.

Kusic, G.L., and Lavi, A. (1968), "Stability of Difference Methods for Initial-Value Type Partial Differential Equations", *Journal of Computational Physics*, Vol. 3, pp. 358-378.

Kusic, G.L. (1969), "On Stability of Numerical Methods for Systems of Initial-Value Partial Differential Equations", *Journal of Computational Physics*, Vol. 4, pp. 272-275.

Laasonen, P. (1949), "Über eine Methode zur Lösung der Warmeleitungsgleichung", *Acta Mathematica*, Vol. 81, p. 309.

Lax, P.D., and Richtmyer, R.D. (1956), "Survey of the Stability of Linear Finite Difference Equations", *Communications on Pure and Applied Mathematics*, Vol. 9, pp. 267-293.

Lax, P.D., and Wendroff, B. (1960), "Systems of Conservation Laws", *Communications on Pure and Applied Mathematics*, Vol. 13, pp. 217-237.

Lax, P.D., and Wendroff, B. (1964), "Difference Schemes with High Order of Accuracy for Solving Hyperbolic Equations", *Communications on Pure and Applied Mathematics*, Vol. 17, pp. 381-398.

Levy, H., and Lessman, F. (1959), *Finite Difference Equations*, Sir Isaac Pitman & Sons, Ltd., London.

MacCormack, R.W. (1971), "Numerical Solution of the Interaction of a Shock Wave with a Laminar Boundary Layer", *Proceedings of Second International Conference on Numerical Methods in Fluid Dynamics*, ed. M. Holt, Lecture Notes in Physics, Vol. 8, Springer-Verlag, New York.

Marchuk, G.I. (1975), *Methods of Numerical Mathematics*, Springer-Verlag, New York.

Moore, A.H., Kaplan, B., and Mitchell, D.B. (1975), "A Comparison of Crandall and Crank-Nicolson Methods for Solving a Transient Heat Conduction Problem", *International Journal for Numerical Methods in Engineering*, Vol. 9, pp. 938-943.

Moretti, G., and Salas, M.D. (1970), "Numerical Analysis of Viscous One-Dimensional Flows", *Journal of Computational Physics*, Vol. 5, pp. 487-506.

Motzkin, T.S., and Wasow, W. (1953), "On the Approximation of Linear Elliptic Differential Equations by Difference Equations with Positive Coefficients", *Journal of Mathematics & Physics*, Vol. 31, pp. 253-259.

Myers, G.E. (1971), *Analytical Methods in Conduction Heat Transfer*, McGraw-Hill Book Company.

Noye, B.J. (1977), "Wind-Induced Circulation and Water Level Changes in Lakes", *Proceedings of International Conference on Applied Numerical Modelling*, University of Southampton, 11-15 July, 1977, Pentech Press, pp. 135-145.

Noye, B.J. (1978), "Finite Difference Schemes for the Solution of Linearised Wind Effect Equations", *Numerical Simulation of Fluid Motion*, ed. John Noye, North-Holland, Amsterdam, pp. 435-451.

O'Brien, G.G., Hyman, M.A., and Kaplan, S. (1950), "A Study of the Numerical Solution of Partial Differential Equations", *Journal of Mathematics and Physics*, Vol. 29, pp. 223-251.

Osher, S.J. (1970), "Mesh Refinements for the Heat Equation", *S.I.A.M. Journal of Numerical Analysis*, Vol. 7, pp. 219-227.

Pao, Y.-H., and Daugherty, R.J. (1969), *Time-Dependent Viscous Incompressible Flow Past a Finite Flat Plate*, Boeing Scientific Research Laboratories, D1-82-0822, January, 1969.

Payne, R.B. (1958), "Calculation of Unsteady Viscous Flow Past a Circular Cylinder", *Journal of Fluid Mechanics*, Vol. 4, p. 81.

Peaceman, D.W., and Rachford, H.H., Jr. (1955), "The Numerical Solution of Parabolic and Elliptical Differential Equations", *Journal Society Indust. Applied Mathematics*, Vol. 1, No. 1, March 1955, pp. 28-41.

Pearson, C.E. (1964), *A Computational Method for Time-Dependent Two-Dimensional Incompressible Viscous Flow Problems*, Sperry Rand Research Centre Report, SRRC-RR-64-17, February, 1964.

Pearson, C.E. (1965a), "A Computational Method for Viscous Flow Problems", *Journal of Fluid Mechanics*, Vol. 21, Part 4, pp. 611-622.

Pearson, C.E. (1965b), "Numerical Solutions for the Time-Dependent Viscous Flow Between Two Rotating Coaxial Disks", *Journal of Fluid Mechanics*, Vol. 21, Part 4, pp. 623-633.

Richardson, L.F. (1910), "The Approximate Arithmetical Solution by Finite Differences of Physical Problems Involving Differential Equations, with an Application to the Stresses in a Masonry Dam", *Transactions of the Royal Society of London*, Series A., Vol. 210, pp. 307-357.

Richardson, L.F., and Gaunt, J.A. (1927), "The Deferred Approach to the Limit", *Philosophical Transactions of the Royal Society, London*, Series A, Vol. 226, pp. 299-361.

Richtmyer, R.D., and Morton, K.W. (1967), *Difference Methods for Initial-Value Problems*, Second Edition, Interscience Publishers, J. Wiley and Sons, New York, New York.

Roache, P.J. (1974), *Computational Fluid Dynamics*, Hermosa Publishers, Albuquerque.

Sadourny, R., and Morel, P. (1969), "A Finite-Difference Approximation of the Primitive Equations for a Hexagonal Grid on a Plane", *U.S. Monthly Weather Review*, Vol. 97, No. 6, pp. 439-445.

Salvadori, M.G. (1951), "Extrapolation Formulas in Linear Difference Operators", *Proceedings 1st Congress Applied Mechanics*, New York, pp. 15-18.

Samarskii, A.A. (1964), "Local One-Dimensional Difference Schemes for Multi-Dimensional Hyperbolic Equations in an Arbitrary Region", *Z. Vycisl. Mat. i Mat. Fiz.*, Vol. 4, pp. 21-35.

Smith, G.D. (1969), *Numerical Solution of Partial Differential Equations*, Oxford University Press, London.

Stommel, H. (1948), "The Westward Intensification of Wind-Driven Ocean Currents", *Transactions of American Geophysical Union*, Vol. 29, pp. 202-206.

Sundquist, H., and Veronis, G. (1970), "A Simple Finite-Difference Grid with Non-Constant Intervals", *Tellus*, Vol. 22, pp. 26-31.

Thom, A., and Apelt, C.J. (1961), *Field Computations in Engineering and Physics*, D. Van Nostrand Company, Ltd., London.

Thoman, D., and Szewczyk, A.A. (1969), "Time Dependent Viscous Flow over a Circular Cylinder", *The Physics of Fluids Supplement II*, pp. 76-87.

Thomas, G.B. (1968), *Calculus and Analytic Geometry*, 4th Edition, Addison-Wesley Publishing Co., Reading, Massachusetts.

Thomas, L.H. (1949), *Elliptic Problems in Linear Difference Equations over a Network*. Watson Scientific Computing Laboratory, Columbia University, New York.

Torrance, K.E. (1968), "Comparison of Finite-Difference Computations of Natural Convection", *Journal of Research of the National Bureau of Standards*, Vol. 72B, No. 4, pp. 281-301.

Trapp, J.A., and Ramshaw, J.D. (1976), "A Simple Heuristic Method for Analyzing the effect of Boundary Conditions on Numerical Stability", *Journal of Computational Physics*, Vol. 20, pp. 238-242.

Venit, S. (1973), "Mesh Refinements for Parabolic Equations of Second Order", *Mathematics of Computation*, Vol. 27, No. 124, pp. 745-754.

Viswanathan, R.V. (1957), "Solution of Poisson's Equation by Relaxation Method - Normal Gradient Specified on Curved Boundaries", *Mathematical Tables and other Aids to Computation*, Vol. 11, pp. 67-78.

Wade, T.L. (1951), *The Algebra of Vectors and Matrices*, Addison-Wesley Publishing Company Inc., Reading, Massachusetts, U.S.A.

Wasow, W. (1955), "Discrete Approximations to Elliptic Differential Equations", *Zeitschrift Angew. Mathematical Physics*, Vol. 6, pp. 81-97.

Williamson, D. (1969), "Numerical Integration of Fluid Flow Over Triangular Grids", *Monthly Weather Review*, Vol. 97, No. 12, pp. 885-895.

Winslow, A.M. (1966), "Numerical Solution of the Quasilinear Poisson Equation in a Nonuniform Triangle Mesh", *Journal of Computational Physics*, Vol. 1, No. 2, p. 149.

Yanenko, N.N. (1971), *The Method of Fractional Steps: The Solution of Problems of Mathematical Physics in Several Variables*, English Translation edited M. Holt, Springer-Verlag, New York.

Young, D. (1954), "Iterative Methods for Solving Partial Differential Equations of Elliptic Type", *Transactions American Mathematical Society*, Vol. 76, pp. 92-111.

Zienkiewicz, O.C. (1967), *The Finite Element Method in Structural and Continuum Mechanics*, McGraw-Hill, London.

Numerical Simulation of Fluid Motion, ed. John Noye
pp. 113-170

THE GALERKIN METHOD: AN INTRODUCTION

C.A.J. Fletcher

Weapons Research Establishment

Salisbury, South Australia

SUMMARY

The traditional Galerkin method is described in detail. Four worked examples using the traditional Galerkin method are described in full. The examples are used to show the limitations of the traditional method. The connection between the traditional method and the following computer-oriented Galerkin methods is demonstrated:

(i) The Galerkin finite element method
(ii) The Galerkin finite difference method
(iii) The spectral method

A brief description, a worked example and some typical applications to fluid flows are given for each computer-oriented method. Finally the Galerkin framework is used to compare finite element, finite difference and spectral methods.

TABLE OF CONTENTS

1. INTRODUCTION

The word Galerkin arises in a number of, apparently unconnected, computational methods. Galerkin computational methods have been applied in such diverse areas as dynamics of structures, heat and mass transfer, fluid flow, convective and hydrodynamic stability, nuclear engineering and electrical circuit design, etc. Galerkin methods have been applied to ordinary differential equations, partial differential equations of elliptic, parabolic and hyperbolic type, to integral equations and to steady, eigenvalue and unsteady problems. The width of application and the variety of methods has led to considerable interpretation as to what constitutes a Galerkin method. In the present paper the traditional definition of a Galerkin method will be given and where various conditions have been relaxed for particular applications, these will be indicated.

Galerkin, a Russian engineer, first published the method(ref.1) in 1915 as a technique for calculating the elastic behaviour of rods and plates. It appears possible(ref.2) that Galerkin was formulating the method during an enforced stay in prison in 1906 because of his anti-Tsarist views. The method gained wider dissemination from the work of Duncan. References 3 and 4 give a readable introduction to the method from an engineering viewpoint. Duncan and his co-workers applied the method to problems associated with the dynamics of aeronautical structures. An introductory treatment of the mathematical aspects of the method, with particular reference to the close link with the variational method, is given by Kantorovich and Krylov(ref.5). A more advanced treatment of the mathematical aspects of the method is given by Mikhlin(ref.6).

A modern treatment would consider the Galerkin method as an important member of the wider class of methods of weighted residuals (MWR). The name, methods of weighted residuals, was first used by Crandall(ref.7) although Collatz(ref.8) put forward the same collective idea by considering the various members as examples of different "error distribution principles". Subsequently Finlayson and Scriven(ref.9) published a comprehensive review of the application of MWR to problems in chemical engineering, fluid flow, nuclear engineering and many others. Since then additional applications of MWR have been given by Finlayson(ref.10) and Ames(ref.11). Finlayson gives comparisons of results obtained from the Galerkin method with results obtained from other members of MWR. Ames stresses the nonlinear applications of MWR.

Four major computational techniques in which the Galerkin method is involved may be identified. These are

(a) the traditional Galerkin method

(b) the Galerkin finite element formulation

(c) the Galerkin finite difference formulation

(d) spectral methods.

Categories (b), (c) and (d) are developments of the traditional Galerkin method designed to achieve more economic and accurate solutions when using a computer. The purpose of this paper is to describe the traditional Galerkin method in detail, to describe briefly the Galerkin finite element method, the Galerkin finite difference method and the spectral method and to describe the way in which these are connected with, and have developed from, the traditional method. Finally the Galerkin framework is used to compare the finite element method, the finite difference method and the spectral method.

The layout of the rest of this paper is as follows. In Section 2 a description of the traditional Galerkin method is given. It is introduced by first considering the wider class of methods of weighted residuals. The steps required in applying the Galerkin method are laid out and a number of examples, drawn mainly from fluid dynamics, are presented to illustrate the method in practice. The connection between the Galerkin method and other methods is discussed with emphasis on the close link with variational methods. Mathematical aspects of the Galerkin method, mainly concerning convergence properties, are touched on. The various modifications to the method required to obtain categories (b), (c) and (d) are described briefly. Discussion of these modifications is taken up in later sections.

Section 3 contains four worked examples (in excruciating detail) to illustrate various practical aspects of the method. The first example is a simple ordinary differential equation. The second and third examples concern two dimensional flows governed by elliptic partial differential equations. The fourth example concerns two dimensional flow governed by a parabolic partial differential equation. No worked example is given of the application of the Galerkin method to a hyperbolic partial differential equation; however a worked example for the one dimensional wave equation is given by Orszag and Israeli(ref.12). The examination of these worked examples throws light on certain deficiencies of the traditional method as a technique for solving large scale problems on a computer; these are discussed.

In Section 4 the Galerkin formulation of the finite element method is described and the links with the traditional Galerkin method are stressed. The Galerkin finite element method is applied to problem 3.3 and the results are compared with those obtained by the traditional method. Applications of the Galerkin finite element method to fluid dynamics are reviewed briefly.

In Section 5 the Galerkin formulation of the finite difference method is described. The Galerkin finite difference method is applied to problem 3.4 and comparisons with results given by the traditional method are discussed. Examples of the application of the Galerkin finite difference method are cited with emphasis on problems governed by parabolic partial differential equations.

In Section 6 spectral methods are discussed in relation to the traditional Galerkin method. An elementary spectral method is applied to problem 3.4 and the results compared with the results of previous sections. Some applications of spectral methods to problems of fluid dynamics are cited.

Based on the common Galerkin framework a brief comparison is made in Section 7 of the relative strengths and weaknesses of finite element methods, finite difference methods and spectral methods.

2. THE TRADITIONAL GALERKIN METHOD

This section begins with a description of the method of weighted residuals which includes the Galerkin method as a special case. The choice of the name, method of weighted residuals, by Crandall(ref.7), was inspired by remarks made by Courant in discussion following the presentation of the subdomain method by Biezeno(ref.13). The method of weighted residuals is an approximate method which seeks a solution that is a good approximation to the exact solution over the whole domain. This may be contrasted to a Taylor expansion which seeks an accurate solution in the small region surrounding a single point in the domain. Thus, following Collatz, the error between the exact and the approximate solutions may be thought of as being distributed, hopefully evenly, throughout the whole domain.

2.1 The method of weighted residuals

A solution of

$$L(u) = 0 \tag{1}$$

is sought where L is a differential operator and equation (1) may be an ordinary differential equation or a partial differential equation of elliptic, parabolic or hyperbolic type. The solution $u = u(\overline{x},t)$ is sought within some domain D and boundary conditions are specified on S, the boundary of D, namely

$$N(u) = f(\bar{x},t), \tag{2}$$

where N is a differential operator but of lower order than L.

The first step in the application of MWR is to introduce a trial solution

$$v(\bar{x},t) = v_o(\bar{x},t) + \sum_{j=1}^{n} a_j(t) \cdot \varphi_j(\bar{x}), \tag{3}$$

which, it is hoped, is close to u in some sense or can be made so if n is large enough. The trial solution is chosen to satisfy the boundary conditions exactly. This is often done by making v_o satisfy the boundary conditions exactly; then the analytic functions, $\varphi_j(\bar{x})$ satisfy homogenous boundary conditions. Since the approximate solution should be capable of converging to the exact solution as n approaches infinity it is important that the $\varphi_j(\bar{x})$ are linearly independent and chosen from a set of functions which is complete in the domain of interest. That is the complete set of functions, $\varphi_j(\bar{x})$, j=1. . ∞ is defined such that there exists no exact solution that cannot be expanded in terms of $\varphi_j(\bar{x})$. In equation (3), $\varphi_j(\bar{x})$ are prescribed analytic functions e.g. polynomials or trigonometric functions, whose form may be suggested by the symmetry of the problem, or a boundary condition or the exact solution of a related problem. Thus an initial value problem might suggest the use of exponential functions. The solution of Laplace's equation might suggest the use of circular and hyperbolic functions.

In equation (3), $a_j(t)$ are coefficients to be determined. The form of equation (3) is typical of initial value problems in which the governing equation (1) is to be reduced to a system of ordinary differential equation for $a_j(t)$. For a steady state problem φ_j might depend on all the independent variables and the a_j's would be constants to be determined from algebraic relations.

The method proceeds by substituting equation (3) into equation (1). In general equation (1) is no longer satisfied and a residual, r, results, i.e.

$$L(v) = r(\bar{x},t,a_j). \tag{4}$$

It is to be expected that r will be small, in some sense, but non-zero throughout, the domain in which a solution is sought. Since r = 0 throughout the domain when the exact solution is obtained, r may be considered a measure of the error. Since the exact solution is not available in general, the size and distribution of r in the domain can be used to assess the accuracy of the solution. Thus if a solution for a particular value of n has been obtained $r(\bar{x},t)$ can be evaluated. The effect of obtaining a new solution with increased n should cause a reduction in r in some average sense.

The method proceeds by imposing the condition

$$\int_D r(\bar{x},t,a_j) \cdot W_k(\bar{x}) \cdot d\bar{x} = 0, \; k = 1 \; . \; . \; . \; n \tag{5}$$

where $W_k(\bar{x})$ represents one member of a family of weighting functions which depend on the same independent variables as the integrating variables and the analytic functions, $\varphi_j(\bar{x})$ in equation (3). Repeated application of equation (5) with different weighting functions, $W_k(\bar{x})$ produces enough equations to permit solution for the a_j's. Since each equation must be independent so each $W_k(\bar{x})$ must be an independent function. As n, the number of unknown coefficients in equation (3), is increased the approximate solution is expected to approach the exact solution. Consequently, the residual, r given by equation (4), is expected to reduce as n increases. In the limit of n approaching infinity the approximate solution will converge to the exact solution, r will be zero and equation (5) will be satisfied for any choice of the weighting function.

Different choices of the weighting function give rise to the different methods which collectively are known as methods of weighted residuals. These methods are

(a) Subdomain method

The domain D is split up into n subdomains, D_j, which may overlap. Then

$$\begin{aligned} W_k &= 1 \quad \text{in } D_k \\ W_k &= 0 \quad \text{outside } D_k. \end{aligned} \tag{6}$$

The application of the Method of Integral Relations to the blunt body problem in supersonic inviscid flow is an example of a subdomain method. A lucid description is given by Belotserkovskii and Chushkin(ref.14). Solutions of laminar boundary layer flow problems have also been obtained by the subdomain method(refs.15,16).

(b) Collocation method

$$W_k(\bar{x}) = \delta(\bar{x} - \bar{x}_k) \tag{7}$$

where δ is the Dirac delta. This choice of $W_k(\bar{x})$ reduces equation (5) to evaluating $r(\bar{x},t)$ at $\bar{x} = \bar{x}_k$. Jain(ref.17) has used collocation to obtain solutions to flow problems with similarity solutions. Villadsen and Stewart(ref.18) choose the collocation points as the zeros of the Jacobi polynomials and call the method orthogonal collocation. They have demonstrated the technique for a number of problems including viscous flow in a channel.

(c) Least squares

$$W_k(\bar{x}) = \partial r/\partial a_k \tag{8}$$

This is equivalent to replacing equation (5) with the requirement that $\int_D r^2 \, d\bar{x}$ is a minimum. Narasimha and Deshpande(ref.19) have used a least-squares formulation of the Botlzmann equation to examine shock structure. Lynn(ref.20) has obtained solutions to the boundary layer equations using a least squares formulation of the finite element method.

(d) Method of moments

$$W_k(\bar{x}) = \sum_{m=0}^{k} b_m \cdot x^k \qquad (9)$$

Often just x^k is used for the weighting function (hence the name "method of moments"). The development of solutions to boundary layer flows (particularly in two dimensions) has produced many moment methods. Here mention is made of the first true moment method due to Yamada(ref.21), the Method of Integral Relations(ref.22) and the use of the moment of the momentum equations by Truckenbrodt(ref.23).

(e) Galerkin method

$$W_k(\bar{x}) = \varphi_k(\bar{x}) \qquad (10)$$

i.e. the weighting functions are chosen from the same family as the trial functions in equation (3). Since the trial functions are chosen from a linearly independent set of functions, complete in the domain of interest, equation (5) and equation (10) may be interpreted as requiring that the residual must be orthogonal to every member of the complete set up to order n. Thus in the limit of n tending to infinity the residual must tend to zero. If the weighting function used in the method of moments is chosen from a complete set then the above remarks hold for the method of moments. Collatz(ref.8) combines the Galerkin method and the method of moments under the single name of orthogonalisation methods. Other authors (e.g. Murphy in reference 24) have grouped the method of moments and the Galerkin method under the collective name of generalised Galerkin method. This is not done here because one of the unique features of the Galerkin method is its equivalence, under certain conditions, to a variational formulation (see Section 2.4). No examples are given here since applications of the Galerkin method to flow problems are treated at greater length in Section 2.3.

Sometimes the Karman - Pohlhausen Integral Method(ref.25) is given as an example of the method of weighted residuals. It uses a weighting function of unity throughout the domain which means it can either be considered as a subdomain method or as a method of moment(s). However it is not a very good example because equation (5), from which the method of weighted residuals takes its name, is only used to find one of the unknown coefficients in equation (3); auxiliary boundary conditions are introduced to obtain the rest.

The description of the method of weighted residuals given above has required that the boundary conditions are satisfied exactly and that the differential equation is satisfied approximately. Ames(ref.26) calls this an interior method. The converse is also possible, in a boundary method the differential equation is satisfied exactly but the boundary conditions are satisfied approximately. Shuleshko(ref.27) has even put forward a general method in which neither differential equation nor boundary conditions are satisfied exactly. The method is rather complicated and does not appear to have been used to date.

2.2 An example of the traditional Galerkin method

The purpose of this section is to lay out the step-by-step procedure required to implement the method. Consider a problem governed by the unsteady heat equation

$$\frac{\partial T}{\partial t} - \frac{\partial^2 T}{\partial x^2} = 0, \tag{11}$$

subject to the boundary conditions

$$\begin{aligned} T(x,o) &= T_a + T_o(x) \\ T(o,t) &= T_a \\ T(1,t) &= T_a . \end{aligned} \tag{12}$$

An analytic representation is postulated

$$T - T_a = T_o(x) + \sum_{j=1}^{n} a_j(t) \cdot \varphi_j(x) . \tag{13}$$

This representation is chosen deliberately so that $\varphi_j(x)$ satisfy homogeneous boundary conditions. Also $\varphi_j(x)$ are chosen from a complete set of functions e.g. x^j. Substitution of equation (13) into equation (11) produces some residual, r, given by

$$r = \sum_{j=1}^{n} \frac{\partial a_j}{\partial t} \cdot \varphi_j(x) - \frac{\partial^2 T_o}{\partial x^2} - \sum_{j=1}^{n} a_j \cdot \frac{\partial^2 \varphi_j}{\partial x^2} (x) . \tag{14}$$

Imposing the Galerkin condition (i.e. equation (5) and (10)) gives

$$\sum_{j=1}^{n} \frac{\partial a_j}{\partial t} \left\{ \int_0^1 \varphi_j(x) \cdot \varphi_k(x)\, dx \right\} - \int_0^1 \frac{\partial^2 T_o}{\partial x^2} \cdot \varphi_k(x) \cdot dx - \sum_{j=1}^{n} a_j \left\{ \int_0^1 \frac{\partial^2 \varphi_j}{\partial x^2} \cdot \varphi_k(x)\, dx \right\}$$

$$= 0, \; k = 1 \ldots n. \tag{15}$$

Equations (15) can be rewritten as

$$[P][\dot{A}] = [Q][A] + [R], \tag{16}$$

where $[A]$ is a column matrix with elements a_j

$[\dot{A}]$ is a column matrix with elements $\partial a_j/\partial t$

$[P]$ has elements p_{kj} given by

$$p_{kj} = \int_0^1 \varphi_j(x) \cdot \varphi_k(x)\, dx. \tag{17}$$

$[Q]$ has elements q_{kj} given by

$$q_{kj} = \int_0^1 \frac{\partial^2 \varphi_j(x)}{\partial x^2} \cdot \varphi_k(x)\, dx. \tag{18}$$

and $[R]$ is a column matrix with elements r_k given by

$$r_k = \int_0^1 \frac{\partial^2 T_0(x)}{\partial x^2} \cdot \varphi_k(x)\, dx. \tag{19}$$

Clearly $[P]$, $[Q]$ and $[R]$ can be evaluated once and for all and equation (16) can be manipulated to give

$$[\dot{A}] = [P]^{-1}\,[Q][A] + [P]^{-1}\,[R]. \tag{20}$$

Equation (20) represents a system of ordinary differential equations for the a_j's which can be solved by any of the standard means. The partial differential equation (11) is solved (in detail) for a viscous flow problem in Section 3.4.

At this point it is worth re-emphasising that the following conditions are required in applying the traditional Galerkin method:

(1) The weighting functions, W_k, used in equation (5) are chosen from the same family as the analytic functions, φ_j, used in equation (3).

(2) The analytic functions (and hence the weighting functions) must be linearly independent.

(3) The analytic functions should be chosen from a complete set of functions.

(4) The analytic functions should satisfy the boundary conditions exactly.

As can be seen from the example above the analytic representation is often manipulated so that condition (4) can be satisfied. Of the above conditions, (1) defines the Galerkin method and (2) follows from the need to generate independent equations to determine the unknown coefficients, a_j. Condition (3) is imposed as a necessary condition for ensuring convergence to the exact solution. Violation of conditions (3) and (4) would not necessarily be disastrous rather it would lead to solutions that were less accurate, for a given number of unknown coefficients, a_j, than solutions which satisfied conditions (3) and (4).

2.3 Applications of the traditional Galerkin method

The work of Galerkin(ref.1) and Duncan(refs.3,4) has already been referred to. Bickley(ref.28) gives a very readable account of the solution of the unsteady heat conduction equation (same equation as considered in Section 3.4) but applied to problems in electrical circuit design. Bickley considered different ways of applying the Galerkin method and made comparisons with the collocation method and the method of least squares.

The earliest application of the Galerkin method to a fluid dynamic problem appears to be the investigation by Kawaguti(ref.29) of low Reynolds number flow past a sphere with the intention of finding the critical Reynolds number above which unsteady solutions could be expected. The Galerkin method was used to provide the steady state solution. For steady axi-symmetric flow of a viscous fluid the governing equation (Navier-Stokes) is

$$\psi_r \left\{ \frac{D^2 \psi}{r^2 \sin^2 \theta} \right\}_\theta - \psi_\theta \left\{ \frac{D^2 \psi}{r^2 \sin^2 \theta} \right\}_r - \frac{2}{Re \, . \, \sin \theta} \, . \, D^4 \psi = 0, \tag{21}$$

where

$$D^2 \equiv \frac{\partial}{\partial r^2} + \frac{\sin \theta}{r^2} \, . \, \frac{\partial}{\partial \theta} \left(\frac{1}{\sin \theta} \, . \, \frac{\partial}{\partial \theta} \right). \tag{22}$$

In equations (21) and (22) ψ is the stream function, (r,θ,φ) are spherical polar coordinates with corresponding velocity components (u,v,w) given by

$$u = \frac{1}{r^2 \sin \theta} \, . \, \frac{\partial \psi}{\partial \theta}, \quad v = -\frac{1}{r \sin \theta} \, . \, \frac{\partial \psi}{\partial r} \text{ and } w = 0. \tag{23}$$

Re is the Reynolds number based on the diameter of the sphere. The boundary conditions for the problem are

$$\psi = \frac{r^2}{2} \, . \, \sin^2 \theta \, . \text{ at } r = \infty \tag{24}$$

$$\psi = \frac{\partial \psi}{\partial r} = 0 \text{ at } r = 1 \text{ (surface of the sphere).} \tag{25}$$

Kawaguti introduced two different analytic representations for the stream function which were appropriate to different Reynolds number ranges. For larger Reynolds numbers (up to 100) Kawaguti used

$$\psi = \left\{ \frac{r^2}{2} + \frac{A1}{r} + \frac{A2}{r^2} + \frac{A3}{r^3} + \frac{A4}{r^4} \right\} \sin^2 \theta + \left\{ \frac{B1}{r} + \frac{B2}{r^2} + \frac{B3}{r^3} + \frac{B4}{r^4} \right\} \sin^2 \theta \cos \theta . \tag{26}$$

The form of equation (26) automatically satisfies equation (24). Four of the unknown coefficients in equation (26) may be eliminated (leaving four unknowns) by requiring that equation (25) must hold for all θ. Substitution of equation (26) into equation (21) leaves some residual, α. Kawaguti obtained two relations by requiring that

$$\alpha(1,\theta) = 0, \tag{27}$$

i.e. that the differential equation should hold exactly on the surface of the sphere. The other two relations amongst the unknown coefficients were obtained by requiring that

$$\int_1^\infty \int_{-1}^{1} r^{-1} \{P_o - P_2(Z)\} \,.\, \alpha \,.\, dZ\, dr = 0 \tag{28}$$

$$\int_1^\infty \int_{-1}^{1} r^{-2} \{P_1(Z) - P_3(Z)\} \,.\, \alpha \,.\, dZ\, dr = 0. \tag{29}$$

Equations (28) and (29) are weighted residual integrals of the Galerkin type. In equations (28) and (29), $P_n(Z)$ is the Legendre function of order n with $Z = \cos\theta$.

Kawaguti obtained good agreement for the drag coefficient but the pressure distribution and vorticity distribution at the surface were not so good, probably due to the coarseness of the representation, equation (26).

Krueger and Di Prima(ref.30) have used a conventional Galerkin technique to consider the stability of a viscous fluid between rotating cylinders. By solving a simpler approximate equation they obtained eigenfunctions for the axial variation of the perturbation velocity. The eigenfunctions were then used in the analytic representation of the more complex problem.

Snyder, Spriggs and Stewart(ref.31) have described the application of the Galerkin method to general problems involving viscous, incompressible flow. As a model problem they considered the similarity problem of two dimensional flow between inclined walls. In a related study(ref.32), Snyder and Stewart applied the Galerkin method to the problem of steady creeping flow through a densely packed bed of spheres. For creeping flow the inertial (convective) terms can be ignored and the momentum equations, now linear, express a balance between the pressure gradient and the viscous diffusion.

Snyder and Stewart used symmetry arguments to derive additional boundary conditions. They introduced trial solutions that were periodic in x, y and z and satisfied all the boundary conditions and symmetry conditions. It was found that after obtaining the usual weighted residual equations (equivalent to equation (5)) additional orthogonality conditions had to be imposed in order to obtain enough equations to solve for the unknown parameters. The integrals (in equations like (5)) were evaluated numerically using Guassian quadrature. For 28 unknown coefficients in expansions for the three velocity components and pressure, the predicted flow rate was within 5% of the experimentally measured value.

Radin and Mintzer(ref.33) have applied an unnamed Galerkin technique to the problem of obtaining solutions for the structure of a strong shock wave by solving the Boltzman equation. An expansion for the distribution function in terms of orthogonal polynomials is used as the analytic representation and the same orthogonal polynomials are used as weighting functions. The orthogonal polynomials are generated in terms of the collision velocities.

A comparable technique has been used by Fletcher and Holt(ref.34) to improve Dorodnitsyn's formulation of the method of integral relations for boundary layer problems. Dorodnitsyn's formulation uses weighting functions of the form $(1-u)^k$ where u is the steamwise velocity component (which is treated as an independent variable). The improvement of Fletcher and Holt is to generate orthonormal functions from the Dorodnitsyn weighting function and to use the orthonormal functions as both weighting functions and as analytic functions in the trial solution i.e. a Galerkin method. In fact of all the variations of the method of integral relations previously proposed only the orthonormal method of integral relations is a true Galerkin method. The technique has been applied to the boundary layer flow around cones at large angles of attack in supersonic flow(ref.35). A description of the method which treats it as a Galerkin method is given in these proceedings(ref.36).

2.4 Connections with other methods

The close link between the Galerkin method and the other methods which are also methods of weighted residuals has already been indicated in Section 2.1. In this section situations are described for which the Galerkin formulation produces solutions that are identical with those produced by other methods.

The first situation concerns problems for which a solution by separation of variables is possible. If the same problem is solved by the Galerkin method with a trial solution (equation (3)) based on the eigenfunctions of the problem (as found by the separation of variables solution, for instance) then precisely the same solution will be obtained if sufficient terms are included in the trial solution. As Finlayson and Scriven(ref.9) point out this simply means that if the trial solution contains the exact solution the Galerkin formulation will find it.

The Galerkin method can be shown to be equivalent to the orthogonal collocation method of Villadsen and Stewart(ref.18) under certain circumstances. It is necessary for the Galerkin method to use a trial solution of degree n in x^2, for the governing equation to be linear and for the residual, given by equation (4), to be a polynomial of degree n or less in x^2.

The numerical coefficients that multiply the unknown parameters, a_j in the algebraic equations are obtained by evaluating equation (5). Traditionally the integrals in equation (5) are evaluated algebraicly, however they may be evaluated numerically. For nonlinear problems, if a Galerkin formulation uses the same trial solution as the orthogonal collocation method and also performs the integrations numerically, identical results are obtained if the integration formula is of the same order as the number of unknowns in the analytic representation, equation (3).

The most important link with other techniques is that between the Galerkin method and variational methods. It is important because the variational methods often give an upper or lower bound on the exact solution. Where exact equivalence to a Galerkin method can be shown, the Galerkin method also gives an upper or lower bound. If a problem can be put in a form that permits a solution to be found by a variational method e.g. Rayleigh-Ritz method, then it can always be solved by a Galerkin method; and the Galerkin method will produce exactly the same solution. This will be illustrated by an example after a brief description of the Rayleigh-Ritz method.

The Rayleigh-Ritz method starts with the requirement that some integral, say

$$I(u) = \int\int F(x,y,u'_x,u'_y)\, dxdy, \tag{30}$$

with appropriate boundary conditions, has an exact solution $u = u_e(x,y)$, and at $u = u_e$, $I(u_e)$ is a minimum. The method proceeds by assuming a trial solution for u i.e.

$$u_a = \sum_{k=1}^{n} a_k \cdot \varphi_k(x,y). \tag{31}$$

The problem then becomes one of finding the minimum of $I(u_a)$ which will be close to $I(u_e)$ in some sense if n is sufficiently large. Since $I_{min}(u_a)$ must be a minimum for the approximate system it is possible to set up sufficient equations to solve for the a_k's. This is done by requiring that

$$\frac{\partial I_{min}}{\partial a_j}(u_a) = 0, \; j = 1 \ldots n. \tag{32}$$

To demonstrate the connection between the Rayleigh-Ritz method and the Galerkin method a problem governed by Poisson's equation is considered. That is

$$\frac{\partial^2 u}{\partial x^2} + \frac{\partial^2 u}{\partial y^2} = f, \tag{33}$$

subject to the condition u = 0 on the boundary. The equivalent variational problem requires that

$$I(u) = \iint \left[\left(\frac{\partial u}{\partial x}\right)^2 + \left(\frac{\partial u}{\partial y}\right)^2 + 2fu\right] dxdy \tag{34}$$

has a minimum corresponding to the exact solution, subject to the same condition u = 0 on the boundary. Substitution of equation (31) into equation (34) gives

$$I(u_a) = \iint \left[\left\{\sum_{k=1}^{n} a_k \frac{\partial \varphi_k}{\partial x}\right\}^2 + \left\{\sum_{k=1}^{n} a_k \frac{\partial \varphi_k}{\partial y}\right\}^2 + 2f \sum_{n=1}^{k} a_k \varphi_k\right] dxdy. \tag{35}$$

Imposing conditions equation (32) produces the result

$$\frac{\partial I}{\partial a_j}(u_a) = \iint \left[\frac{2\partial \varphi_j}{\partial x} \cdot \left\{\sum_{k=1}^{n} a_k \cdot \frac{\partial \varphi_k}{\partial x}\right\} + \frac{2\partial \varphi_j}{\partial y} \left\{\sum_{k=1}^{n} a_k \frac{\partial \varphi_k}{\partial y}\right\} + 2f \cdot \varphi_j\right] dxdy = 0 \tag{36}$$

or

$$\iint \left[\frac{\partial \varphi_j}{\partial x} \cdot \frac{\partial u_a}{\partial x} + \frac{\partial \varphi_j}{\partial y} \cdot \frac{\partial u_a}{\partial y} + f \cdot \varphi_j\right] dxdy = 0. \tag{37}$$

Application of the Galerkin method to equation (33) gives

$$\iint \varphi_j \left[\frac{\partial^2 u_a}{\partial x^2} + \frac{\partial^2 u_a}{\partial y^2} - f \right] dxdy = 0. \tag{38}$$

Applying Green's theorem to the first two terms leads to

$$\int \varphi_j \left[\frac{\partial u_a}{\partial x} \cdot 1_x + \frac{\partial u_a}{\partial y} \cdot 1_y \right] ds - \iint \left[\frac{\partial \varphi_j}{\partial x} \cdot \frac{\partial u_a}{\partial x} + \frac{\partial \varphi_j}{\partial y} \cdot \frac{\partial u_a}{\partial y} + f \cdot \varphi_j \right] dxdy = 0. \tag{39}$$

Since φ_j is chosen to satisfy homogeneous boundary conditions the first term in equation (39) disappears and equation (39) reduces to equation (37). Thus the two techniques are equivalent.

2.5 Convergence properties and error bounds

In the Galerkin method the trial functions should be linearly independent and members of a complete set. Since the weighting functions are chosen from the same set, the weighted residual equation (5) ensures that the residual is orthogonal to all members of the complete set of functions used. As n approaches infinity the residual must approach zero as requires. Although it has been suggested(ref.32) that choosing the trial functions as members of a complete set is sufficient for convergence, Finlayson and Scriven(ref.9) quote Mikhlin(ref.6) to caution that this may not always be so.

Convergence proofs for the application of the Galerkin method to some linear elliptic partial differential equations have been shown by Mikhlin (ref.6). Where specific problems possess a variational formulation, convergence proofs and error bounds obtained for the Rayleigh-Ritz method are also appropriate to the Galerkin method applied to the same problem. For examples see Kantorovich and Krylov(ref.5). A readable discussion of convergence proofs and error bounds applicable to the Galerkin method and other methods of weighted residuals may be found in Finlayson(ref.10).

2.6 Improvements to the traditional Galerkin method

The description of the traditional Galerkin method given so far, and many of the applications cited, have been concerned with obtaining solutions of low accuracy (perhaps around 5 to 10% error) but with minimal computational effort i.e. essentially hand or pre-computer calculations.

One type of improvement that has been commonly used in conjunction with the Galerkin method is the introduction of ancilliary conditions. Thus Duncan(ref.3) and Kawaguti(ref.29) requires that the approximating functions satisfy the differential equation exactly on certain boundaries.

Another improvement is the reduction of the original partial differential equation to a set of ordinary differential equations rather than to a set of algebraic equations. This technique was originally proposed for elliptic partial differential equations by Kantorovich and Krylov(ref.5). The advantage is that ordinary differential equations can often be solved analytically or are amenable to very efficient numerical algorithms(ref.37). A comparable idea of reducing the original partial differential equation by a single order and then solving the resultant partial differential equation by finite difference methods has been used by Kaplan et al(ref.38) to obtain economical solutions to nuclear reactor problems.

The widespread availability of computers and the desire to obtain accurate (say less than 1% error) solutions using the Galerkin method poses an immediate problem. If the number of terms, n, in the trial solution is increased arbitrarily then the inversion of matrix [P] in equation (16), or [Q] in a steady state problem, becomes inefficient and inaccurate. This is because, for conventional applications of the traditional Galerkin method, [P] and [Q] are dense, that is they have no obvious zero elements, and are ill-conditioned for large n.

Improvements to the traditional Galerkin method to obtain accurate solutions have pursued two different directions. Firstly the unknown coefficients a_j have been replaced by nodal values of the unknown. Thus equation (3) can be replaced with

$$v = \sum_{j=1}^{n} \varphi_j(x) \,.\, \bar{v}_j . \tag{40}$$

$\bar{v}_j$ are the nodal values of v and become the unknowns of the problem. The $\varphi_j(x)$, which could be just Lagrange interpolation coefficients, are chosen from piecewise polynomials. φ_j is chosen to be equal to unity at the jth node and to fall to zero at adjacent nodes and is zero beyond the adjacent nodes. The result of deriving equations like equation (16), but based on equation (40), produces matrices [P] and [Q] that are sparse. That is they have few non-zero elements and they are clustered around the main diagonal. Thus the matrices can be inverted economically and accurately. The Galerkin formulation of the finite element method belongs to this group. This is described in Section 4. If the nodes are part of a regular rectangular grid then the final equations simplify and the Galerkin finite difference formulation is obtained. This is described in Section 5.

The second direction has been to introduce orthogonal functions for the trial functions in equation (3) and for the weighting functions in equation (5). Typically these might be orthogonal polynomials or Fourier series and are either introduced directly as in reference 32 or are generated from a system of independent functions connected with the problem as in references 33 and 34. The result is that the matrices [P] and [Q] only have entries on the diagonal and hence are trivial to invert. Spectral methods are of this type and these will be considered in Section 6.

3. WORKED EXAMPLES USING THE TRADITIONAL GALERKIN METHOD

In this section four examples are treated in sufficient depth for the details of the method to become apparent. The different examples are carried through to completion and illustrate different aspects of the method. Solutions to some of the examples are also obtained using the finite element, finite difference and spectral methods in Sections 4, 5 and 6 respectively. The first example is a first order ordinary differential equation (taken from reference 3). The second and third examples consider the steady viscous flow in a channel which leads to elliptic partial differential equations. The fourth example considers the viscous flow between moving plates with a prescribed velocity distribution at the left hand boundary. The development of the flow in the right hand direction is governed by a parabolic partial differential equation. The examples are used to illustrate the drawbacks of trying to extend the traditional Galerkin method to high order so that solutions of higher accuracy may be obtained using a computer.

3.1 An ordinary differential equation

The governing equation is

$$L(y) = \frac{\partial y}{\partial x} - y = 0, \tag{41}$$

subject to the boundary condition $y = 1$ at $x = 0$.
The exact solution is $y = e^x$. Following the procedure given in Section 2.2 an approximate analytic representation is introduced, e.g. with $n = 3$,

$$y_a = 1 + \sum_{j=1}^{3} a_j \, . \, x^j . \tag{42}$$

The unknown coefficients a_j are constants and equation (42) satisfies the boundary condition exactly. Substitution of equation (42) into equation (41) produces a residual, r.

$$r = L(y_a) = -1 + a_1 \, . \, (1 - x) + a_2 \, . \, (2x - x^2) + a_3 \, . \, (3x^2 - x^3).$$

Setting the integral of the weighted residual equal to zero gives

$$\int_0^1 r \, . \, x^j \, . \, dx = 0, \quad j = 1, 2, 3, \tag{43}$$

which leads to

$$\left[\int_0^1 (1 - x) \, . \, x^j \, dx\right] . \, a_1 + \left[\int_0^1 (2x - x^2) \, x^j \, dx\right] . \, a_2 + \left[\int_0^1 (3x^2 - x^3) \, . \, x^j \, dx\right] . \, a_3$$

$$= \int_0^1 1 \, . \, x^j \, dx, \quad j = 1, 2, 3. \tag{44}$$

Evaluating the integrals in equation (44) leads to

$$\begin{aligned} \frac{1}{6} a_1 + \frac{5}{12} a_2 + \frac{11}{20} a_3 &= \frac{1}{2} \\ \frac{1}{12} a_1 + \frac{3}{10} a_2 + \frac{13}{30} a_3 &= \frac{1}{3} \\ \frac{1}{20} a_1 + \frac{7}{30} a_2 + \frac{5}{14} a_3 &= \frac{1}{4}. \end{aligned} \tag{45}$$

Equations (45) can be solved for a_1, a_2 and a_3 and substituted into equation (42) to give

$$y_a = 1 + 1.034\, x + 0.388\, x^2 + 0.302\, x^3 . \qquad (46)$$

By discarding appropriate equations and coefficients in equation (45) the first and second approximation can also be obtained. The following table provides an indication of the increase in accuracy with increasing order of approximation.

x	Approximate solution, y_a			Exact solution
	linear	quadratic	cubic	$y = e^x$
0	1.0	1.0	1.0	1.0
0.2	1.6	1.1818	1.2248	1.2214
0.4	2.2	1.4364	1.4952	1.4918
0.6	2.8	1.7636	1.8255	1.8221
0.8	3.4	2.1636	2.2303	2.2255
1.0	4.0	2.6364	2.7241	2.7183

Reference 3 contains a number of other worked examples of similar complexity.

3.2 Symmetric viscous flow in a channel

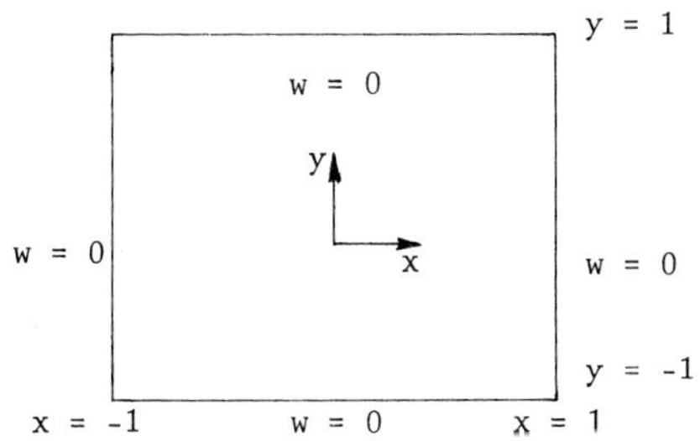

Figure 1

The steady flow of a viscous fluid within a square channel (figure 1) is governed by the 'z' momentum equation (in a conventional cartesian coordinate system).

$$u\frac{\partial w}{\partial x} + v\frac{\partial w}{\partial y} + w\frac{\partial w}{\partial z} + \frac{1}{\rho}\cdot\frac{\partial p}{\partial z} = \nu\left\{\frac{\partial^2 w}{\partial x^2} + \frac{\partial^2 w}{\partial y^2} + \frac{\partial^2 w}{\partial z^2}\right\} . \qquad (47)$$

Far from any entrance or exit the flow does not vary in the z direction and u and v are zero. Equation (47) reduces to

$$\frac{\partial p}{\partial z} = \mu\left\{\frac{\partial^2 w}{\partial x^2} + \frac{\partial^2 w}{\partial y^2}\right\} . \qquad (48)$$

If $\partial p/\partial Z$, which is a constant, and μ are absorbed into a non-dimensional velocity then equation (48) can be written

$$\frac{\partial^2 w}{\partial x^2} + \frac{\partial^2 w}{\partial y^2} - 1 = 0. \tag{49}$$

An approximate solution of equation (49), subject to the boundary conditions $w = 0$ on x, $y = \pm 1$, is sought using the Galerkin technique. An approximate solution is introduced as

$$w_a = \sum_{j=1}^{4} a_j (1 - x^2)^j \,.\, (1 - y^2)^j. \tag{50}$$

The form of equation (50) is dictated by the boundary conditions and by the symmetry of the problem. Substitution of equation (50) into equation (49) produces some residual, r, which may be written in the form

$$r = -1 + \sum_{j=1}^{4} a_j \,.\, k_j(x,y), \tag{51}$$

where

$$k_1 = -2(1 - x^2) -21 - y^2)$$

$$k_2 = 4(1 - y^2)^2.(-1 + 3x^2) + 4(1 - x^2)^2.(-1 + 3y^2)$$

$$k_3 = 6(1 - y^2)^3.(-1 + 6x^2 - 5x^4) + 6(1 - x^2)^3.(-1 + 6y^2 - 5y^4), \text{ and}$$

$$k_4 = 8(1 - y^2)^4.(-1 + 9x^2 - 15x^4 + 7x^6) + 8(1 - x^2)^4.(-1 + 9y^2 - 15y^4 + 7y^6).$$

Applying the weighted residual condition gives

$$\int_{-1}^{1}\int_{-1}^{1} r.\, \varphi_i(x,y)\, dxdy = 0, \tag{52}$$

where

$$\varphi_i(x,y) = (1 - x^2)^i\, (1 - y^2)^i. \tag{53}$$

Substituting for r in equation (52) from equation (51) gives

$$\sum_{j=1}^{4} a_j \int_{-1}^{1}\int_{-1}^{1} k_j(x,y) \,.\, \varphi_i(x,y)\, dxdy = \int_{-1}^{1}\int_{-1}^{1} \varphi_i(x,y)\, dxdy,\; i = 1 \ldots 4, \tag{54}$$

or

$$[C] \cdot [A] = [B], \tag{55}$$

where an element of [C] is given by

$$c_{ij} = \int_{-1}^{1}\int_{-1}^{1} k_j(x,y) \cdot \varphi_i(x,y)\, dxdy, \tag{56}$$

and an element of [B] is given by

$$b_i = \int_{-1}^{1}\int_{-1}^{1} \varphi_i(x,y)\, dxdy. \tag{57}$$

Traditionally the coefficients c_{ij} and b_i are evaluated analytically, however here they have been evaluated numerically using Gauss quadrature. A sufficiently high order numerical scheme was used to ensure that the above integrations were evaluated exactly (within the round-off error of the computer). The Gauss weights and function evaluation points are given by Abramowitz and Stegun(ref.39).

The matrix [C] is

$$\begin{bmatrix} -5.68889 & -3.90095 & -2.97215 & -2.40174 \\ -3.90095 & -3.96287 & -3.60261 & -3.22471 \\ -2.97215 & -3.60261 & -3.62780 & -3.47277 \\ -2.40174 & -3.22471 & -3.47277 & -3.48639 \end{bmatrix}$$

The matrix $[B]^T$ is

$$[-1.77778 \quad -1.13778 \quad -0.83592 \quad -0.66048]$$

The matrix $[A]^T$, which has coefficients a_j, is

$$[A]^T = [0.36994 \quad -0.13154 \quad 0.08833 \quad -0.03173]. \tag{58}$$

substitution of equation (58) into equation (50) produces an approximate solution that can be compared with the exact solution given by Dryden et al (ref.40). Comparisons will be made for the non-dimensional centre line velocity, $w_{C/L}$ and the non-dimensional flow rate $\dot{q}$, which is given by

$$\dot{q} = \int_{-1}^{1}\int_{-1}^{1} w(x,y)\, dxdy.$$

	Approximate solution (n terms expansion, equation (50))			Exact solution
	n = 2	n = 3	n = 4	
$w_{C/L}$	0.29267	0.29588	0.29501	0.29469
$\dot{q}$	0.56068	0.56088	0.56090	0.56231

The results indicate that quite an accurate result is obtained with just two terms in equation (50). Since the centre line velocity is approaching the exact result with increasing n with little corresponding change in the flow rate, it would appear that increasing n is essentially causing an improvement in the velocity profile.

3.3 Asymmetric viscous flow in a channel

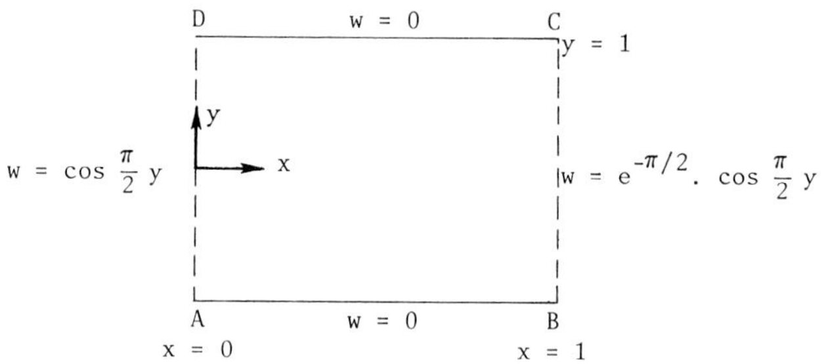

Figure 2

This example concerns the viscous flow between two parallel plates AB and CD with prescribed velocity distributions on AD and BC (figure 2). This example is similar to the symmetric problem considered in Section 3.2. The governing equation is

$$\frac{\partial^2 w}{\partial x^2} + \frac{\partial^2 w}{\partial y^2} = 0, \tag{59}$$

and the boundary conditions are

$$w(0,y) = \cos\frac{\pi}{2}y, \quad w(1,y) = e^{-\pi/2}\,.\,\cos\frac{\pi}{2}y, \quad w(x,\pm 1) = 0. \tag{60}$$

The solution is simplified if a new velocity, w_d is introduced as

$$w = w_d + e^{-\pi/2}\,.\,\cos\frac{\pi}{2}y. \tag{61}$$

The governing equation for w_d is

$$\frac{\partial^2 w_d}{\partial x^2} + \frac{\partial^2 w_d}{\partial y^2} - e^{-\pi/2}\,.\,\frac{\pi^2}{4}\,.\,\cos\frac{\pi}{2}\,y = 0, \tag{62}$$

and the boundary conditions are

$$w_d(0,y) = (1 - e^{-\pi/2}) \cdot \cos\frac{\pi}{2}y,\quad w_d(1,y) = 0,\quad w_d(x,\pm 1) = 0. \tag{63}$$

An approximate solution for w_d is introduced as

$$w_a = (1 - e^{-\pi/2}) \cdot \cos\frac{\pi}{2}y \cdot (1 - x) + x(1 - x)(1 - y^2)$$
$$(a_1 + a_2 x + a_3 y^2 + a_4 x^2 + a_5 xy^2). \tag{64}$$

The first term in equation (64) is chosen to satisfy the non-zero boundary condition; consequently the second term must be chosen to satisfy homogeneous boundary conditions. Equation (64) may be rewritten as

$$w_a = \varphi_o(x,y) + \sum_{j=1}^{5} a_j \cdot \varphi_j(x,y), \tag{65}$$

$$\begin{aligned}
\text{where } \varphi_o(x,y) &= (1 - e^{-\pi/2}) \cdot \cos\frac{\pi}{2}y \cdot (1 - x)\\
\varphi_1(x,y) &= x(1 - x)(1 - y^2)\\
\varphi_2(x,y) &= x^2(1 - x)(1 - y^2)\\
\varphi_3(x,y) &= x(1 - x)\, y^2(1 - y^2)\\
\varphi_4(x,y) &= x^3(1 - x)(1 - y^2)\\
\varphi_5(x,y) &= x^2(1 - x)\, y^2(1 - y^2).
\end{aligned} \tag{66}$$

Substitution of equation (65) into equation (62) produces a residual given by

$$r = k_o(x,y) + \sum_{j=1}^{5} a_j \cdot k_j(x,y), \tag{67}$$

$$\begin{aligned}
\text{where } k_o(x,y) &= -\frac{\pi^2}{4} \cdot \cos\frac{\pi}{2}y \cdot \lfloor 1 - (1 - e^{\pi/2})x \rfloor\\
k_1(x,y) &= -2(1 - y^2) - 2x(1 - x)\\
k_2(x,y) &= 2(1 - 3x)(1 - y^2) - 2x^2(1 - x)\\
k_3(x,y) &= -2(y^2 - y^4) + 2x(1 - x)(1 - 6y^2)\\
k_4(x,y) &= (6x - 12x^2)(1 - y^2) - 2x^3(1 - x)\\
k_5(x,y) &= (1 - 3x)(2y^2 - 2y^4) + 2x^2(1 - x)(1 - 6y^2).
\end{aligned}$$

Applying the weighted residual condition gives

$$\int_{-1}^{1}\int_{0}^{1} r \cdot \varphi_i(x,y)\, dxdy = 0,\quad i = 1 \ldots . 5. \tag{68}$$

Substitution for r from equation (67) and rearrangement allows equation (68) to be rewritten as

$$[C] \cdot [A] = [B], \tag{69}$$

where an element of [C] is given by

$$c_{ij} = \int_{-1}^{1}\int_{0}^{1} k_j(x,y) \cdot \varphi_i(x,y)\, dxdy$$

and an element of [B] is given by

$$b_i = -\int_{-1}^{1}\int_{0}^{1} \varphi_o(x,y) \cdot \varphi_i(x,y)\, dxdy.$$

Once [C] and [B] are formed equation (69) can be solved for [A] which has as elements the coefficients a_j. As in Section 3.2 c_{ij} and b_i are evaluated numerically. The matrix [C] is

$$\begin{bmatrix} -0.44444 & -0.22222 & -0.06857 & -0.13206 & -0.03429 \\ -0.22222 & -0.16762 & -0.03429 & -0.12254 & -0.02540 \\ -0.06857 & -0.03429 & -0.04487 & -0.02032 & -0.02243 \\ -0.13206 & -0.12254 & -0.02032 & -0.10201 & -0.01841 \\ -0.03429 & -0.02540 & -0.02243 & -0.01841 & -0.01475 \end{bmatrix}.$$

The matrix $[B]^T$ is

$$[\,0.25632 \quad 0.11135 \quad 0.03500 \quad 0.06009 \quad 0.01520\,],$$

and the solution $[A]^T$ is

$$[A]^T = [\,-0.77441 \quad 0.42326 \quad 0.16694 \quad -0.11574 \quad -0.06877\,]. \tag{70}$$

Substitution of equation (70) into equations (64) and (61) gives an approximate solution. This can be compared with the exact solution

$$w = e^{-\frac{\pi}{2}x} \cdot \cos\frac{\pi}{2}y. \tag{71}$$

Location		Approximate solution (n term, equation (65))				Exact solution, w
x	y	n = 2	n = 3	n = 4	n = 5	
0.25	0	0.67989	0.67605	0.67572	0.67525	0.67523
0.50	-0.50	0.31891	0.32130	0.32232	0.32232	0.32240
0.50	0	0.45976	0.45464	0.45601	0.45601	0.45594
0.50	0.50	0.31891	0.32130	0.32232	0.32232	0.32240
0.75	0	0.31172	0.30788	0.30755	0.30802	0.30786

The results indicate that an accurate solution can be obtained with very few terms in equation (65), that the accuracy increases with n and that the error is of the same order throughout the domain.

3.4 A parabolic problem

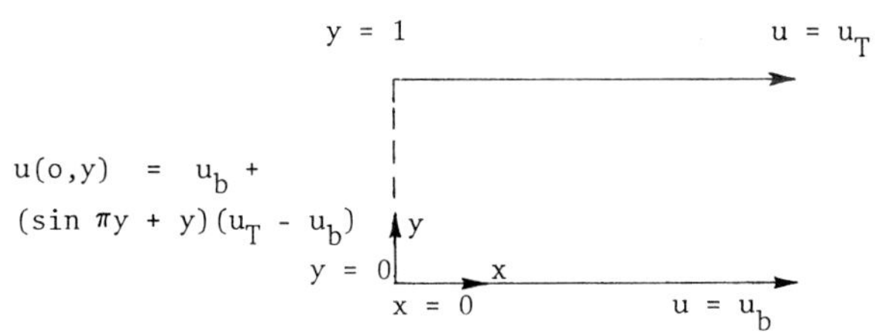

Figure 3

This problem concerns the flow of an incompressible viscous fluid between two plates moving with velocities u_b and u_T (figure 3). The fluid has a prescribed 'inlet' velocity distribution at x = 0. The 'x' momentum equation reduces to

$$u\frac{\partial u}{\partial x} = \nu\frac{\partial^2 u}{\partial y^2}\ . \tag{72}$$

A new variable ψ is introduced by

$$\psi = \frac{u - u_b}{u_T - u_b} \tag{73}$$

Substitution of equation (73) into equation (72) gives

$$[u_b + \psi(u_T - u_b)]\frac{\partial \psi}{\partial x} = \nu\frac{\partial^2 \psi}{\partial x^2}\ . \tag{74}$$

If it is assumed that $u_T \approx u_b$ then the problem can be linearised to give

$$\frac{u_b}{\nu}\cdot\frac{\partial \psi}{\partial x} = \frac{\partial^2 \psi}{\partial y^2} \tag{75}$$

If a new variable $x' = \nu x/u_b$ is introduced and the superscript ' is dropped the result is

$$\frac{\partial \psi}{\partial x} - \frac{\partial^2 \psi}{\partial y^2} = 0, \tag{76}$$

subject to the boundary conditions

$$\psi(o,y) = \sin \pi y + y, \; \psi(x,o) = 0, \; \psi(x,1) = 1. \tag{77}$$

An approximate solution is introduced as

$$\psi_a(x,y) = \varphi_o(y) + \sum_{j=1}^{5} a_j(x) \, . \, \varphi_j(y), \tag{78}$$

where

$$\varphi_o(y) = \sin \pi y + y$$

and

$$\varphi_j(y) = (1 - y) \, . \, y^j .$$

The form of equation (78) is dictated by the parabolic nature of equation (76) i.e. it is convenient to make the trial functions depend on y alone and to allow the unknown coefficients to be functions of x. Substitution of equation (78) into equation (76) produces a residual

$$r = M_o(y) + \sum_{j=1}^{5} \left[\varphi_j(y) \, . \, \frac{\partial a_j}{\partial x} + M_j(y) \, . \, a_j \right], \tag{79}$$

where

$$M_o(y) = \pi^2 \, . \, \sin \pi \, y$$

$$M_j(y) = (j + 1) \, . \, j \, . \, y^{j-1} - j \, . \, (j - 1) \, . \, y^{j-2} .$$

Applying the weighted residual integral gives

$$\int_0^1 r \, . \, \varphi_i(y) \, dy = 0, \; i = 1 \ldots 5, \tag{80}$$

and substitution for r leads to

$$\sum_{j=1}^{5} \frac{\partial a_j}{\partial x} \left[\int_0^1 \varphi_j(y) \cdot \varphi_i(y)\, dy\right] + \sum_{j=1}^{5} a_j \left[\int_0^1 M_j(y) \cdot \varphi_i(y)\, dy\right] +$$

$$\int_0^1 M_o(y) \cdot \varphi_i(y)\, dy = 0, \quad i = 1 \ldots 5. \tag{81}$$

Equation (81) can be rewritten in matrix form as

$$[P]\,[\dot{A}] = [Q]\,[A] + [R], \tag{82}$$

where an element of [P] is

$$p_{ij} = \int_0^1 \varphi_j(y) \cdot \varphi_i(y)\, dy,$$

an element of [Q] is

$$q_{ij} = -\int_0^1 M_j(y) \cdot \varphi_i(y)\, dy,$$

an element of [R] is

$$r_i = -\int_0^1 M_o(y) \cdot \varphi_i(y)\, dy,$$

an element of $[\dot{A}]$ is $\partial a_j/\partial x$ and an element of [A] is a_j. Equation (82) represents a system of ordinary differential equations for the a_j's as functions of x. Once [P], [Q] and [R] have been evaluated, equation (82) can be manipulated to give

$$[\dot{A}] = [S] \cdot [A] + [T] \tag{83}$$

where $[S] = [P]^{-1}\,[Q]$
and $[T] = [P]^{-1}\,[R]$.

The matrices [P], [Q] and [R] have been evaluated numerically.

[P] is

$$\begin{bmatrix} 0.03333 & 0.01667 & 0.00952 & 0.00595 & 0.00397 \\ 0.01667 & 0.00952 & 0.00595 & 0.00397 & 0.00278 \\ 0.00952 & 0.00595 & 0.00397 & 0.00278 & 0.00202 \\ 0.00595 & 0.00397 & 0.00278 & 0.00202 & 0.00152 \\ 0.00397 & 0.00278 & 0.00202 & 0.00152 & 0.00117 \end{bmatrix}$$

[Q] is

$$\begin{bmatrix} -0.33333 & -0.16667 & -0.10000 & -0.06667 & -0.04762 \\ -0.16667 & -0.13333 & -0.10000 & -0.07619 & -0.05952 \\ -0.10000 & -0.10000 & -0.08571 & -0.07143 & -0.05952 \\ -0.06667 & -0.07619 & -0.07143 & -0.06349 & -0.05556 \\ -0.04762 & -0.05952 & -0.05952 & -0.05556 & -0.05051 \end{bmatrix}$$

and $[R]^T$ is

$$[-1.27324 \quad -0.63662 \quad -0.36179 \quad -0.22437 \quad -0.14831].$$

Equation (83) has been integrated using a first order scheme i.e.

$$a_j^{n+1} = a_j^n + \frac{\partial a_j}{\partial x} \cdot \Delta x. \tag{84}$$

The results shown in the table were obtained with $\Delta x = 0.001$.

x location (y = 0.50)	Approximate solution (n term, equation (78))				Exact solution, ψ	Approximate solution, ψ_b
	n = 2	n = 3	n = 4	n = 5		
0	1.50000	1.50000	1.50000	1.50000	1.50000	1.50000
0.02	1.32611	1.32020	1.32020	1.32007	1.32087	1.32006
0.04	1.18389	1.17278	1.17278	1.17251	1.17383	1.17251
0.06	1.06757	1.05188	1.05188	1.05150	1.05312	1.05150
0.08	0.97242	0.95274	0.95274	0.95227	0.95404	0.95226
0.10	0.89461	0.87144	0.87144	0.87089	0.87271	0.87089
0.12	0.83096	0.80477	0.80477	0.80416	0.80594	0.80415
0.14	0.77890	0.75009	0.75009	0.74943	0.75114	0.74942
0.16	0.73632	0.70526	0.70526	0.70455	0.70615	0.70454
0.18	0.70150	0.66849	0.66849	0.66775	0.66923	0.66774
0.20	0.67301	0.63833	0.63833	0.63756	0.63891	0.63756

The exact solution is

$$\psi = \sin \pi y \cdot e^{-\pi^2 x} + y. \tag{85}$$

Physically the initial conditions ($x = 0$) may be viewed as a perturbation $\sin \pi y$ on the linear variation y. The perturbation dies away for large x leaving just the linear velocity variation. Because of the particular trial functions used the results for n = 3 and 4 are the same. The tabulated results indicate that the solutions for n = 3, 4 agree more closely with the exact solution than does the solution for n = 5. However, the superior agreement is deceptive. The approximate solutions shown in the table contain errors from two sources; firstly the error in the Galerkin approximation i.e. equation (78), and secondly the error in numerically integrating equation (83). These errors could be split up by integrating equation (83) analytically. However in order to make comparisons with the results of solving the present problem with a Galerkin finite difference method (Section 4), the two sources of error will be separated in another way.

The governing equation (76) will be solved by a separation of variables technique. Thus if

$$\psi = \chi(x) \sin \pi y + y \tag{86}$$

is substituted into the governing equation, the result is

$$\frac{\partial \chi}{\partial x} + \pi^2 \chi = 0 \tag{87}$$

which clearly has the solution $\chi = e^{-\pi^2 x}$. However if equation (87) is solved using the numerical scheme, equation (84), the result χ_a will be different from χ. Substituting χ_a into equation (86) will give a result ψ_b which only differs from the exact solution due to the numerical integration. ψ_b is given in the last column of the table. Thus errors associated with the Galerkin approximation can be isolated is the approximate results, equation (78), are compared with the approximate results, ψ_b. In this case the accuracy of the solution increases with n and for n = 5 is very high.

3.5 Limitations of the traditional Galerkin method

All the solutions show an increase in accuracy as the order, n, of the approximation is increased. However, in some cases, the accuracy of the highest order of approximation is not very great. There is a strong temptation to code a traditional Galerkin formulation to be evaluated on a computer, to input a large value of n and to sit back and wait for some accurate answers.

An immediate problem is that, for large n, there is very little difference between two consecutive weighting functions, w_{n-1} and w_n in equation (5). Consequently, the $(n - 1)^{th}$ and the nth algebraic relations formed using equation (5) would be almost linearly dependent. This can be illustrated by considering problem 3.1 with n = 100. In this case the last two algebraic equations, corresponding to $w_{99} = x^{99}$ and $w_{100} = x^{100}$, are given by equation (44) with j = 99, 100. These are

$$0.000099a_1 + 0.009998a_2 + 0.019703a_3 \; .... + 0.494975a_{99} + 0.497513a_{100} = 0.01,$$

and

$$0.000097a_1 + 0.009899a_2 + 0.019511a_3 \; .... + 0.492487a_{99} + 0.495025a_{100} = 0.0099.$$

Clearly such equations would lead to a very ill-conditioned matrix. The factorisation of the matrix would be very inaccurate.

Another problem is that the matrices, like [C] in problems 3.2 and 3.3, which have to be factorised are always dense. For large n, such problems will require considerable time to evaluate the coefficients, considerable storage and, most important, will require an increase of execution time like n^3 for solution by a direct method. Clearly this would be an inefficient usage of both CPU time and storage.

Another limitation of the traditional Galerkin method is apparent from a consideration of the weighted residual integral (e.g. equation (5)) used to obtain the algebraic relations between the coefficients. In all the examples considered the limits of integration of the residual integral have coincided with coordinate lines. That is they have been independent of the other coordinate(s). This means that the boundary of the domain must lie on coordinate lines. This is clearly a considerable restriction. If the boundary of the domain is reasonably close to the coordinate lines it would be feasible to map the physical domain onto a regular domain with boundaries that coincided with coordinate lines. However this technique increases the complexity of the governing equations and hence the algebraic effort of setting up the problem. Also it may cause a degradation in accuracy since the mapping is equivalent to distorting the coordinate lines in the real domain until they fit the domain boundary.

The way that Galerkin finite element methods, Galerkin finite difference methods and spectral methods overcome some or all of these limitations is described in Sections 4, 5 and 6 respectively.

4. THE GALERKIN FINITE ELEMENT METHOD

Since a general description of the finite element method is given elsewhere in the proceedings(ref.41), the description in this section will be brief and will concentrate on those aspects that are connected with the traditional Galerkin method. The application of the present method to the asymmetric viscous flow problem (Section 3.3) provides results that illustrate the differences between the present method and the traditional Galerkin method. Finally, some of the applications of the Galerkin finite element method to fluid dynamic problems are referred to.

4.1 Description

As with the traditional Galerkin method the starting point is a differential equation, e.g.,

$$\frac{\partial^2 u}{\partial x^2} + \frac{\partial^2 u}{\partial y^2} - f = 0, \tag{88}$$

for which a solution is sought within some domain D subject to appropriate boundary conditions on S the boundary of D. An approximate solution, defined by

$$v = \sum_{j=1}^{n} N_j(x,y) \cdot \bar{v}_j, \tag{89}$$

is introduced. It would be possible to start from an analytic representation like equation (3), with $\varphi_j(x,y)$ as ascending powers of x,y and xy, and to invert the relationship(ref.42, p.105) to obtain the representation shown in equation (89). In equation (89), $\bar{v}_j$ are the nodal values of v and they become the unknowns of the problem, that is they play the same role as the unknown coefficients, a_j, in equation (3). $N_j(x,y)$ are referred to as shape functions and under certain circumstances they are the same as Lagrange interpolation coefficients. The shape functions have the property,

$$N_i = 1 \text{ at the } i^{th} \text{ node,}$$
$$= 0 \text{ at all other nodes.} \qquad (90)$$

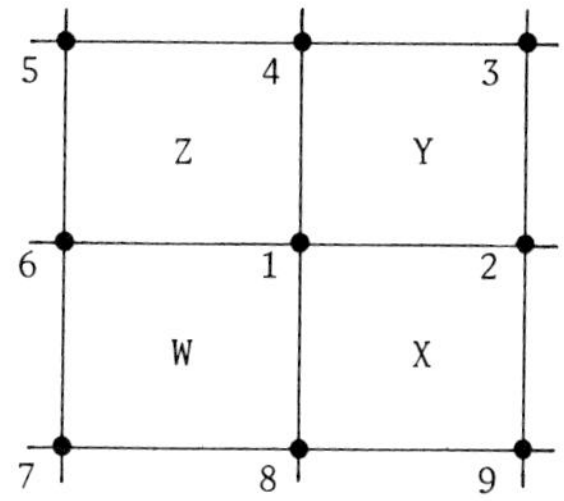

Figure 4

In figure 4 N_1 equals one at node 1, equals zero at nodes 2 to 9, takes values between zero and one in elements W, X, Y and Z and is zero outside elements W, X, Y and Z. If element coordinates

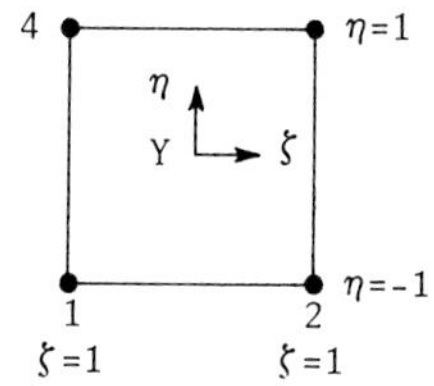

Figure 5

are defined as in figure 5, then a linear shape function N_1 would have the following form in element Y

$$N_1(\zeta,\eta) = \tfrac{1}{4}(1 - \zeta)(1 - \eta). \qquad (91)$$

Comparable formulae are available(ref.42) for the variation of N_1 in elements W, X and Z. It is apparent from equation (91) that N_1 demonstrates the correct behaviour at nodes 1, 2, 3 and 4. The nature of equation (89) makes it inappropriate to choose trial functions $(N_j(x,y))$ that satisfy the boundary conditions directly.

As with the traditional method, the Galerkin finite element method proceeds by substituting equation (89) into equation (88), which results in a residual given by

$$r = \frac{\partial^2 v}{\partial x^2} + \frac{\partial^2 v}{\partial y^2} - f. \qquad (92)$$

Application of the weighted residual integral gives

$$\iint_D r \,.\, N_i \,.\, dxdy = 0, \; i = 1 \ldots n, \qquad (93)$$

$$\iint_D \left[N_i \left\{ \frac{\partial^2 v}{\partial x^2} + \frac{\partial^2 v}{\partial y^2} \right\} - N_i f \right] dxdy = 0, \; i = 1 \ldots n. \tag{94}$$

Because of inter-element continuity requirements it is desirable to reduce the maximum order of the derivatives appearing in equation (94). This can be done by applying Green's theorem to the Laplacian in equation (94). The result is

$$\int_S N_i \left\{ \frac{\partial v}{\partial x} \cdot 1_x + \frac{\partial v}{\partial y} \cdot 1_y \right\} ds - \iint_D \left\{ \frac{\partial N_i}{\partial x} \cdot \frac{\partial v}{\partial x} + \frac{\partial N_i}{\partial y} \cdot \frac{\partial v}{\partial y} \; dxdy \right\}$$

$$- \iint_D N_i \; f \; dxdy = 0, \; i = 1 \ldots n. \tag{95}$$

The first integral only makes a contribution on the boundary S and only then if the ith node is on the boundary. Substituting for v from equation (89) into equation (95) produces the result

$$\sum_{j=1}^{n} \left[\int_S N_i \; \frac{\partial N_j}{\partial x} \cdot 1_x + \frac{\partial N_j}{\partial y} \cdot 1_y \; ds - \iint_D \left\{ \frac{\partial N_i}{\partial x} \cdot \frac{\partial N_j}{\partial x} + \frac{\partial N_i}{\partial y} \cdot \frac{\partial N_j}{\partial y} \right\} dxdy \right] \bar{v}_j$$

$$= \iint N_i f \; dxdy, \; i = 1 \ldots n. \tag{96}$$

Equation (96) is just an algebraic equation which can be written as

$$\sum_{j=1}^{n} c_{ij} \cdot \bar{v}_j = b_i, \; i = 1 \ldots n. \tag{97}$$

The whole system of equations can be written as

$$[C] \cdot [V] = [B], \tag{98}$$

where an element of [C] is c_{ij}, defined by

$$c_{ij} = \int_S N_i \left\{ \frac{\partial N_j}{\partial x} \cdot 1_x + \frac{\partial N_j}{\partial y} \cdot 1_y \right\} ds - \iint_D \left\{ \frac{\partial N_i}{\partial x} \cdot \frac{\partial N_j}{\partial x} + \frac{\partial N_i}{\partial y} \cdot \frac{\partial N_j}{\partial y} \right\} dxdy, \tag{99}$$

and an element of [B] is b_i, defined by

$$b_i = \iint_D N_i \; . \; f \;\; dxdy. \tag{100}$$

The form of equation (98) is precisely the same as equation (55), Section 3.2 obtained from the application of the traditional Galerkin method. However, the evaluation of the coefficients, c_{ij} demonstrates a significant difference from the traditional Galerkin method. Because of the local nature of N_i the only contributions to c_{ij} come from the elements surrounding the ith node. For example, in obtaining the Galerkin equation at node 1 (see previous figure), the integrations shown in equations (99) and (100) only have contributions from the elements W, X, Y and Z. This is because N_1 is zero outside those four elements. For i = 1 the only non-zero c_{ij}'s will be those corresponding to nodes 1 to 9 i.e., $c_{1,1}$ to $c_{1,9}$.

If linear shape functions are used and only one dependent variable appears in the governing equation, then the ith row in the [C] matrix can only have a maximum of nine non-zero entries, irrespective of how many nodes are in the total domain. With a sensible numbering system the non-zero elements are clustered around the main diagonal. Since [C] is sparse, equation (98) can be solved economically if the sparse character of [C] is taken advantage of. This advantage can be traced back directly to the local nature of the shape function N_j in equation (89).

Another advantage of the finite element method comes from reducing the integration over the whole domain to the sum of the integrations over individual elements. Although the elements shown here have been rectangular this is not generally necessary. Using an isoparametric formulation (ref.42,p.129) the elements can be of any reasonable shape. Consequently by distorting the elements within the domain it is possible to handle arbitrary boundary contours and the corresponding boundary conditions without the need for special procedures. This situation may be contrasted with the need for boundaries to follow coordinate lines in the traditional Galerkin method. The integration over the distorted elements is usually done numerically after mapping the distorted element in the (x,y) plane into a regular element in the (ξ,η) plane.

Thus the Galerkin finite element formulation has two advantages over the traditional Galerkin formulation:

(i) it leads to a sparse solution matrix [C] that can be accurately and economically factorised.

(ii) by reducing the integration over the whole domain to the numerical integration over discrete, distortable elements arbitrary boundaries are easily handled.

A more detailed description of the Galerkin finite element method using a similar notation to the above may be found in reference 43.

4.2 A worked example

The Galerkin finite element method is applied to asymmetric viscous flow in a channel, the same problem as considered in Section 3.3 A solution is sought for the intermediate function w_d which is governed by the equation

$$\frac{\partial^2 w_d}{\partial x^2} + \frac{\partial^2 w_d}{\partial y^2} - e^{-\pi/2} \; . \; \frac{\pi^2}{4} \; . \; \cos \frac{\pi}{2} \; y \; = \; 0. \tag{101}$$

The boundary conditions are

$$W_d = 0 \text{ on } y = \pm 1 \text{ and on } x = 1$$
$$= (1 - e^{-\pi/2}) \cos \frac{\pi}{2} y \quad \text{on } x = 0. \qquad (102)$$

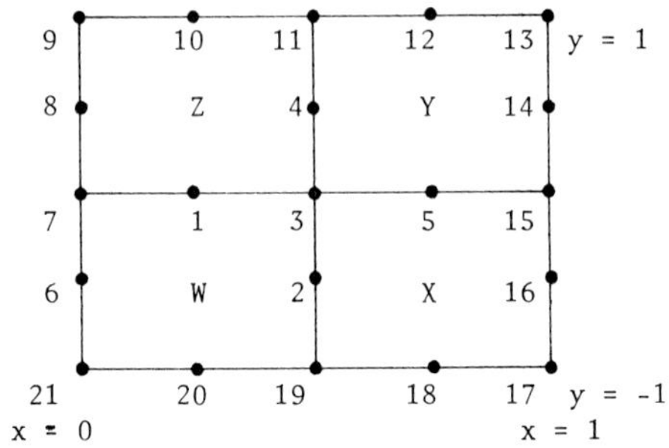

Figure 6

An approximate solution is defined by

$$U_a = \sum_{j=1}^{21} N_j(x,y) \cdot \overline{U}_{aj} \qquad (103)$$

$\overline{U}_{aj}$ are the nodal values of the velocity and these become the unknowns of the problem. Strictly only $\overline{U}_{aj}$, j = 1 to 5 are unknown. $\overline{U}_{aj}$, j = 6 to 8, are known from the boundary condition on x = 0; and $\overline{U}_{aj}$, j = 9 to 21, are zero due to appropriate boundary conditions. The domain (figure 6) is split into four elements W, X, Y and Z and the analytic variation within the elements is determined by shape functions $N_j(x,y)$ in equation (103). For this particular problem quadratic shape functions of the Serendipity family (ref.42, p.109) have been used for the N_j's. These were chosen to obtain five unknowns with four elements; this may be contrasted with the use of linear shape functions and four elements which would lead to only one unknown.

The quadratic shape functions are most easily defined using an element-based coordinate system (ζ,η) (figure 7)

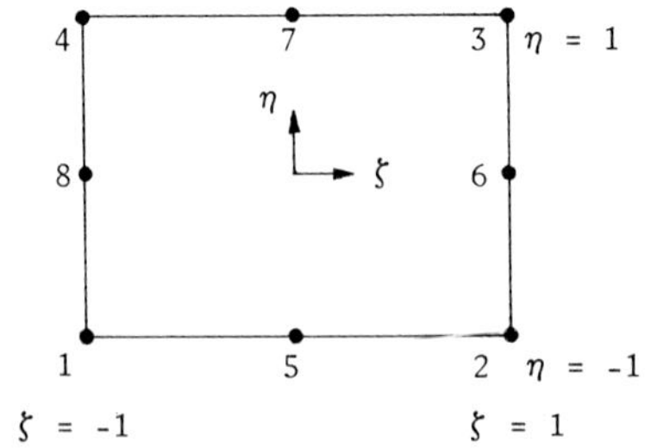

Figure 7

For corner nodes 1, 2, 3 and 4,

$$N_i = \tfrac{1}{4}(1 + \zeta . \zeta_i)(1 + \eta . \eta_i)(\zeta . \zeta_i + \eta . \eta_i - 1). \tag{104}$$

For midside nodes 5 and 7,

$$N_i = \tfrac{1}{2}(1 - \zeta^2)(1 + \eta . \eta_i). \tag{105}$$

For midside nodes 6 and 8,

$$N_i = \tfrac{1}{2}(1 + \zeta . \zeta_i)(1 - \eta^2). \tag{106}$$

As in Section 4.1 the method proceeds by substituting equation (103) into equation (101) which produces a residual

$$r = \frac{\partial^2 U_a}{\partial x^2} + \frac{\partial^2 U_a}{\partial y^2} - e^{-\pi/2} . \frac{\pi^2}{4} . \cos\frac{\pi}{2} y. \tag{107}$$

Application of the weighted residual integral gives

$$\int_{-1}^{1}\int_{0}^{1} r . N_i(x,y)\, dxdy = 0, \; i = 1 \ldots 5. \tag{108}$$

Since there are five unknowns $\overline{U}_{aj}$, j = 1 to 5, equation (108) is applied at the corresponding nodes 1 to 5 (figure 6). Substitution of equation (107) into equation (108) gives

$$\int_{-1}^{1}\int_{0}^{1} N_i \left\{\frac{\partial^2 U_a}{\partial x^2} + \frac{\partial^2 U_a}{\partial y^2}\right\} dxdy - e^{-\pi/2} . \frac{\pi^2}{4} . \int_{-1}^{1}\int_{0}^{1} N_i . \cos\frac{\pi}{2} y .\, dxdy = 0,$$

$$i = 1 \ldots 5. \tag{109}$$

Application of Green's theorem gives

$$\int_S N_i \left\{\frac{\partial U_a}{\partial x} . 1_x + \frac{\partial U_a}{\partial y} . 1_y\right\} ds - \int_{-1}^{1}\int_{0}^{1} \left\{\frac{\partial N_i}{\partial x} . \frac{\partial U_a}{\partial x} + \frac{\partial N_i}{\partial y} . \frac{\partial U_a}{\partial y}\right\} dxdy$$

$$- e^{-\pi/2} . \frac{\pi^2}{4} \int_{-1}^{1}\int_{0}^{1} N_i . \cos\frac{\pi}{2} y . dxdy = 0, \; i = 1 \ldots 5. \tag{110}$$

The line integral can only contribute on the boundary of the domain. However, for i = 1 to 5, N_i is zero on all exterior boundaries; therefore the line integral makes no contribution to equation (110). Substitution of equation (103) into equation (110) produces the result

$$\sum_{j=1}^{8} \left[\int_{-1}^{1}\int_{0}^{1} \left\{\frac{\partial N_i}{\partial x} \cdot \frac{\partial N_j}{\partial x} + \frac{\partial N_i}{\partial y} \cdot \frac{\partial N_j}{\partial y}\right\} dxdy\right] \overline{U}_{aj}$$

$$= e^{-\pi/2} \cdot \frac{\pi^2}{4} \cdot \int_{-1}^{1}\int_{0}^{1} N_i \cdot \cos\frac{\pi}{2}\, y\, dxdy, \; i = 1 \ldots 5. \tag{111}$$

or

$$\sum_{j=1}^{8} c_{ij} \cdot \overline{U}_{aj} = b_i, \; i = 1 \ldots 5, \tag{112}$$

where

$$c_{ij} = \int_{-1}^{1}\int_{0}^{1} \left\{\frac{\partial N_i}{\partial x} \cdot \frac{\partial N_j}{\partial x} + \frac{\partial N_i}{\partial y} \cdot \frac{\partial N_j}{\partial y}\right\} dxdy,$$

and

$$b_i = -e^{-\pi/2} \cdot \frac{\pi^2}{4} \int_{-1}^{1}\int_{0}^{1} N_i \cdot \cos\frac{\pi}{2}\, y \,.\, dxdy.$$

The integrations to evaluate c_{ij} and b_i are performed most easily in the (ζ,η) plane, either numerically or analytically. Since $x = x(\zeta)$ only and $y = y(\eta)$ only, the conversion to the (x,y) plane is trivial.

In equation (112) $\overline{U}_{aj}$, $j = 6, 7, 8$ are known, therefore it is convenient to manipulate equation (112) to give

$$\sum_{j=1}^{5} c_{ij} \cdot \overline{U}_{aj} = b_i - \sum_{j=6}^{8} c_{ij} \cdot \overline{U}_{aj}, \; i = 1 \ldots 5. \tag{113}$$

Equations (113) can be written in matrix form as

$$[C] \,.\, [U] = [D], \tag{114}$$

where an element of [C] is c_{ij}, an element of [U] is $\overline{U}_{aj}$ and an element of [D] is d_i, given by

$$d_i = b_i - \sum_{j=6}^{8} c_{ij} . \overline{U}_{aj} . \qquad (115)$$

Equation (114) is equivalent to equations (98) and (69). In evaluating the coefficients c_{ij} the number of elements that contribute depend on the locations of nodes i and j. Thus to obtain $c_{1,1}$ contributions are obtained from elements W and Z, to obtain $c_{2,3}$ contributions are obtained from W and X, to obtain $c_{3,5}$ contributions are obtained from X and Y, to obtain $c_{3,3}$ contributions are obtained from W, X, Y and Z (figure 6), and so on. In equation (114) the matrix [C] is

$$\begin{bmatrix} 7.6444 & 0 & -3.4889 & 0 & 0 \\ 0 & 3.9111 & -0.6222 & 0 & 0 \\ -3.4889 & -0.6222 & 5.7778 & -0.6222 & -3.4889 \\ 0 & 0 & -0.6222 & 3.9111 & 0 \\ 0 & 0 & -3.4889 & 0 & 7.6444 \end{bmatrix}$$

and $[D]^T$ is

$$[2.6250 \quad 0.51656 \quad -0.69276 \quad 0.51656 \quad -0.13859]$$

The solution obtained from equation (114) is an appriximation to W_d. To facilitate a comparison with the traditional Galerkin Method a solution for W is obtained from

$$W_a = U_a + e^{-\pi/2} . \cos \frac{\pi}{2} y. \qquad (116)$$

Here U_a is the finite element solution from equation (114). W_a can be compared with the exact solution

$$W = e^{-\frac{\pi}{2} x} . \cos \frac{\pi}{2} y. \qquad (117)$$

Location		Approximate solution, W_a	Exact solution, W
x	y		
0.25	0	0.66682	0.67523
0.50	-0.50	0.31935	0.32240
0.50	0	0.46105	0.45594
0.50	0.50	0.31935	0.32240
0.75	0	0.30530	0.30786

A comparison of the above results with the results shown in Section 3.3 indicates that, for the same number of unknown coefficients, the traditional Galerkin method gives more accurate results than the Galerkin finite element method. This result can be traced to restricting the region of the domain over which the ith shape function operates. Alternatively, if the number of unknowns is held constant more accurate results are obtained as the order of the shape functions is increased. This implies the elements are made larger (and fewer) to include more nodes which, in turn, makes the solution matrix [C] denser. In the limit of treating the whole domain as a single element, [C] could be completely dense, the approximating and weighting functions would span the whole domain and accuracies comparable to using the traditional Galerkin method would be obtained.

However, the comparisons just noted have taken no account of the economy of the different methods. If a comparison were made of the accuracy for the same execution time it is likely that the Galerkin finite element method would be more accurate than the traditional Galerkin method, particularly for complex problems requiring accurate solutions.

The sparsity of [C] above is deceptive; it is only partly due to restricting the domain of influence of the shape functions. This is the reason why $c_{1,5}$, $c_{2,4}$, $c_{4,2}$ and $c_{5,1}$ are zero. However, the other zeros are due to the regular nature of the grid (figure 6). If a solution were sought with many more nodes but still using quadratic rectangular elements of the Serendipity family the maximum number of non-zero coefficients c_{ij} in the ith row would be 21. This can be compared with 9 if linear rectangular elements were used.

4.3 Applications

There is a strong correlation between the development of the Galerkin finite element method and the number of applications of the finite element method to fluid dynamics. Traditionally the finite element method has been treated as a variational method. However, although many inviscid flows and a few viscous flows possess a variational formulation the general fluid flow problem governed by the Navier-Stokes equations does not possess a useful variational formulation.

Consequently, it was not until Oden put forward a theory of finite elements divorced from a variational formulation in 1969(ref.44) that non-variational formulations were used with a clear conscience. The first application to a fluid dynamic problem appears to be that due to Leonard (ref.45) who considered compressible inviscid flow. Of the many later applications of the Galerkin method the paper by Oden and Wellford(ref.46) is worth mentioning because it gave a general treatment of the Navier-Stokes equations using a Galerkin formulation and presented results for a number of model viscous flows.

A number of applications to creeping flows have been made using a Galerkin formulation. For creeping flows the inertial (convective) terms can be ignored and it is fairly straight forward to borrow the notation, formulation and computer codes from essentially structural applications; this approach is illustrated in the review article by Zienkiewicz(ref.47).

As the Reynolds number is increased the convective terms (which are highly nonlinear) play a dominant role and most Galerkin formulations have not been based so closely on the structural background of the method. Baker has developed a Galerkin finite element formulation based on a continuum mechanics approach and has written a general finite element code for fluid flow problems(ref.48). Applications have been made to two dimensional unsteady flow at moderate Reynolds numbers(ref.49), to three dimensional boundary layer flow(ref.50) and to low speed flows of environmental significance(ref.51,52), amongst others.

5. THE GALERKIN FINITE DIFFERENCE METHOD

The Galerkin finite difference method can be considered as a special case of the Galerkin finite element method. That is if the Galerkin finite element method is applied on a rectangular grid many of the terms in the solution matrix, e.g. [C] in equation (98), are zero. The sparsity of [C] is then of the same order and distribution as for many conventional finite difference techniques. This often implies that the finite difference algorithms, e.g. the tridiagonal algorithm, are available for solution. Galerkin finite difference methods constitute a very restricted class of finite difference methods.

To date most of the Galerkin finite difference techniques have been proposed for solving parabolic partial differential equations and a few have been proposed for solving hyperbolic partial differential equations. In this section a simple Galerkin finite difference scheme will be derived for the model parabolic equation, the unsteady heat equation. The Galerkin finite difference method is applied to the parabolic problem of Section 3.4 and the results are compared with the traditional Galerkin method and with conventional finite difference techniques. Finally, a number of papers that consider Galerkin finite difference schemes are discussed.

5.1 Description

The method can be understood most easily by applying it to an example. Here the same example that was used in Section 2.2 to describe the traditional Galerkin method is used.

The governing equation for the unsteady temperature distribution problem is

$$\frac{\partial T}{\partial t} - \frac{\partial^2 T}{\partial x^2} = o, \tag{116}$$

subject to the boundary conditions

$$\begin{aligned} T(x,o) &= T_o(x) + T_a \\ T(o,t) &= T_a \\ T(1,t) &= T_a. \end{aligned} \tag{117}$$

The temperature difference $T_d = T - T_a$ is treated as the unknown since this simplifies the boundary conditions. As with the Galerkin finite element method an approximate representation is introduced as

$$T_d = \sum_{j=1}^{n} N_j(x) \,.\, \bar{T}_j(t). \tag{118}$$

N_j are one dimensional shape functions that will be defined later. $\bar{T}_j$ are the nodal values of T_d and are functions of t. Substitution of equation (118) into the governing equation (116) produces a residual

$$r = \sum_{j=1}^{n} \left\{ N_j \,.\, \frac{\partial \bar{T}_j}{\partial t} - \frac{\partial^2 N_j}{\partial x^2} \,.\, \bar{T}_j \right\}. \tag{119}$$

The weighted residual integral condition is

$$\int_0^1 r \,.\, N_i \, dx = 0. \tag{120}$$

Substitution for r gives

$$\sum_{j=1}^{n} \int_0^1 N_i \left\{ N_j \,.\, \frac{\partial \overline{T}_j}{\partial x} - \frac{\partial^2 N_j}{\partial x^2} \,.\, \overline{T}_j \right\} dx = 0, \; i = 1 \ldots n \tag{121}$$

or

$$\sum_{j=1}^{n} p_{ij} \,.\, \frac{\partial \overline{T}_j}{\partial x} = \sum_{j=1}^{n} q_{ij} \,.\, \overline{T}_j, \tag{122}$$

where

$$p_{ij} = \int_0^1 N_i \, N_j \, dx, \tag{123}$$

and

$$q_{ij} = \int_0^1 N_i \,.\, \frac{\partial^2 N_j}{\partial x^2} dx = \left[N_i \,.\, \frac{\partial N_j}{\partial x} \right]_0^1 - \int_0^1 \frac{\partial N_i}{\partial x} \,.\, \frac{\partial N_j}{\partial x} dx. \tag{124}$$

To evaluate p_{ij} and q_{ij} a choice for the shape function must be made. Here the one dimensional equivalents of the linear shape functions defined by equation (91) are used (figure 8).

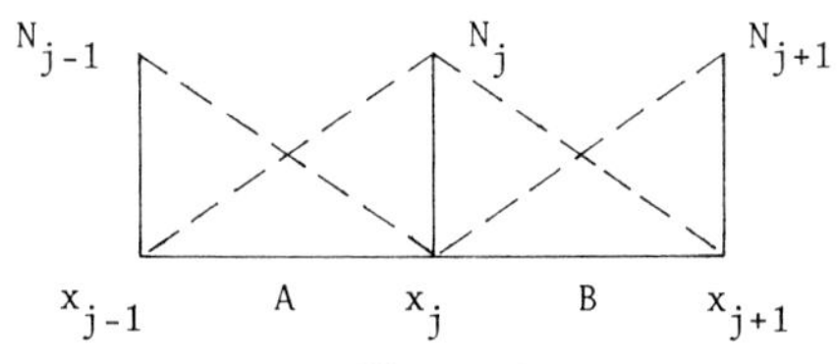

Figure 8

For a regular grid, $x_{j+1} - x_j = x_j - x_{j-1} = \Delta x$.
In element A

$$N_{j-1} = (x_j - x)/\Delta x, \; N_j = (x - x_{j-1})/\Delta x. \tag{125}$$

In element B

$$N_j = (x_{j+1} - x)/\Delta x, \quad N_{j+1} = (x - x_j)/\Delta x. \tag{126}$$

When i = j only three p_{ij}'s are non zero, they are $p_{j,j-1}$, $p_{j,j}$ and $p_{j,j+1}$. These can be evaluated using equation (123).

$$p_{j,j-1} = \int_0^1 N_j \cdot N_{j-1}\, dx = \int_{x_{j-1}}^{x_j} \frac{(x - x_{j-1})}{\Delta x} \cdot \frac{(x_j - x)}{\Delta x} \cdot dx.$$

Introducing

$$\zeta = (x - x_{j-1})/\Delta x$$

leads to

$$p_{j,j-1} = \Delta x \int_0^1 \zeta \cdot (1 - \zeta)\, d\zeta = \Delta x/6.$$

$$p_{j,j} = \int_{x_{j-1}}^{x_j} \left[\frac{(x - x_{j-1})}{\Delta x}\right]^2 dx + \int_{x_j}^{x_{j+1}} \left[\frac{(x_{j+1} - x)}{\Delta x}\right]^2 dx$$

$$= \Delta x/3 + \Delta x/3 = 2\Delta x/3.$$

$$p_{j,j+1} = \int_{x_{j-1}}^{x_j} \frac{(x_{j+1} - x)}{\Delta x} \cdot \frac{(x - x_j)}{\Delta x}\, dx = \Delta x/6.$$

Similarly only three of the q_{ij}'s are non zero. Since the Galerkin equations are not required on the boundary, equation (124) gives

$$q_{ij} = -\int_0^1 \frac{\partial N_i}{\partial x} \cdot \frac{\partial N_j}{\partial x}\, dx. \tag{127}$$

With i = j the following are obtained

$$q_{j,j-1} = \int_{x_{j-1}}^{x_j} -\frac{1}{\Delta x^2}\, dx = 1/\Delta x$$

$$q_{j,j} = -\int_{x_{j-1}}^{x_j} \frac{1}{\Delta^2 x}\,dx - \int_{x_j}^{x_{j+1}} \frac{1}{\Delta^2 x}\,dx = -2/\Delta x$$

and

$$q_{j,j+1} = -\int_{x_j}^{x_{j+1}} -\frac{1}{\Delta^2 x}\,dx = 1/\Delta x.$$

Substituting into equation (122) and rearranging gives

$$\frac{1}{6}\cdot\frac{\partial \overline{T}_{j-1}}{\partial t} + \frac{2}{3}\cdot\frac{\partial \overline{T}_j}{\partial t} + \frac{1}{6}\frac{\partial \overline{T}_{j+1}}{\partial t} = \frac{(\overline{T}_{j+1} - 2\,\overline{T}_j + \overline{T}_{j-1})}{\Delta^2 x}. \tag{128}$$

In the present form this is one equation out of a system of ordinary differential equations for the $\overline{T}_j$'s. Following conventional finite difference practice the derivative terms are replaced by

$$\frac{\partial \overline{T}_j}{\partial t} = \frac{\overline{T}_j^{n+1} - \overline{T}_j^n}{\Delta t},$$

and the right hand side of equation (128) is evaluated as a weighted $(1 - \lambda:\lambda)$ average of the values at the nth and $(n+1)^{th}$ steps. Thus

$$\frac{1}{6}\cdot\frac{(\overline{T}_{j-1}^{n+1} - \overline{T}_{j-1}^n)}{\Delta t} + \frac{2}{3}\cdot\frac{(\overline{T}_j^{n+1} - \overline{T}_j^n)}{\Delta t} + \frac{1}{6}\cdot\frac{(\overline{T}_{j+1}^{n+1} - \overline{T}_{j+1}^n)}{\Delta t}$$

$$= \lambda\,.\,\frac{(\overline{T}_{j+1}^{n+1} - 2\,\overline{T}_j^{n+1} + \overline{T}_{j-1}^{n+1})}{\Delta^2 x} + (1-\lambda)\,.\,\frac{(\overline{T}_{j+1}^n - 2\,\overline{T}_j^n + \overline{T}_{j-1}^n)}{\Delta^2 x}. \tag{129}$$

Equation (129) can be re-arranged to give

$$\frac{\overline{T}_j^{n+1} - \overline{T}_j^n}{\Delta t} = \theta\,.\,\overline{T}_{xx,j}^{n+1} + (1 - \theta)\,.\,\overline{T}_{xx,j}^n \tag{130}$$

where $\theta = (\lambda - \Delta^2 x/6\Delta t)$

and

$$\overline{T}_{xx,j} = (\overline{T}_{j+1} - 2\,\overline{T}_j + \overline{T}_{j-1})/\Delta^2 x.$$

Equation (130) is a classic finite difference result. Different choices of θ (and hence λ) give the fully implicit, Crank-Nicolson and explicit schemes as special cases. Swartz and Wendroff first gave the above result in a paper(ref.53) that appears to pre-date the first Galerkin finite element formulation. Swartz and Wendroff note that "Galerkin's method, using a local basis, provides unconditionally stable, implicit generalised finite difference schemes for a large class of linear and nonlinear problems". Since the Galerkin finite difference formulation can be derived from a finite element formulation the treatment of boundary conditions should be superior to the essentially ad-hoc treatment that conventional finite difference formulations use. It was stated above that application of Galerkin finite difference schemes has been made to parabolic and hyperbolic equations. If a Galerkin finite element formulation using linear triangular elements on a rectangular grid is made to a model elliptic equation, e.g. Laplaces equation, the result is

$$4\,\psi_{ij} - (\psi_{i-1,j} + \psi_{i,j-1} + \psi_{i+1,j} + \psi_{i,j+1}) = 0 \tag{131}$$

which is a standard finite difference result.

5.2 A worked example

The formulae derived above are applied to problem 3.4. The governing equation is

$$\frac{\partial\psi}{\partial x} - \frac{\partial^2\psi}{\partial y^2} = 0, \tag{132}$$

and the boundary conditions are

$$\psi(o,y) = \sin\pi y + y,\ \psi(x,o) = 0,\ \psi(x,1) = 1. \tag{133}$$

In terms of problem 3.4 equation (130) becomes

$$\frac{(\bar{\psi}_j^{n+1} - \bar{\psi}_j^n)}{\Delta x} = \theta\,.\,\bar{\psi}^{n+1}_{yy,j} + (1-\theta)\,.\,\bar{\psi}^n_{yy,j}, \tag{134}$$

where

$$\theta = \lambda - \Delta^2 y/6\Delta x.$$

To facilitate comparison with the traditional Galerkin method equation (134) is converted back into an ordinary differential equation and rearranged to give

$$\left(\frac{1}{6} - \frac{\lambda\Delta x}{\Delta^2 y}\right).\frac{\partial\psi_{j-1}}{\partial x} + \left(\frac{2}{3} + 2\lambda\frac{\Delta x}{\Delta^2 y}\right).\frac{\partial\psi_j}{\partial x} + \left(\frac{1}{6} - \lambda\frac{\Delta x}{\Delta^2 y}\right).\frac{\partial\psi_{j+1}}{\partial x} = \psi^n_{yy,j}. \tag{135}$$

or

$$-\theta\,.\frac{\partial\psi_{j-1}}{\partial x} + (2\theta + \Delta^2 y/\Delta x)\frac{\partial\psi_j}{\partial x} - \theta\,.\frac{\partial\psi_{j+1}}{\partial x} = \left(\frac{\Delta^2 y}{\Delta x}\right).\,\psi^n_{yy,j}. \tag{136}$$

Equation (136) is equivalent to equation (82) i.e.,

$$[P]\,[\dot{\psi}] = [Q]\,[\psi] + [R] \tag{137}$$

where [P] is

$$\begin{bmatrix} (2\theta + \Delta^2 y/\Delta x) & -\theta & 0 & 0 & 0 \\ -\theta & (2\theta + \Delta^2 y/\Delta x) & -\theta & 0 & 0 \\ 0 & -\theta & (2\theta + \Delta^2 y/\Delta x) & -\theta & \\ 0 & 0 & -\theta & (2\theta + \Delta^2 y/\Delta x) & -\theta \\ 0 & 0 & 0 & -\theta & (2\theta + \Delta^2 y/\Delta x) \end{bmatrix}$$

[Q] is

$$\begin{bmatrix} -2 & 1 & 0 & 0 & 0 \\ 1 & -2 & 1 & 0 & 0 \\ 0 & 1 & -2 & 1 & 0 \\ 0 & 0 & 1 & -2 & 1 \\ 0 & 0 & 0 & 1 & -2 \end{bmatrix}$$

and $[R]^T$ is $[0 \quad 0 \quad 0 \quad 0 \quad 1/\Delta x]$.

As in Section 3.4 [P], [Q] and [R] are evaluated and equation (137) is used to obtain $[\dot{\psi}]$. $[\dot{\psi}]$ is integrated using the same first order scheme as is Section 3.4, i.e.,

$$\bar{\psi}_j^{n+1} = \bar{\psi}_j^n + \frac{\partial \bar{\psi}_j}{\partial x} \,.\, \Delta x. \tag{138}$$

Five nodal values $\bar{\psi}_j$ (equivalent to $\Delta y = 1/6$) have been integrated simultaneously using equation (138) for various values of θ. For all the results shown $\Delta x = 0.001$. Setting $\theta = -\Delta^2 y/6\Delta x = -4.63$ corresponds to setting $\lambda = 0$ in the equation equivalent to (129), i.e., an explicit right hand side and the original Galerkin left hand side. Setting $\theta = 0$ corresponds to setting $\lambda = \Delta^2 y/6\Delta x$, i.e. a conventional explicit formula. Setting $\theta = 0.5$ gives a Crank-Nicolson scheme. The results at $y = 0.5$ for these three values of θ, are shown in the table. For comparison the exact solution is included as is the approximate solution, ψ_b which only has errors due to the numerical integration scheme.

x location (y = 0.5)	Approximate solution of equation (137)			Exact solution, ψ	Approximate solution, ψ_b
	Galerkin LHS θ = -4.63	Explicit θ = 0	Crank-Nicolson θ = 0.5		
0	1.50000	1.50000	1.50000	1.50000	1.50000
0.02	1.31630	1.32377	1.32454	1.32087	1.32006
0.04	1.16635	1.17860	1.17987	1.17383	1.17251
0.06	1.04395	1.05902	1.06059	1.05312	1.05150
0.08	0.94403	0.96050	0.96223	0.95404	0.95226
0.10	0.86246	0.87935	0.88113	0.87271	0.87089
0.12	0.79588	0.81250	0.81426	0.80594	0.80415
0.14	0.74153	0.75743	0.75912	0.75114	0.74942
0.16	0.69716	0.71206	0.71365	0.70615	0.70454
0.18	0.66094	0.67469	0.67617	0.66923	0.66774
0.20	0.63138	0.64391	0.64526	0.63891	0.63756

None of the results shown above are as accurate as those obtained using the traditional Galerkin method and the same number of unknown coefficients i.e. five. This is consistent with the lower accuracy of the finite element results shown in Section 4.2. The sparsity of [P], [Q] and [R] may be noted and the comments about accuracy versus economy in Section 4.2 also apply here. The above results indicate that θ (or λ) could be adjusted to give good agreement with ψ or ψ_b but, almost certainly, such a solution would suffer from the stability restriction $\Delta x < 0.5\ \Delta^2 y$.

5.3 Applications

Most of the papers that have examined the Galerkin finite difference formulation have been written by numerical analysts and have been concerned with error estimates and devising optimum schemes. To date the technique only appears to have been applied to model problems; no-one seems to have used a Galerkin finite difference technique on a 'real' problem. This could change in the future.

After the initial paper by Swartz and Wendroff(ref.53), Douglas and Dupont have published a number of papers(refs.54-56) dealing with improvements to the Swartz-Wendroff schemes, alternating-direction Galerkin schemes and the effect of non-linear boundary conditions. Gourlay(ref.57) has reviewed various finite difference techniques for the solution of time-dependent partial differential equations and includes Galerkin techniques as a special class. Gourlay notes that Galerkin schemes require more setting up than conventional finite difference schemes but should produce answers more efficiently (same accuracy in less computing time). Gourlay explores an alternating direction Galerkin formulation and suggests other combinations e.g. Lax-Wendroff-Galerkin.

Dendy and Fairweather(ref.58) have extended the alternating direction Galerkin methods of Douglas and Dupont(ref.55) to parabolic and hyperbolic problems in rectangular polygons. Wahlbin(ref.59) has considered the use of a dissipative Galerkin technique for the solution of one dimensional hyperbolic problems.

Swartz(ref.60) has taken the parabolic problem governed by $U_t = U_{xx}$ as a model problem and used it to compare a large number of higher order finite difference techniques. For the spatial terms Swartz considers explicit centered differencing schemes, super convergent smooth spline-Galerkin schemes due to Thomée(ref.61) and high order implicit schemes. Seven different time discretisations are considered.

6. THE SPECTRAL METHOD

The main difference between a spectral method and the traditional Galerkin method is the use of orthogonal functions for the trial and weighting functions. In this sense references 32, 33 and 34 may be interpreted as spectral methods. The chief difficulty with spectral methods is in choosing orthogonal functions appropriate to the problem. For the purposes of description the spectral method is applied to Poisson's equation in a rectangle and to the nonlinear vorticity convection equation. An elementary spectral method is applied to the parabolic problem of Section 3.4. A number of references are considered which illustrate the application of the spectral method to fluid flow.

6.1 Description

The spectral method is illustrated by considering Poisson's equation in a rectangle bounded by $0 \leqslant x \leqslant 1$, $0 \leqslant y \leqslant 1$. The governing equation is

$$\frac{\partial^2 u}{\partial x^2} + \frac{\partial^2 u}{\partial y^2} - f = 0. \tag{139}$$

An approximate solution is introduced as

$$u_a(x,y) = u_o(x,y) + \sum_{i=1}^{N} \sum_{j=1}^{N} a_{ij} \sin i\pi x \,.\, \sin j\pi y. \tag{140}$$

$u_o(x,y)$ is included to take care of the boundary conditions; then the trial functions are chosen to satisfy homogeneous boundary conditions. The trial functions are eigenfunctions of the related equation

$$\frac{\partial^2 u}{\partial x^2} + \frac{\partial^2 u}{\partial y^2} = \lambda \,.\, u, \tag{141}$$

in a rectangular domain. The ploy of choosing the trial functions from eigenfunctions of a related problem is often used in spectral methods. In order to facilitate comparison with the traditional Galerkin method, the approximate solution, equation (140), will be expanded to give

$$u_a = u_o + b_1 \sin \pi x \,.\, \sin \pi y + b_2 \sin 2\pi x \,.\, \sin \pi y + b_3 \sin \pi x \,.\, \sin 2\pi y \ldots$$

or

$$u_a = u_o + \sum_{j=1}^{N^2} b_j \,.\, \varphi_j(x,y). \tag{142}$$

Substituting equation (142) into the governing equation (139) gives a residual

$$r = \left[\frac{\partial^2 u_o}{\partial x^2} + \frac{\partial^2 u_o}{\partial y^2} - f\right] + \sum_{j=1}^{N^2} b_j \left\{\frac{\partial^2 \varphi_j}{\partial x^2} + \frac{\partial^2 \varphi_j}{\partial y^2}\right\}. \tag{143}$$

Application of the weighted integral condition gives

$$\int_0^1\int_0^1 r \cdot \varphi_i(x,y)\, dxdy = 0, \quad i = 1 \ldots N^2. \tag{144}$$

Substitution of equation (143) into equation (144) gives

$$\int_0^1\int_0^1 \left[\frac{\partial^2 u_o}{\partial x^2} + \frac{\partial^2 u_o}{\partial y^2} - f\right] \cdot \varphi_i(x,y)\, dxdy$$

$$+ \sum_{j=1}^{N^2} b_j \int\int_0^1 \left\{\frac{\partial^2 \varphi_j}{\partial x^2} + \frac{\partial^2 \varphi_j}{\partial y^2}\right\} \cdot \varphi_i\, dxdy = 0, \quad i = 1 \ldots N^2. \tag{145}$$

Since the φ_j's are orthogonal functions the only contribution from the sum occurs when $j = i$. Thus equation (145) can be written as

$$b_i \cdot \left[\int_0^1\int_0^1 \left\{\frac{\partial^2 \varphi_i}{\partial x^2} + \frac{\partial^2 \varphi_i}{\partial y^2}\right\} \varphi_i\, dxdy\right] = -\int_0^1\int_0^1 \left\{\frac{\partial^2 u_o}{\partial x^2} + \frac{\partial^2 u_o}{\partial y^2} - f\right\} \cdot \varphi_i\, dxdy,$$

$$i = 1 \ldots N^2. \tag{146}$$

Clearly b_i is given explicitly by equation (146). This may be contrasted with the application of the traditional Galerkin method to Poisson's equation described in Sections 3.2 and 3.3. There a matrix (see equations (55) and (69)) had to be inverted before the unknown coefficients in the approximate solution could be obtained. A similar situation occurred with the Galerkin finite element formulation (see equation (80)). The equivalent matrix of coefficients multiplying the b_j's for a spectral formulation would only have non-zero entries on the main diagonal. Once the b_j's have been obtained from equation (146) the solution u_a follows from substitution into equation (142).

Up to now all the examples used to illustrate the various formulations have been linear. This is because the linearity, or not, of the governing equation is independent of the Galerkin formulation. However, the treatment of non-linear equations is more cumbersome with a traditional Galerkin or spectral formulation than it is with a Galerkin finite element or difference formulation. The reason is that the unknowns in the spectral formulation, b_j in equation (142), have a global character whereas the unknowns in a finite element (or difference) formulation, e.g. the nodal values $\bar{v}_j$ in equation (73), have a purely local character.

The increased complexity for a non-linear spectral formulation can be appreciated by considering the two-dimensional convection of vorticity in an inviscid, incompressible fluid. This is governed by the equations

$$\frac{\partial w}{\partial t} - \frac{\partial \psi}{\partial y} \cdot \frac{\partial w}{\partial x} + \frac{\partial \psi}{\partial x} \cdot \frac{\partial w}{\partial y} = 0 \tag{147}$$

and

$$w = - \left\{ \frac{\partial^2 \psi}{\partial x^2} + \frac{\partial^2 \psi}{\partial y^2} \right\} , \tag{148}$$

where w is the vorticity and ψ is the stream function. Approximate solutions solutions for w and ψ are introduced as

$$w_a = \sum_{i=1}^{N} \sum_{j=1}^{N} a_{ij} \sin i\pi x \, . \, \sin j\pi y. \tag{149}$$

and

$$\psi_a = \sum_{i=1}^{N} \sum_{j=1}^{N} b_{ij} \sin i\pi x \, . \, \sin j\pi y. \tag{150}$$

Equation (148) can be used to relate the coefficients a_{ij} and b_{ij}. From equation (149)

$$\frac{\partial w_a}{\partial x} = \sum_{i=1}^{N} \sum_{j=1}^{N} a_{ij} \cdot (i\pi) \, . \, \cos i\pi x \, . \, \sin j\pi y, \tag{151}$$

and from equation (150)

$$\frac{\partial \psi_a}{\partial y} = \sum_{i=1}^{N} \sum_{j=1}^{N} b_{ij} \cdot (j\pi) \, . \, \sin i\pi x \, . \, \cos j\pi y. \tag{152}$$

The weighted residual integral equation, equivalent to equation (144), is

$$\int_0^1 \int_0^1 r \, . \, \sin k\pi x \, . \, \sin l\pi y \, . \, dxdy = 0, \quad k = 1 \ldots N, \quad l = 1 \ldots N. \tag{153}$$

The evaluation of the second term in equation (147) leads to

$$\int_0^1 \int_0^1 \frac{\partial \psi_a}{\partial y} \cdot \frac{\partial w}{\partial x} \, . \, \sin k\pi x \, . \, \sin l\pi y \, dxdy, \quad k = 1 \ldots N, \quad l = 1 \ldots N$$

$$= \int_0^1 \int_0^1 \left[\sum_{i=1}^{N} \sum_{j=1}^{N} b_{ij} \,.\, (j\pi) \,.\, \sin i\pi x \,.\, \cos j\pi y \right] .$$

$$\left[\sum_{m=1}^{N} \sum_{n=1}^{N} a_{mn} \, (m\pi) \,.\, \cos m\pi x \,.\, \sin n\pi y \right]$$

$$.\, \sin k\pi x \,.\, \sin l\pi y \; dxdy.$$

Since the trial functions are orthogonal one might hope the only non-zero coefficient would multiply the $b_{kl}.a_{kl}$ term. However in the above integral terms of the form sin $i\pi x$. cos $m\pi x$ can be replaced with $\frac{1}{2}[\sin(i+m)\,\pi x + \sin(i-m)\,\pi x]$ and terms of the form cos $j\pi y$. sin $n\pi y$ can be replaced with $\frac{1}{2}[\sin(n+j)\,\pi y + \sin(n-j)\,\pi y]$. Therefore, if $i+m = k$ or $i-m = k$ there will be a non-zero contribution to $\int dx$ and if $n+j = 1$ or $n-j = 1$, there will be a non-zero contribution to $\int dy$. Thus in each equation formed using equation (153) there will be $(N-1)^2$ non-zero b_{ij} . a_{mn} terms. Orszag(ref.62) indicates that if $N > 10$ the computation time required to handle the interaction terms for equations (147) and (148) in a spherical coordinate system becomes prohibitive unless special procedures are introduced. The special procedures are referred to in Section 6.3.

The choice of trial functions is very important in the spectral method. Orszag and Israeli(ref.12) suggest the following:

(i) for periodic boundary conditions use finite Fourier series.

(ii) for discontinuous boundary conditions in planar geometries use Chebyshev series.

(iii) for problems described in spherical coordinates, e.g. large scale atmospheric calculation, use spherical-harmonic expansions.

6.2 A worked example

In order to make comparisons with the traditional Galerkin method an elementary spectral method is applied to the parabolic problem of Section 3.4. The governing equation (76) is

$$\frac{\partial \psi}{\partial x} - \frac{\partial^2 \psi}{\partial y^2} = 0, \tag{154}$$

subject to the boundary conditions

$$\psi(o.y) = \sin \pi y + y, \; \psi(x,o) = 0, \; \psi(x,1) = 1. \tag{155}$$

ψ is a non-dimensional velocity defined by equation (73). An approximate solution is introduced as

$$\psi_a = \sin \pi y + y + \sum_{j=1}^{5} a_j(x) \,.\, \sin j\pi y. \tag{156}$$

substitution of equation (156) into equation (154) produces a residual

$$r = m_o(y) + \sum_{j=1}^{5} \left[\sin j\pi y \cdot \frac{\partial a_j}{\partial x} + m_j(y) \cdot a_j \right], \tag{157}$$

where

$$m_o(y) = \pi^2 \sin \pi y$$

and

$$m_j(y) = j^2 \pi^2 \sin j\pi y.$$

The weighted residual integral condition gives

$$\int_0^1 r \cdot \sin i\pi y \cdot dy = 0, \quad i = 1 \ldots 5, \tag{158}$$

and this leads to

$$\sum_{j=1}^{5} \left[\int_0^1 \sin j\pi y \cdot \sin i\pi y \, dy \right] \frac{\partial a_j}{\partial x} + \sum_{j=1}^{5} \left[\int_0^1 m_j(y) \cdot \sin i\pi y \, dy \right]$$
$$+ \int_0^1 m_o(y) \cdot \sin i\pi y \, dy = 0, \quad i = 1 \ldots 5. \tag{159}$$

The choice of $\sin j\pi y$ as trial function allows a considerable simplification of equation (159). It becomes

$$0.5 \frac{\partial a_i}{\partial x} + 0.5\,(i\pi)^2 \cdot a_i + r_i = 0, \quad i = 1 \ldots 5 \tag{160}$$

where

$$\begin{aligned} r_i &= \pi^2/2 \text{ if } i = 1 \\ &= 0 \quad \text{ if } i \neq 1 \end{aligned} \tag{161}$$

This is equivalent to obtaining just diagonal terms in matrices like [P] and [Q] in equation (82) obtained from a traditional Galerkin formulation. Since a_i, $i = 1 \ldots 5$ are zero at $x = 0$ equation (160) and (161) have as solutions

$$a_1 = e^{-\pi^2 x} - 1 \tag{162}$$

and

$$a_i = 0, \; i = 2 \; . \; . \; . \; 5.$$

Substitution of equations (162) into equation (156) gives

$$\psi_a = \sin \pi y \; . \; e^{-\pi^2 x} + y$$

which is the exact solution. This has occurred because the trial solutions contain the exact solution and the boundary conditions are expressed in terms of the exact solution. A more interesting situation develops if the initial conditions

$$\psi(o,y) = 5y - 4y^2, \tag{163}$$

are introduced. The only change in the above development is that $5y - 4y^2$ replaces $\sin \pi y + y$ in equation (155) and $m_o = 8$ in equation (157). Then r_i, which is defined by

$$r_i = \int_0^1 m_o \; . \; \sin i\pi y \; . \; dy, \tag{164}$$

leads to $[R]^T$ as

$$\frac{16}{\pi}[1 \quad 0 \quad \frac{1}{3} \quad 0 \quad \frac{1}{5}].$$

To permit a comparison to be made with the traditional Galerkin method, equation (160) is evaluated numerically using a first order integration scheme.

$$a_i^{n+1} = a_i^n + \frac{\partial a_i}{\partial x} \; . \; \Delta x. \tag{165}$$

Results are presented in the table for one, three and five terms in the approximate solution given by equation (156). All results shown have been obtained with $\Delta x = 0.001$. Also shown in the table is the exact solution, ψ and an approximate solution, ψ_b which only contains errors due to the numerical integration formula, equation (165).

x location (y = 0.50)	Approximate solution, ψ_a (n term, equation (156))			Exact solution, ψ	Approximate solution, ψ_b
	n = 1	n = 3	n = 5		
0	1.50000	1.50000	1.50000	1.5000	1.5000
0.02	1.31430	1.34657	1.33835	1.3408	1.3404
0.04	1.16201	1.19931	1.19105	1.1943	1.1931
0.06	1.03712	1.07520	1.06695	1.0707	1.0690
0.08	0.93471	0.97291	0.96465	0.96856	0.96673
0.10	0.85072	0.88894	0.88069	0.88465	0.88277
0.12	0.78185	0.82007	0.81182	0.81575	0.81389
0.14	0.72537	0.76359	0.75533	0.75919	0.75741
0.16	0.67905	0.71727	0.70902	0.71276	0.71110
0.18	0.64107	0.67929	0.67103	0.67465	0.67311
0.20	0.60992	0.64814	0.63988	0.64337	0.64196

The exact solution can be obtained from a separation of variables solution as

$$\psi = y + 4 \sum_{j=1}^{\infty} \left[\frac{2}{(2j - 1)\pi} \right]^3 . \sin(2j - 1)\pi y . \exp[-\{(2j - 1)\pi\}^2 x] . \tag{166}$$

The approximate solution ψ_b is obtained from

$$\psi_b = y + 4 \sum_{j=1}^{\infty} \left[\frac{2}{(2j - 1)\pi} \right]^3 . \sin(2j - 1)\pi y . \chi_j(x), \tag{167}$$

with $\chi_j(x)$ evaluated numerically, using the equivalent of equation (165) applied to

$$\frac{\partial \chi_j(x)}{\partial x} = -\{(2_j - 1)\pi\}^2 \chi_j(x) . \tag{168}$$

Because of the form of equation (164) only odd terms in equation (156) contribute. Thus the results shown above for n = 5 actually have only three unknown coefficients. Noting the remarks made in Section 3.4, the approximate solution ψ_a should be compared with the approximate solution ψ_b in order to gauge the error of the Galerkin approximation. The results indicate a reduction in the error as n is increased. The spectral (and traditional) method implicitly fits the initial data with the series of trial functions in the assumed approximate solution. Thus the difference between ψ_a and ψ_b is independent of x and can be traced back to the error in fitting the initial data with a truncated series of the trial functions.

Orszag and Israeli(ref.12) have considered the wave equation $u_t + u_x = 0$

as a model problem in the region $-1 \leqslant x < 1$ with boundary conditions $u(x,o) = 0$ and $u(-1,t) = \sin m\pi t$. They obtain results using a Galerkin formulation based on Chebyshev functions in x. They compare results with second and fourth order finite difference schemes and demonstrate that fewer unknown coefficients are required with a spectral formulation, to obtain comparable accuracy.

6.3 Applications

Applications of the spectral method to fluid flows may be conveniently divided into meteorological and non-meteorological applications. Recently, a review of mainly non-meteorological applications has appeared(ref.12).

Orszag and Kruskal(ref.63) have considered the unsteady Navier-Stokes equations in a cube with periodic boundary conditions. A Galerkin formulation (unnamed) with discrete Fourier series has been introduced and used to formulate a theory of homogeneous, isotropic turbulence. In a subsequent paper, Orszag(ref.64) has put forward numerical schemes for integrating the unsteady Navier-Stokes equations. This paper is very important for spectral techniques because in it Orszag introduced a scheme which greatly economises the computation associated with the non-linear terms (see Section 6.1). The essence of this scheme is to transform, using a fast Fourier transform, from the spectral domain (e.g., working in terms of the unknown coefficients a_{ij} in equation (149)) to the physical domain. The non-linear multiplications are performed in terms of nodal values in the physical domain and an inverse transform is applied to the result, to get back into the spectral domain. In a later paper, Orszag(ref.62) applied the transform technique using spherical harmonics to the equivalent of the vorticity convection equations (147) and (148), and in so doing reduced the number of computations per time step from $O(N^5)$ to $O(N^3)$.

Orszag(ref.65) has applied a spectral method based on a Fourier series to a scalar-convection test problem and made comparisons with a number of finite difference schemes applied to the same problem. In addition Orszag applied a spectral method to the decay of a vortex by solving the unsteady Navier-Stokes equations in three dimensions. Comparisons are made with some finite difference schemes. Orszag concludes that finite difference formulations require 2^p times as many grid points as there are unknowns in the spectral representation for the same problem; p is the number of spatial dimensions in the problem.

An interesting treatment of the vorticity convection equations (147) and (148) is given by Lorenz(ref.66). Lorenz obtained the "minimum hydrodynamic equations" by introducing three term Fourier expansions for the vorticity and stream function. Platzman(ref.67) considered the vorticity convection equations in spherical coordinates and developed a general spectral method using spherical harmonics as trial functions. In a related paper Platzman (ref.68) considered the solution of the vorticity convection equations using the spectral method but for low-order systems. Bourke(ref.69) took the spectral equations of Platzman, generalised them to allow for atmospheric divergence, introduced a transform technique to deal with the non-linear terms in a manner similar to Orszag, and presented results that showed convincingly the greater economy of the transform technique in comparison with the older method of interaction coefficients in which the complete solution is obtained in the spectral domain.

7. OVERALL COMPARISON

The purpose of this section is to consider real applications, e.g., the unsteady Navier-Stokes equations, and to see what types of problems the different computational techniques are best suited for. In comparing finite element methods, finite difference methods and spectral methods three gross features would appear to be important: accuracy, economy and ease of coding.

Any Galerkin method allows a reciprocal influence between neighbouring points in the domain. Consequently Galerkin methods are unsuitable for situations where the influence should only be in one direction. The most obvious example is the time dimension. Thus applications to unsteady problems use finite difference representations for the time derivatives. Thus a comparison of the three Galerkin methods really becomes a comparison of the spatial representation. This could be reinterpreted as a comparison between a global (spectral methods) and a local (finite element/difference methods) representation.

There is strong evidence in references 12 and 65 that where spectral methods can be used they are more accurate than finite difference techniques for comparable execution times but require considerably greater effort to code. This effort comes mainly from the need to transform to a local representation to handle the non-linear terms. Thus spectral methods would appear to be better suited to situations where the code is to be re-used many times rather than for solving one-off problems. They also suffer from the need to have very simple domain boundaries. Thus spectral methods find application for problems where the complexity comes from the physics rather than the boundary of the problem. Simulation of atmospheric phenomena on a global scale is an area where spectral methods are particularly effective.

Where the problem is dominated by the complexity of the domain boundary then finite element methods are the most suitable. Even where the boundary geometry is simple enough to allow non-Galerkin finite difference techniques to be used, finite element techniques can achieve results of comparable accuracy with superior economy because of the greater flexibility in refining the grid in areas where the flow variables are changing rapidly. However, finite element methods require more effort to code than do finite difference methods. Although, once a general finite element code has been written it usually has a wide range of applicability. General purpose finite element codes are commonplace in the structural field but only one general code for fluid dynamic applications(ref.48) is known to the author. Finite element methods should prove most effective for the problem of obtaining flow fields about complex bodies such as aircraft and rockets, where the numerical technique becomes a design tool for obtaining the changing force and moment distribution on the body due to detailed changes in the body geometry.

The main virtues of non-Galerkin finite difference methods are their ease of coding, economy and wide applicability. However, considerable refinement is required to obtain high accuracy or to cope with complex boundaries or boundary conditions.

Up to the present time much of computational fluid dynamics has had an essentially build-it-yourself flavour. It seems likely that the future will bring more general purpose codes that will be treated as tools in the sense that wind tunnels represent tools for experimentalists. If this is the case then the general purpose codes may well combine the virtues of the finite element, finite difference and spectral methods.

REFERENCES

No.	Author	Title
1	Galerkin, B.G.	"Rods and Plates. Series Occurring in Some Problems of Elastic Equilibrium of Rods and Plates". Vestnik Inzhenerov, Tech., Vol.19, 1915, pp 897-908
2	Anon.	"On the Seventieth Anniversary of the Birth of B.T. Galerkin". Prikl. Mat. Mech., Vol.5, 1941, pp 337-341
3	Duncan, W.J.	"Galerkin's Method in Mechanics and Differential Equations". ARC R&M 1798, 1937
4	Duncan, W.J.	"The Principles of Galerkin's Method". ARC R&M 1848, 1938
5	Kantorovich, L.V. and Krylov, V.I.	"Approximate Methods of Higher Analysis". Wiley, New York, 1958
6	Mikhlin, S.G.	"Variational Methods in Mathematical Physics". Pergamon, Oxford, 1964
7	Crandall, S.H.	"Engineering Analysis". McGraw-Hill, New York, 1956
8	Collatz, L.	"The Numerical Treatment of Differential Equations". Springer-Verlag, Berlin, 1960
9	Finlayson, B.A. and Scriven, L.E.	"The Method of Weighted Residuals - A Review". Applied Mechanics Reviews, Vol.19, 1966, pp 735-748
10	Finlayson, B.A.	"The Method of Weighted Residuals and Variational Principles". Academic Press, New York, 1972
11	Ames, W.F.	"Non linear Partial Differential Equations in Engineering". Vol.II, Academic Press, New York, 1972
12	Orszag, S.A. and Israeli, M.	"Numerical Simulation of Viscous Incompressible Flow" in "Annual Review of Fluid Mechanics, Vol.6". Annual Review Inc., Palo Alto Ca., 1974, pp 281-318
13	Biezeno, C.B.	"Graphical and Numerical Methods for Solving Stress Problems". Proc. 1st Int. Congr. Appl. Mech., Delft, 1924, pp 3-17

No.	Author	Title
14	Belotserkovskii, O.M. and Chushkin, P.I.	"The Numerical Solution of Problems in Gas Dynamics", in "Basic Developments in Fluid Dynamics". M. Holt (ed.), Academic Press, New York, pp 1-126
15	Pallone, A.	"Nonsimilar Solutions of the Compressible Laminar Boundary Layer Equations with Applications to the Upstream-Transpiration Cooling Problem". Jour. Aero. Sci., Vol.28, No.6, 1961, pp 449-456
16	Bartlett, E.P. and Kendall, R.M.	"Nonsimilar Solutions of the Multicomponent Laminar Boundary Layer by an Integral Matrix Method". NASA CR-1062, 1968
17	Jain, M.K.	"An Extremal Point Collocation Method for Fluid Flow Problems". Appl. Sci. Res., Vol.A11, 1962, pp 177-188
18	Villadsen, J.V. and Stewart, W.E.	"Solution of Boundary-Value Problems by Orthogonal Collocation". Chem. Eng. Sci., Vol.22, 1967, pp 1483-1501
19	Narasimha, R. and Deshpande, S.M.	"Minimum-Error Solutions of the Boltzmann Equations for Shock Structure". J. Fluid Mech., Vol.36, 1969, pp 555-570
20	Lynn, P.P.	"Least Squares Finite Element Analysis of Laminar Boundary Layer Flows". Int. J. Num. Meth. Eng., Vol.8, 1974, pp 865-876
21	Yamada, H.	"An Approximate Method of Integration of Laminar Boundary-Layer Equations". Rept. Res. Inst. Fluid Engng., Kyushu Univ., Vol.4, 1948, pp 27-42
22	Dorodnitsyn, A.A.	"General Method of Integral Relations and Its Application to Boundary Layer Theory". in "Advances in Aeronautical Sciences". Vol.3, Pergamon Press, New York, 1962, pp 207-219
23	Truckenbrodt, E.	"An Approximate Method for the Calculation of the Laminar and Turbulent Boundary Layer by Simple Quadrature for Two-Dimensional and Axially Symmetric Flow". J. Aero. Sci., Vol.19, 1952, pp 428-429
24	Murphy, J.D.	"Application of the Generalised Galerkin Method to the Computation of Fluid Flows". Proc. 1st AIAA Conf. on Computational Fluid Dynamics, Palm Springs, Ca., 1973

No.	Author	Title
25	Rosenhead, L.	"Laminar Boundary Layers". Clarendon Press, Oxford, 1963
26	Ames, W.F.	"Nonlinear Partial Differential Equations in Engineering". Academic Press, New York 1965, p 262
27	Shuleshko, P.	"A New Method of Solving Boundary-Value Problems of Mathematical Physics; Generalisations of Previous Methods of Solving Boundary-Value Problems of Mathematical Physics". Australian J. Appl. Sci., Vol.10, 1959, pp 1-7, 8-16
28	Bickley, W.G.	"Experiments in Approximating to Solutions of a Partial Differential Equation". Phil. Mag. (7), Vol.32, 1941, pp 50-66
29	Kawaguti, M.	"The Critical Reynolds Number for the Flow Past a Sphere". J. Phys. Soc. of Japan, Vol.10, No.8, 1955, pp 694-699
30	Krueger, E.R. and Di Prima, R.C.	"The Stability of a Viscous Fluid between Rotating Cylinders with an Axial Flow". J. Fluid Mech., Vol.19, 1964, pp 528-538
31	Snyder, L.J., Spriggs, T.W. and Stewart, W.E.	"Solution of the Equations of Change by Galerkin's Method". Am. Inst. Chem. Eng. J., Vol.10, No.4, 1964, pp 535-540
32	Snyder, L.J. and Stewart, W.E.	"Velocity and Pressure Profiles for Newtonian Creeping Flow in Regular Packed Beds of Spheres". Am. Inst. Chem. Eng. J., Vol.12, No.1, 1966, pp 167-173
33	Radin, S.H. and Mintzer, D.	"Orthogonal Polynomial Solution of the Boltzmann Equation for a Strong Shock Wave". Phys. of Fluids, Vol.9, No.9, 1966, pp 1621-1633
34	Fletcher, C.A.J. and Holt, M.	"An Improvement to the Method of Integral Relations". J. Comp. Phys., Vol.18, 1975, pp 154-164
35	Fletcher, C.A.J. and Holt, M.	"Supersonic Viscous Flow over Cones at Large Angles of Attack". J. Fluid Mech., Vol.74, Part 3, pp 561-591
36	Fletcher, C.A.J.	"Application of an Improved Method of Integral Relations to the Supersonic Boundary Layer Flow about Cones at Large Angles of Attack". Numerical Simulation of Fluid Motion, ed. John Noye, North-Holland, 1978, p 537

No.	Author	Title
37	Gear, C.W.	"The Automatic Integration of Ordinary Differential Equations". Comm. ACM., Vol.14, 1971, pp 176-179
38	Kaplan, S., Marlowe, O.J. and Bewick, J.	"Application of Synthesis Techniques to Problems Involving Time Dependence". Nuc. Sci. Engng., Vol.18, 1964, pp 163-176
39	Abramowitz, M. and Stegun, I.A. (eds).	"Handbook of Mathematical Functions". National Bureau of Standards, Washington D.C., 1964, p 916
40	Dryden, H.L. Murnaghan, F.P. and Bateman, H.	"Hydrodynamics". Dover, New York, 1956, p 197
41	Steven, G.P.	"On the Application of the Finite Element Method to Problems in Fluid Flow". Numerical Simulation of Fluid Motion, ed. John Noye, North-Holland, 1978, p 171.
42	Zienkiewicz, O.C.	"The Finite Element Method in Engineering Science". McGraw-Hill, London, 1971
43	Fletcher, C.A.J.	"The Application of the Finite Element Method to Inviscid Flow". WRE-TN-1606 (WR&D), May 1976
44	Oden, J.T.	"A General Theory of Finite Elements, I. Topological Consideration". Int. J. Num. Meth. Eng., Vol.1, 1969, pp 205-221
45	Leonard, J.W.	"Finite Element Analysis of Perturbed Compressible Flow". Int. J. Num. Meth. Eng., Vol.4, 1972, pp 123-132
46	Oden, J.T. and Wellford, L.C.	"Analysis of the Flow of Viscous Fluids by the Finite Element Method". AIAA J., Vol.10, 1972, pp 1590-1599
47	Zienkiewicz, O.C.	"Some Linear and Nonlinear Problems in Fluid Mechanics: FEM Formulation", in "Numerical Solution of Partial Differential Equations". J.G. Gram (ed.), Reidel Publ. Co., Holland, 1973, pp 173-194
48	Baker, A.J. and Zelazny, S.W.	"COMOC: Three Dimensional Boundary Region Variant, Theoretical Manual and Users Guide". NASA-CR-132450, 1974
49	Baker, A.J.	"A Finite Element Solution Algorithm for the Navier-Stokes Equations". NASA-CR-2391, 1974

No.	Author	Title
50	Baker, A.J.	"Finite Element Solution Theory for Three-dimensional Boundary Flows". Comp. Meth. App. Mech. Eng., Vol.4, 1974, pp 367-386
51	Baker, A.J.	"Predictions of Environmental Hydrodynamics Using the Finite Element Method - I - Theoretical Development". AIAA J., Vol.13, 1975, pp 36-42
52	Baker, A.J. and Zelazny, S.W.	"Predictions of Environmental Hydrodynamics Using the Finite Element Method - II - Applications". AIAA J., Vol.13, 1975, pp 43-46
53	Swartz, B. and Wendroff, B.	"Generalised Finite-Difference Schemes". Math. Comput., Vol.23, 1969, pp 37-49
54	Douglas, J. and Dupont, T.	"Galerkin Methods for Parabolic Equations". SIAM J. Num. Anal., Vol.4, 1970, pp 575-626
55	Douglas, J. and Dupont, T.	"Alternating Direction Galerkin Methods on Rectangles". Proc. Symposium on Numerical Solution of Partial Differential Equations, II, B. Hubbard (ed.), Academic Press, New York, 1971, pp 133-214
56	Douglas, J. and Dupont, T.	"Galerkin Methods for Parabolic Equations with Non-linear Boundary Conditions". Numer. Math., Vol.20, 1973, pp 213-237
57	Gourlay, A.R.	"Some Recent Methods for the Numerical Solution of Time Dependent Partial Differential Equations". Proc. Roy. Soc., Series A, Vol.323, 1971, pp 219-235
58	Dendy, J.E. and Fairweather, G.	"Alternating Direction Galerkin Methods for Parabolic and Hyperbolic Problems on Rectangular Polygons". SIAM J. Num. Anal., Vol.12, 1975, pp 144-163
59	Wahlbin, L.B.	"A Dissipative Galerkin Method for the Numerical Solution of First Order Hyperbolic Equations", in "Mathematical Aspects of Finite Elements in Partial Differential Equations". C. de Boor (ed.), Academic Press, New York, 1974, pp 147-169
60	Swartz, B.K.	"The Construction and Comparison of Finite Difference Analogues of Some Finite Element Schemes", in "Mathematical Aspects of Finite Elements in Partial Differential Equations". C. de Boor (ed.), Academic Press, New York, 1974, pp 279-213

No.	Author	Title
61	Thomee, V.	"Convergence Estimates for Semi-discrete Galerkin Methods for Initial Value Problems", in "Lecture Notes in Mathematics, No.333". Springer-Verlag, New York, 1973, pp 243-262
62	Orszag, S.A.	"Transform Method for the Calculation of Vector-Coupled Sums : Application to the Spectral Form of the Vorticity Equation". J. Atmos. Sci., Vol.27, 1970, pp 890-895
63	Orszag, S.A. and Kruskal, M.D.	"Formulation of the Theory of Turbulence". Phys. Fluids, Vol.11, 1968, pp 43-60
64	Orszag, S.A.	"Numerical Methods for the Simulation of Turbulence". Phys. Fluids Supplement II, Vol.12, pp 250-257
65	Orszag, S.A.	"Numerical Simulation of Incompressible Flows within Simple Boundaries; Accuracy". J. Fluid Mech., Vol.49, 1971, pp 75-112
66	Lorenz, E.N.	"Maximum Simplification of the Dynamic Equations". Tellus, Vol.12, 1960, pp 243-254
67	Platzman, G.W.	"The Spectral Form of the Vorticity Equation". J. Meteor., Vol.17, 1960, pp 635-644
68	Platzman, G.W.	"The Analytical Dynamics of the Spectral Vorticity Equation". J. Meteor., Vol.19, 1962, pp 313-328
69	Bourke, W.	"An Efficient, One-Level, Primitive Equation Spectral Model". Mon. Weather Rev., Vol.100, 1972, pp 683-689

Numerical Simulation of Fluid Motion, ed. John Noye
pp. 171-197

ON THE APPLICATION OF THE FINITE ELEMENT METHOD TO PROBLEMS IN FLUID FLOW

G.P. Steven

University of Sydney

New South Wales, Australia

SUMMARY

The purpose of this paper is to demonstrate the various means by which the differential equations governing incompressible viscous flow can be moulded into a suitable form for solution by the finite element process. It examines various solution criteria available at present and discusses the way that the independent variables are represented both in the solution domain and in subdivisions of that domain namely the finite elements.

To illustrate the process from start to finish two examples are given for simple flow processes where it is possible to put more on paper rather than in the computer.

1. INTRODUCTION

It is not the author's intention to give historical background to the finite element method (F.E.M.); the reader is referred to the bibliographic details attached to the reference listing at the end of this paper. The ground covered in this present work gives details of how field distributions of the various independent variables in a flow problem, such as velocity components and pressure are represented on a local basis within a subdivision of the solution domain, section 3. Section 4 presents some of the various governing principles for approximate solution and section 5 shows for two of these principles, namely Galerkin and Least Squares, how the mathematical formulation may be reduced to a format suitable for machine computation.

The path to any solution of the Navier-Stokes equations for incompressible viscous flow using the FEM contains many stages at which a choice between several alternatives need be made. It is hoped to indicate what the choices are and their merits and demerits. It is important to note that most of the information and mathematics is of a general nature and may be directly applied to any physical process which is governed by differential equations.

2. BASIC EQUATIONS

Although, as mentioned in the introduction, the nature of the work presented herein is of a general nature, we are concerned primarily at this conference with fluid flow and the equations that shall be looked at in this work are those governing unsteady two dimensional incompressible flow of a viscous fluid, namely the Navier-Stokes equations. These may be presented in two non-dimensionalised forms. Firstly in velocity and pressure format, where (u,v) are the cartesian (x,y) velocity components, p the pressure, t is time, R the Reynolds number and the subscript denotes differentiation with respect to the indicated variable.

$$\frac{1}{R}(u_{xx} + u_{yy}) - p_x - uu_x - vu_y - u_t = 0 \tag{1}$$

$$\frac{1}{R}(v_{xx} + v_{yy}) - p_y - uv_x - vv_y - v_t = 0 \tag{2}$$

$$u_x + v_y = 0. \tag{3}$$

Secondly in stream function (ψ) and vorticity (w) format,

$$\frac{1}{R}(w_{xx} + w_{yy}) - \psi_y w_x + \psi_x w_y - w_t = 0 \tag{4}$$

$$(\psi_{xx} + \psi_{yy}) + w = 0 \tag{5}$$

$$(p_{xx} + p_{yy}) - 2\psi_{xx}\psi_{yy} + 2\psi_{xy}\psi_{xy} = 0, \tag{6}$$

where $u = \psi_y, \quad v = -\psi_x, \quad w = v_x - u_y.$ (7)

3. ELEMENT INTERPOLATION

Before proceeding to elaborate on the various means by which the governing differential equations are reduced to a suitable form for computation of solutions, it would facilitate things if first we mention something about the "element" and what goes on inside it.

Ideas related to the representation of continuous distributions in a domain, be it one, two or multi dimensional, by a discrete number of subregions each with their own simple internal representation are familiar to all. For instance, a curve represented by a series of straight lines goes back many thousand years. In the finite element method we generalise these concepts of interpolation and derive the necessary associated algebra. It is not intended to present a full coverage of all that can be done to represent the variation of independent field variables within elements of the solution field, merely to give some of the basic ideas. At one end of the scale of representation we could have an element with many curved sides of arbitrary shape with an internal representation of very high order; while at the other end there could be a simple triangle with only linear variation within it. In the extreme, the high order element could be the whole solution domain and the solution process would be identical to classical methods for the problem under consideration. For the simple element, many hundreds of elements may be required to give adequate approximation to the true solution. In practice it has been found that many simple elements give better, quicker solutions than using a few complex ones.

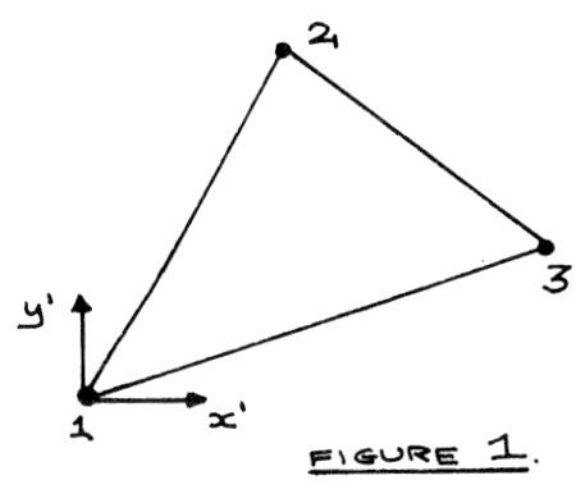

FIGURE 1.

Let us consider such a simple element, a triangle, figure 1, with corners numbered 1, 2 and 3, local coordinates x'y' and origin at 1. The independent variable χ will be given only linear variation within the element.

$$\text{i.e.}\quad \chi = a + bx' + cy' = [1\ x'\ y'] \begin{bmatrix} a \\ b \\ c \end{bmatrix} = [1\ x'\ y']\ \{T\} \tag{8}$$

Apply the boundary conditions of known χ values χ_1, χ_2 and χ_3 at the respective nodes, then,

$$\begin{bmatrix} \chi_1 \\ \chi_2 \\ \chi_3 \end{bmatrix} = \begin{bmatrix} 1 & 0 & 0 \\ 1 & x'_2 & y'_2 \\ 1 & x'_3 & y'_3 \end{bmatrix} \begin{bmatrix} a \\ b \\ c \end{bmatrix}$$

i.e. $\{\chi\}^e = [c]\{T\} \qquad \{T\} = [c]^{-1}\{\chi\}^e.$

Finally,

$$\chi = [1\ x'\ y']\ [c]^{-1}\{\chi\}^e = [N]\{\chi\}^e. \tag{9}$$

In words, a linear variation of χ can be represented by an interpolation matrix [N] (also referred to as the shape function) operating on a vector of nodal values $\{\chi\}^e$. The vector $\{\chi\}^e$ is a sub set of larger vector $\{\chi\}$ which represents the distribution of the variable χ throughout the whole solution domain. Any adjacent triangular element would have a similar shape function and consequently continuity of χ across element boundaries is ensured but there is no gradient compatibility. It should be noted that χ could by multi-dimensional, say in the case of the Navier Stokes equations χ would be $\{u\ v\ p\}$ or $\{\psi\ w\}$; the interpolation process would be the same.

All elements, into which the solution region is divided have their own unique vector $\{\chi\}^e$, although it is possible to select different shape functions from element to element. Besides simple polynomials there are other forms that the interpolating functions may have, among these are Legendre polynomials and area functions. The area functions, alternatively called natural coordinates are based on the ability to define any point within a triangle in terms of the areas it subtends with the nodes. Two dimensional solutions are implied here but the general principles are valid in any system of dimensions. Denoting $L_i = A_i/A$, figure 2, A being the total area, then it can be shown that;

$$\chi = L_1\chi_1 + L_2\chi_2 + L_3\chi_3 = [N]\{\chi\}^e \quad ..(10)$$

provides a linear interpolation function. In this the relationship between the area coordinates and the cartesian coordinates can be expressed as,

$$\Sigma L_i x'_i = x', \quad \Sigma L_i y'_i = y', \quad \Sigma L_i = 1. \quad ..(11)$$

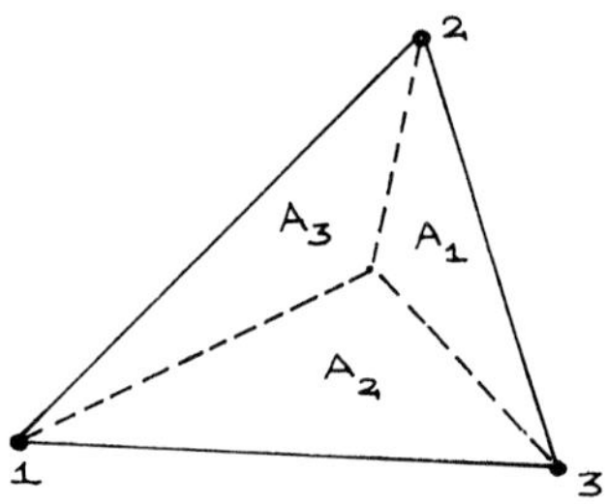

FIGURE 2

Solving these three equations in (11) would give the L's in terms of x' and y'.

As mentioned previously linear interpolation provides only a stepwise linear approximation to any continuous distribution. If a quadratic polynomial is used then there are six arbitrary coefficients in the quadratic expansion. Six element boundary conditions are obtained from values at the corners and another three from the mid points of the sides. For a single variable problem such an element is said to have six degrees of freedom (d.o.f.). To obtain the shape functions a similar process to that presented by equations (8) to (9) is conducted. In area coordinates the six components of the shape function matrix are;

$$N_i = L_i(2L_i - 1) \qquad i = 1, 2, 3 \text{ for corner nodes}$$

$$N_{i+3} = 4L_iL_{i+1} \qquad i = 1, 2, 3 \text{ if } i + 1 = 4 \Rightarrow i + 1 = 1 \text{ for mid side nodes.} \qquad (12)$$

It is possible to go on to higher order shape functions, but unless it is necessary to do so, such higher order functions have been shown to offer little advantage. One reason for using a higher order could be that there may be certain dependent variables which are higher order derivatives of independent variables and therefore the independent variable must have sufficiently high an order to be capable to producing the dependent variable. For instance the Navier-Stokes equations in stream function vorticity format, equations (4) and (5) could be solved. To determine pressure values it is necessary that at least second derivatives of stream function be non zero and consequently the stream function must be at least second order. Another reason for high order shape functions being adopted is that the criterion used for the solution of the governing differential equations may contain high order derivatives. Finally if gradient continuity across element boundaries were necessary or desirable then with a triangle there would be nine boundary conditions of one value of the variable and two gradients at each of three nodes giving nine d.o.f. This would require at least a cubic representation. The full cubic has ten coefficients and it is unwise to discard a term in the cubic polynomial just to suit a nine d.o.f. element. Instead an additional node at the centre of the element is introduced to give a ten d.o.f. element. Such an element is shown in figure 3.

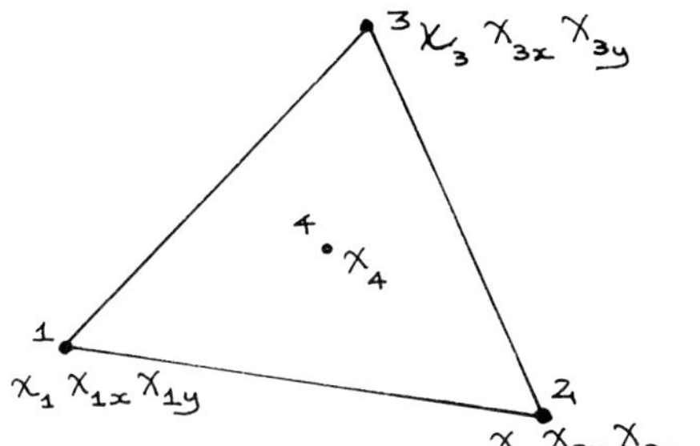

FIGURE 3

In conclusion, the contents of this section may be summarised by stating that, regardless of the nature of the problem under consideration or the shape of element used, it is possible to represent an approximation to the internal behaviour of the independent variable vector χ in terms of the shape function matrix and a vector of nodal values for each degree of freedom, i.e. $\chi = [N]\{\chi\}^e$.

4. SOLUTION CRITERIA

In this section we examine various criteria that have been presented or suggest themselves as ways of translating the equations of section 2 into computer programmes. The following techniques apply to the full N-S equations and consequently any simplified form of these equations for very high or very low speed could also be solved by them. The criteria fall into three main categories, namely, quasi-variational, weighted residual and other sensible schemes. We shall examine each in turn.

4.1 QUASI VARIATIONAL

Millikan [1] has shown that the Navier-Stokes equations for viscous, incompressible flow cannot be equivalent (except in some exceptional cases) to a true variational problem. However, Deshpande [2] obtains a system of functional differential equations which are equivalent to the N-S equations. If we define the following two integrals;

$$C(u,v,p) = \int_A \left[\frac{1}{2R}\left(u_x^{\,2} + u_y^{\,2} + v_x^{\,2} + v_y^{\,2}\right) + up_x + vp_y + \frac{1}{2}\left(u^2\right)_t + \frac{1}{2}\left(v^2\right)_t\right] dA, \tag{13}$$

$$D(u,v,p) = \int_A \left[\frac{1}{2}\left(u^2 v_x - v^2 u_y\right)\right] dA. \tag{14}$$

where A is the whole solution domain, then Deshpande has shown that the following variational form of these equations yields the N-S equations.

$$\frac{\partial C}{\delta p} = 0, \quad \frac{\delta C}{\delta u} = \frac{\delta D}{\delta v}, \quad \frac{\delta C}{\partial v} = -\frac{\delta D}{\delta u}. \tag{15}$$

It is to be noted that there are several forms that the quasi-functionals C and D may have. Using equations (13), (14) and (15) investigations are currently under way to establish the suitability of this approach.

Another quasi-variational technique is that of De Vries et al [3] [4] [5] where the vorticity is eliminated between (4) and (5) to yield,

$$\nabla^4\psi = R\left[\psi_x\nabla^2\psi_y - \psi_y\nabla^2\psi_x\right]. \tag{16}$$

The non-linearity is removed by setting the right hand side at that value from the previous iteration, where an iterative solution procedure is implied.

$$\nabla^4\psi^n = R\left[\psi_x\nabla^2\psi_y - \psi_y\nabla^2\psi_x\right]^{n-1} \tag{17}$$

On the basis of standard variational calculus where at the nth iteration the right hand side of (17) is a constant, the functional $I^n(\psi)$ associated with (17) is

$$I^n(\psi) = \int_A \left[\left(\psi_{xx} - \psi_{yy}\right)^2 + 4\psi^2_{xy}\right]^n dA - R\int_A \psi^n\left[\psi_x\nabla^2\psi_y - \psi_y\nabla^2\psi_x\right]^{n-1} dA. \tag{18}$$

It would be appropriate at some stage in this section to give some indication of how one proceeds from the solution criteria towards eventual computer solution and this equation (18) will be taken as an example. Dropping the superscript n, the quasi-functional, being a scalar, can be expressed as the sum of the contributions from a number of finite elements.

$$I = \sum_{\text{elements}} I^e. \tag{19}$$

Within the solution domain the independent variable can be expressed as a vector of point values $\{\psi\}$. For an approximate solution the best that can be done is to simultaneously minimise I with respect to $\{\psi\}$ appreciating that $\{\psi\}$ is not necessarily the exact solution.

$$\frac{\delta I}{\delta\{\psi\}} = \sum_{\text{elements}} \frac{\delta I^e}{\delta\{\psi\}} = 0 \tag{20}$$

But, in this, I^e is a function of that subset of $\{\psi\}$, namely $\{\psi\}^e$, only, and consequently,

$$\sum_{\text{elements}} \frac{\delta I^e}{\delta\{\psi\}^e} = 0 \tag{21}$$

The summation in (21) is termed "assembly" and the summed quantity for each element is the element characteristic which in this case will consist of an element "stiffness" matrix $[K]^e$, and an element "force" vector $\{F\}^e$ and thus (21) may be expressed as

$$\sum_{\text{elements}} [K]^e\{\psi\}^e = \sum_{\text{elements}} \{F\}^e \tag{22}$$

This last equation is in a suitable form for machine calculation. To go from (21) to (22) we substitute $\psi = [N]\{\psi\}^e$ into the expression for I, equation (18), and carry out the differentiation of (21) followed by integration over the area of the element. This yields expressions for $[K]^e$ and $\{F\}^e$ and,

$$[K]^e = \int_{A_e} \Big[[N_{xx} - N_{yy}]^T \; [N_{xx} - N_{yy}] \; + 4[N_{xy}]^T \; [N_{xy}] \Big] dA_e \tag{23}$$

$$\{F\}^e = \int_{A_e} [N]^T R\{\psi^{n-1}\}^{eT} \Big[[N_x]^T [N_{xx} + N_{yy}]_y - [N_y]^T [N_{xx} + N_{yy}]_x \Big] \{\psi^{n-1}\}^e dA_e \tag{24}$$

where the superscript T denotes the transpose of the matrix enclosed in square brackets. Note that the level of differentiation in (24) implies that ψ must be represented by, at least, a cubic polynomial within the element.

A third quasi-variational approach is that of Bratanow et al [6] [7] [8] [9] which is based on the following three functionals,

$$I_1(w,\psi) = \int_A [w \cdot w_t + \tfrac{1}{2}\{\psi_y w_x - \psi_x w_y\} w + \frac{1}{2R}\{w_x^{\,2} + w_y^{\,2}\}] dA \tag{25}$$

$$I_2(w,\psi) = \int_A [\tfrac{1}{2}\{\psi_x^{\,2} + \psi_y^{\,2}\} - w\psi] \; dA \tag{26}$$

$$I_3(w,\psi,p) = \int_A [\tfrac{1}{2}\{p_x^{\,2} + p_y^{\,2}\} + 2p\{\psi_{yy}\psi_{xx} - \psi_{xy}\}^2]. \tag{27}$$

The first two of these are independent of pressure and could be solved initially for w and ψ while the pressure could be determined afterwards using the third. The first function is not an exact function but is based upon a linearisation of the convective term in the vorticity transport equation (4) whereby a Taylor series expansion of the velocity field is expressed at any point of time as,

$$\psi - \psi_0 + (\delta\psi/\delta w)_0 dw + \tfrac{1}{2}(\delta^2\psi/\delta w^2)_0 dw^2 + \ldots \tag{28}$$

There are some doubts about the validity of the first functional but nevertheless this technique has produced a vast range of results for velocity and pressure fields around two and three dimensional (suitably amended) steady or unsteady lifting surfaces.

4.2 WEIGHTED RESIDUAL

There is a large range of procedures for the approximate solution to field problems which come under the title of weighted residual. The reader should refer to Finlayson and Scriven [10] for a review of the topic. Two of the more common weighted residual techniques are Least Squares and Galerkin and these have found extensive usage in attempted solutions to the Navier-Stokes equations.

If the governing differential equations are abbreviated to,

$$L(\chi) - f = 0, \tag{29}$$

where L is a matrix differential operator, χ a vector of independent variables and f some "load" vector, then for an approximate solution χ^*, the residual R is defined as,

$$R = L(\chi^*) - f. \tag{30}$$

The least squares criterion is such that,

$$\int_A R^2 dA = \sum_{\text{elements}} \int_{A_e} R^{e^2} dA_e \tag{31}$$

be a minimum. This approach with some modification has been used by Lynn and Arya [11] [12] and Milthorpe [13].

In the Galerkin approach to finite element solutions we use the local representation of the approximate solution within an element described in section §3, i.e. $\chi^* = [N]\{\chi^*\}^e$. Applying this to the residual on an elemental basis the Galerkin criterion is,

$$\sum_{\text{elements}} \int_{A_e} R^e [N] \, dA_e = 0. \tag{32}$$

Here the weighting function is the shape function rather than the residual as in the least squares criterion. This approach was suggested by Zienkiewicz [12] and has been used with some success by Leonard [11], Oden [15] [16], Tanner et al [17] and Baker [18] [19] [20], and many more.

4.3 OTHER TECHNIQUES

Taylor and Hood [22] have used a solution procedure which appears to give good results. Having prescribed initial velocity boundary conditions and guessed field values of vorticity the equation,

$$\nabla^2\psi = - w,$$

is solved and the boundary values of w are updated. Then $\nabla^2 w = 0$ is solved and the cycle given by these two equations is allowed to carry on until it converges. Upon convergence another iterative cycle is started using,

$$\nabla^2\psi = - w$$

$$\frac{1}{R} \nabla^2 w - \psi_y w_x + \psi_x w_y = 0,$$

until eventually a solution is reached. This process has yielded results for cavity flow and flow around a cylinder at low Reynolds number.

5. DETAILED EXAMPLES

Here we select two of the techniques mentioned in the previous section and consider them in more detail. Also by simplifying the governing equations a little it is possible to take this illustrative process further on paper than would normally be the case. Firstly we look at the Galerkin approach.

5.1 GALERKIN APPROACH

In equation (28) the vector {f} is zero for the Navier-Stokes equations and we shall look at the terms involved in the assembly process of equation (32); namely,

$$\int_{A_e} [N]^T L(\chi^*)\, dA_e. \tag{33}$$

The vector of the independent variable for the velocity, pressure formulation can be represented within an element as,

$$\{\chi^*\} = \begin{bmatrix} u \\ v \\ p \end{bmatrix} = [N]\{\chi^*\}^e = [N] \begin{bmatrix} \{u\}^e \\ \{v\}^e \\ \{p\}^e \end{bmatrix} \tag{34}$$

Galerkin's technique has the inherent simplicity that the weighting functions can be made the same as the shape functions. Matrix algebra dictates that the transpose of N be used in (33).

Green's theorem may now be applied to (33) to reduce the order of differentiation in L and consequently reduce the order of the interpolating functions. For brevity, matrix notation has been partially dropped in the following equations where N represents [N] and u^e represents $\{u\}^e$, the subscript still denotes differentiation.

$$\int_{A_e} \left\{ N^T N u^e N_x u^e + N^T N v^e N_y u^e + N^T N \dot{u}^e \right\} dA_e$$
$$+ \int_{A_e} \left[\frac{1}{R}\left(N_x^T N_x + N_y^T N_y\right) \quad (0) \quad \left(N^T N_x\right)\right] \chi^e dA_e = \int_s \frac{1}{R} N^T \frac{\delta u}{\delta n}\, ds. \tag{35}$$

$$\int_{A_e} \left\{ N^T N u^e\, N_x v^e + N^T N v^e N_y v^e + N^T N \dot{v}^e \right\} dA_e$$
$$+ \int_{A_e} \left[(0) \quad \frac{1}{R}\left(N_x^T N_x + N_y^T N_y\right) \quad \left(N^T N_y\right)\right] \chi^e dA_e = \int_s \frac{1}{R} N^T \frac{\delta v}{\delta n}\, ds. \tag{36}$$

$$\int_{A_e} \left[\left(N^T N_x\right) \quad \left(N^T N_y\right) \quad (0) \right] \chi^e dA_e = 0. \tag{37}$$

In (35) and (36) the first integrals are the non linear and transient terms while the second integrals together with (37) are the linear parts in matrix form. The right hand sides come from the application of Green's theorem and represent some boundary loading where the coordinate s is around the contour and n is the normal to the boundary. However, in the majority of problems in fluid mechanics, the boundaries are selected naturally such that either $\delta u/\delta n$ and $\delta v/\delta n$ are zero or u and v are known and thereby these right hand sides need not be established in the formulation of the solution routine.

The presence of the non-linear terms would lead to a set of quadratically non linear algebraic equations once the form of [N] were decided and the integrations over the element carried out. It is the opinion of the author that the use of available means of solving these non-linear equations, such as Newton-Raphson, conjugate gradient or search techniques are not valid in this present context. There may be several thousand degrees of freedom for the whole solution domain. This would involve, in the Newton-Raphson technique, the evaluation of a Jacobian of great size with prohibitive calculation time and high probability of poor accuracy.

Thus the use of an iterative scheme is indicated whereby the equations are linearised a prioré. Here again there is a range of ways by which this linearisation may be achieved. One such scheme could be to express the non linear terms, in equation (35) say, as

$$\int_{A_e} \left\{ N^T N u^e_{n-1} N_x u_n^e + N^T N v^e_{n-1} N_y u_n^e \right\} dA_e, \tag{38}$$

where the subscript n or $n-1$ tells which cycle of the iterative process is being used. Consequently the non-linearity is removed by having certain nodal velocity vectors set constant at the level of the previous iteration. For an unsteady problem the linearisation above can be incorporated into a simple differencing time stepping scheme whereby,

$$\{\dot{u}\}^e = \left(\{u\}_n^e - \{u\}^e_{n-1} \right) / \Delta t, \tag{39}$$

where Δt is a time increment and each iterative cycle represents a point of time starting from some initial condition.

Alternatively, it is possible to take the whole of the expression in (38) as based on the values at the $(n-1)$th iteration. For simplicity of the example presented herein this alternative proposal is adopted, but it should be noted that this is not as good as the original one.

The time stepping scheme mentioned above is of the simplest form. More complex temporal representations are available and go up to cubic time interpolation functions with "leap-frog" facilities which enable much larger time increments to be used, see Desai and Abel [B1]. It has been shown in the case of transient heat conduction by Wood and Lewis [21] that simple time stepping algorithms are more effective. Finally on this point concerning time, Oden [B3] suggests an approach whereby time is treated as another field dimension and incorporated in the shape functions.

Returning to the example in hand, we are now at the stage of deciding on the shape of the finite element to be used and the form of its shape function. The element adopted is a quadrilateral. Within the element linear changes of the field variables are used and an "isoparametric" transformation is utilised to transform the local x'y' plane into a $\zeta\eta$ plane, see figure 4 on the next page. These isoparametric transformations are such that the shape functions within the element are identical to the transformation from the x'y' plane to $\zeta\eta$ plane. The $\zeta\eta$ plane is a square bounded by $-1 < (\zeta\eta) < 1$. Clearly if the evaluation of the integrals involved in (35) (36) and (37) is very complex then it would be much easier to do them numerically in the transformed $\zeta\eta$ plane; this is

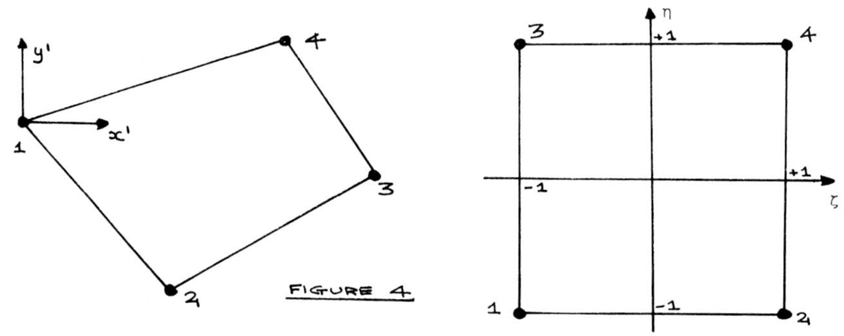

FIGURE 4

one of the main reasons for isoparametric transformations. The other main reason for the transformation is that when a quadratic or higher order shape function is used the sides of the element in the x'y' plane can be curved.

The appropriate linear shape functions in this example are,

$$\begin{aligned} N_1 &= \tfrac{1}{4}(1-\zeta)(1-\eta) \\ N_2 &= \tfrac{1}{4}(1+\zeta)(1-\eta) \\ N_3 &= \tfrac{1}{4}(1-\zeta)(1+\eta) \\ N_4 &= \tfrac{1}{4}(1+\zeta)(1+\eta) \end{aligned} \tag{40}$$

with $u = [N]\{u\}^e$, $v = [N]\{v\}^e$, $p = [N]\{p\}^e$ or $\chi^* = [N]\{\chi^*\}^e$. The characteristics of the shape functions so described, and indeed all shape functions of this type are,

$$\sum_{i=1}^{4} N_i = 1, \quad \sum_{i=1}^{4} N_i x'_i = x', \quad \sum_{i=1}^{4} N_i y'_i = y' \tag{41}$$

In order to develop expression for the element characteristic matrices as indicated by the integrations in (34) (35) and (36) it is required to evaluate various first derivatives of [N]. It can be shown that,

$$\begin{bmatrix} \dfrac{\delta N_r}{\delta\zeta} \\ \dfrac{\delta N_r}{\delta\eta} \end{bmatrix} = \begin{bmatrix} \dfrac{\delta x}{\delta\zeta} & \dfrac{\delta y}{\delta\zeta} \\ \dfrac{\delta x}{\delta\eta} & \dfrac{\delta y}{\delta\eta} \end{bmatrix} \begin{bmatrix} \dfrac{\delta N_r}{\delta x} \\ \dfrac{\delta N_r}{\delta y} \end{bmatrix} = [J] \begin{bmatrix} \dfrac{\delta N_r}{\delta x} \\ \dfrac{\delta N_r}{\delta y} \end{bmatrix}, \tag{42}$$

where using (40) and (41) the Jacobian is,

$$J = \tfrac{1}{4}\begin{bmatrix} \{(1-\eta)(x_2-x_1) + (1+\eta)(x_4-x_3)\} & \{(1-\zeta)(x_3-x_1) + (1+\zeta)(x_4-x_2)\} \\ \{(1-\eta)(y_2-y_1) + (1+\eta)(y_4-y_3)\} & \{(1-\zeta)(y_3-y_1) + (1+\zeta)(y_4-y_2)\} \end{bmatrix}. \tag{43}$$

The integrations involve terms such as,

$$\int_{A_e} \left(\frac{\delta N_i}{\delta x} \cdot \frac{\delta N_j}{\delta x} + \frac{\delta N_i}{\delta y} \cdot \frac{\delta N_j}{\delta y} \right) dA_e. \tag{44}$$

Transforming to the $\zeta\eta$ plane this integral becomes,

$$\int_{-1}^{+1}\int_{-1}^{+1}\left\{\frac{\delta N_i}{\delta\zeta}\cdot\frac{\delta N_j}{\delta\zeta}\;[J_{22}{}^2 + J_{21}{}^2] + \frac{\delta N_i}{\delta\eta}\cdot\frac{\delta N_j}{\delta\eta}\;[J_{12}{}^2 + J_{11}{}^2]\right.$$

$$\left. - \left[\frac{\delta N_i}{\delta\zeta}\cdot\frac{\delta N_j}{\delta\eta} + \frac{\delta N_i}{\delta\eta}\cdot\frac{\delta N_j}{\delta\zeta}\right]\cdot[J_{22}J_{12} + J_{11}J_{21}]\right\}\frac{d\zeta d\eta}{|J|}\;. \tag{45}$$

Despite the apparent complexity of this last expression and the corresponding ones for the other terms involved in (35) (36) and (37), it can be calculated numerically using a Gaussian scheme. Nine Guassian points have been shown to give sufficient accuracy.

To illustrate the above process through to a logical conclusion of solving some physical flow situation a simple square element of side unity is taken. Under these circumstances the stiffness matrix for the linear terms is shown below in equation (46).

$$\left[\begin{array}{cccc|cccc|cccc}
8 & -2 & -2 & -4 & & & & & -2R & 2R & -R & R \\
-2 & 8 & -4 & -2 & & 0 & & & -2R & 2R & -R & R \\
-4 & -4 & 8 & -2 & & & & & -R & R & -2R & 2R \\
-4 & -2 & -2 & 8 & & & & & -R & R & -2R & 2R \\
\hline
 & & & & 8 & -2 & -2 & 4 & -2R & -R & 2R & R \\
 & 0 & & & -2 & 8 & -4 & -2 & -R & -2R & R & 2R \\
 & & & & -2 & -4 & 8 & -2 & -2R & -R & 2R & R \\
 & & & & -4 & -2 & -2 & 8 & -R & -2R & R & 2R \\
\hline
-2R & 2R & -R & R & -2R & -R & 2R & R & & & & \\
-2R & 2R & -R & R & -R & -2R & R & 2R & & 0 & & \\
-R & R & -2R & 2R & -2R & -R & 2R & R & & & & \\
-R & R & -2R & 2R & -R & -2R & R & 2R & & & &
\end{array}\right]\left[\begin{array}{c} u_1 \\ u_2 \\ u_3 \\ u_4 \\ \hline v_1 \\ v_2 \\ v_3 \\ v_4 \\ \hline p_1 \\ p_2 \\ p_3 \\ p_4 \end{array}\right] \tag{46}$$

The contributions from the non-linear terms are presented in appendix A, and the transient terms will be mentioned later.

If we consider steady Couette-Poiseville flow between two parallel surfaces using eight elements, figure 5 opposite, then $v = \frac{\delta u}{\delta x} = \frac{\delta p}{\delta y} = 0$, $\frac{\delta p}{\delta x} = \Delta p$. The non linear terms disappear from the governing differential equations and the stiffness matrix for the linear terms can be extracted from (46) by the removal of all rows and columns associated with v. Carrying out the assembly process indicated in equation (32), the final assembled system of equations for solution are,

$$[H]\{u\} = -\{I\}\frac{R}{2}\Delta p + \{M\}\frac{U}{2} = \{F\}\;, \tag{47}$$

where $\{I\}$ is a unit vector and $\{M\}$ is a zero vector except for unit values associated with nodes adjacent to the moving surface, i.e. nodes 8 and 17 in figure 5. These are solved for various Δp and the values of

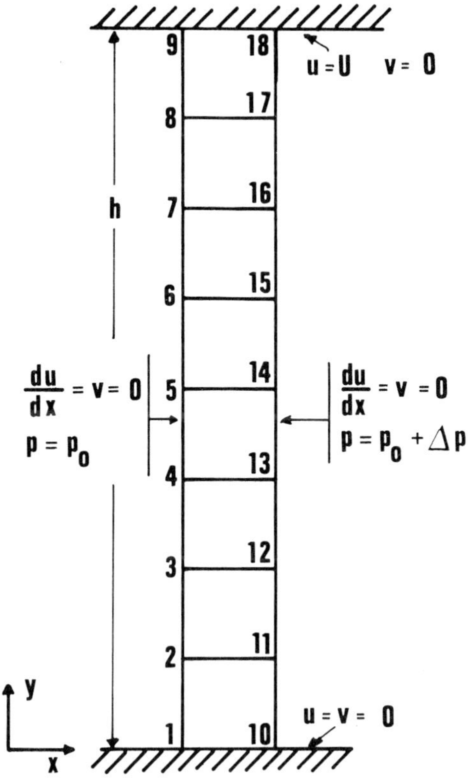

Figure 5 - Arrangement of eight elements and boundary conditions for flow between parallel surfaces.

$\{u\}$ together with the exact solution,

$$u = U\frac{y}{h} - \frac{R}{2}\ \Delta p\ (hy - y^2)\ ,$$

are presented in figure 6 from which the good correspondence is noted.

Dropping the pressure gradient term but including the transient term for this problems enables us to study transient Couette flow. The contribution from the transient term associated with u can be expressed from equation (35) as,

$$\int_{A_e} N^T N\dot{u}^e dA_e = \int_{-1}^{+1}\int_{-1}^{+1} N^T N\dot{u}^e |J|\, d\zeta d\eta = [K]^e\ \{\dot{u}\}^e\ , \tag{48}$$

and for the simple square element with shape functions as described in (40) the expression $[K]^e\{\dot{u}\}^e$ is

$$\frac{1}{36}\begin{bmatrix} 4 & 2 & 2 & 1 \\ 2 & 4 & 1 & 2 \\ 2 & 1 & 4 & 2 \\ 1 & 2 & 2 & 4 \end{bmatrix}\begin{bmatrix} \dot{u}_1 \\ \dot{u}_2 \\ \dot{u}_3 \\ \dot{u}_4 \end{bmatrix} \tag{49}$$

Adopting the time stepping scheme of equation (39) whereby,

$$\{\dot{u}\} = \left(\{u\}_{t+\Delta t} - \{u\}_t\right)\ \Delta t,$$

it is reasonable to represent $\{u\}$ as,

$$\{u\} = \left(\{u\}_{t+\Delta t} + \{u\}_t\right)\ 2,$$

thus upon assembly, the system of linear equations for this flow system can be presented as,

$$\left[\frac{[K]}{\Delta t} + \frac{[H]}{2}\right]\{u\}_{t+\Delta t} = \{F\} + \left[\frac{[K]}{\Delta t} - \frac{[H]}{2}\right]\{u\}_t . \tag{50}$$

If one surface has the initial condition at $t = 0$ that $u = U$ while the rest of the fluid is at rest then the starting vector of $\{u\}$ is all zero. Also $\{F\}$ is the term $\{M\}U/2$ in equation (47). Using eight elements a sample of the transient behaviour is presented in figure 7 and from this the expected behaviour is observed. For small time steps the solution presented herein is very close to the analytical solution for such a case.

5.2 LEAST SQUARES APPROACH

In this example to simplify things it shall be taken, a prioré, that we are looking at steady inviscid irrotational flow, i.e. potential flow. From equations (4) and (5) we have,

$$\nabla^2\psi = 0. \tag{51}$$

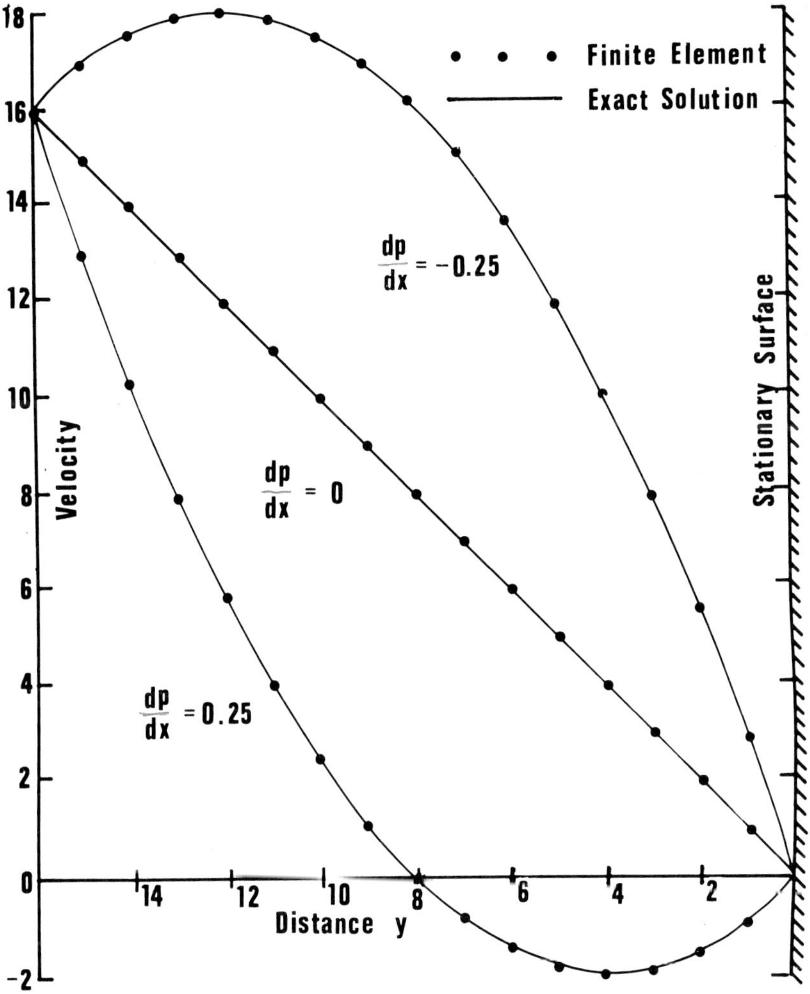

Figure 6 - Solution for Couette-Poiseuille flow, steady state, 16 elements used, various pressure gradients.

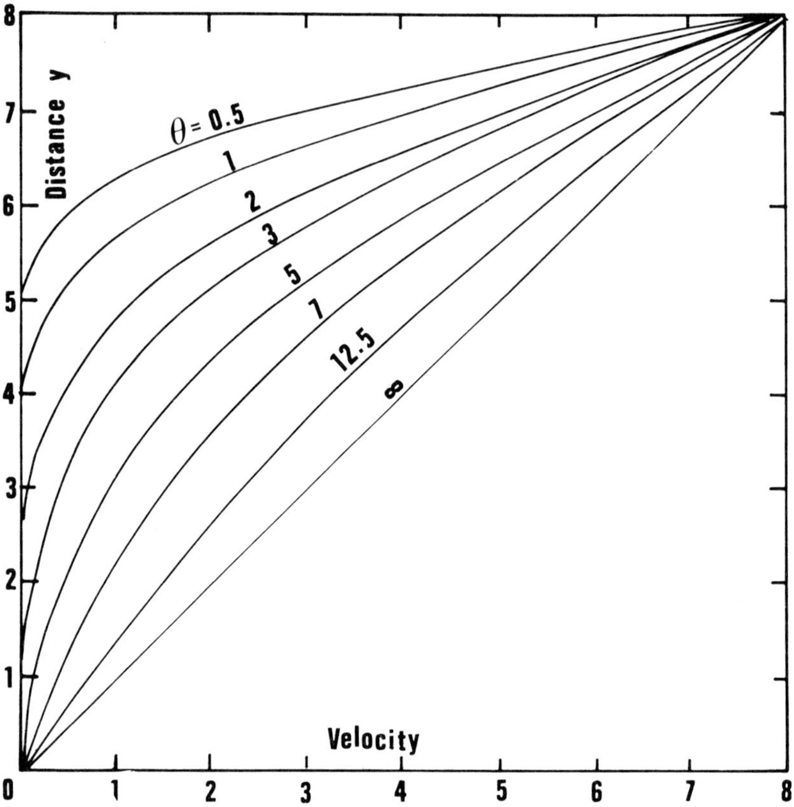

Figure 7 - Solution for transient Couette flow, $\theta = t/R$, 8 elements used.

For illustrative purposes the auxiliary functions $\alpha = \delta\psi/\delta x$ and $\beta = \delta\psi/\delta y$ are introduced so that (51) becomes three first order equations rather than one second order equation.

$$\alpha - \frac{\delta\psi}{\delta x} = 0$$

$$\beta - \frac{\delta\psi}{\delta y} = 0$$

$$\frac{\delta\alpha}{\delta x} + \frac{\delta\beta}{\delta y} = 0$$

i.e.

$$\begin{bmatrix} 1 & 0 & -\frac{\delta}{\delta x} \\ 0 & 1 & -\frac{\delta}{\delta y} \\ \frac{\delta}{\delta x} & \frac{\delta}{\delta y} & 0 \end{bmatrix} \begin{bmatrix} \alpha \\ \beta \\ \psi \end{bmatrix} = [L]\{\chi\} = 0 \tag{52}$$

The least squares criterion tells us that if χ^* is an approximation to χ then for the best solution possible,

$$\int_A ([L]\{\chi^*\})^2 dA, \tag{53}$$

shall be a minimum. Clearly such a minimum may be achieved by simultaneously differentiating (53) with respect to a vector of field values of χ^*.

$$\frac{\delta(53)}{\delta\chi^*} = \frac{\delta}{\delta\chi^*} \sum_{\text{elements}} \int_{A_e} ([L]\{\chi^*\}^e)^2 dA_e$$

$$= \sum_{\text{elements}} \frac{\delta}{\delta\chi^*} \int_{A_e} ([L]\{\chi^*\}^e)^2 dA_e$$

$$= \sum_{\text{elements}} \frac{\delta}{\delta\chi^{*e}} \int_{A_e} ([L]\{\chi^*\}^e)^2 dA_e \tag{54}$$

Since only the subset χ^{*e} of the total field vector for χ^* is involved in representing the residual within an element, then the last step towards (54) should be clear. If the variable χ^{*e} is interpolated as $\chi^{*e} = [N]\{\bar{\chi}^*\}^e$ then let,

$$[L]\ [N]\{\bar{\chi}^*\}^e = [K]\ \{\bar{\chi}^*\}^e,$$

and (54) becomes,

$$\sum_{\text{elements}} \int_{A_e} [K]^T[K]\{\bar{\chi}^*\}^e dA_e = 0 \tag{55}$$

The expression,

$$\int_{A_e} [K]^T[K]dA_e = [S] \tag{56}$$

is the element characteristic or stiffness matrix.

As in the previous example it is now time to select an element and associated shape functions. Once again the well favoured triangle is adopted within which the stream function is represented by quadratic area functions,

$$\psi = L_1(2L_1-1)\psi_1 + L_2(2L_2-1)\psi_2 + L_3(2L_3-1)\psi_3 + 4L_1L_2\psi_4 + 4L_2L_3\psi_5 + 4L_3L\psi_6 \ ,$$

and α and β are represented by linear area functions,

$$\alpha = L_1\alpha_1 + L_2\alpha_2 + L_3\alpha_3,$$

$$\beta = L_1\beta_1 + L_2\beta_2 + L_3\beta_3.$$

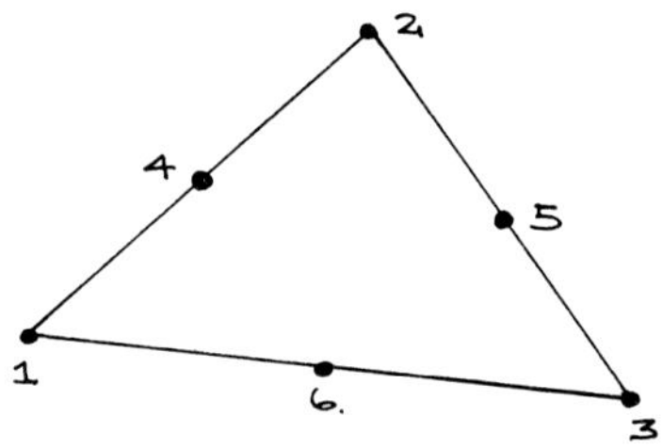

i.e.

$$\begin{bmatrix} \alpha \\ \beta \\ \psi \end{bmatrix} = \begin{bmatrix} L_1 & 0 & 0 & L_2 & 0 & 0 & L_3 & 0 & 0 & 0 & 0 & 0 \\ 0 & L_1 & 0 & 0 & L_2 & 0 & 0 & L_3 & 0 & 0 & 0 & 0 \\ 0 & 0 & L_1(2L_1-1) & 0 & 0 & L_2(2L_2-1) & 0 & 0 & L_3(2L_3-1) & 4L_1L_2 & 4L_2L_3 & 4L_3L_1 \end{bmatrix} \begin{bmatrix} \alpha_1 \\ \beta_1 \\ \psi_1 \\ \alpha_2 \\ \beta_2 \\ \psi_2 \\ \alpha_3 \\ \beta_3 \\ \psi_3 \\ \psi_4 \\ \psi_5 \\ \psi_6 \end{bmatrix}$$

i.e. $\{\chi^*\} = [N]\{\bar{\chi}^*\}^e.$

In order to evaluate [S], equation (56), we require from equation (11),

$$\frac{\delta L_i}{\delta x} = \frac{b_i}{2\Delta} = y'_{i+1} - y'_{i+2} \ ,$$

$$\frac{\delta L_i}{\delta y} = \frac{c_i}{2\Delta} = x'_{i+2} - x'_{i+1} \ ,$$

where if $i + 1$ or $i + 2$ are greater than 3 then $i + 1 = i + 1 - 3$, $i + 2 = i + 2 - 3$, and

$$2\Delta = \begin{vmatrix} 1 & x'_1 & y'_1 \\ 1 & x'_2 & y'_2 \\ 1 & x'_3 & y'_3 \end{vmatrix}.$$

The whole process of integration over the element is greatly facilitated by use of the expression,

$$\int_{A_e} L_1^{\,a}\, L_2^{\,b}\, L_3^{\,c} = \frac{a!b!c!}{(a+b+c+2)!} \cdot 2\Delta \,. \qquad (57)$$

Indeed, this last equation is so useful that it is one of the main reasons for the adoption of area coordinates.

Without details of the algebraic process, the lower triangular portion of the stiffness matrix is shown on the next page. The symmetrical nature of the stiffness matrices produced by the least squares procedure is an advantage over the unsymmetrical ones produced by the Galerkin technique especially in terms of computer storage space.

The assembly process indicated in equation (55) consists of overlaying each element matrix onto the global stiffness matrix and is simply handled in the computer. Boundary conditions are treated as follows. Consider that we have assembled all the element stiffness matrices to give a global stiffness matrix $[S^A]$. Let this be partitioned as,

$$\left[\begin{array}{c|c} S^A_{pp} & S^A_{pq} \\ \hline S^A_{qp} & S^A_{qq} \end{array}\right] \left[\begin{array}{c} \chi^*_p \\ \hline \chi^*_q \end{array}\right] = 0 \,,$$

where p terms are unknown values and q terms are known boundary values. Multiply the first row of this to get,

$$\left[S^A_{pp}\right] \{\chi^*_p\} = - \left[S^A_{pq}\right] \{\chi^*_q\} \,,$$

and it is a simple matter to solve for $\{\chi^*_p\}$. In the computer it is unnecessary to interchange rows and columns and the following brief segment of FORTRAN illustrates the assembly process.

```
        DO    400    M = 1, 12
        L = IEROW(M)
420     IF  (L)   400, 400, 421
421     DO  500  I = 1, 12
        K = IEROW(N)
        IF  (K)   430, 430, 440
430     RHS(L) = RHS(L) - ESTIF(M,N) * BVALUE(K)
        GOTO  500
440     IF (K.LT.L)   GO TO 500
        K = K - L + 1
        GSTIF (L,K)  =  GSTIF(L,K) + ESTIF(M,N)
500     CONTINUE
400     CONTINUE
```

$$
\Delta \cdot \left[\begin{array}{cccccccccccc}
\frac{1}{6}+b_1^2 & & & & & & & & & & & \\
b_1c_1 & \frac{1}{6}+c_1^2 & & & & & & & & & & \\
-b_1/3 & -c_1/3 & b_1^2+c_1^2 & & & & & & \text{SYMMETRICAL.} & & & \\
\frac{1}{12}+b_1b_2 & b_2c_1 & 0 & \frac{1}{6}+b_2^2 & & & & & & & & \\
b_1c_2 & \frac{1}{12}+c_1c_2 & 0 & b_2c_2 & \frac{1}{6}+c_2^2 & & & & & & & \\
0 & 0 & -\frac{1}{3}(b_1b_2+c_1c_2) & -b_2/3 & -c_2/3 & b_2^2+c_2^2 & & & & & & \\
\frac{1}{12}+b_1b_3 & b_3c_1 & 0 & \frac{1}{12}+b_2b_3 & c_2b_3 & 0 & \frac{1}{6}+b_3^2 & & & & & \\
b_1c_3 & \frac{1}{12}+c_1c_3 & 0 & b_2c_3 & \frac{1}{12}+c_2c_3 & 0 & b_3c_3 & \frac{1}{6}+c_3^2 & & & & \\
0 & 0 & -\frac{1}{3}(b_1b_3+c_1c_3) & 0 & 0 & -\frac{1}{3}(b_2b_3+c_2c_3) & -b_3/3 & -c_3/3 & b_3^2+c_3^2 & & & \\
-(\frac{2}{3}b_2+\frac{1}{3}b_1) & -(\frac{2}{3}c_2+\frac{1}{3}c_1) & \frac{4}{3}(b_1b_2+c_1c_2) & -(\frac{1}{3}b_2+\frac{2}{3}b_1) & -(\frac{1}{3}c_2+\frac{2}{3}c_1) & \frac{4}{3}(b_1b_2+c_1c_2) & -(\frac{1}{3}b_2+\frac{1}{3}b_1) & -(\frac{1}{3}c_2+\frac{1}{3}c_1) & 0 & \begin{array}{c}\frac{8}{3}(b_1^2+b_2^2+b_1b_2\\+c_1^2+c_2^2+c_1c_2)\end{array} & & \\
-(\frac{1}{3}b_3+\frac{1}{3}b_2) & -(\frac{1}{3}c_3+\frac{1}{3}c_2) & 0 & -(\frac{2}{3}b_3+\frac{1}{3}b_2) & -(\frac{2}{3}b_3+\frac{1}{3}c_2) & \frac{4}{3}(b_2b_3+c_2c_3) & -(\frac{1}{3}b_3+\frac{2}{3}b_2) & -(\frac{1}{3}c_3+\frac{2}{3}c_2) & \frac{4}{3}(b_2b_3+c_2c_3) & \begin{array}{c}\frac{8}{6}(b_2b_3+b_2^2+2b_1b_3\\+b_1b_2+c_2c_3+c_2^2\\+2c_1c_3+c_1c_2)\end{array} & \begin{array}{c}\frac{8}{3}(b_2^2+b_3^2+b_2b_3\\+c_2^2+c_3^2+c_2c_3)\end{array} & \\
-(\frac{1}{3}b_1+\frac{2}{3}b_3) & -(\frac{1}{3}c_1+\frac{2}{3}c_2) & \frac{4}{3}(b_1b_3+c_1c_3) & -(\frac{1}{3}b_1+\frac{1}{3}b_3) & -(\frac{1}{3}c_1+\frac{1}{3}c_3) & 0 & -(\frac{2}{3}b_1+\frac{1}{3}b_3) & -(\frac{2}{3}c_1+\frac{1}{3}c_3) & \frac{4}{3}(b_1b_3+c_1c_3) & \begin{array}{c}\frac{8}{6}(b_2b_1+b_1^2+2b_2b_3\\+b_1b_3+c_2c_1+c_1^2\\+2c_2c_3+c_1c_3)\end{array} & \begin{array}{c}\frac{8}{6}(b_3b_1+b_3^2+2b_1b_2\\+b_2b_3+c_3c_1+c_3^2\\+2c_1c_2+c_2c_3)\end{array} & \begin{array}{c}\frac{8}{3}(b_3^2+b_1^2+b_3b_1\\+c_3^2+c_1^2+c_3c_1)\end{array}
\end{array}\right]
$$

IEROW - vector of row numbers in glodal stiffness matrix associated with every d.o.f. If boundary value IEROW is negative, if unknown value IEROW is positive.

RHS - vector of right hand sides for solution subroutine.

BVALUE - vector containing boundary value associated with IEROW negative.

ESTIF - 12 x 12 element stiffness matrix.

GSTIF - upper triangular part of global stiffness matrix which has been skewed so that diagonal becomes first column.

Figure 8 shows a grid of elements for some region with boundary conditions and plots streamlines for this problem. This sample result was taken at random from a selection and is not meant to represent a problem of any physical significance.

6. GENERAL CONSIDERATIONS ON SOLUTION PROCEDURES

To obtain a result for any problem using the finite element method there are five main stages to the solution procedure

i Point input, element specification, boundary condition specification.

ii Generation of element stiffness matrix and force vector.

iii Assembly of above to give global stiffness matrix and right hand side vector.

iv Solution of resulting equations.

v Past solution calculations and output.

There may be an iterative cycle involving ii to v for non-linear problems.

In previous sections of this paper it has been noted that various decisions have to be made, these are listed below.

i Overall solution criteria.

ii Type of element.

iii Form of shape function.

iv Form of linearisation for non-linear equations.

v Iterative scheme in computer.

vi Time stepping algorithm.

vii Linear equation solver, Cholesky, Gauss-Jordon etc.

Most of these items have been mentioned herein, others are straightforward programming tasks. The reader is referred to the bibliography for further reading.

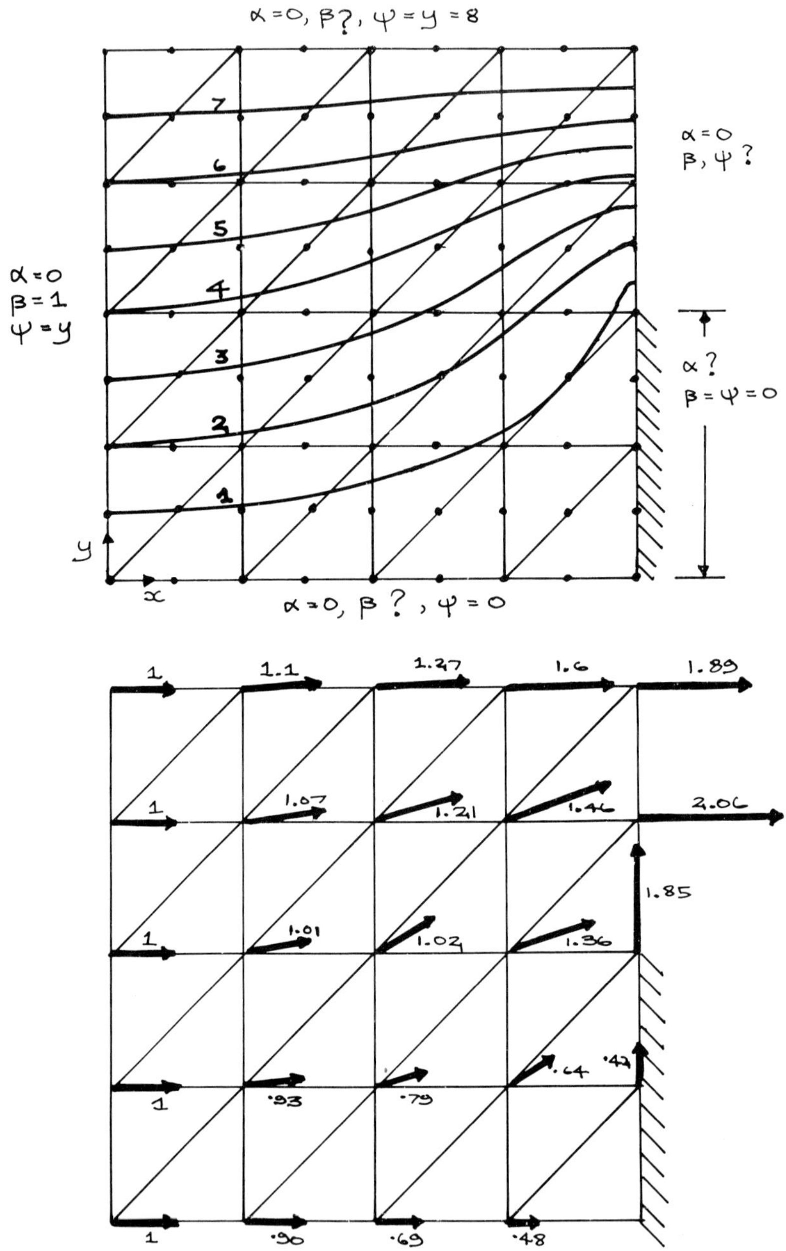

Figure 8 - Finite element grid, boundary conditions, streamlines and velocity vectors for a sample problem using least squares approach.

7. CONCLUDING REMARKS

It is hoped that the work in this paper provides some sort of insight to the finite element method. Understanding and appreciation of the method come with time and effort. It dominates the solution to problems in solid mechanics, but in fluid mechanics there is still some way to go before effective and efficient schemes are developed to cater for the types of fluid flow problems that are of interest to us all.

APPENDIX A

The evaluation of the characteristic matrices for the non linear terms involve integrals over the element of the following type.

$$\int_e [N]^T[N]\{p\}\frac{\partial[N]}{\partial x}\{q\}dA, \qquad \int_e [N]^T[N]\{p\}\frac{\partial[N]}{\partial y}\{q\}\,dA.$$

using the iterative scheme whereby $\{u\}^e$ and $\{v\}^e$ are known from the previous iteration, then $\{p\}$ and $\{q\}$ in the above expressions are known, as either $\{u\}^e$ or $\{v\}^e$ and these integrals can be evaluated explicity. Let the first integral be denoted by $\alpha[k,i,j]$ and the second by $\beta[k,i,j]$ where $k,i,j = 1 \rightarrow 4$. Thus typical terms in α and β are,

$$\alpha[k,i,j] = p_i q_j \int_e N_K N_i \frac{\partial N_j}{\partial x}\,dA = p_i q_j \alpha'[k,i,j]$$

$$\beta[k,i,j] = p_i q_j \int_e N_k N_i \frac{\partial N_j}{\partial y}\,dA = p_i q_j \beta'[k,i,j].$$

Transforming the required integrals to the $\zeta\eta$ plane and remembering equations (40) and (42), α' and β' can be expressed as follows,

$$\alpha'[k,i,j] = \int_e N_k N_i \frac{\partial N_j}{\partial x}\,dA, \quad \beta'[k,i,j] = \int_e N_k N_i \frac{\partial N_j}{\partial y}\,dA,$$

where

$$\begin{bmatrix} \dfrac{\partial N_j}{\partial x} \\[2ex] \dfrac{\partial N_j}{\partial y} \end{bmatrix} = [J]^{-1} \begin{bmatrix} \dfrac{\partial N_j}{\partial \zeta} \\[2ex] \dfrac{\partial N_j}{\partial \zeta} \end{bmatrix}$$

$$dA = dxdy = [J]\;d\zeta d\eta.$$

Thus

$$\alpha'[k,i,j] = \int_{-1}^{+1}\int_{-1}^{+1} N_k N_i \left(J_{22}\frac{\partial N_j}{\partial \zeta} - J_{12}\frac{\partial N_j}{\partial \eta} \right) d\zeta d\eta,$$

$$\beta'[k,i,j] = \int_{-1}^{+1}\int_{-1}^{+1} N_k N_i \left(-J_{21}\frac{\partial N_j}{\partial \zeta} + J_{11}\frac{\partial N_j}{\partial \eta} \right) d\zeta d\eta.$$

For the simple square element of side unity the α' and β' matrices work out as follows.

$$\alpha'[1,i,j] = \frac{1}{72}\begin{bmatrix} -6 & 6 & -2 & 2 \\ -3 & 3 & -1 & 1 \\ -4 & 4 & -2 & 2 \\ -1 & 1 & -1 & 1 \end{bmatrix} \quad \beta'[1,i,j] = \frac{1}{72}\begin{bmatrix} -6 & -2 & 6 & 2 \\ -2 & -2 & 2 & 2 \\ -3 & -1 & 3 & 1 \\ -1 & -1 & 1 & 1 \end{bmatrix}$$

$$\alpha'[2,i,j] = \frac{1}{72}\begin{bmatrix} -3 & 3 & -1 & 1 \\ -6 & 6 & -2 & 2 \\ -1 & 1 & -1 & 1 \\ -2 & 2 & -2 & 2 \end{bmatrix} \quad \beta'[2,i,j] = \frac{1}{72}\begin{bmatrix} -2 & -2 & 2 & 2 \\ -2 & -6 & 2 & 6 \\ -1 & -1 & 1 & 1 \\ -1 & -3 & 1 & 3 \end{bmatrix}$$

$$\alpha'[3,i,j] = \frac{1}{72}\begin{bmatrix} -4 & 4 & -2 & 2 \\ -1 & 1 & -1 & 1 \\ -2 & 2 & -6 & 6 \\ -1 & 1 & -3 & 3 \end{bmatrix} \quad \beta'[3,i,j] = \frac{1}{72}\begin{bmatrix} -3 & -1 & 3 & 1 \\ -1 & -1 & 1 & 1 \\ -6 & -2 & 6 & 2 \\ -2 & -2 & 2 & 2 \end{bmatrix}$$

$$\alpha'[4,i,j] = \frac{1}{72}\begin{bmatrix} -1 & 1 & -1 & 1 \\ -2 & 2 & -2 & 2 \\ -1 & 1 & -3 & 3 \\ -2 & 2 & -6 & 6 \end{bmatrix} \quad \beta'[4,i,j] = \frac{1}{72}\begin{bmatrix} -1 & -1 & 1 & 1 \\ -1 & -3 & 1 & 3 \\ -2 & -2 & 2 & 2 \\ -2 & -6 & 2 & 6 \end{bmatrix}$$

REFERENCES

[1] C.B. MILLIKAN, On the steady motion of viscous, incompressible fluids; with particular reference to a variational principle, Phil. May. 7 641 (1929).

[2] S.M.DESHPANDE, On the connection between the alliptic equations of the Navier-Stokes type and the theory of harmonic functionals, Q. Appl. Maths. 43 (1973).

[3] G.de VRIES, R.BALASUBRAMANIAN and D.H. NORRIE, The application of the pseudo-functional finite element method to viscous flow problems, Proc. 5th Austasia. Conf. on Hyd. and Fluid Mech. 2 561 (1974). Also Mech. Eng. Dept. Rpt. 63, University of Calgary (1974).

[4] G. de VRIES and D.H. NORRIE, Application of the pseudo-functional finite element method to compressible flow problems, Univ. Calgary, Mech. Eng. Dept. Rpt. 51 (1974).

[5] G.de VRIES and D.H. NORRIE, Application of the pseudo-functional finite element method to visco-plastic torsion, Univ. Calgary, Mech. Eng. Dept. Rpt. 46 (1973).

[6] T. BRATANOW, A. ECER and M. KOBISKE, Finite element analysis of unsteady incompressible flow around an oscillating obstacle of arbitrary shape, AIAA Jol. 11 1471 (1973).

[7] T. BRATANOW and A. ECER, On the application of the finite element method in unsteady aerodynamics, AIAA Jol. 12 503 (1974).

[8] T. BRATANOW and A. ECER, Analysis of three dimensional unsteady flow around oscillating wings, AIAA paper 74-184, Washington,D.C. (1974).

[9] T. BRATANOW and A. ECER, Analysis of moving body problems in aerodynamics, Conf. Finite Element Methods in Flow Problems, University of Wales 225 (1974).

[10] B.A. FINLAYSON and L.E. SCRIVEN, The method of weighted residuals - a review, AMR 19 735 (1966).

[11] P.P. LYNN, Least squares finite element analysis of laminar boundary layer flows, Int. J. Num. Math. Engng. 8 865 (1974).

[12] P.P. LYNN and S.K. ARYA, Use of the least squares criterion in the finite element formulation, Int. J. Num. Math. Engng. 6 75 (1973).

[13] J.F. MILTHORPE, The use of the finite element method in numerical integration of the Navier-Stokes equations, M.Eng.Sci. thesis, University of Sydney, (1975).

[14] J.W. LEONARD, Finite element analysis of perturbed compressible flow, Int. J. Num. Math. Engng. 4,123 (1972).

[15] J.T. ODEN and D. SOMOGYI, Finite element applications in fluid dynamics, Jnl. Eng. Mech. Div. ASCE 95, EM3, 821 (1969).

[16] J.T. ODEN and L. CARTER-WELLFORD, Analysis of flow of viscous fluids by the finite-element method, AIAA Jol. 10, 1590 (1972).

[17] R.I. TANNER, R.E. NICKELL and R.W. BILGER, Finite element methods for the solution of some incompressible non-Neutonian fluid mechanics problems with free surfaces, Comp. Math. Appl. Mech. and Engng, 1,1 (1975).

[18] A.J. BAKER, Finite element solution algorithm for viscous incompressible fluid dynamics Int. J. Num. Math. Engng. 6,89 (1973).

[19] A.J. BAKER, Predictions of envioronmental hydrodynamics using the finite element method - 1 - Theoretical development, AIAA J. 13,36 (1975).

[20] A.J. BAKER and S.W. ZELAZNY, Predictions in enviornmental hydrodynamics using the finite element method - 2 - Applications, AIAA J. 13, 43 (1975).

[21] W.L. WOOD and R.W. LEWIS, A comparison of time marching schemes for the transient heat conduction equations, Int. J. Num. Math. Engng. 9, 679 (1975).

[22] C. TAYLOR and P. HOOD, A numerical solution of the Navier-Stokes equations using the finite element technique, Computers and Fluids,I,73(1973).

BIBLIOGRAPHY

[B1] C.S. DESAI and J.F. ABEL, Introduction to the Finite Element Method, Van Nostrand-Reinhold (1972).

[B2] O.C. ZIENKIEWICZ, The Finite Element Method in Engineering Science, McGraw-Hill (1971).

[B3] J.T. ODEN, Finite Elements of Non-Linear Continua, McGraw-Hill (1972).

[B4] J.T. ODEN, O.C. ZIENKIEWICZ, R.H. GALAGHER and C. TAYLOR editors, Finite Element Methods in Flow Problems, Collected papers from the International Conf. at University of Wales, Swansea, (1974).

[B5] D.H. NORRIE and G. de VRIES, A Finite Element Bibliography, Mechanical Engineering Department, University of Calgary, Report Numbers 57, 58, 59 (1974).

[B6] D. H. NORRIE and G. de VRIES, The Finite Element Method, Academic Press (1973).

[B7] G. STRANG and G.J. FIX, An Analysis of the Finite Element Method, Prentice-Hall (1973).

Numerical Simulation of Fluid Motion, ed. John Noye
pp. 199-222

HYBRID FINITE ELEMENT METHODS

G.J. Fix

Carnegie-Mellon University

Pittsburgh, U.S.A.

SUMMARY

The purpose of this paper is to show how Lagrange multipliers can be used with finite elements to achieve a number of desirable properties in the underlying approximation. For elliptic boundary value problems variational principles can be developed in which all boundary conditions are natural. In fluid flow problems one can endow the approximations with physically essential conservation laws.

1. INTRODUCTION

In recent years a hybrid finite element has typically meant a variational method using finite elements where Lagrange multipliers are used to treat troublesome constraints. T. Pian (1972) and his colleagues have been using these techniques for several years to solve the fourth order elliptic problems arising in solid mechanics. In the latter the "troublesome constraint" is the condition of inter-element compatibility, i.e., the requirement that the test and trial functions have continuous derivatives. They alleviate the need to impose these constraints by in effect introducing Lagrange multipliers over each element boundary.

I. Babuska (1972) has also used these techniques for second order elliptic problems. His goal was to develop a variational method in which all boundary conditions are natural, i.e., need not be satisfied by the test and trial functions. Such a principle is important since it permits the use of rectangular elements such as cubic splines for problems which have complicated geometries.

The fundamental goal of this paper is to use the hybrid ideas for incompressible viscous fluid problems. The purpose of the Lagrange multipliers is to build in physically essential conservation laws that are not normally satisfied in standard finite element (or for that fact finite difference) schemes. This algorithm plus the associated time integration scheme are developed in Sections 4-5 for flows in polygonal regions.

To prepare for this material, generalizations of Babuska's work are given in Section 2. The latter treated the "strongly elliptic case"

$$\Delta u - cu = f \quad \text{in} \quad \Omega ,$$

where $c > 0$ in $\overline{\Omega}$. We develop an alternate analysis that covers the case $c = 0$, a rather essential point for our work since part of the fluid equations is the relation

$$\Delta\psi = \zeta$$

between the stream function ψ and vorticity ζ.

Finally, we propose and study an apparently new iterative method for solving the algebraic equations arising from the hybrid method as applied to second order elliptic problems. The latter will play a central role in the algorithm for fluid problems discussed in Section 4. It should be emphasized that the development of such algorithms is perhaps the most important item on our agenda. For example, the normal finite element schemes for $\Delta\psi = \zeta$ produce banded positive definite matrices. Gaussian elimination has been a work horse for such systems since the band structure and the profile of the bands can be fully exploited. The hybrid method, on the other hand, is a "stationary technique" -- it does not minimize a quadratic function -- hence the matrices are not positive definite. If Gaussian elimination were to be used, pivoting would be required and this would lead to unacceptable storage and computation costs. Thus the ultimate practicality of the method is fundamentally linked to development of efficient "equation solvers" for the associated algebraic problems.

2. SECOND ORDER ELLIPTIC PROBLEMS

Let Ω be a planar region and consider

$$\Delta u + f = 0 \quad \text{in} \quad \Omega \tag{2.1}$$

$$u = g \quad \text{on} \quad \Gamma , \tag{2.2}$$

where Γ is the boundary of Ω. The hybrid formulation of (2.1) - (2.2) (see Babuska, 1972) involves the Sobolev spaces $H^1(\Omega)$ and $H^{\frac{1}{2}}(\pi)$ where the now standard notation of Lions and Magenes (1968) is used. In particular, the goal is to compute an ordered pair

$$U = (u, \lambda) \in H_E = H_E(\Omega) \oplus H_E(\Gamma)$$

such that

$$a\big((u, \lambda), (v, \mu)\big) = F(v, \mu) \tag{2.3}$$

for all (v, μ) in H_E. The bilinear form $a(U, V)$ on $H_E \times H_E$ is defined by

$$a\big((u, \lambda), (v, \mu)\big) = \iint_\Omega \nabla u \cdot \nabla v - \int_\Gamma (\lambda v + \mu u), \tag{2.4}$$

while the functional F on H_E is given by

$$F(v, \mu) = \iint_\Omega f\, v - \int_\Gamma g\, \mu \tag{2.5}$$

Observe that H_E is a Hilbert space with norm

$$|||(v, \mu)||| = \left(\|v\|_1^2 + |\mu|_{-\frac{1}{2}}^2 \right)^{\frac{1}{2}}, \tag{2.6}$$

where in the sequel we shall consistently let $\|v\|_\ell$ denote the norm on $H^\ell(\Omega)$ and $|\mu|_r$ denote the norm on $H^r(\Gamma)$. In addition, $a(U, V)$ is a bounded linear form on $H_E \times H_E$. Indeed,

$$\begin{aligned} |a(U, V)| &\leq \iint_\Omega |\nabla u|\,|\nabla v| + \left|\int_\Gamma \lambda v\right| + \left|\int_\Gamma \mu u\right| \\ &\leq \|u\|_1 \|v\|_1 + |\lambda|_{-\frac{1}{2}} |v|_{\frac{1}{2}} + |\mu|_{-\frac{1}{2}} |u|_{\frac{1}{2}} \\ &\leq C_1 \left(\|u\|_1^2 + |\lambda|_{-\frac{1}{2}}^2 \right)^{\frac{1}{2}} \left(\|v\|_1^2 + |\mu|_{-\frac{1}{2}}^2 \right)^{\frac{1}{2}} \end{aligned} \tag{2.7}$$

In this calculation we have used the fact that

$$\left|\int_\Omega \sigma w\right| \leq |\sigma|_{-\frac{1}{2}} |w|_{\frac{1}{2}} \tag{2.7a}$$

as well as

$$|w|_{\frac{1}{2}} \leq C_2 \|w\|_1 . \tag{2.7b}$$

A similar calculation shows that F is a bounded linear functional on H_E if for example $f \in H^0(\Omega)$ and $g \in H^{\frac{1}{2}}(\Gamma)$.

It is an easy calculation (see Babuska, 1972) to show that if u_0 is a classical solution of (2.1) - (2.2) (for example, u_0 has two continuous derivatives in $\overline{\Omega}$ and satisfies (2.1) at each $\underline{x} \in \Omega$ and (2.2) at each $\underline{x} \in \Gamma$) and if

$$\lambda_0 = \partial u_0 / \partial n ,$$

then

$$a\big((u_0, \lambda_0), (v, \mu)\big) = F(v, \mu) \tag{2.8}$$

for all $(v, \mu) \in H_E$. A more elaborate calculation shows that (u_0, λ_0) is the only solution of the variational problem (2.3).

This variational principle can be used for approximation by introducing finite dimensional spaces

$$S^h(\Omega) \subset H_E(\Omega), \quad S^h(\Gamma) \subset H_E(\Gamma) \tag{2.9}$$

where in the sequel we shall let

$$h_\Omega^{-2}, \quad h_\Gamma^{-1}, \quad h^{-2} = h_\Omega^{-2} + h_\Gamma^{-1}$$

denote the respective dimensions of $S^h(\Omega)$, $S^h(\Gamma)$ and $S^h = S^h(\Omega) \oplus S^h(\Gamma)$. (Thus, in applications where finite element approximations are used, h_Ω and h_Γ can be interpreted as average mesh spacings.) An approximate solution

$$u_h \in S^h(\Omega), \quad \lambda_h \in S^h(\Gamma)$$

is determined by requiring

$$a\big((u_h, \lambda_h), (v^h, \mu^h)\big) = F(v^h, \mu^h) \tag{2.10}$$

for all (v^h, μ^h) in S^h. The latter is equivalent to a set of linear algebraic equations. Indeed, let $\phi_1, \ldots, \phi_N$ $(N = h_\Omega^{-2})$ be a basis for $S^h(\Omega)$ and $\psi_1, \ldots, \psi_M$ $(M = h_\Gamma^{-1})$ be a basis for $S^h(\Gamma)$. Then

$$u_h(\underline{x}) = \sum_{j=1}^{N} u_j \phi_j(\underline{x}), \quad \lambda_h(\underline{x}) = \sum_{j=1}^{M} \lambda_j \psi_j(\underline{x}) \tag{2.11}$$

and the vector of weights $\underline{u} = (u_j)$, $\underline{\lambda} = (\lambda_j)$ are computed from

$$\begin{pmatrix} K & M \\ M^* & 0 \end{pmatrix} \begin{pmatrix} \underline{u} \\ \underline{\lambda} \end{pmatrix} = \begin{pmatrix} \underline{f} \\ \underline{g} \end{pmatrix} \tag{2.12}$$

The entries of the $N \times N$ matrix K are

$$\iint_\Omega \nabla\phi_i \cdot \nabla\phi_j \qquad (i, j = 1, \ldots, N);$$

the entries of the $N \times M$ matrix M are

$$-\int_\Gamma \phi_i \, \psi_j \qquad (i = 1, \ldots, N) \; (j = 1, \ldots, M);$$

and the entries of $\underline{f}$, $\underline{g}$ are

$$\iint_\Omega f\,\phi_i \quad (i = 1, \ldots, N), \; -\int_\Gamma g\,\psi_j \quad (j = 1, \ldots, M).$$

Our goal is to study the errors

$$u_0 - u_h, \quad \lambda_0 - \lambda_h$$

as $h \to 0$ (i.e., as the dimension of S^h goes to ∞). To do this we make the usual assumption (Strang and Fix, 1973) that asserts in effect that smooth functions can be approximated by appropriate functions on S^h. More precisely, we assume there is a positive constant $C_3 > 0$(*) and numbers $k > 1$, $m > -\frac{1}{2}$ for which the following is true. Given any $v \in H^{\ell}(\Omega)$, $1 \leq \ell \leq k$, there is a $\tilde{v}_h \in S^h(\Omega)$ satisfying

$$\|v - v_h\|_r \leq C_3 h^{\ell-r} \|v\|_\ell ; \tag{2.13}$$

and given any $\mu \in H^n(\Gamma)$, $n \leq m$, there is a $\tilde{\mu}_h \in S^h(\Gamma)$ for which

$$|\mu - \mu_h|_s \leq C_3 h^{n-s} |\mu|_s \quad . \tag{2.13a}$$

For technical reasons we shall also need to assume

$$S^h(\Gamma) \subset H^{-\frac{1}{2}+\varepsilon}(\Gamma) \tag{2.14}$$

for some $\varepsilon > 0$ and that the inverse inequality

$$|\mu^h|_{-\frac{1}{2}+\varepsilon} \leq C_4 h_\Gamma^{-\varepsilon} |\mu^h|_{-\frac{1}{2}} \tag{2.14a}$$

is valid.

Such a setting can be obtained for example if we introduce a grid (rectangular or triangular) covering Ω and let $S^h(\Omega)$ denote the space of continuous piecewise linear functions on this grid (Strang and Fix, 1973). The integer k in (2.13) will equal 2 for this choice. As we shall see momentarily, an appropriate boundary space $S^h(\Gamma)$ to use with the linear elements is piecewise constant functions defined on a grid covering Γ. For this choice $m = 1$ in (2.14). Higher order combinations are obtained for example by using quadratics in $\Omega(k = 3)$ and linear elements on $\Gamma(m = 2)$.

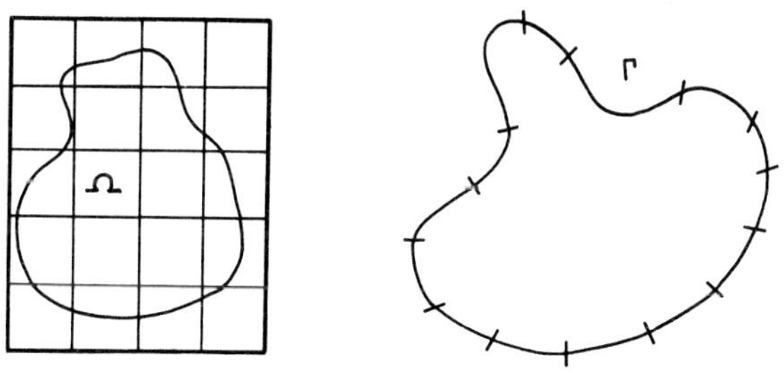

FIGURE 2.1. GRIDS COVERING Ω AND Γ .

(*) The constants that appear in our analysis will be more important than is usually the case in "finite element analysis". Thus we shall index them sequentially so we can keep track of their origin. So far we have $C_1 > 0$, $C_2 > 0$ from (2.7) and the approximation constant $C_3 > 0$.

The basic convergence theorem for this method is:

THEOREM 1.

Let (2.13) - (2.14) be true, and let,

$$h_\Omega / h_\Gamma \to 0 \quad \text{as} \quad h \to 0 \tag{2.15}$$

Then the matrix in (2.12) is invertible for all sufficiently small $h > 0$. Thus the variational problem (2.10) has a unique solution (u_h, λ_h) for $h > 0$ as $h \to 0$. In addition, assume that the solution (u_0, λ_0) of (2.3) satisfies

$$\| u_0 \|_k < \infty \qquad | \lambda_0 |_m < \infty \tag{2.16}$$

Then

$$\| u_0 - u_h \|_1 + | \lambda_0 - \lambda_h |_{-\frac{1}{2}} \leq C_0 \, \{ h_\Omega^{k-1} + h_\Gamma^{m+\frac{1}{2}} \} \tag{2.17}$$

where the constant C_0 depends only on the constants $C_1 - C_4$; the ratio (2.15); and (2.16). In particular, it is bounded independent of h as $h \to 0$ if (2.15) is satisfied.

Remarks (i). The condition of (2.15) asserts in effect that the grid on Ω must be finer than the grid on Γ. To see that this condition is necessary let Ω be a star-shaped region to be chosen momentarily and let $S^h(\Omega)$ be the space of Hermit cubics (Strang and Fix, 1973), although any rectangular element will do. In addition, let $S^h(\Gamma)$ be the functions on Ω restricted to the boundary Γ. Then $k = m = 4$ in (2.13) - (2.14), however (2.15) is violated.

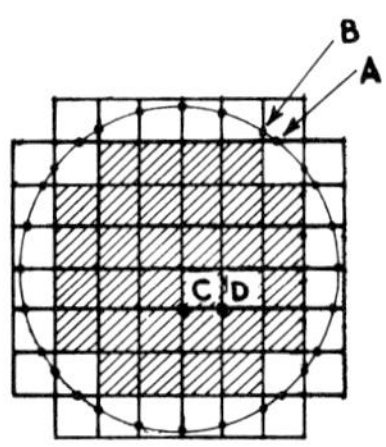

FIGURE 2.2. REGION AND GRID FOR COUNTEREXAMPLE.

(Compare the distance between the points A, B and C, D for example). We assert that

$$\| u - u_h \|_1 \geq C \, h_\Omega$$

as $h_\Omega \to 0$. Indeed, putting $v^h \equiv 0$ in (2.10) we have for the case $g = 0$

$$\int_\Gamma \mu^h \, u_h = 0$$

for all μ^h in $S^h(\Gamma)$. But such a μ^h is for example u_h restricted to Γ. Hence

$$\int_\Gamma (u_h)^2 = 0$$

and so $u_h \equiv 0$ on Γ. But as yet no assumptions have been made about Γ. Thus if Γ were described by say $r(x, y) = 0$ and if r were not a polynomial in x and y, then the only way u_h could vanish on Γ is for it to vanish outside the shaded region in Figure 2.2.

(ii) It is interesting to note that this calculation shows that if Ω were say a rectangle then u_h is zero on Γ (for the case $g = 0$). That is, the second equation in (2.12)

$$M^* \underline{u} = \underline{0}$$

is in fact equivalent to the standard procedure for forcing u_h to satisfy the essential boundary conditions.

(iii) If for example piecewise polynomials of degree $k - 1$ were used for $S^h(\Omega)$, then the appropriate boundary space would be piecewise polynomials of degree $k - 2$. The error would then behave like

$$O(h_\Omega^{k-1}) + O(h_\Gamma^{k-½})$$

The terms are balanced if

$$h_\Gamma = O(h_\Omega^{(k-1)/(k-½)})$$

and in addition (2.15) is valid.

Proof. The key step is to show that there is a positive number σ for which

$$\text{SUP } |a(U^h, V^h)| \geq \sigma \ |||U^h||| \tag{2.18}$$

for all $V^h \varepsilon S$. The sup is taken over all $V^h \varepsilon S^h$ with $|||V^h||| \leq 1$.

Indeed, if (2.18) were true then matrix problem (2.12) is certainly non-singular. To see this we note that if (2.12) had a nontrivial solution

$$\begin{pmatrix} \underline{u} \\ \underline{\lambda} \end{pmatrix} \neq \underline{0}$$

with $\underline{f} = \underline{0}$, $\underline{g} = \underline{0}$, then the functions

$$u(\underline{x}) = \Sigma u_j \phi_j(\underline{x}), \quad \lambda(\underline{x}) = \Sigma \lambda_j \psi_j(\underline{x})$$

would be in $S^h(\Omega)$, $S^h(\Gamma)$ and satisfy (2.3) for $(v, \mu) \varepsilon S^h$ with $f = g = 0$. The inequality (2.18) would then give $|||(u, \lambda)||| = 0$ and so $\underline{u} = \underline{0}$, $\underline{\lambda} = \underline{0}$.

To prove the error estimate (2.17) put

$$U_0 = (u_0, \lambda_0) \qquad U_h = (u_h, \lambda_h)$$

Choose

$$\tilde{U}_h = (\tilde{u}_h, \tilde{\lambda}_h)$$

such that

$$|||U_0 - \tilde{U}_h||| \leq C_3 [h_{\Omega}^{k-1} ||u_0||_k + h_{\Gamma}^{m+\frac{1}{2}} |\partial u_0/\partial n|_m] \tag{2.19}$$

Then for any $V^h \in S^h$ we have from (2.3) and (2.10)

$$0 = a(U_0 - U_h, V^h) = a(U_0 - \tilde{U}_h, V^h) + a(\tilde{U}_h - U_h, V^h) ;$$

that is

$$a(U_0 - \tilde{U}_h, V^h) = -a(\tilde{U}_h - U_h, V^h) .$$

Taking the sup over $V^h \in S^h$, $|||V^h||| \leq 1$, gives (in light of (2.7) and (2.18))

$$\sigma|||\tilde{U}_h - U_h||| \leq C_1|||U_0 - \tilde{U}_h|||$$

This with (2.19) gives (2.17).

Let us now estimate the number σ in (2.18). Suppose

$$U^h = (u^h, \lambda^h), \quad |||U^h||| = 1,$$

is given. We shall construct a $V^h = (v^h, \mu^h)$ and a number C independent of U^h, V^h such that

$$a(U^h, V^h) = 1, \quad |||V^h||| \leq C .$$

Note that in such a case we may take $\sigma = 1/C$, and we shall show that σ is bounded as $h \to 0$ if (2.15) holds. Also for convenience we shall let

$$||w||_1 = \{\iint_{\Omega} |\nabla w|^2 + \int_{\Gamma} w^2\}^{\frac{1}{2}} ,$$

the latter being equivalent (Necas, 1967) to the usual form

$$\{\iint_{\Gamma} |\nabla w|^2 + w^2\}$$

for $||w||_1$.

Let $v^h = u^h - w^h$. Then

$$a(U^h, V^h) = \iint_{\Omega} \nabla u^h \cdot \nabla u^h - \iint_{\Omega} \nabla u^h \nabla w^h + \int_{\Gamma} \lambda^h w^h - \int_{\Gamma} (\lambda^h + \mu^h)u^h .$$

We start by solving the boundary value problem

$$\Delta w = 0 \quad \text{in} \quad \Omega ,$$

$$\partial w/\partial m + w = \lambda \quad \text{on} \quad \Gamma ,$$

i.e.,

$$\iint_{\Omega} \nabla w \cdot \nabla\theta + \int_{\Gamma} w\theta = \int_{\Gamma} \lambda\theta \quad \text{all} \quad \theta \in H^1(\Omega) \tag{2.20}$$

Since $\lambda^h \in H^{-\frac{1}{2}+\varepsilon}(\Gamma)$ for some $\varepsilon > 0$, w is in $H^{1+\varepsilon}(\Omega)$ for some $\varepsilon > 0$ [3] and

$$||w||_{1+\varepsilon} \leq C_5 |\lambda|_{-\frac{1}{2}+\varepsilon} \tag{2.21}$$

for some $0 < C_5 < \infty$ which depends only on Ω (Strang and Fix, 1973). Also

(2.20) gives

$$\iint_\Omega \nabla w \cdot \nabla u^h + \int_\Gamma w\, u^h = \int_\Gamma \lambda^h\, u^h \tag{2.22}$$

$$\iint_\Omega \nabla w \cdot \nabla w + \int_\Gamma w^2 = \int_\Gamma \lambda^h\, w \tag{2.23}$$

$$a\big((u^h, \lambda^h), (u^h - w, \mu)\big)$$

$$= \|u^h\|_1^2 + \|w^h\|_1^2 + \int_\Gamma (w - 2\lambda^h - \mu - u^h)u^h .$$

We let

$$\mu = w - u^h - 2\lambda^h \tag{2.24}$$

to eliminate the boundary integral. Also, we note that (2.20) gives

$$|\lambda|_{-\frac{1}{2}} = \sup_{\|\theta\|_1 \leq 1} \left|\int_\Gamma \lambda\, \theta\right| \leq \|w\|_1$$

Thus

$$a\big((u^h, \lambda^h), (u^h - w, \mu)\big) \geq \|u^h\|_1^2 + |\lambda^h|_{-\frac{1}{2}}^2 = 1. \tag{2.24a}$$

In addition

$$\|v\|_1 = \|u^h - w\| \leq \|u^h\|_1 + \|w\| \leq \|u^h\|_1 + |\lambda^h|_{-\frac{1}{2}}^2 \leq 2,$$

$$|\mu|_{-\frac{1}{2}} \leq |w|_{-\frac{1}{2}} + |u^h|_{-\frac{1}{2}} + 2|\lambda^h|_{-\frac{1}{2}} \tag{2.25}$$

Of course, we are not finished since w is not in $S^h(\Omega)$ and μ is not in $S^h(\Gamma)$. However, to complete the proof we approximate these functions. Indeed, choose a $w^H \in S^h(\Omega)$ for which

$$\|w - w^h\|_1 \leq C_4\, h_\Omega^\varepsilon\, \|w\|_{1+\varepsilon}$$

Using (2.21) we obtain

$$\|w - w^h\|_1 \leq C_3 C_5\, h_\Omega^\varepsilon\, |\lambda^h|_{-\frac{1}{2}+\varepsilon}$$

With the inverse inequality (2.14a), this becomes

$$\|w - w^h\|_1 \leq \varepsilon_h\, |\lambda^h|_{-\frac{1}{2}} \tag{2.25a}$$

where

$$\varepsilon_h = C_3 C_4 C_5 (h_\Omega / h_\Gamma)^\varepsilon$$

In addition, choose $u_\Gamma^h \in S^h(\Gamma)$ and $w_\Gamma^h \in S^h(\Gamma)$ for which

$$|u^h - u^h|_{-\frac{1}{2}} \leq C_3\, h_\Gamma\, |u^h|_{\frac{1}{2}} \leq C_2 C_3\, h_\Gamma \|u^h\|_1$$

$$|w^h - w^h|_{-\frac{1}{2}} \leq C_3\, h_\Gamma\, |w^h|_{\frac{1}{2}} \leq C_2 C_3\, h_\Gamma \|w^h\|_1 .$$

We then let

$$v^h = u^h - w^h$$

$$\mu^h = w^h_\Gamma - u^h_\Gamma - 2\lambda^h$$

Then (2.24) will be true if the unit right hand side is replaced with $1 - C(\varepsilon_h + h_\Gamma)$ where C depends C_j $(j = 1, 5)$. Moreover,

$$|||V^h||| = |||(v^h, \mu^h)||| \leqslant C$$

where C depends on the same constants.

3. SOLUTION OF THE ALGEBRAIC EQUATIONS.

Our problems with the system (2.12) are not with conditioning provided

$$h_\Omega/h_\Gamma \to 0 \quad \text{as} \quad h \to 0 . \tag{3.1}$$

Indeed, calculations similar to the one given in this section indicate that the condition number of the system is the same order of magnitude as that for the usual stiffness matrix. However, unlike the latter, Gaussian elimination would be a very inefficient method for solving (2.12). The zero entries in the block (2.2) position indicates that pivoting will be required throughout the calculation. This of course would be entirely unacceptable both in terms of storage and operation costs.

In this section we propose an iterative method which in effect is an approximation to the time dependent problem

$$\begin{aligned} -\Delta u &= f \quad \text{in} \quad \Omega \\ \frac{\partial u}{\partial t} + \frac{\partial u}{\partial n} - \lambda &= 0 \quad \text{on} \quad \Gamma \\ u &= g \quad \text{on} \quad \Gamma \end{aligned} \tag{3.1a}$$

More precisely we shall compute

$$u_h = u_h(\underline{x}, t) \qquad \lambda_h = \lambda_h(\underline{x}, t)$$

satisfying

$$\int_\Gamma \partial u_h/\partial t \; v^h + \iint_\Gamma \nabla u_h \cdot \nabla v^h - \int_\Gamma [\lambda_h v^h + \mu^h u_h] = \iint_\Omega f v^h - \int_\Gamma g\mu^h \tag{3.2}$$

for each $t > 0$ and $(v^h, \mu^h) \in S^h$. We shall show that (3.2) is a set of dissipative ordinary differential equations and that the steady state can be reached with the same amount of work as that required to solve the standard finite element system.

To fix ideas let us restrict attention to the normal finite element spaces where $S^h(\Omega)$ and $S^h(\Gamma)$ consist of piecewise polynomial functions. We let

$$\Omega = \Omega_{STRIP} \cup \overline{\Omega}_{CORE} \tag{3.3}$$

where Ω_{CORE} satisfies

$$\sqrt{2}\; 2h_\Omega \geqslant \text{distance}\; (\Omega_{CORE}, \Gamma) \geqslant h_\Omega \sqrt{2} \tag{3.3a}$$

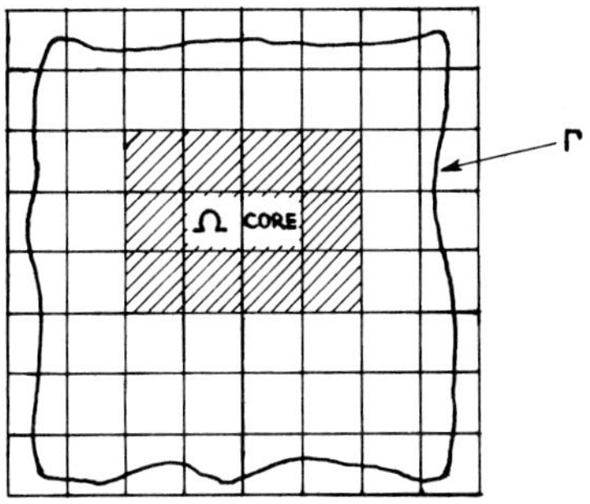

FIGURE 3.1 Ω_{CORE}

For each $t \geq 0$ we write

$$u_h(\underline{x}, t) = u_h^{\Omega}(\underline{x}, t) + u_h^{\Gamma}(\underline{x}, t), \tag{3.4}$$

where

$$u_h^{\Omega} = 0 \quad \text{on} \quad \Gamma , \qquad u_h^{\Gamma} = 0 \quad \text{in} \quad \Omega_{CORE}$$

In addition, we select the basis for S^h such that

$$\phi_1, \ldots, \phi_{N_{CORE}}$$

vanish on Γ and the remaining $N_\Gamma = N - N_{CORE}$ functions φ_j vanish in Ω_{CORE}. Then (3.2) becomes a matrix system

$$\begin{pmatrix} 0 & 0 & 0 \\ 0 & M_{11} & 0 \\ 0 & 0 & 0 \end{pmatrix} \begin{pmatrix} \dot{\underline{u}}_\Omega \\ \dot{\underline{u}}_\Gamma \\ \dot{\underline{\lambda}} \end{pmatrix} \qquad \begin{pmatrix} K_{00} & K_{01} & 0 \\ K_{01}^* & K_{11} & M_{01} \\ 0 & M_{01}^* & 0 \end{pmatrix} \begin{pmatrix} \underline{u}_\Omega \\ \underline{u}_\Gamma \\ \underline{\lambda} \end{pmatrix} \qquad \begin{pmatrix} \underline{f}_\Omega \\ \underline{f}_\Gamma \\ g \end{pmatrix} \tag{3.5}$$

Introduce a time grid $t_j = j\Delta t$ and put

$$\underline{u}^{(n)} = \underline{u}(n\Delta t), \qquad u_h^{(n)} = u_h(\underline{x}, n\Delta t) .$$

We integrate (3.5) with

$$M_{11}(\underline{u}_\Gamma^{(n+1)} - \underline{u}_\Gamma^{(n)}) + \Delta t[K_{01}^* \underline{u}_\Omega^{(n)} + K_{11} \underline{u}_\Gamma^{(n)}] + \Delta t\, M_{01} \underline{\lambda}^{(n+1)} = \Delta t\, \underline{f}_\Gamma \tag{3.6}$$

$$M_{01}^* \underline{u}_\Gamma^{(n+1)} = \underline{g} \tag{3.7}$$

$$K_{00} \underline{u}_\Omega^{(n+1)} + K_{01} \underline{u}_\Gamma^{(n+1)} = \underline{f}_\Omega \tag{3.8}$$

or what is the same,

$$\int_\Gamma \left(\frac{u_\Gamma^{(n+1)} - u_\Gamma^{(n)}}{\Delta t} \right) \phi_j + \iint_\Omega u_\Gamma^{(n)} \cdot \nabla\phi_j + \iint \nabla u_\Omega^{(n)} \cdot \nabla\phi_j$$

$$- \int \lambda^{(n+1)} \phi_j = \iint_\Omega + \phi_j \ , \ N_\Omega \geq j > N_{CORE} \tag{3.9}$$

$$\int_\Gamma u_\Gamma^{(n+1)} \psi_j^h = \int_\Gamma g \, \psi_j^h \qquad 1 \leq j \leq M_\Gamma \tag{3.10}$$

$$\iint_\Omega \nabla u_\Omega^{(n+1)} \cdot \nabla\phi_j + \iint_\Omega \nabla u_\Gamma^{(n+1)} \cdot \nabla\phi_j = \iint_\Omega f\phi_j \ , \quad 1 \leq j \leq N_{CORE} \tag{3.11}$$

Let us first show that the matrix

$$\begin{pmatrix} M_{11} & \Delta t \, M_{01} \\ \Delta t \, M_{01} & 0 \end{pmatrix} \tag{3.12}$$

to be inverted at each time step is nonsingular. Indeed if we are to map a vector

$$\begin{pmatrix} \underline{u}_\Gamma \\ \underline{\lambda} \end{pmatrix}$$

into zero, we would have

$$\int_\Gamma u_\Gamma \phi_j + \Delta t \int_\Gamma \lambda\phi_j = 0 \ , \qquad j > N_{CORE} \tag{3.13}$$

$$\int_\Gamma u_\Gamma \psi_j = 0 \ , \qquad 1 \leq j \leq M_\Gamma \tag{3.14}$$

u_Γ by construction is a linear combination of ϕ_j $(j > N_{CORE})$ so (3.13) can be written

$$\int_\Gamma u_\Gamma^2 = - \Delta t \int_\Gamma \lambda u_\Gamma$$

But replacing ψ_j with λ in (3.14) we get

$$\int_\Gamma u_\Gamma^2 = 0 \ , \quad \int_\Gamma \lambda\phi_j = 0 \ , \qquad j > N_{CORE}$$

The first of these gives $u_\Gamma = 0$ on Γ and hence $\underline{u}_\Gamma = \underline{0}$ (since $u_\Gamma \equiv 0$ in $\overline{\Omega}_{CORE}$). The second relation involves M unknowns (the weight $\lambda_1, \ldots \lambda_M$ of λ) and N_Γ equations. It is easy to see that if N_Γ exceeds M we can conclude $\lambda = 0$ on Γ, and this condition is implied by our basic assumption

$$h_\Omega / h_\Gamma \to 0 \quad \text{as} \quad h \to 0 \ .$$

We factor (3.12) at the start of the time integration by Gaussian elimination with pivoting. This requires

$$(h_\Omega^{-1} + h_\Gamma^{-1})^3 \simeq h^{-3}$$

operations. In addition we also factor the stiffness matrix K_{00} at this time, the cost being h_Ω^{-4} operations. The backsolves on (3.12) require $\simeq h_\Omega^{-2}$ operations per time step while the stiffness matrix requires h_Ω^{-3}. Thus the total work will be of order h_Ω^{-4} provided we can get to an "acceptable" steady state in h_Ω^{-1} time steps.

To show that this is indeed the case we replace $\dot{\underline{u}}_\Gamma$ with $e^{-\xi t}\, \underline{u}_\Gamma$ in (3.5) and study the eigenvalues ξ of

$$\iint_\Omega \nabla u^h \cdot \nabla\phi^h - \int [\lambda^h\phi^h + \mu^h u^h] = \xi \int_\Gamma u^h\phi^h \tag{3.15}$$

More precisely, we show that if ξ is a number (necessarily real) for which there is a

$$(u^h, \lambda^h) \in S^h \ , \quad \iint_\Omega \nabla u^h \cdot \nabla u^h + \int_\Gamma {u^h}^2 + |\lambda|_{-\frac{1}{2}}^2 = 1,$$

and (3.15) holds for all

$$(\phi^h, \mu^h) \in S^h \ ,$$

then

$$\gamma_1 \leqslant \xi \leqslant \gamma_2\, h_\Omega^{-1} \ , \tag{3.16}$$

where the constants γ_1 and γ_2 depend only on the ratio (3.1).

To prove the first inequality we follow the same general pattern of the proof in Section 2. In particular, ϕ^h is replaced with $u^h - w$, where w is the solution of

$$\iint_\Omega \nabla w\, \nabla\theta + \int_\Gamma w\theta = \int_\Gamma \lambda\theta \quad \text{all} \quad H(\Omega)$$

We then replace w with a w^h satisfying

$$\|w - w^h\|_1 \leqslant (h_\Omega/h_\Gamma)^\varepsilon\ |\lambda|_{-\frac{1}{2}}$$

as before. The left hand side of (3.15) will equal

$$\iint_\Omega \nabla u^h \cdot \nabla u^h + \int_\Gamma {u^h}^2 + \iint_\Omega \nabla w^h \cdot \nabla w^h + \int_\Gamma {w^h}^2$$

plus the boundary integral

$$\int_\Gamma (2\lambda^h + \mu + u^h - w^h)u^h \quad ,$$

the right hand side is

$$\xi \int {u^h}^2 - \xi \int u^h\, w^h \quad .$$

Thus putting

$$\mu^h = -2\lambda^h - u_\Gamma^h + w_\Gamma^h + \xi\, w_\Gamma^h \quad ,$$

where as before

$$|u^h - u^h|_{-\frac{1}{2}} \leqslant h_\Gamma |u^h|_{\frac{1}{2}} \ , \quad |w_\Gamma^h - w^h|_{-\frac{1}{2}} \leqslant h_\Gamma |w^h|_{\frac{1}{2}} \quad ,$$

$$\xi \geqslant \xi \int_\Gamma (u^h)^2 \geqslant \gamma_1$$

where σ depends only on h_Ω/h_Γ .

For the upper bound we put $\mu^h = -\lambda^h$ and $\phi^h = u^h$ to get

$$\xi \int_\Gamma {u^h}^2 = \iint_\Omega \nabla u^h \nabla u^h , \quad \iint_\Omega \nabla u^h \cdot \nabla\phi_j = 0 \quad (1 \leq j \leq N_{CORE}) \tag{3.17}$$

The result will then follow from the inverse inequality

$$C h_\Omega^{-1} \int_\Gamma {u^h}^2 \geq \iint_\Omega \nabla u^h \cdot \nabla u^h \tag{3.18}$$

First, observe that the scales in the latter are correct, i.e., $\int_\Gamma w^2$ is dominated by $\|w\|_{1/2}^2$

$$\{\iint_\Omega |\nabla w^h|^2\}^{1/2} \leq C h_\Omega^{-1/2} \|w^h\|_{1/2}^2$$

is the usual inverse inequality.

The inequality (3.17) does not follow from this since u^h could be zero on Γ with ∇u^h being zero in Ω . In fact, we would have only that $u_\Gamma = 0$ or $u = u_\Omega$. However, the second of (3.17) gives

$$0 = \iint_\Omega \nabla u^h \nabla u_\Omega^h = \iint_\Omega \nabla u_\Omega^h \cdot \nabla u_\Omega^h \quad .$$

4. VISCOUS INCOMPRESSIBLE FLOW.

For simplicity we shall assume that the flow region has solid boundaries. Thus using a stream function ψ and vorticity ζ we have

$$\partial\zeta/\partial t = J(\zeta, \psi) + \nu\Delta\zeta \quad \text{in } \Omega \ , \tag{4.1}$$

$$\Delta\psi = \zeta \quad \text{in } \Omega \ , \tag{4.2}$$

$$\psi = 0 \ , \ \partial\psi/\partial n = 0 \quad \text{on } \Gamma \ . \tag{4.3}$$

where $J(v, w)$ is the nonlinear convective term

$$\frac{\partial v}{\partial x} \cdot \frac{\partial w}{\partial y} - \frac{\partial v}{\partial y} \cdot \frac{\partial w}{\partial x} \quad .$$

To complete the statement of the problem we specify ζ,ψ at $t = 0$.

There are two physical laws implicit in (4.1) - (4.3). The first is the conservation of vorticity

$$\iint_\Omega \zeta = 0 \tag{4.4}$$

a relation that follows directly from (4.2) - (4.3). The second is even more significant and asserts

$$\int_\Gamma \partial\zeta/\partial n = 0 \tag{4.5}$$

Indeed, the vorticity production at the solid walls Γ is the dominant physical mechanism in (4.1) - (4.3), and (4.5) is a constraint on the way this can be done.

To derive (4.5) we integrate (4.1) over Ω . In light of (4.4) this becomes

$$\begin{aligned} 0 &= \iint_\Omega J(\zeta,\psi) + \nu \iint_\Omega \Delta\zeta \\ &= \iint_\Omega J(\zeta,\psi) - \nu \int_\Gamma \partial\zeta/\partial n \end{aligned}$$

But an integration by parts gives

$$\iint_\Omega J(\zeta,\psi) = \int_\Gamma \partial\psi/\partial\tau\ \zeta \quad , \tag{4.6}$$

where $\partial/\partial\tau$ denotes the tangential derivative on Γ. Since $\psi = 0$ on Γ, (4.5) is valid (provided the viscosity ν is nonzero).

In the usual finite element schemes (e.g. Taylor and Hood, 1973, Baker, 1975) neither (4.4) or (4.5) are satisfied by the approximation. The problem of course is that we have two boundary conditions on the stream function ψ and no boundary conditions on the vorticity ζ. The latter, however, is needed to solve the parabolic problem (4.1) (for a given ψ). Thus one is normally forced (Roache, 1973) to require that (4.2) hold on the boundary and use the second of (4.3) to calculate ζ on Γ. In this paper we shall use Lagrange multipliers to avoid these ad hoc modifications.

The hybrid formulation is

$$\iint_\Omega \partial\zeta/\partial t\ \theta - \iint_\Omega J(\zeta,\psi)\theta + \nu \iint_\Omega \nabla\zeta \cdot \nabla\theta - \nu \int_\Gamma \sigma\theta = 0 \tag{4.7}$$

$$\iint_\Omega \nabla\psi \cdot \nabla\phi - \int_\Gamma \mu\psi + \iint_\Omega \zeta\phi = 0 \quad , \tag{4.8}$$

where $\sigma = \partial\zeta/\partial n$ is the "Lagrange multiplier". More precisely we seek a family

$$\zeta(\cdot,t)\ ,\ \psi(\cdot,t) \in H^1(\Omega) \quad \sigma(\cdot,t) \in H^{-1/2}(\Gamma)$$

such that (4.7) - (4.8) holds for all

$$\theta,\phi \in H^1(\Omega)\ ,\quad \mu \in H^{-1/2}(\Gamma)\ .$$

To approximate we introduce finite dimensional subspaces

$$S^h(\Omega) \subset H^1(\Omega)\ ,\quad S^h(\Gamma) \subset H^{-1/2}(\Gamma)\ ,$$

and require that (4.7) - (4.8) hold on the latter.

We shall take here a slightly simpler setting and assume that Ω is a polygon. In this context

$$\psi = 0 \quad \text{on} \quad \Gamma$$

can be satisfied for finite element approximations. To be precise let $S^h(\Omega) \subset H^1(\Omega)$ be a finite element space, and let

$$S_0^h(\Omega) = \{v^h \in S^h(\Omega)\ |\ v^h = 0 \quad \text{on} \quad \Gamma\}$$

$$S^h(\Gamma) = \{v^h \ \text{restricted to}\ \Gamma\ |\ v^h \in S^h(\Omega)\}\ .$$

The semi-discrete approximation is to find

$$\zeta^h(\cdot,t) \in S^h(\Omega)\ ,\quad \psi^h(\cdot,t) \in S_0^h(\Omega)\ ,$$

$$\sigma^h(\cdot,t) \subset S^h(\Gamma)$$

such that

$$\iint_\Omega \partial\zeta^h/\partial t\ \theta^h - \iint_\Omega J(\zeta^h,\psi^h)\theta^h + \nu \iint_\Omega \nabla\zeta^h \cdot \nabla\theta^h - \nu \int_\Gamma \sigma^h\ \theta^h = 0 \tag{4.9}$$

for all θ^h in $S^h(\Omega)$;

$$\iint_\Omega \nabla\psi^h \cdot \nabla\phi + \iint_\Omega \zeta^h \phi^h = 0 \tag{4.10}$$

for all ϕ^h in $S^h(\Omega)$; and

$$\psi^h = 0 \quad \text{on} \quad \Gamma \ . \tag{4.11}$$

The initial conditions are determined from

$$\iint_\Omega \zeta^h(\cdot,0)\theta^h = \iint_\Omega \zeta(\cdot,0)\theta^h \quad \text{all} \quad \theta^h \in S^h(\Omega) \tag{4.12}$$

$$\iint_\Omega \psi^h(\cdot,0)\phi^h = \iint \psi(\cdot,0)\phi^h \quad \text{all} \quad \phi^h \in S_0^h(\Omega) \tag{4.13}$$

$$\int_\Gamma \sigma^h(\cdot,0)^h = \int \partial\zeta/\partial n\ (\cdot,0)^h \quad \text{all} \quad \phi^h \in S^h(\Gamma) \tag{4.14}$$

The status of the conservation laws (4.4) - (4.5) is covered in

THEOREM 4.1.

Let Ω be a polygon and let $(\zeta^h, \psi^h, \sigma^h)$ be the solution of (4.9) - (4.14). Then

$$\left.\begin{aligned} &\iint_\Omega \zeta^h(\cdot,t) = 0 \\ &\int_\Gamma \sigma^h(\cdot,t) = 0 \end{aligned}\right\} \quad \text{all} \quad t > 0 \tag{4.15}$$

Proof We put $\phi^h = 1$ in (4.10) to get

$$\iint_\Omega \zeta^h(\cdot,t) = -\iint \nabla\psi^h\ \nabla 1 = 0 \quad . \tag{4.16}$$

Similarly, putting $\theta^h = 1$ in (4.12) gives

$$\partial/\partial t \iint_\Omega \zeta^h = \iint_\Omega J(\zeta^h, \psi^h) + \nu \int_\Gamma \sigma \tag{4.17}$$

The term on the left is 0 by (4.16). An integration by parts shows

$$\iint_\Omega J(\zeta^h,\ \psi^h) = \int_\Gamma \partial\psi^h/\partial\tau\ \zeta^h$$

But $\psi^h = 0$ on Γ and hence this term is zero.

Care must be taken with the time integrations to preserve these properties; and the usual "simplification" or "crimes" (Strang and Fix, 1973) must be avoided. To give one example let us show that if the "lumped mass" approximation is used for (4.9) - (4.14) with an explicit time scheme, then the system reduces to the usual finite element schemes where the wall vorticities are in effect calculated by requiring $\Delta\psi = \zeta$ to hold on Γ .

To see this let Ω be a rectangle and let $S^h(\Omega)$ be the space of piecewise linear functions on a uniform grid (jh, kh) $(h = h_\Omega = h_\Gamma)$. For each node let ϕ_{jk} denote the hill function that is one at (jh, kh) and zero at the other nodes. Then for example

$$\zeta^h(x,\ y,\ n\Delta t) = \sum_{j,k} \zeta_{jk}^n\ \phi_{jk}(x,\ y)$$

where

$$\zeta^n_{jk} = \zeta^h(jh, kh, n\Delta t) .$$

From (4.9) we have for each interior node

$$\iint \partial\zeta^h/\partial t \ \phi_{jk} = \iint_\Omega J(\zeta^h, \psi^h)\phi_{jk} - \nu \iint_\Omega \nabla\zeta^h \ \nabla\phi_{jk} .$$

In the explicit lumped mass approximation the term on the left is replaced with

$$\left(\frac{\zeta^{n+1}_{jk} - \zeta^n_{jk}}{\Delta t} \right) h^2 , \quad (*)$$

giving the difference scheme

$$\frac{\zeta^{n+1}_{jk} - \zeta^n_{jk}}{\Delta t} = J(\zeta^n, \psi^n)_{jk} + \nu(\Delta_h \ \zeta^n)_{jk} . \tag{4.18}$$

The first term on the right is the "modified Arakawa approximation" to the Jacobian (Jespersen, 1974) and the last term is the usual five point difference star. Note that (4.18) holds for each interior node and this determines ζ^{n+1} at all such points. Given the latter we can compute ψ^{n+1} from (4.10) - (4.11). The former becomes with $\phi^h = \phi_{jk}$

$$(\Delta_h \ \psi)_{jk} = \zeta_{jk} , \quad (jk, kh) \in \Omega , \tag{4.19}$$

and of course

$$\psi_{jk} = 0 \quad \text{if} \quad (jk, kh) \in \Gamma . \tag{4.20}$$

To compute the values of ζ on the wall we put $\phi^h = \phi_{jk}$ in (4.10) for

$$(jh, kh) \in \Gamma .$$

If for example (jh, kh) were on a horizontal wall (Figure 4-1) then (4.10) reduces to

$$\zeta^{n+1}_{jk} = 2/h^2 \ \psi^{n+1}_{jk+1} , \tag{4.21}$$

which is precisely the Thom formula (Roache, 1973) used in finite element approximations described above.

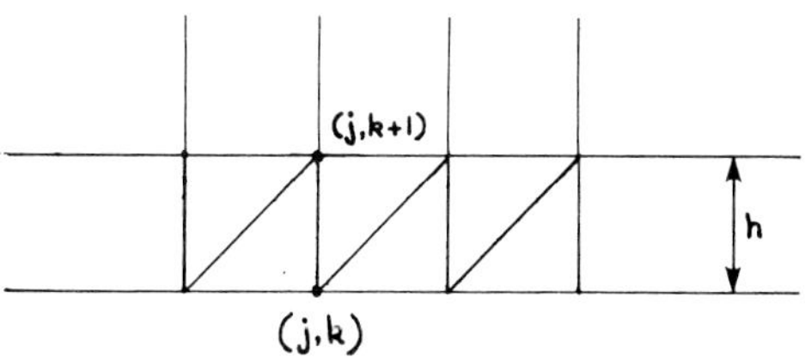

FIGURE 4-1

(*) This lumping invalidates the calculations in the proof of Theorem 4.1, hence (4.15) are not necessarily valid.

From our point of view the flaw with such a scheme is the passive role played by the Lagrange multiplier σ. Indeed, (4.18) - (4.21) uniquely determines ζ^{n+1}, ψ^{n+1} in Ω and Γ. The only independent relations remaining are (4.19) with $\theta^h = \phi_{jk}$, $(jh, kh) \in \Gamma$. The latter determine σ^{n+1}, however these values have no effect whatsoever on the calculation of ζ^{n+2}, ψ^{n+2}. In fact, in the standard finite element scheme σ is not computed at all.

In our time integration, on the other hand, σ will play a central role and the values of ψ^{n+1}, ζ^{n+1} will depend explicitly on the values of this Lagrange multiplier. Recall that $\sigma \simeq \partial\zeta/\partial n$, so this apparently acts as a control of the vorticity production at the walls. Indeed, returning to the general setting we propose the following implicit scheme

$$\iint_\Omega [\zeta^{(n+1)} - \zeta^{(n-1)}]\,\theta^h - 2\Delta t \iint_\Omega J(\zeta^{(n)}, \psi^{(n)})\theta^h$$

$$+ \nu\Delta t \iint_\Omega \nabla[\zeta^{(n+1)} + \zeta^{(n-1)}]\nabla\theta^h - \nu\Delta t \int_\Gamma [\sigma^{(n+1)} + \sigma^{(n-1)}]\,\theta^h = 0 \tag{4.22}$$

$$\iint_\Omega \nabla\psi^{(n+1)}\,\nabla\phi^h + \iint_\Omega \zeta^{(n+1)}\,\phi^h = 0 \tag{4.23}$$

$$\psi^{(n+1)} = 0 \quad \text{on} \quad \Gamma . \tag{4.24}$$

The first two equations are required to hold for all θ^h, $\phi^h \in S^h(\Omega)$.

To specify the associated matrix problem, let $\phi_1, \ldots, \phi_N$ be a basis for $S^h(\Omega)$ such that $\phi_1, \ldots, \phi_{N_0}$ is a basis for $S_0^h(\Omega)$ and $\phi_{N_0+1}, \ldots, \phi_N$ (restricted to Γ) is a basis for $S^h(\Gamma)$.(*) Then

$$\zeta = \zeta_\Omega + \zeta_\Gamma , \quad \psi = \psi_\Omega , \tag{4.25}$$

where

$$\begin{aligned} \zeta_\Omega^{(n)}(x, y) &= \sum_{j=1}^{N_0} \zeta_j^{(n)}\,\phi_j(x, y) \\ \zeta_\Gamma^{(n)}(x, y) &= \sum_{j=N_0+1}^{N} \zeta_j^{(n)}\,\phi_j(x, y) \\ \psi_\Omega^{(n)}(x, y) &= \sum_{j=1}^{N_0} \psi_j^{(n)}\,\phi_j(x, y) \end{aligned} \tag{4.26}$$

In addition,

$$\sigma^{(n)}(x,y) = \sum_{j=N_0+1}^{N} \sigma_j^{(n)}\,\phi_j(x, y)$$

for (x, y) on Γ. The associated matrix problem is obtained by substituting the above into (4.22) - (4.23). The first is required to hold for $\theta^h = \phi_j$, $1 \leq j \leq N$ and the second for $\phi^h = \phi_j$, $1 \leq j \leq N$. This represents a set of 2N equations in 2N unknowns: $\zeta_j^{(n+1)}$ $(1 \leq j \leq N)$, $\psi_j^{(n+1)}$ $(1 \leq j \leq N_0)$, $\sigma_j^{(n+1)}$ $(N_0 + 1 \leq j \leq N)$.

The inversion of the associated coefficient matrix by Gaussian elimination would not be very practical since extensive pivoting would be required. We thus consider iterations along the lines of those introduced in Section 3. First note that (4.22) - (4.23) is in effect a spatial approximation to

$$\begin{aligned} \zeta - \nu\Delta t\,\Delta\zeta &= f \quad \text{in} \quad \Omega \\ \Delta\psi - \zeta &= 0 \quad \text{in} \quad \Omega \\ \psi &= 0 \quad \text{on} \quad \Gamma \\ \partial\zeta/\partial n - \sigma &= 0 \quad \text{on} \quad \Gamma \end{aligned} \tag{4.27}$$

(*) In terms of our previous notation $M_\Gamma = N - N_0$. In addition, since Ω is a polygon and homogeneous boundary conditions can be satisfied exactly, the condition (3.1) is no longer applicable.

We are thus led to an artificial time s and

$$\begin{aligned} \zeta - \nu\Delta t\ \Delta\zeta &= f \quad \text{in } \Omega \\ \Delta\psi - \zeta &= 0 \quad \text{in } \Omega \\ \psi &= 0 \quad \text{on } \Gamma \\ \partial\zeta/\partial s + \partial\zeta/\partial n - \sigma &= 0 \quad \text{on } \Gamma \end{aligned} \tag{4.28}$$

In variational form this is

$$\begin{aligned} \int_\Gamma \partial\zeta/\partial s\ \phi_j + \nu\Delta t \iint_\Omega \nabla\zeta \cdot \nabla\phi_j - \nu\Delta t \int_\Gamma \sigma\phi_j &= \iint_\Omega f\phi_j \\ \iint_\Omega \nabla\psi \cdot \nabla\phi_j + \iint_\Omega \zeta\phi_j &= 0 \end{aligned} \tag{4.29}$$

for $1 \leqslant j \leqslant N$. The iterative method is defined by a specific time integration of (4.29). In particular, put $\zeta = \zeta_\Omega + \zeta_\Gamma$ as before and

$$\zeta^{(\ell)}(x, y) = \zeta(x, y, \ell\Delta s)\ .$$

Then we use

$$\int_\Gamma \left(\frac{\zeta_\Gamma^{(\ell+1)} - \zeta_\Gamma^{(\ell)}}{\Delta s} \right) \phi_j + \nu\Delta s \iint_\Omega \nabla\zeta_\Gamma^{(\ell+1)} \nabla\phi_j + \nu\Delta s \iint_\Omega \nabla\zeta_\Omega^{(\ell)} \nabla\phi_j$$

$$- \nu\Delta s \int_\Gamma \sigma^{(\ell+1)} \phi_j = \iint_\Omega f\phi_j \tag{4.30}$$

$$\iint_\Omega \zeta_\Gamma^{(\ell+1)} \phi_j + \iint_\Omega \zeta_\Omega^{(\ell)} \phi_j + \iint_\Omega \nabla\psi_\Omega^{(\ell)} \nabla\phi_j = 0 \tag{4.31}$$

for $N_0 + 1 \leqslant j \leqslant N$. These $2(N - N_0)$ equations determine $\zeta_\Gamma^{(\ell+1)}$, $\sigma^{(\ell+1)}$. The same calculations as in Section 3 show that the coefficient matrix is non-singular, is of order $O(h^{-1}) \times O(h^{-1})$ (as compared to $O(h^{-2}) \times O(h^{-2})$ for the stiffness matrix), and can be factorized at the very start of the calculation by Gaussian elimination. The final set of equations are

$$\begin{aligned} \nu\Delta t \iint_\Omega \nabla\zeta_\Omega^{(\ell+1)} \nabla\phi_j &= \iint_\Omega f\phi_j - \nu\Delta t \iint_\Omega \nabla\zeta_\Gamma^{(\ell+1)} \nabla\phi_j \\ \iint_\Omega \nabla\psi_j^{(\ell+1)} \nabla\phi_j &= - \iint_\Omega \zeta^{(\ell+1)} \phi_j \end{aligned} \tag{4.32}$$

for $1 \leqslant j \leqslant N_0$.

Let us review the entire algorithm and estimate the work that is involved.

STEP 1 (factorizations). We factor the standard stiffness matrix

$$\left[\iint_\Omega \nabla\phi_j \cdot \nabla\phi_k \right]_{1 \leqslant j,\ k \leqslant N_0} \tag{4.33}$$

by Gaussian elimination. This costs $O(h^{-4})$ operations. We then factor the matrix

$$\begin{pmatrix} M_{00} + \nu\Delta t K_{11} & -\nu\Delta t M_{00} \\ R_{00} & 0 \end{pmatrix} \tag{4.34}$$

needed in (4.30) - (4.31) by Gaussian elimination with pivoting. The entries of M_{00} are

$$\int_\Gamma \phi_j \ \phi_k , \qquad N_0 + 1 \leq j,k \leq N ;$$

the entries of K_{11} are

$$\iint_\Omega \Delta\phi_j \cdot \Delta\phi_k , \quad N_0 + 1 \leq j,k \leq N ;$$

and the entries of R_{00} are

$$\iint_\Omega \phi_j \ \phi_k , \qquad N_0 + 1 \leq j,k \leq N .$$

As in Section 2, the work required by this factorization is $O(h^{-3})$.

STEP 2 (Iteration). For the sequence $n = 0, 1, 2, \ldots$ in time $t = n\Delta t$ we want to solve (4.22) - (4.24). We do this by introducing at each $t = n\Delta t > 0$ an artificial time $s = \ell\Delta s$ and considering the iteration (4.30) - (4.32). We start the iteration by letting the initial values of $\zeta_\Omega, \zeta_\Gamma, \psi_\Omega, \sigma$ be those of the previous time step, i.e., $\zeta_\Omega^{(n)}, \zeta_\Gamma^{(n)}, \psi_\Omega^{(n)}, \sigma^{(n)}$. Then for $\ell = 0, 1, 2, \ldots$ we compute $\zeta_\Gamma^{(\ell+1)}, \sigma^{(\ell+1)}$ from (4.30) - (4.31) by backsolving (4.34). The functions $\psi_\Omega^{(\ell+1)}, \zeta_\Omega^{(\ell+1)}$ are computed from (4.32) by backsolving the stiffness matrix (4.33). Using the arguments in Section 3, it can be shown that for any given $\varepsilon > 0$

$$\| \zeta^{(\ell)} - \zeta^{(n+1)} \|_1 , \ \| \psi^{(\ell)} - \psi^{(n+1)} \|_1 , \ |\sigma^{(\ell)} - \sigma^{(n+1)}|_{-\frac{1}{2}}$$

are less than ε provided $\ell = O(h^{-1})$. Since the backsolves on (4.32) - (4.33) require $O(h^{-3})$ operations, a total of $O(h^{-4})$ operations are required per real time step. This is typical of the work requirement in the inviscid case (Fix, 1975a,b).

5. FLOWS WITH INFLOW AND OUTFLOW REGIONS

In this section we generalize the model proposed in the previous section to include flow regions Ω that have not only solid walls but also inflow and outflow regions. The channel flow in Figure 5.1 is an example.

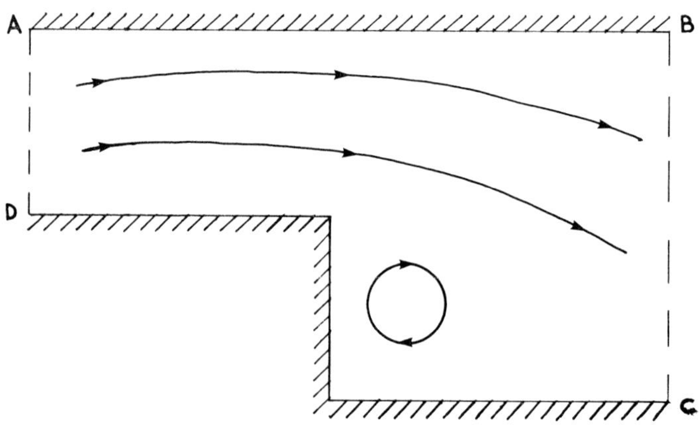

FIGURE 5.1 THE REGION Ω

The dynamic equations remain the same, namely (4.1) - (4.2), however the boundary conditions change. To be precise, let Γ_s denote the solid wall (AB and CD in Figure 5.1) so that $\Gamma - \Gamma_s$ represents the inflow and outflow regions (BC and DA in Figure 5.1). As usual the stream-function, or what is the same the normal flow, must be specified everywhere on Γ:

$$\psi = \psi_0 \quad \text{on} \quad \Gamma . \tag{5.1}$$

(Note that ψ_0 is a constant on Γ_s .) In addition, on the solid walls there is no tangential flow, i.e.,

$$\partial\psi/\partial n = 0 \quad \text{on} \quad \Gamma_s \quad . \tag{5.2}$$

Finally the vorticity ζ must be specified on the inflow and outflow

$$\zeta = \zeta_0 \quad \text{on} \quad \Gamma - \Gamma_s \quad . \tag{5.3}$$

Of course, if $\Gamma = \Gamma_s$ we reduce to the case (4.3).

With this change in the boundary conditions the conservation laws (4.4) - (4.5) need to be modified to include the effects of vorticity production and transport on the inflow and outflow regions. In particular, (4.4) is replaced by

$$\iint_\Omega \zeta = \int_\Gamma \partial\psi/\partial n = \int_{\Gamma-\Gamma_s} \partial\psi/\partial n \tag{5.4}$$

and (4.5) is replaced with

$$\nu \int_\Gamma \partial\zeta/\partial n = -\int_{\Gamma-\Gamma_s} \partial\psi/\partial t \; \zeta_0 + \partial/\partial t \iint_\Omega \zeta \; . \tag{5.5}$$

The term on the right of (5.4) is the vorticity transported into (out of) Ω through the inflow (outflow) regions. The terms on the right of (5.5) are the resulting vorticity production mechanisms.

The normal hybrid formulation with inhomogeneous data is similar to the one introduced in Section 4 but with extra terms. In particular, vorticity transport is approximated with

$$\iint_\Omega \{\partial\zeta/\partial t \; \theta - J(\zeta,\psi)\theta + \nu\nabla\zeta \cdot \nabla\theta\} - \nu\int_\Gamma \sigma\theta - \nu \int_{\Gamma-\Gamma_s} \xi(\zeta-\zeta_0) = 0 \tag{5.6}$$

and the elliptic equation is replaced with

$$\iint_\Omega \nabla\psi \cdot \nabla\varphi - \int_\Gamma [\lambda\varphi + \mu(\psi-\psi_0)] + \iint_\Omega \zeta\varphi = 0 \; . \tag{5.7}$$

For each time $t > 0$ the idea is to seek functions $\zeta(\cdot,t)$, $\psi(\cdot,t)$ in $S^h(\Omega)$ and Lagrange multipliers $\lambda(\cdot,t)$, $\sigma(\cdot,t)$ in $S^h(\Gamma)$ approximating $\partial\varphi/\partial n$ and $\partial\sigma/\partial n$ such that (5.6) - (5.7) are valid for all θ, φ in $S^h(\Omega)$ and μ, ξ in $S^h(\Gamma)$. This in turn is equivalent to a set of nonlinear ordinary differential equations once a basis has been chosen.

This system -- like (4.7) - (4.8) -- will not necessarily be conservative (in the sense that (5.4) - (5.5) hold) as we shall see in the sequel. To obtain the latter it will be necessary to assume (as we shall do from this point on) that Ω is a polygon and

$$S^h(\Gamma) = \{v^h \text{ restricted to } \Gamma | v^h \in S^h(\Omega)\} \; .$$

In this setting we can satisfy homogeneous boundary conditions, and in particular the normal hill function basis $\varphi_1,\ldots,\varphi_N$ for $S^h(\Omega)$ can be chosen so that $\varphi_1,\ldots,\varphi_{N_0}$ is a basis for

$$S_0^h(\Omega) = \{v^h \in S^h(\Omega) \,|\, v^h = 0 \text{ on } \Gamma\} \quad ;$$

$\varphi_1,\ldots,\varphi_{K_0}$ $(K_0 > N_0)$ is a basis for

$$S_{00}^h(\Omega) = \{v^h \in S^h(\Omega) \,|\, v^h = 0 \text{ on } \Gamma_S\} \quad ,$$

and finally $\varphi_{N_0+1},\ldots,\varphi_N$ is a basis for $S^h(\Omega)$.(*)

We shall use (5.6) - (5.7) except that some of the terms will be omitted by forcing boundary conditions. To be precise, for each $t > 0$ we seek an approximate stream function of the form

$$\psi^h(x,y,t) = \sum_{j=1}^{K_0} \psi_j^h(t)\, \varphi_j(x,y) + \psi_0^h(x,y,t) \tag{5.8}$$

where $\psi_0^h = \Sigma_{j=K_0+1}^{N} \psi_j^0(t)\varphi_j(x,y)$ is the interpolant of ψ_0 on Γ_S ; an approximate vorticity of the form

$$\zeta^h(x,y,t) = \sum_{\substack{1 \leq j \leq N_0 \\ K_0 \leq j \leq N}} \zeta_j^h(t)\, \varphi_j(x,y) + \zeta_0^h(x,y,t), \tag{5.9}$$

where $\zeta_0^h = \Sigma_{N_0+1}^{K_0} \zeta_j^0(t)\,\varphi_j(x,y)$ is the interpolant of ζ_0 on $\Gamma - \Gamma_S$; and finally Lagrange multipliers

$$\lambda^h(x,y,t) = \sum_{j=N_0+1}^{K_0} \lambda_j^h(t)\, \varphi_j(x,y) \tag{5.10}$$

$$\sigma^h(x,y,t) = \sum_{j=N_0+1}^{N_0} \sigma_j^h(t)\, \varphi_j(x,y) \quad . \tag{5.11}$$

Observe that the approximation λ^h to $\partial\psi/\partial n$ vanishes on Γ_S and that $\psi^h = \psi_0^h$, $\zeta^h = \zeta_0^h$ on Γ_S, $\Gamma - \Gamma_S$. These conditions have $2N + K_0 - N_0$ unknown weights. The latter are determined by the variational principle. In particular, we require

$$\iint_\Omega [\partial\zeta^h/\partial t\, \theta^h - J(\zeta^h, \psi^h)\theta^h + \nu\nabla\zeta^h \cdot \nabla\theta^h] - \nu \int_\Gamma \sigma^h \theta^h = 0 \tag{5.12}$$

for $\theta^h = \varphi_j^h$, $1 \leq j \leq N$. (i.e. all $\theta^h \in S^h(\Omega)$)

$$\iint_\Omega \nabla\psi^h \cdot \nabla\varphi^h - \int_{\Gamma-\Gamma_S} [\lambda^h\varphi^h + \mu^h(\psi^h - \psi_0)] + \iint_\Omega \zeta\varphi^h = 0 \tag{5.13}$$

for $\varphi^h = \varphi_j^h$, $1 \leq j \leq N$, $\mu^h = 0$, and $\varphi^h = 0$, $\mu^h = \varphi_j^h$, $N_0 + 1 \leq j \leq K_0$ (i.e., all $\varphi^h \in S^h(\Omega)$ and $\mu^h \in S^h(\Gamma)$ which vanishes on Γ_S).

The system (5.12) - (5.13) (together with initial conditions (4.12) - (4.14) is equivalent to a set of $2N + K_0 - N_0$ nonlinear ordinary differential equations in $2N + K_0 - N_0$ unknowns. Moreover, if $\Gamma = \Gamma_S$ (the entire boundary is solid), then (5.9) - (5.13) reduces to the algorithm (4.9) - (4.11) (taking $\psi_0 = 0$ on Γ_S).

(*) This is achieved with a nodal finite element method for example, if the nodes $\underline{z}_1,\ldots,\underline{z}_{N_0}$ associated with $\varphi_1,\ldots,\varphi_{N_0}$ are in Ω and the nodes nodes $\underline{z}_{N_0+1},\ldots,\underline{z}_N$ are on Γ with $\underline{z}_{N_0+1},\ldots,\underline{z}_{K_0}$ being on $\Gamma - \Gamma_S$.

Time integrations of (5.12) - (5.13) cause no problems, the algorithm of Section 4 being directly applicable. We thus turn our attention to the vorticity production and transport laws (5.4) - (5.5).

Theorem 5.1. Let Ω be a polygon and let ζ^h, ψ^h, λ^h be the approximation defined by (5.9) - (5.13). Then

$$\iint_\Omega \zeta^h = \int_{\Gamma-\Gamma_s} \lambda^h \tag{5.14}$$

and

$$\nu \int_\Gamma \sigma^h = - \int_{\Gamma-\Gamma_s} \frac{\partial\psi^h}{\partial\tau} \zeta_0^h + \frac{\partial}{\partial t} \iint_\Omega \zeta^h . \tag{5.15}$$

Proof. Putting $\varphi^h \equiv 1$(*) and $\mu^h = 0$ in (5.13) we obtain (5.14). Similarly, putting $\theta^h \equiv 1$ in (5.12) we obtain

$$\nu \int_\Gamma \sigma^h = - \iint_\Omega J(\zeta^h, \psi^h) + \frac{\partial}{\partial t} \iint_\Omega \zeta^h .$$

An integration by parts on the first term on the right of this equation gives

$$\nu \int_\Gamma \sigma^h = - \int_\Gamma \frac{\partial\psi^h}{\partial\tau} \zeta^h + \frac{\partial}{\partial t} \iint_\Omega \zeta^h . \tag{5.16}$$

Since $\psi^h = \psi_0^h$ is constant on Γ, $\frac{\partial\psi^h}{\partial\tau} = 0$ there, and hence (5.16) becomes (5.15).

Note that if we were unable to make ψ^h constant on Γ_s (which would happen if Ω were not polygonal, then (5.15) would fail).

ACKNOWLEDGEMENT

This work was supported in part by the Office of Naval Research under contract NR-044-468 and in part by NASA Grant NGR 47-102-001 while the author was in residence at I.C.A.S.E. at NASA Langley Research Center.

REFERENCES

Babuska, I. (1972). Mathematical Foundations of the Finite Element Method. K. Aziz, ed., Academic Press.

Baker, A.J. (1975). Nonlinear Initial Boundary Value Problems. Comp. Methods in Nonlinear Mechanics, J. Oden, ed., Univ. of Texas Press.

Fix, G. (1975a) Finite Elements and Fluid Dynamics. Comp. Methods in Nonlinear Mechanics, J. Oden ed., Univ. of Texas Press.

Fix, G. (1975b) Finite Element Models for Ocean Circulation Problems. S.I.A.M. Jour. Appl. Math. 29, 371-387.

Jespersen, D. (1974). Arakawas' Method is a Finite Element Method. Journal of Comp. Phys. 16, 383 - 390.

Lions, J.L. & Magenes, E. (1968). Problems aux Limites non Homogenes et Applications. Dunod.

(*) Since there are no boundary restrictions for functions v^h in $S^h(\Omega)$, $v^h \equiv 1$ is a member of this space.

Necas, J. (1967). Les methodes directes. Academia Editions de L'Academie Tchecoslovaque des Science, Prague.

Pian, T.H.T. & Mau, S.T. (1972). Some recent studies in assumed stress models. Adv. in Comp. Methods, J. Oden, ed., U.A.H. Press.

Roache, P. (1973). Computational Fluid Dynamics. Hermosa Press.

Strang, G. & Fix, G. (1973). An Analysis of the Finite Element Method. Prentice Hall.

Taylor, C. & Hood, P. (1973). A Numerical Solution of the Navier-Stokes Equations by use of the Finite Element Technique. Computers and Fluids, Vol. 1, 73-100.

Numerical Simulation of Fluid Motion, ed. John Noye
pp. 223-247

THE MARKER AND CELL TECHNIQUE

L.W.B. Browne

The University of Newcastle

New South Wales, Australia

SUMMARY

The development of the marker and cell Eulerian technique for the computer solution of viscous, transient, free-surface fluid flow problems is given. Refinements and extensions of the original technique are discussed and particular attention is paid to the instability problems inherent in this and similar finite difference techniques. Reference is made to later work involving incompressible flow and a wide range of flow speeds. Some details of computer experiments carried out by the author are given.

CONTENTS

1. HISTORICAL BACKGROUND

The Marker and Cell (M A C) computing technique was first described in 1965 by a group of workers at the University of California's Los Alamos Scientific Laboratory. (Harlow, Shannon and Welch (1965), Harlow and Welch (1965), Welch et al (1965)). The technique was used by them to solve incompressible, viscous, transient, free-surface fluid flow problems.

Particularly involved in this work was F.H. Harlow who in 1955 had published at Los Alamos the first proposal of the Particle in Cell (P I C) computing technique. This technique was for solving compressible, non-viscous, transient confined fluid flow. A comprehensive report and bibliography of the P I C method was probided by Amsden (1966).

The first published work of the Los Alamos group on the computer solution of incompressible, viscous, transient but confined flows was Fromm (1963), Fromm and Harlow (1963), and Harlow and Fromm (1964). These approaches used vorticity and the stream function as the primary dependent variables. It was out of this work, in order to deal with the free-surface problem, that the M A C method evolved. The M A C method uses velocity and pressure as the primary physical variables.

Since 1965 there have been some 40 publications dealing with or related to the M A C technique and these are listed in the bibliography. The major contribution has been from the Los Alamos group and practically all publications have been by workers in the U.S.A.

2. BRIEF DESCRIPTION OF THE M A C TECHNIQUE

The region of interest is divided into cells, and particles (markers) are placed in those cells which initially contain fluid. Typically 10 to 20 particles are evenly distributed through each fluid cell. Initial velocities, pressures and temperatures are specified for each fluid cell, together with appropriate boundary conditions. Finite difference forms of the continuity, momentum and energy equations are solved iteratively to obtain, for some small interval of time later, new values of velocity, pressure and temperature. Each particle is then moved by allotting to it an average of the velocities in surrounding cells and then multiplying this velocity by the time interval to obtain the displacement in both magnitude and direction. All cells are then examined and reflagged as full, empty or surface so that a new configuration of the fluid is determined. This process is repeated as often as is necessary to obtain the required flow information. The method is thus Eulerian (uses a non-moving reference frame) and discretizes both the fluid and time.

3. THE DIFFERENTIAL EQUATIONS

The continuity, momentum and energy differential equations applicable to flow problems can be expressed in many different forms. Vector forms are most general (see for example Hughes and Gaylord (1964) and Welch et al (1965)) followed by those which can be used for either cartesian or cylindrical coordinate systems depending on the value of an exponent included in the formulae (see for example Amsden and Harlow (1970b) and Viecelli (1971)).

For ease of illustration and also since a major portion of the actual work reported in the literature is of this form, the energy equation will be omitted and the 2-dimensional cartesian coordinate form of the constant density continuity and momentum equations will be considered. These have been used in 3 different groupings -

SET 1 Standard Navier-Stokes equations

$$\frac{\partial u}{\partial x} + \frac{\partial v}{\partial y} = 0 \tag{3.1}$$

$$\frac{\partial u}{\partial t} + u\,\frac{\partial u}{\partial x} + v\,\frac{\partial u}{\partial y} = -\,\frac{\partial \phi}{\partial x} + \nu\,\left(\frac{\partial^2 u}{\partial x^2} + \frac{\partial^2 u}{\partial y^2}\right) + g_x \tag{3.2}$$

$$\frac{\partial v}{\partial t} + u\,\frac{\partial v}{\partial x} + v\,\frac{\partial v}{\partial y} = -\,\frac{\partial \phi}{\partial y} + \nu\,\left(\frac{\partial^2 v}{\partial x^2} + \frac{\partial^2 v}{\partial y^2}\right) + g_y \tag{3.3}$$

Where u & v are the velocities in the x and y direction

$\phi = \frac{\text{Pressure}}{\text{Density}}$

ν = kinematic viscosity

and g_x & g_y are the gravity components in the x and y direction.

The M A C method was applied to this set of equations by Chan and Street (1970b) and Browne (1973) while Deville (1974) used a simplified version.

SET 2

The convective terms in (3.2) and (3.3) can be expressed in conservative form by adding u x (3.1) to (3.2) and v x (3.1) to (3.3) -

$$\frac{\partial u}{\partial x} + \frac{\partial v}{\partial y} = 0 \tag{3.4}$$

$$\frac{\partial u}{\partial t} + \frac{\partial u^2}{\partial x} + \frac{\partial uv}{\partial y} = -\,\frac{\partial \phi}{\partial x} + \nu\,\left(\frac{\partial^2 u}{\partial x^2} + \frac{\partial^2 u}{\partial y^2}\right) + g_x \tag{3.5}$$

$$\frac{\partial v}{\partial t} + \frac{\partial uv}{\partial x} + \frac{\partial v^2}{\partial y} = -\,\frac{\partial \phi}{\partial y} + \nu\,\left(\frac{\partial^2 v}{\partial x^2} + \frac{\partial^2 v}{\partial y^2}\right) + g_y \tag{3.6}$$

The M A C method was applied to this set of equations by Welch et al (1965), Nakayama and Romero (1971), and Hirt and Cook (1972).

SET 3

A complete conservative form can be obtained by subtracting $\frac{\partial}{\partial x}$ (3.4) from the terms in the bracket in (3.5) and $\frac{\partial}{\partial y}$ (3.4) from the terms in the bracket in (3.6) -

$$\frac{\partial u}{\partial x} + \frac{\partial v}{\partial y} = 0 \tag{3.7}$$

$$\frac{\partial u}{\partial t} + \frac{\partial^2 u}{\partial x} + \frac{\partial uv}{\partial y} = -\,\frac{\partial \phi}{\partial x} + \nu\,\frac{\partial}{\partial y}\,\left(\frac{\partial u}{\partial y} - \frac{\partial v}{\partial x}\right) + g_x \tag{3.8}$$

$$\frac{\partial v}{\partial t} + \frac{\partial uv}{\partial x} + \frac{\partial v^2}{\partial y} = -\,\frac{\partial \phi}{\partial y} - \nu\,\frac{\partial}{\partial x}\,\left(\frac{\partial u}{\partial y} - \frac{\partial v}{\partial x}\right) + g_y \tag{3.9}$$

The M A C method was applied to this set of equations of Viecelli (1971).

The inclusion of the energy equation and its subsequent use with any of these sets of equations is a straightforward process. See for example Welch et al (1965) and Hirt and Cook (1972).

The inclusion of variable density is covered by Young and Hirt (1972) and Rivard et al (1973). The latter also include the energy equation.

4. FINITE DIFFERENCE FORM OF THE FLOW EQUATIONS - ORIGINAL M A C

The field is discretized into cells of size $\delta x \times \delta y$ with cell centres being designated by i in the x direction and j in the y. The u values are located at the side faces of a cell, v values at the upper and lower faces while ϕ values are located at the cell centre. This arrangement is illustrated in Figure 1.

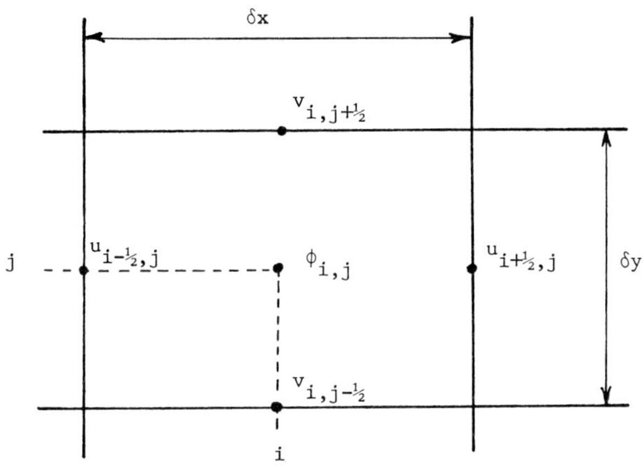

Figure 1

Location of velocities and pressures in M A C cell

If values are required at other points they are obtained by simple interpolation.

e.g. $$u_{i,j} = \frac{1}{2}(u_{i+\frac{1}{2},j} + u_{i-\frac{1}{2},j})$$

$$(uv)_{i+\frac{1}{2},j-\frac{1}{2}} = \frac{1}{4}(u_{i+\frac{1}{2},j} + u_{i+\frac{1}{2},j-1})(v_{i,j-\frac{1}{2}} + v_{i+1,j-\frac{1}{2}})$$

The cells do not change in size or position and the calculations are designed to find appropriate velocities and pressures for each cell at any point in time. Thus the M A C approach is Eulerian.

Original MAC used the Set 2 equations and expressed them in finite difference form using central difference formulae and explicit forms for all the spatial derivatives except for the pressure derivatives.

Thus, for the point $i+\frac{1}{2},j$ (3.5) becomes -

$$\frac{1}{\delta t}(u^{n+1}_{i+\frac{1}{2},j} - u^{n}_{i+\frac{1}{2},j}) + \frac{1}{\delta x}[(u^{n}_{i+1,j})^2 - (u^{n}_{i,j})^2]$$

$$+ \frac{1}{\delta y}[(uv)^{n}_{i+\frac{1}{2},j+\frac{1}{2}} - (uv)^{n}_{i+\frac{1}{2},j-\frac{1}{2}}] = -\frac{1}{\delta x}(\phi^{n+1}_{i+1,j} - \phi^{n+1}_{i,j})$$

$$+ \frac{\nu}{(\delta x)^2}(u^{n}_{i+\frac{3}{2},j} - 2u^{n}_{i+\frac{1}{2},j} + u^{n}_{i-\frac{1}{2},j})$$

$$+ \frac{\nu}{(\delta y)^2}(u^{n}_{i+\frac{1}{2},j+1} - 2u^{n}_{i+\frac{1}{2},j} + u^{n}_{i+\frac{1}{2},j-1}) + g_x \qquad (4.1)$$

Where superscript n indicates current time values and n+1 the values at time interval δt later.

This can be written -

$$u^{n+1}_{i+\frac{1}{2},j} = u^{n}_{i+\frac{1}{2},j} + \delta t\,\{\frac{1}{\delta x}[(u^{n}_{i,j})^2 - (u^{n}_{i+1,j})^2]$$

$$+ \frac{1}{\delta y}[(uv)^{n}_{i+\frac{1}{2},j-\frac{1}{2}} - (uv)^{n}_{i+\frac{1}{2},j+\frac{1}{2}}] + \frac{1}{\delta x}(\phi^{n+1}_{i,j} - \phi^{n+1}_{i+1,j})$$

$$+ \frac{\nu}{(\delta x)^2}(u^{n}_{i+\frac{3}{2},j} - 2u^{n}_{i+\frac{1}{2},j} + u^{n}_{i-\frac{1}{2},j})$$

$$+ \frac{\nu}{(\delta y)^2}(u^{n}_{i+\frac{1}{2},j+1} - 2u^{n}_{i+\frac{1}{2},j} + u^{n}_{i+\frac{1}{2},j-1}) + g_x\} \qquad (4.2)$$

Similarly (3.6) can be expressed in finite difference form at the point $i,j+\frac{1}{2}$ to give -

$$v^{n+1}_{i,j+\frac{1}{2}} = v^{n}_{i,j+\frac{1}{2}} + \delta t\,\{\frac{1}{\delta x}[(uv)^{n}_{i-\frac{1}{2},j+\frac{1}{2}} - (uv)^{n}_{i+\frac{1}{2},j+\frac{1}{2}}]$$

$$+ \frac{1}{\delta y}[(v^{n}_{i,j})^2 - (v^{n}_{i,j+1})^2] + \frac{1}{\delta y}(\phi^{n+1}_{i,j} - \phi^{n+1}_{i,j+1})$$

$$+ \frac{\nu}{(\delta x)^2}(v^{n}_{i+1,j+\frac{1}{2}} - 2v^{n}_{i,j+\frac{1}{2}} + v^{n}_{i-1,j+\frac{1}{2}})$$

$$+ \frac{\nu}{(\delta y)^2}(v^{n}_{i,j+\frac{3}{2}} - 2v^{n}_{i,j+\frac{1}{2}} + v^{n}_{i,j-\frac{1}{2}}) + g_y\} \qquad (4.3)$$

5. SOLUTION OF THE FINITE DIFFERENCE EQUATIONS - ORIGINAL M A C

To evaluate (4.2) and (4.3), ϕ^{n+1} values are required and original M A C found these by firstly manipulating (3.5) and (3.6). Operating on (3.5) with $\frac{\partial}{\partial x}$ and (3.6) with $\frac{\partial}{\partial y}$ and adding (in effect taking the divergence of the momentum equations) leads to -

$$\frac{\partial^2\phi}{\partial x^2} + \frac{\partial^2\phi}{\partial y^2} = - \frac{\partial D}{\partial t} - \frac{\partial^2 u^2}{\partial x^2} - 2\,\frac{\partial^2 uv}{\partial x\partial y} - \frac{\partial^2 v^2}{\partial y^2} + \nu\,\left(\frac{\partial^2 D}{\partial x^2} + \frac{\partial^2 D}{\partial y^2}\right) \qquad (5.1)$$

Where $$D = \frac{\partial u}{\partial x} + \frac{\partial v}{\partial y} \qquad (5.2)$$

Expressing (5.1) in finite difference form at point i,j gives -

$$\frac{1}{(\delta x)^2}\,(\phi^{n+1}_{i+1,j} - 2\,\phi^{n+1}_{i,j} + \phi^{n+1}_{i-1,j}) + \frac{1}{(\delta y)^2}\,(\phi^{n+1}_{i,j+1} - 2\,\phi^{n+1}_{i,j} + \phi^{n+1}_{i,j-1})$$
$$= - \frac{1}{\delta t}\,(D^{n+1}_{i,j} - D^{n}_{i,j}) - \frac{1}{(\delta x)^2}\,[(u^{n}_{i+1,j})^2 - 2\,(u^{n}_{i,j})^2 + (u^{n}_{i-1,j})^2]$$
$$- \frac{2}{\delta x\delta y}\,[(uv)^{n}_{i+\frac{1}{2},j+\frac{1}{2}} + (uv)^{n}_{i-\frac{1}{2},j-\frac{1}{2}} - (uv)^{n}_{i+\frac{1}{2},j-\frac{1}{2}} - (uv)^{n}_{i-\frac{1}{2},j+\frac{1}{2}}]$$
$$- \frac{1}{(\delta y)^2}\,[(v^{n}_{i,j+1})^2 - 2\,(v^{n}_{i,j})^2 + (v^{n}_{i,j-1})^2]$$
$$+ \frac{\nu}{(\delta x)^2}\,(D^{n}_{i+1,j} - 2\,D^{n}_{i,j} + D^{n}_{i,j-1})$$
$$+ \frac{\nu}{(\delta y)^2}\,(D^{n}_{i,j+1} - 2\,D^{n}_{i,j} + D^{n}_{i,j-1}) \qquad (5.3)$$

Where $$D^{n}_{i,j} = \frac{1}{\delta x}\,(u^{n}_{i+\frac{1}{2},j} - u^{n}_{i-\frac{1}{2},j}) + \frac{1}{\delta y}\,(v^{n}_{i,j+\frac{1}{2}} - v^{n}_{i,j-\frac{1}{2}}) \qquad (5.4)$$

Here the spatial derivatives of the velocities are again in explicit form. Setting $D^{n+1}_{i,j}$ to zero and writing (5.3) in the form -

$$\phi^{n+1}_{i,j} = \frac{1}{2\,\left[\frac{1}{(\delta x)^2} + \frac{1}{(\delta y)^2}\right]}\,\left\{ \frac{1}{(\delta x)^2}\,(\phi^{n+1}_{i+1,j} + \phi^{n+1}_{i-1,j}) + \frac{1}{(\delta y)^2}\,(\phi^{n+1}_{i,j+1} + \phi^{n+1}_{i,j-1}) + \text{D, u \& v terms evaluated at n}\right\} \qquad (5.5)$$

- allows an iterative procedure to be used to find each ϕ^{n+1} value.

In fact, with $D^{n+1}_{i,j} = 0$, (5.3) is obviously the finite difference form of a Poisson equation -

$$\Delta^2\phi = \text{Constant}$$

- and the iterative solution procedure for these is well known.

The convergence criterion was that percentage ϕ changes between iterations were to be negligible.

Once ϕ^{n+1} values are obtained, (4.2) and (4.3) can be used to obtain u^{n+1} and v^{n+1} values.

6. SOLUTION OF THE FINITE DIFFERENCE EQUATIONS - CURRENT M A C

There has been a fundamental change in the procedure to solve (4.2) and (4.3) although the results are similar. Instigated by Chorin (1968), elaborated on by Viecelli (1971) and used subsequently by Hirt and Cook (1972), Nichols and Hirt (1973) and Browne (1973), the method involves a simultaneous iteration on pressure and velocity components.

Chorin (1968) and Viecelli (1971) showed that an iterative solution to (4.2) and (4.3) was given by -

$$(\phi^{n+1}_{i,j})^{m+1} = (\phi^{n+1}_{i,j})^{m} - \lambda\,(D^{n+1}_{i,j})^{m} \tag{6.1}$$

where m indicates the m th. iteration and λ is a relaxation parameter which for stability is given by -

$$\lambda \leqslant \frac{1}{\delta t\,\left[\frac{1}{(\delta x)^2} + \frac{1}{(\delta y)^2}\right]} \tag{6.2}$$

and to minimise the number of iterations

$$\lambda\text{optimum} \simeq 0.9\ \lambda\text{maximum} \tag{6.3}$$

The iterative loop is then -

1. Calculate $(D^{n+1}_{i,j})^m$ using (5.4).

   Note that for m = 1, u^{n+1} & v^{n+1} will be u^n & v^n.

2. Calculate $(\phi^{n+1}_{i,j})^{m+1}$ using (6.1).

   Note that for m = 1, ϕ^{n+1} will be ϕ^n.

3. Calculate $(u^{n+1}_{i+\frac{1}{2},j})^{m+1}$ using (4.2) and the latest iterate value for the ϕ's.

Assuming the field is swept with both i & j increasing, this gives -

$$(u^{n+1}_{i+\frac{1}{2},j})^{m+1} = \eta^{n}_{i+\frac{1}{2},j} + \frac{\delta t}{\delta x}\left[(\phi^{n+1}_{i,j})^{m+1} - (\phi^{n+1}_{i+1,j})^{m}\right]$$

Where $\eta^{n}_{i+\frac{1}{2},j}$ contains velocity values that do not change during the iteration.

4. Similarly calculate

$$(u^{n+1}_{i-\frac{1}{2},j})^{m+1} = \eta^{n}_{i-\frac{1}{2},j} + \frac{\delta t}{\delta x} [(\phi^{n+1}_{i-1,j})^{m+1} - (\phi^{n+1}_{i,j})^{m+1}]$$

$$(v^{n+1}_{i,j+\frac{1}{2}})^{m+1} = \xi^{n}_{i,j+\frac{1}{2}} + \frac{\delta t}{\delta y} [(\phi^{n+1}_{i,j})^{m+1} - (\phi^{n+1}_{i,j+1})^{m}]$$

$$(v^{n+1}_{i,j-\frac{1}{2}})^{m+1} = \xi^{n}_{i,j-\frac{1}{2}} + \frac{\delta t}{\delta y} (\phi^{n+1}_{i,j-1})^{m+1} - (\phi^{n+1}_{i,j})^{m+1}]$$

Where ξ is similar to η.

5. $(D^{n+1}_{i,j})^{m+1}$ can now be calculated and the whole process repeated to convergence.

Convergence here is when $(D^{n+1}_{i,j})_{max}$ is less than some small value - typically 1×10^{-4}.

Note that both pressures and velocities are updated during the iterative process and no further calculations of these are required once convergence is achieved.

A working computer program incorporating these ideas is listed in Browne (1973). This is written in FORTRAN IV and contains 1593 statements. It was written to be run on an ICL - 1904A computer as an OVERLAY program and with DUMP facilities. Single precision was used and storage requirements for problem using 544 cells and 1568 particles are approximately 28K. Computer time per time cycle is approximately 135 s. The Los Alamos group have available a CDC 7600 with a 64K fast core memory giving 1 to 2 s computer time for each time cycle calculation.

7. DEVELOPMENTS IN THE FINITE DIFFERENCE FORM OF THE FLOW EQUATIONS

The most recent work involves a move towards more implicitness in the space derivatives of the velocity terms rather than the pure explicit approach used in original M A C.

For example, Deville (1974), in considering very low Reynolds number flows, was able to neglect the convective terms of (3.2) in comparison with the diffusion. Thus for flows with $\frac{\partial \phi}{\partial x} \simeq 0$

$$\frac{\partial u}{\partial t} = \nu \left(\frac{\partial^2 u}{\partial x^2} + \frac{\partial^2 u}{\partial y^2}\right)$$

- and it was with this equation that he experimented with various explicit-implicit schemes for the space derivatives.

Harlow and Amsden (1975) in dealing with multiphase flow calculations use a purely Eulerian mesh of computational cells. Referring to their Equations (28) and (32) they say - "The essence of our new methodology lies in the implicitness expressed in these equations ......".

Experiments with M A C applied to single phase flow and incorporating some of these ideas may be worthwhile.

8. KEEPING TRACK OF FREE SURFACES

At the end of the calculations for the velocities and pressures in each cell the marker particles are moved using an average of 4 new velocities surrounding each particle.

e.g. if u_k and v_k are the averages of the velocities surrounding particle K then -

$$x_k^{n+1} = x_k^n + \delta t . u_k$$
$$y_k^{n+1} = y_k^n + \delta t . v_k \qquad (8.1)$$

The averages used are area averages. For example, consider particle K in the 1st quadrant of a cell. A cell area is superimposed with K at the centre, as in Figure 2 -

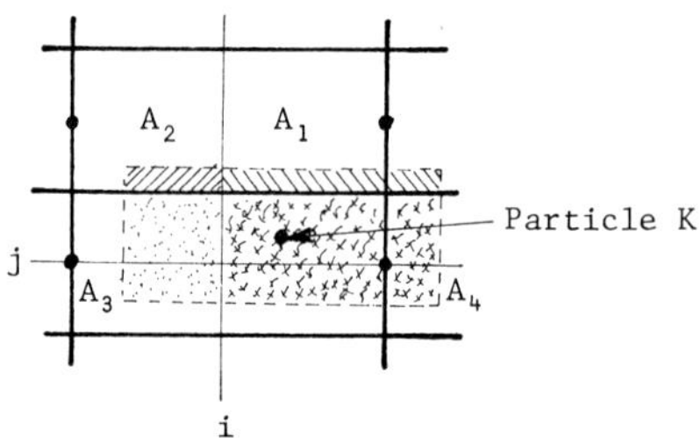

Figure 2

Areas used in determining average velocities

Then
$$u_k = \frac{1}{\delta x \delta y} (A_1 . u_{i+\frac{1}{2},j+1}^{n+1} + A_2 . u_{i-\frac{1}{2},j+1}^{n+1} + A_3 . u_{i-\frac{1}{2},j}^{n+1} + A_4 . u_{i+\frac{1}{2},j}^{n+1}) \qquad (8.2)$$

Details of the areas to be used when the particle is in other quadrants, and when v_k values are required are in Browne (1973).

The new particle positions gives the new configuration of the fluid. Cells can also then be reflagged e.g. as full or empty or surface and so on. This flagging is particularly important when considering the boundary conditions that are discussed in the next section.

9. BOUNDARY CONDITIONS

The application of appropriate boundary conditions are important for the satisfactory working of any finite difference technique.

One of the main problems is that in applying the finite difference equations, velocities are referenced outside the fluid. For example, in calculating velocities for a surface cell, velocities in a cell which does not contain any fluid will be referenced. A similar problem occurs at a solid boundary. The M A C approach is to use virtual fluid cells and determine appropriate velocities for these cells.

For example, consider the surface cell shown in Figure 3.

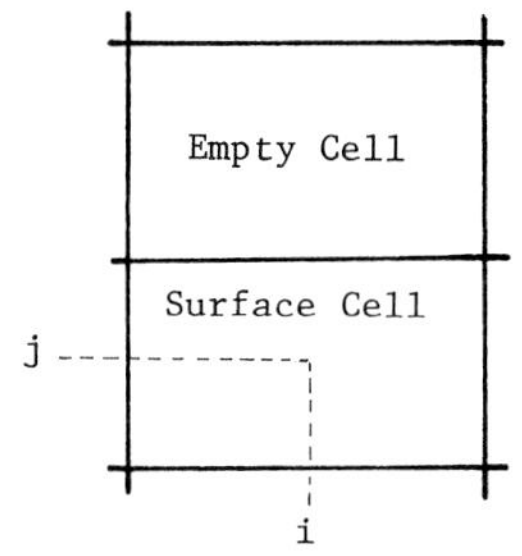

Figure 3

Setting of surface cell velocities

Rather than calculate $v_{i,j+\frac{1}{2}}$ values by the usual techniques, the 3 other velocities of cell i, j are calculated and then using D = 0 (5.4),

$$v_{i,j+\frac{1}{2}} = v_{i,j-\frac{1}{2}} - \frac{\delta y}{\delta x}(u_{i+\frac{1}{2},j} - u_{i-\frac{1}{2},j})$$

Then $u_{i+\frac{1}{2},j+1}$ in the virtual cell can be found by considering a pseudo cell with its centre at $i+\frac{1}{2}$, $j+\frac{1}{2}$ -

$$u_{i+\frac{1}{2},j+1} = u_{i+\frac{1}{2},j} - \frac{\delta x}{\delta y}(v_{i+1,j+\frac{1}{2}} - v_{i,j+\frac{1}{2}})$$

At a solid boundary, as in Figure 4, an averaging approach is used.

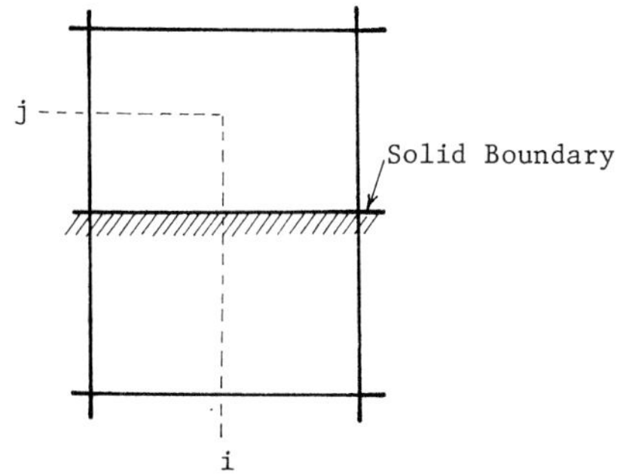

Figure 4

Treatment of cells in solid boundary

If the flow is viscous then for the virtual cell i,j-1 $u_{i+\frac{1}{2},j-1}$ is set equal to $-u_{i+\frac{1}{2},j}$ so that $u_{i+\frac{1}{2},j-\frac{1}{2}} = 0$ as required.

Pressure boundary conditions are generally straightforward. For example, the ϕ values of all surface cells are set equal to the ϕ in the non-fluid domain. Details of the setting of boundary conditions are given by Welch et al (1965) and Browne (1973).

10. REFINEMENTS AND EXTENSIONS OF THE M A C TECHNIQUE

A number of authors have suggested refinements and extensions of the M A C technique.

The use of vorticity and stream function as the primary dependent variable was, as pointed out in Section 1, the approach out of which M A C grew. The Los Alamos group reverted to this at one stage in the so-called S M A C (Simplified M A C) technique (Amsden and Harlow (1970a) and (1970b)). These ideas however appear to have been dropped.

Forester (1973) suggested that particles could be moved with greater accuracy by using an average of the old and new velocities surrounding a particle. Thus (8.2) would become -

$$\begin{aligned} u_k = \frac{1}{\delta x \delta y} \Big[& \frac{A_1}{2} (u^n_{i+\frac{1}{2},j+1} + u^{n+1}_{i+\frac{1}{2},j+1}) \\ & + \frac{A_2}{2} (u^n_{i-\frac{1}{2},j+1} + u^{n+1}_{i-\frac{1}{2},j+1}) \\ & + \frac{A_3}{2} (u^n_{i-\frac{1}{2},j} + u^{n+1}_{i-\frac{1}{2},j}) \\ & + \frac{A_4}{2} (u^n_{i+\frac{1}{2},j} + u^{n+1}_{i+\frac{1}{2},j}) \Big] \end{aligned} \tag{10.1}$$

This approach requires extra computer storage to retain the u^n values.

Hirt and Shannon (1968), Nichols and Hirt (1971) and Easton (1972) make suggestions for improving the free surface boundary conditions.

Daly (1969) and Foote (1973) include surface tension effects in the M A C method.

Variable density is covered by Young and Hirt (1972) and Rivard et al (1973).

Three dimensional fields have been considered by Nakayama and Romero (1971) and Hirt and Cook (1972).

The applicability of M A C to turbulent flow is considered by Amsden and Harlow (1968) and Hirt (1969).

Most workers using the M A C method have only used boundaries that coincide with the cells i.e. typically the solid boundaries are rectangular. A method of incorporating arbitrary boundary shapes with M A C is given by Viecelli(1969) and Viecelli(1971).

11. INSTABILITIES IN THE M A C TECHNIQUE

Instabilities constitute the major problem of the M A C technique.

11.1 The Need for High Viscosity Values

Hirt (1968) considered the stability of the original M A C explicit finite difference equations and was able to show that as a useful rule-of-thumb the M A C method is considered stable if -

(a) $$\nu > \frac{\delta t\ u^2}{2}$$

Where u is the average maximum fluid speed

and (b) $$\nu > \frac{(\delta x)^2\ \frac{\partial u}{\partial x}}{2}$$

Where $\frac{\partial u}{\partial x}$ is the average maximum velocity gradient in the direction of flow. The first condition is needed for linear stability. The second condition is a nonlinear stability requirement.

In general this leads to the requirement that a large value for the kinematic viscosity be used. Hirt and Cook (1972) (p.334) point out that this restriction is not unique to the M A C method but is a necessary feature of all finite-difference methods.

Just how large the viscosity needs to be be can be illustrated by experiments carried out by the author on the "broken-dam" problem. In this problem fluid is contained behind a wall which is removed at zero time - as illustrated in Figure 5.

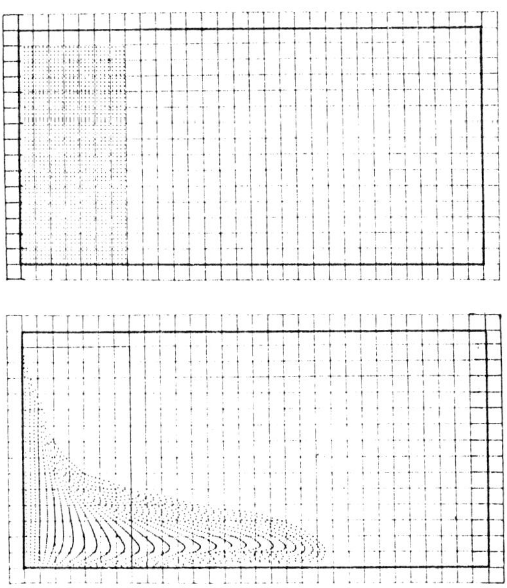

Zero Time

After 90 Time Intervals

Figure 5 Broken-dam problem.

The reason for selecting this particular problem was that information was available in the literature on the results obtained by other workers. (Martin and Moyce (1952), Amsden and Harlow (1970) and Nichols and Hirt (1971)).

Considering the fluid to be in a field of interest of 17.5 cm × 8.75 cm (to coincide with the size of the plotter print-outs used) a value of $\nu = 30 \frac{cm^2}{s}$ was required for stability. This represents a very thick oil

$$(\nu_{water} \simeq 0.01 \frac{cm^2}{s}, \quad \nu_{car\ engine\ oil} \simeq 0.5 \frac{cm^2}{s})$$

Figure 6 shows the onset of instabilities as the viscosity is decreased.

Changing the δt by a factor of 5 had little effect on the instabilities. In fact, because of computational round-off errors, the larger number of calculations led to a more enhanced instability - Figure 6(d). Obviously the non-linear stability requirement of Hirt is applying.

11.2 The Use of Other Finite Difference Formulae

One way to reduce the viscosity and overcome the instabilities is to search for improved finite difference formulae. Following Fromm (1968) and Chan and Street (1970a), this author introduced a more complex finite difference form for the convective velocity terms of equations (3.2) and (3.3). Expressing (3.2) in a partly finite difference form about the point, $i+\frac{1}{2},j$ -

$$\frac{1}{\delta t} (u^{n+1}_{i+\frac{1}{2},j} - u^{n}_{i+\frac{1}{2},j}) = < - u \frac{\partial u}{\partial x} - v \frac{\partial u}{\partial y} > + \text{other terms.}$$

Where the finite difference form of the convective terms $- u \frac{\partial u}{\partial x} - v \frac{\partial u}{\partial y}$ has not yet been stated.

i.e. $$u^{n+1}_{i+\frac{1}{2},j} = u^{*}_{i+\frac{1}{2},j} + \delta t \text{ (other terms)}$$

Where $$u^{*}_{i+\frac{1}{2},j} = \delta t.\ u^{n}_{i+\frac{1}{2},j} - \delta t < - u \frac{\partial u}{\partial x} - v \frac{\partial u}{\partial y} > \qquad (11.2.1)$$

This is obviously a finite difference form of

$$\frac{\partial u^*}{\partial t} = - u \frac{\partial u}{\partial x} - v \frac{\partial u}{\partial y} \qquad (11.2.2)$$

Leith (1965) showed that a second order central difference solution to this diffusion equation was -

$$u^{*}_{i+\frac{1}{2},j} = \tilde{u}_{i+\frac{1}{2},j} + \frac{\beta}{2} (\tilde{u}_{i+\frac{1}{2},j-1} - \tilde{u}_{i+\frac{1}{2},j+1}) + \frac{\beta^2}{2} (\tilde{u}_{i+\frac{1}{2},j-1} + \tilde{u}_{i+\frac{1}{2},j+1} - 2\, \tilde{u}_{i+\frac{1}{2},j}) \qquad (11.2.3)$$

Where -

$$\tilde{u}_{i+\frac{1}{2},j} = u_{i+\frac{1}{2},j} + \frac{\alpha}{2} (u_{i-\frac{1}{2},j} - u_{i+\frac{3}{2},j}) + \frac{\alpha^2}{2} (u_{i+\frac{1}{2},j} + u_{i+\frac{3}{2},j} - 2\, u_{i+\frac{1}{2},j})$$

$$\alpha = \frac{\delta t}{\delta x} u_{i+\frac{1}{2},j} \qquad \beta = \frac{\delta t}{\delta y} v_{i+\frac{1}{2},j}$$

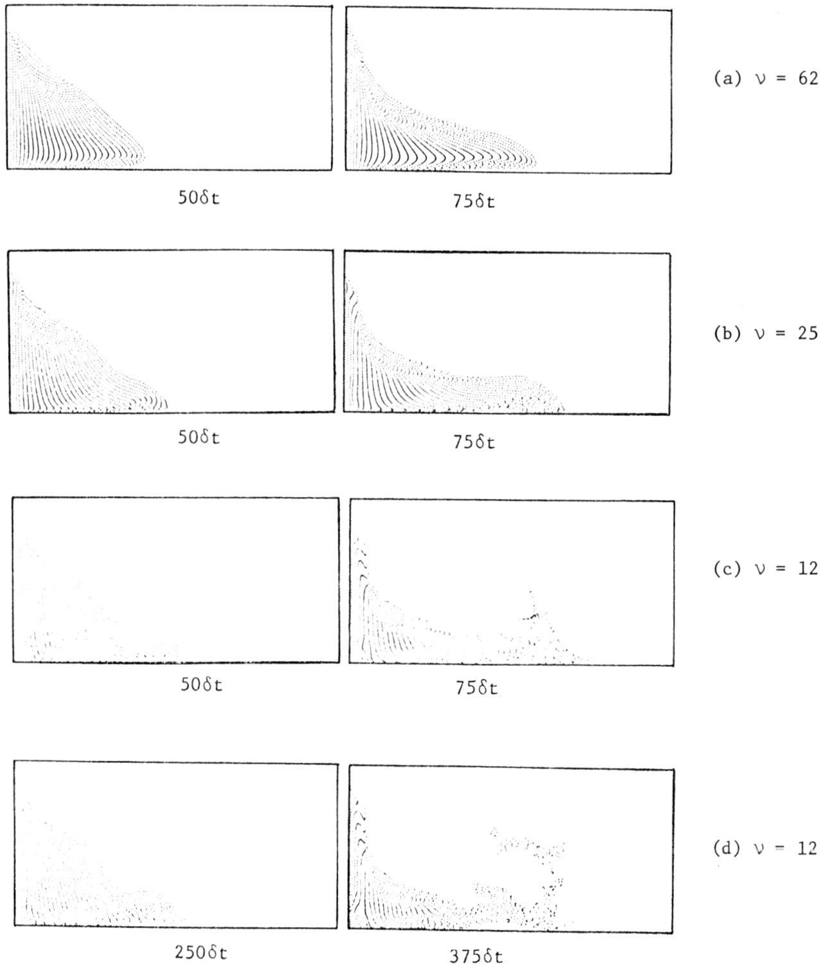

Elapsed time = 0.194 s Elapsed time = 0.291 s

Figure 6

Results of computer experiments on broken-dam problem

ν values are in $\frac{cm^2}{s}$

Fromm (1968a) showed that a second order upstream difference solution to (11.2.2) was (for the velocity vector in the first quadrant) -

$$u^*_{i+\frac{1}{2},j} = \tilde{u}_{i+\frac{1}{2},j-1} + \frac{1-\beta}{2} (\tilde{u}_{i+\frac{1}{2},j} - \tilde{u}_{i+\frac{1}{2},j-2}) + \frac{(1-\beta)^2}{2} (\tilde{u}_{i+\frac{1}{2},j} + \tilde{u}_{i+\frac{1}{2},j-2} - 2\,\tilde{u}_{i+\frac{1}{2},j-1}) \qquad (11.2.4)$$

Where -

$$\tilde{u}_{i+\frac{1}{2},j-1} = u_{i-\frac{1}{2},j-1} + \frac{1-\alpha}{2} (u_{i+\frac{1}{2},j-1} - u_{i-\frac{3}{2},j-1}) + \frac{(1-\alpha)^2}{2} (u_{i+\frac{1}{2},j-1} + u_{i-\frac{3}{2},j-1} - 2\,u_{i-\frac{1}{2},j-1})$$

and α & β are defined above.

The formulae that apply when the velocity vector is in other quadrants and those for $v^*_{i,j+\frac{1}{2}}$ are contained in Browne (1973).

Fromm demonstrated that a simple average of the central difference method and the upstream difference method produces a scheme with sharply reduced dispersion. This method is the one used by the author.

The difference between this method and original M A C is illustrated in Figure 7.

Original M A C uses 5 u-velocities and 4 v-velocities to estimate $u^*_{i+\frac{1}{2},j}$ while the technique which averages the second order central difference and second order upstream difference methods uses a total of 14 u-velocities and 4 v-velocities.

Comparisons between the results obtained by the two approaches can be made by use of the broken-dam problem - see Figure 8.

The surface seems to have taken on a more appropriate profile using the average of second order difference formulae - see the experimental results of Martin and Moyce (1952). However the instability problems were not overcome as the techniques were included in the programs used to produce the results of Figure 6.

11.3 The Use of Variable Cell Sizes

Because of the larger velocity gradients near a wall, instabilities may also be reduced by using smaller cells near the walls. Hirt (1969) indicates that Harlow and Pracht have tried a variable space increment but as far as is known these results have not been reported in the literature. This author is currently experimenting with such an approach. The program complexities are greatly increased - over twice as many FORTRAN statements are required. The junction area between the larger and smaller cells causes major problems and convergence during the iterative part of the process has not yet been obtained.

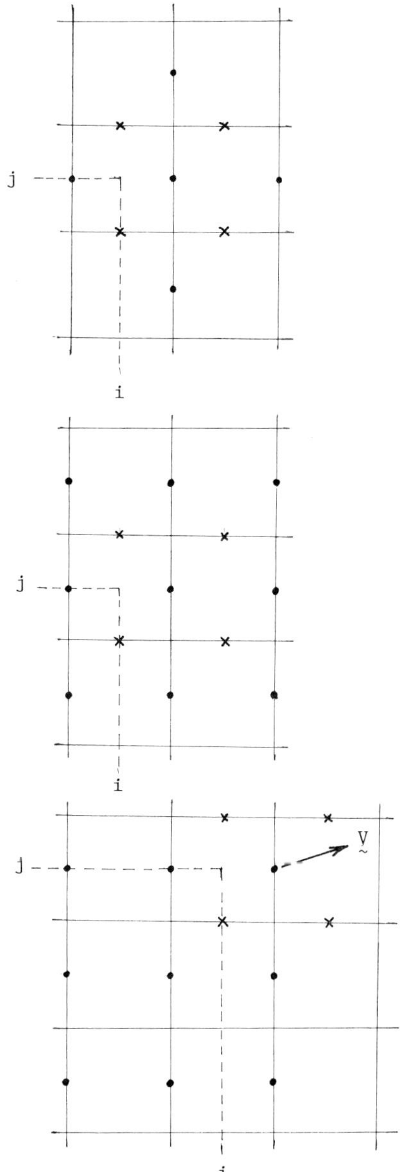

(a) Original M A C

(b) Second order central difference method

(c) Second order upstream difference method

Figure 7

Velocities used by various finite difference methods to obtain the convective velocity at $u_{i+\frac{1}{2},j}$

• u velocities

× v velocities

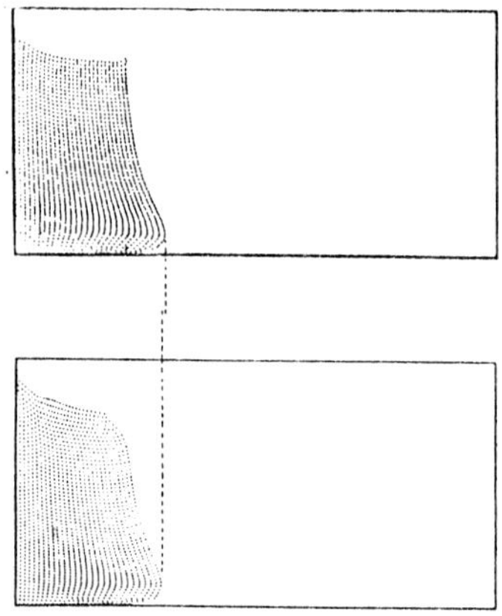

Original M A C solution
(Amsden and Harlow (1970b))

Results obtained by this author

a) Solution after 30 time intervals.

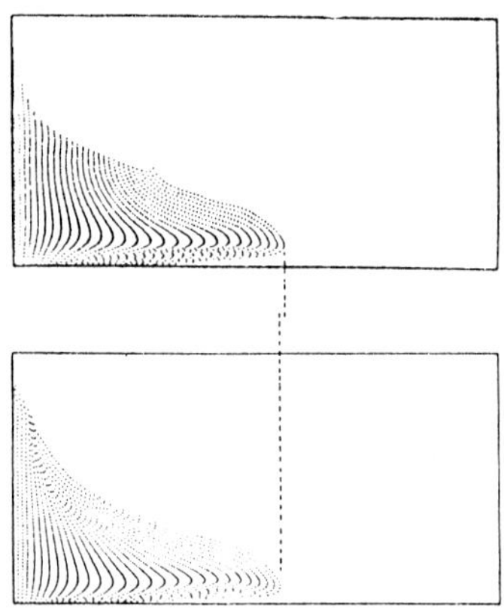

Original M A C solution

Results obtained by this author

b) Solution after 70 time intervals.

Figure 8

The broken-dam problem. Comparison of solutions obtained by 2 methods that differ in their determination of u^* & v^*.

11.4 Cancellation of Truncation Errors

A further technique to reduce instabilities was suggested to the author in a letter from F.C. Harlow. This is the method of local cancellation of low-order diffusional truncation errors as detailed in Rivard et al (1973). They call this the Truncation Error Cancellation (T E C) method. Early work in this field is that of Hirt (1968).

Rivard et al point out that a properly chosen amount of high viscosity (as discussed in Section 11.1) can counter any unreal diffusion arising from the truncation errors of any finite difference form of the differential equations. However one must be extremely careful in the use of artificial viscosity for calculations in more than one space dimension. For example, if the same coefficient is used to diffuse u-momentum in both the x and y directions, excessive diffusion can result in one direction. This will occur when the diffusional truncation errors yield vastly different diffusion coefficients in the two directions, a situation that is quite common. Thus they developed a technique which does not require high viscosity values.

To illustrate the method, consider equation (3.5) written as -

$$DE(u) = \frac{\partial u}{\partial t} + \frac{\partial u^2}{\partial x} + \frac{\partial uv}{\partial y} + \frac{\partial \phi}{\partial x} - \nu \left(\frac{\partial^2 u}{\partial x^2} + \frac{\partial^2 u}{\partial y^2}\right) = 0 \qquad (11.4.1)$$

and the finite difference form of this, about $i+\frac{1}{2},j$ (see Equation (4.2)) -

$$FD(u) = \frac{1}{\delta t}\left(u^{n+1}_{i+\frac{1}{2},j} - u^{n}_{i+\frac{1}{2},j}\right) + \frac{1}{2\delta x}\left[(u^{n}_{i+\frac{3}{2},j})^2 - (u^{n}_{i-\frac{1}{2},j})^2\right]$$

$$+ \frac{1}{\delta y}\left[(uv)^{n}_{i+\frac{1}{2},j+\frac{1}{2}} - (uv)^{n}_{i+\frac{1}{2},j-\frac{1}{2}}\right] + \text{Etc.} \qquad (11.4.2)$$

Now a Taylor series expansion of u^{n+1} about the point $i+\frac{1}{2},j,n$ is -

$$u^{n+1}_{i+\frac{1}{2},j} = u^{n}_{i+\frac{1}{2},j} + \delta t.\frac{\partial u}{\partial t} + \frac{(\delta t)^2}{2}\frac{\partial^2 u}{\partial t^2} + \frac{(\delta t)^3}{6}\frac{\partial^3 u}{\partial t^3} + \ldots\ldots$$

$$\therefore \frac{1}{\delta t}\left(u^{n+1}_{i+\frac{1}{2},j} - u^{n}_{i+\frac{1}{2},j}\right) = \frac{\partial u}{\partial t} + \frac{\delta t}{2}\frac{\partial^2 u}{\partial t^2} + \frac{(\delta t)^2}{6}\frac{\partial^3 u}{\partial t^3} + \ldots\ldots$$

Similarly a Taylor series expansion of $u^2_{i+\frac{3}{2},j}$ about the point $u_{i+\frac{1}{2},j}$ is -

$$u^2_{i+\frac{3}{2},j} = u^2_{i+\frac{1}{2},j} + \delta x\frac{\partial u^2}{\partial x} + \frac{(\delta x)^2}{2}\frac{\partial^2 u^2}{\partial x^2} + \frac{(\delta x)^3}{6}\frac{\partial^3 u^2}{\partial x^3} + \ldots\ldots$$

and similarly -

$$u^2_{i-\frac{1}{2},j} = u^2_{i+\frac{1}{2},j} - \delta x\frac{\partial u^2}{\partial x} + \frac{(\delta x)^2}{2}\frac{\partial^2 u^2}{\partial x^2} - \frac{(\delta x)^3}{6}\frac{\partial^3 u^2}{\partial x^3} + \ldots\ldots$$

$$\therefore \frac{1}{2\delta x}\left(u^2_{i+\frac{3}{2},j} - u^2_{i-\frac{1}{2},j}\right) = \frac{1}{2\delta x}\left(2\,\delta x\frac{\partial u^2}{\partial x} + \frac{(\delta x)^3}{3}\frac{\partial^3 u^2}{\partial x^3} + \ldots\ldots\right)$$

$$= \frac{\partial u^2}{\partial x} + \frac{(\delta x)^2}{6}\frac{\partial^3 u^2}{\partial x^3} + \ldots\ldots$$

Using similar expansions for the other terms of (11.4.2), leads to -

$$FD(u) = DE(u) + \frac{\delta t}{2}\frac{\partial^2 u}{\partial t^2} + \frac{(\delta t)^2}{6}\frac{\partial^3 u}{\partial t^3} + \ldots + \frac{(\delta x)^2}{6}\frac{\partial^3 u^2}{\partial x^3} + \ldots + \text{Etc} \tag{11.4.3}$$

Since $DE(u) = 0$ then -

$FD(u)$ = Truncation Errors

$\therefore$ $u^{n+1}_{i+\frac{1}{2},j}$ is the value calculated previously plus $\delta t \times$ Truncation Errors (see (11.4.2)).

The truncation errors such as those shown in (11.4.3) can be estimated by finite differences.

e.g.
$$\frac{\partial^2 u}{\partial t^2} \simeq \frac{1}{(\delta t)^2}\left(u^{n-1}_{i+\frac{1}{2},j} - 2u^{n}_{i+\frac{1}{2},j} + u^{n+1}_{i+\frac{1}{2},j}\right) \tag{11.4.4}$$

and
$$\frac{\partial^3 u^2}{\partial x^3} \simeq \frac{1}{2(\delta x)^3}\left(-u_{i-\frac{3}{2},j} + 2u_{i-\frac{1}{2},j} - 2u_{i+\frac{3}{2},j} + u_{i+\frac{5}{2},j}\right) \tag{11.4.5}$$

Extra computer storage is required for the n-1 and n values in (11.4.4). Rivard et al (1973) used a scheme that overcame this storage problem by expressing $\frac{\partial^2 u}{\partial t^2}$ in spatial derivatives. From the momentum equation -

$$\frac{\partial u}{\partial t} = -u\frac{\partial u}{\partial x} - v\frac{\partial u}{\partial y} - \frac{\partial \phi}{\partial x} + \nu\left(\frac{\partial^2 u}{\partial x^2} + \frac{\partial^2 u}{\partial y^2}\right)$$

Then

$$\frac{\partial^2 u}{\partial t^2} = -u\frac{\partial^2 u}{\partial x \partial t} - \frac{\partial u}{\partial x}\frac{\partial u}{\partial t} - \nu\frac{\partial^2 u}{\partial y \partial t} - \frac{\partial u}{\partial y}\frac{\partial v}{\partial t} - \frac{\partial^2 \phi}{\partial x \partial t} + \nu\frac{\partial}{\partial t}\left(\frac{\partial^2 u}{\partial x^2} + \frac{\partial^2 u}{\partial y^2}\right)$$

All terms except the one involving ϕ can be expressed in spatial derivatives using the momentum equations for $\frac{\partial u}{\partial t}$ and $\frac{\partial v}{\partial t}$,

e.g. $$\frac{\partial^2 u}{\partial x \partial t} = \frac{\partial}{\partial x}\left(\frac{\partial u}{\partial t}\right), \text{ etc.}$$

The time derivative of ϕ is not expressed in spatial derivatives but since the truncation errors associated with this are assumed negligible, this has no effect on the final formulation.

The derivation of such formulae for the truncation errors is a very lengthy and tedious process.

This author has experimented with the Rivard et al approach but not enough to know how worthwhile the techniques are in overcoming stability problems - see Figure 9.

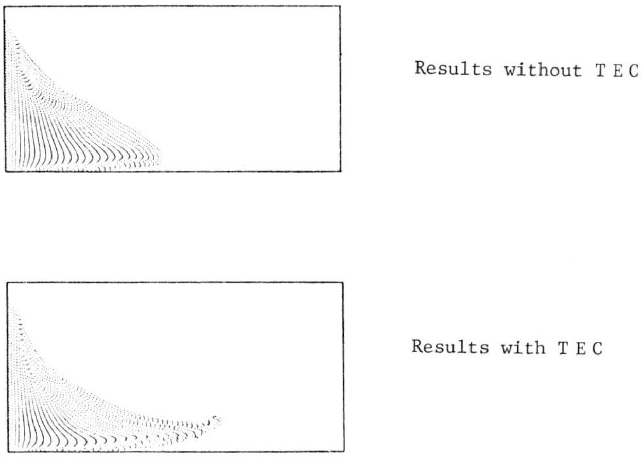

Figure 9

Experiments with Truncation Error Cancellation (T E C) technique. Broken dam problem after 57 time intervals.

11.5 Problems With Very Slow Flows

A further instability problem occurs with M A C at very slow flows. A von Neuman stability analysis of the explicit linearized momentum equations yields the condition -

$$2\,\nu\,\delta t < \frac{(\delta x)^2\,(\delta y)^2}{(\delta x)^2 + (\delta y)^2} \qquad (11.5.1)$$

(See for example the approach of Hirt (1968)).

Pracht (1971) showed that this stability requirement can be related to the Reynold number and yields -

$$R_e > 1$$

For flows with R_e values less than 1 an alternative approach is therefore required. Pracht (1971) and Deville (1974) both provide techniques for overcoming the problem. The techniques involve the use of implicit formulations of the finite difference equations.

11.6 Values for the Time Step

The basic restriction on the size of time step δt, for stability purposes, is that the fluid must not be permitted to flow across more than one computational cell in one time step (Hirt and Cook (1972)).

However for computing efficiency the time step should not be made too small. Pracht (1971) suggests that, for efficiency, δt should be chosen so that the fluid moves about half a mesh interval per time step.

11.7 Number of Cells

Fairly obviously the calculations will tend to stability with a large number of cells. The computational facilities place a limitation on increasing the number of cells indefinitely. Hirt and Cook (1972) use general arguments to show that if a typical dimension, L, in a flow is resolved by N cells then -

$$N > R_e^{\frac{1}{2}}$$

- for the normal M A C calculations. This can be used as a guide in estimating cell requirements.

12. POSSIBLE DEVELOPMENTS OF THE M A C TECHNIQUE

To estimate the direction of developments in this field is difficult. If one looks at the work of the group that developed the M A C technique - the Los Alamos Scientific Laboratory group, then the trends would seem to be -

A. The use of more implicit terms in the finite difference form of the equations - as discussed in Section 7.

B. The use of the Truncation Error Cancellation technique as discussed in Section 11.4.

C. The combination of Lagrangian and Eulerian techniques to handle a wider range of flow speeds. (See Hirt et al (1974) and Amsden and Hirt (1973) and the references given in these publications).

13. CONCLUSION

The M A C method and its developments are useful techniques for investigating fluid flow problems. A study of the methods, the instabilities and developing a computer program to experiment with these allows a more useful evaluation to be made of the various techniques that appear in the literature. Preferably a high speed large core computer should be available for such experiments.

14. REFERENCES

AMSDEN, A.A. (1966), The Particle-in-Cell Method for the Calculation of the Dynamics of Compressible Fluids. Los Alamos Sci. Lab., Rep. No. LA-3466.

AMSDEN, A.A. and HARLOW, F.H. (1968), Transport of Turbulence in Numerical Fluid Dynamics. J. Comp. Phys. 3, 94-110.

AMSDEN, A.A. and HARLOW, F.H. (1970a), A Simplified M A C Technique for Incompressible Fluid Flow Calculations. J. Comp. Phys. 6, 322-325.

AMSDEN, A.A. and HARLOW, F.H. (1970b), The S M A C Method: A Numerical Technique for Calculating Incompressible Fluid Flows. Los Alamos Sci. Lab., Rep. No. LA-4370.

AMSDEN, A.A. and HIRT, C.W. (1973), YAQUI: An Arbitrary Lagrangian-Eulerian Computer Program for Fluid Flow at all Speeds. Los Alamos Sci. Lab., Rep. No. LA-5100.

BROWNE, L.W.B. (1973), The Development of a Computer Program for Solving Fluid Flow Problems Using the Marker-and-Cell Numerical Method. U. of Newcastle, Dept. Mech. Eng., Eng. Bulletin No. ME25.

CHAN, R.K.-C. and STREET, R.L. (1970a), SUMMAC - A Numerical Model for Water Waves. Dept. Civil Eng., Stanford U., Tech. Rep. No. 135.

CHAN, R.K.-C. and STREET, R.L. (1970b), A Computer Study of Finite-Amplitude Water Waves. J. Comp. Phys. 6, 68-94.

CHORIN, A.J. (1968), Numerical Solution of the Navier-Stokes Equations. Math. Comp. 22, 745-762.

DALY, B.J. (1969), A Technique for Including Surface Tension Effects in Hydrodynamic Calculations. J. Comp. Phys. 4, 97-117.

DEVILLE, M.O. (1974), Numerical Experiments on the M A C Code for a Slow Flow. J. Comp. Phys. 15, 362-374.

EASTON, C.R. (1972), Homogeneous Boundary Conditions for Pressure in the M A C Method. J. Comp. Phys. 9, 375-379.

FOOTE, G.B. (1973), A Numerical Method for Studying Liquid Drop Behaviour: Simple Oscillation. J. Comp. Phys. 11, 507-530.

FORESTER, C.K. (1973), A Method of Time-Centering the Lagrangian Marker Particle Computation. J. Comp. Phys. 12, 269-274.

FORESTER, C.K. and EMERY, A.F. (1972), A Computational Method for Low Mach Number Unsteady Compressible Free Convective Flows. J. Comp. Phys. 10, 407-502.

FROMM, J.E. (1963), A Method for Computing Nonsteady, Incompressible, Viscous Fluid Flows. Los Alamos Sci. Lab., Rep. No. LA-2910.

FROMM, J.E. (1968a), Practical Investigations of Convective Difference Approximation of Reduced Dispersion. I.B.M. Res. Lab., Rep. No. RJ.531.

FROMM, J.E. (1968b), A Method of Reducing Dispersion in Convective Difference Schemes. J. Comp. Phys. 3, 176-189.

FROMM, J.E. and HARLOW, F.H. (1963), Numerical Solution of Problem of Vortex Street Development. The Phys. Fluids 6, 975-982.

HARLOW, F.H. (1955), A Machine Calculation Method for Hydrodynamic Problems. Los Alamos Sci. Lab., Rep. No. LAMS-1956.

HARLOW, F.H. (1969), Numerical Methods for Fluid Dynamics, an Annotated Bibliography. Los Alamos Sci. Lab., Rep. No. LA-4281.

HARLOW, F.H. and AMSDEN, A.A. (1968), Numerical Calculation of Almost Incompressible Flow. J. Comp. Phys. 3, 80-93.

HARLOW, F.H. and AMSDEN, A.A. (1971), A Numerical Fluid Dynamics Calculation Method for All Flow Speeds. J. Comp. Phys. 8, 197-213.

HARLOW, F.H. and AMSDEN, A.A. (1975), Numerical Calculation of Multiphase Fluid Flow. J. Comp. Phys. 17, 19-52.

HARLOW, F.H. and FROMM, J.E. (1964), Dynamics and Heat Transfer in the Von Karman Wake of a Rectangular Cylinder. The Phys. Fluids 7, 1147-1156.

HARLOW, F.H., SHANNON, J.P. and WELCH, J.E. (1965), Liquid Waves by Computer. Science 149, 1092-1093.

HARLOW, F.H. and SHANNON, J.P. (1967), Distortion of Splashing Liquid Drop. Science 157, 547-550.

HARLOW, F.H. and WELCH, J.E. (1965), Numerical Calculation of Time-Dependent Viscous Incompressible Flow of Fluid with Free Surface. The Phys. Fluids 8, 2182-2189.

HARLOW, F.H. and WELCH, J.E. (1966), Numerical Study of Large-Amplitude Free-Surface Motions. The Phys. Fluids 9, 842-851.

HIRT, C.W. (1968), Heuristic Stability Theory for Finite-Difference Equations. J. Comp. Phys. 2, 339-355.

HIRT, C.W. (1969), Computer Studies of Time-Dependent Turbulent Flows. The Physics of Fluids 12 Suppl. II, 219-227.

HIRT, C.W., AMSDEN, A.A. and COOK, J.L. (1974), An Arbitrary Lagrangian-Eulerian Computing Method for All Flow Speeds. J. Comp. Phys. 14, 227-253.

HIRT, C.W. and COOK, J.L. (1972), Calculating Three-Dimensional Flows Around Structures and Over Rough Terrain. J. Comp. Phys. 10, 324-340.

HIRT, C.W. and HARLOW, F.H. (1967), A General Corrective Procedure for the Numerical Solution of Initial-Value Problems. J. Comp. Phys. 2, 114-119.

HIRT, C.W. and SHANNON, J.P. (1968), Free-Surface Stress Conditions for Incompressible-Flow Calculations. J. Comp. Phys. 2, 403-411.

HUGHES, W.F. and GAYLORD, E.W. (1964), Basic Equations of Engineering Science. Schaum Publ. Co., New York.

LEITH, C.E., (1965), Numerical Simulation of the Earth's Atmosphere. Methods in Comp. Phys. Vol. 4, Academic Press, New York, 1-28.

MARTIN, J.C. and MOYCE, W.J. (1952). An Experimental Study of the Collapse of Liquid Columns on a Rigid Horizontal Plane. Roy. Soc. Lond., Phil. Trans., Ser. A., A244, 312-320.

NAKAYAMA, P.I. and ROMERO, N.C. (1971), Numerical Method for Almost Three-Dimensional Incompressible Fluid Flow and a Simple Internal Obstacle Treatment. J. Comp. Phys. 8, 230-240.

NICHOLS, B.D. (1970), Recent Extensions to the Marker-and-Cell Method for Incompressible Fluid Flows. Proc. 2nd Int. Conf. Num. Meth. Fl. Dyn., Berkeley, Calif.

NICHOLS, B.D. and HIRT, C.W. (1971), Improved Free Surface Boundary Conditions for Numerical Incompressible Flow Calculations. J. Comp. Phys. 8, 434-448.

NICHOLS, B.D. and HIRT, C.W. (1973), Calculating Three-Dimensional Free Surface Flows in the Vicinity of Submerged and Exposed Structures. J. Comp. Phys. 12, 234-246.

PRACHT, W.E. (1971), A Numerical Method for Calculating Transient Creep Flows. J. Comp. Phys. 7, 46-60.

PRITCHETT, J.W. (1970), The MAC L6 Hydrodynamic Code: A Numerical Method for Calculating Axisymmetric Time-Dependent Free-Surface Fluid Flows at High Reynolds Number. Information Research Associates, Rep. No. IRA-TR-1-70.

RIVARD, W.C., FARMER, O.A., BUTLER, T.D. and O'Rourke, P.J. (1973), A Method for Increased Accuracy in Eulerian Fluid Dynamics Calculations. Los Alamos Sci. Lab., Rep. No. LA-5426-MS.

VIECELLI, J.A. (1969), A Method for Including Arbitrary External Boundaries in the MAC Incompressible Fluid Computing Technique. J. Comp. Phys. 4, 543-551.

VIECELLI, J.A. (1971), A Computing Method for Incompressible Flows Bounded by Moving Walls. J. Comp. Phys. 8, 119-143.

WELCH, J.E., HARLOW, F.H., SHANNON, J.P. and DALY, B.J. (1965), The MAC Method - A Computing Technique for Solving Viscous, Incompressible, Transient Fluid-Flow Problems Involving Free Surfaces. Los Alamos Sci. Lab., Rep. No. LA-3425.

WILLIAMS, J.M. and HOLMES, D.W. (1974), Marker-and-Cell Technique: A Computer Program for Transient Stratified Flows With Free Surfaces. Hyd. Res. Station, Wallingford, Oxfordshire, Eng., Rep. No. INT 134.

YOUNG, J.A. and HIRT, C.W. (1972), Numerical Calculation of Internal Wave Motions. J. Fl. Mech. 56, 265-276.

Numerical Simulation of Fluid Motion, ed. John Noye
pp. 249-259

A DIRECT MACROSCOPIC RELATIVISTIC APPROACH TO THE NUMERICAL SIMULATION OF DYNAMIC SYSTEMS

A.G. Barnett

Ministry of Works and Development

Wellington, New Zealand

SUMMARY

Applications of Galileo's relativity principle to the first law of thermodynamics results in a complete set of macroscopic conservation laws. These general laws are not subject to the limitations introduced in the derivation of the conventional partial differential equations (such as the assumption of continuity) and are therefore more directly related to the requirements of numerical simulation. An example is given which illustrates that the general conservation laws can often be expressed, without approximation, as simple difference equations, although these form a closed numerical model only in the simplest cases.

1 INTRODUCTION

This paper considers the formulation of numerical models of dynamic systems. At present both finite difference methods and finite element methods usually begin from classical systems of partial differential equations such as the Navier-Stokes equations. This tends to focus attention on the consistency of the numerical scheme with the partial differential equations, whereas the real problem is the consistency of the numerical scheme with the information available about the prototype.

It is therefore desirable to seek a direct approach to numerical simulation, discriminating between axiomatic physical concepts and historical conventions of analysis. For instance, is it necessary in fluid dynamics to decide whether matter ultimately acts as particles or a continuum? Why then introduce a continuum hypothesis? Is it necessary to base the analysis on elements fixed in the fluid? Why then work with an Eulerian formulation?

Let us consider the classical thermodynamic approach. This is based on a macroscopic point of view (Zemansky(1968)), involving the consideration of a few fundamental measurable properties of a system. These macroscopic properties are as sure as our senses and will remain unchanged as long as our senses remain the same.

From this viewpoint the first law of thermodynamics for a closed system can be written

$$\Delta E = đW + đQ$$

where E is the internal energy of the system, W is the work done <u>on</u> the system and Q is the cumulative heat entering the system.

E is a state variable, which means that the change ΔE in E is simply the difference between the initial E and the final E, but the "inexact differences" $đW$ and $đQ$ depend on all the intermediate conditions between the initial and final states. Since we are concerned with mechanics, mechanical work only will be considered and another state variable K must be added to represent kinetic energy. The first law of thermodynamics then becomes a general expression of the principle of energy conservation:

$$\Delta(K + E) = đW + đQ \qquad (1.1)$$

This expression is free of assumptions about the microscopic structure of the system and the frame of reference, and no-one will deny that it applies to all mechanical systems of interest. However Equation (1.1) is still too qualitative to form the basis for detailed numerical simulations, so the quantities involved must be precisely defined.

2 PRECISE FORMULATION

The "closed system" can be defined as a region R in physical space enclosed between an initial time t_1 and a final time t_2 by a two sided surface A (i.e. a surface with an "inside" and an "outside"). This surface may move in any way during the time interval provided the velocity u of any surface element dA is finite and single valued at any time, that is, provided the position of the surface never jumps discontinuously.

The total intrinsic magnitude of the material within R is taken as a scalar - the mass of the system. If this is definable it is permissible to represent the mass as an integral of density ρ over the volume of the region regardless of the nature of the microscopic distribution of the density. Even if the mass is literally concentrated in point particles the density is still integrable over any finite region at any instant. Similarly the total kinetic energy of the system can be represented as an integral of $\frac{1}{2}\rho v^2$ over the volume where v is the mass velocity at any point, while the total internal energy is a volume integral of ρe, where e is the point internal energy per unit mass. In other words

$$K + E = \int_R \left(\frac{v^2}{2} + e\right) \rho dR$$

External work may be done first by body forces, which as usual in mechanics, will be taken to be proportional to mass with f the force per unit mass, and second by stresses σ at the surface A of the system. Therefore

$$\Delta W = \int_{t_1}^{t_2}\int_R f.v\ \rho\ dR\ Dt + \int_{t_1}^{t_2}\int_A n.\sigma.v\ dA\ Dt$$

Dt is used here for the infinitesimal time increment to emphasise the relationship with the effective time differencing symbol Δ used in (1.1) and n is the unit outward normal vector to the surface element dA. The contracted product of the vector ndA and the tensor σ is, by convention, the stress force exerted on the material inside the surface by the material outside the surface.

Heat must enter or leave the system through the surface. Therefore, taking q as the heat flux density vector

$$\Delta Q = -\int_{t_1}^{t_2}\int_A n.q\ dA\ Dt$$

Finally, a change in the energy enclosed within the surface A occurs not only as a result of external work and heat flux, but also as a result of the convection of material through the surface, carrying with it kinetic and internal energy. The net energy increase due to convection is therefore

$$\int_{t_1}^{t_2}\int_A n.(v-u)\left(\frac{v^2}{2} + e\right)\ \rho\ dA\ Dt$$

Summing up, Equation (1.1) can be formulated precisely as

$$\left[\int_R \left(\frac{v^2}{2} + e\right)\rho dR\right]_{t_1}^{t_2} + \int_{t_1}^{t_2}\int_A n.(v-u)\left(\frac{v^2}{2} + e\right)\rho dADt$$

$$= \int_{t_1}^{t_2}\int_R f.v\rho dRDt + \int_{t_1}^{t_2}\int_A n.\sigma.vdADt - \int_{t_1}^{t_2}\int_A n.qdADt \tag{2.1}$$

3 THE FRAME OF REFERENCE

So far the frame of reference has not been considered except for a mention of physical space and time. The postulates of Newtonian mechanics are adequate in the present context, so the frame of reference is inertial. According to Galileo's relativity principle (Landau and Lifshitz (1972)) all the laws of nature are the same in every inertial frame. Therefore equations such as (2.1) which express the laws of nature should have the same form when referred to various inertial frames.

In other words all velocities in (2.1) are understood to be relative to some inertial frame, and Equation (2.1) should be equally valid when the velocities are related to some arbitrary inertial frame in which the first inertial frame moves at an arbitrary fixed velocity U. The relativity principle therefore requires

$$\left[\int_R \left(\frac{(v+U).(v+U)}{2} + e\right)\rho dR\right]_{t_1}^{t_2} + \int_{t_1}^{t_2}\int_A n.(v+U-(u+U))\left(\frac{(v+U).(v+U)}{2} + e\right)\rho dADt$$

$$= \int_{t_1}^{t_2}\int_R f.(v+U)\rho dRDt + \int_{t_1}^{t_2}\int_A n.\sigma.(v+U)dADt - \int_{t_1}^{t_2}\int_A n.qdADt$$

Rearranging

$$\left[\int_R \left(\frac{v^2}{2} + e\right)\rho dR\right]_{t_1}^{t_2} + \int_{t_1}^{t_2}\int_A n.(v-u)\left(\frac{v^2}{2} + e\right)\rho dADt$$

$$- \int_{t_1}^{t_2}\int_R f.v\rho dRDt - \int_{t_1}^{t_2}\int_A n.\sigma.vdADt + \int_{t_1}^{t_2}\int_A n.qdADt$$

$$+ U.\left\{\left[\int_R v\rho dR\right]_{t_1}^{t_2} + \int_{t_1}^{t_2}\int_A (n.(v-u))v\rho dADt - \int_{t_1}^{t_2} f\rho dRDt - \int_{t_1}^{t_2}\int_A n.\sigma dADt\right\}$$

$$+ \frac{U^2}{2}\left\{\left[\int_R \rho dR\right]_{t_1}^{t_2} + \int_{t_1}^{t_2}\int_A n.(v-u)\rho dADt\right\} = 0$$

For this equation to be true with U arbitrary in magnitude and direction

$$\left[\int_R \rho dR\right]_{t_1}^{t_2} + \int_{t_1}^{t_2}\int_A n.(v-u)\rho dADt = 0 \qquad (3.1)$$

and since the dot product of an arbitrary vector and a given vector is zero only if the given vector is a null vector

$$\left[\int_R v\rho dR\right]_{t_1}^{t_2} + \int_{t_1}^{t_2}\int_A (n.(v-u))v\rho dADt$$

$$= \int_{t_1}^{t_2}\int_R f\rho dRDt + \int_{t_1}^{t_2}\int_A n.\sigma dADt \qquad (3.2)$$

It can be seen that Equation (3.1) is a mass conservation equation and Equation (3.2) is a vector momentum conservation equation. If both Equations (3.1) and (3.2) are satisfied the transformed equation reverts to Equation (2.1). If Equation (3.2) is transformed in the same way, Equations (3.1) and (3.2) are the result, while Equation (3.1) is independent of the inertial frame.

Equations (2.1), (3.1) and (3.2) therefore form a complete set satisfying Galileo's relativity principle, and it is clear that the conservation of energy, momentum and mass is self-consistent in a very fundamental way.

It is not possible to proceed further without sacrificing generality - for instance it is useless to introduce Cartesian coordinates if the problem involves flow in a cylindrical pipe, or to introduce an Eulerian Formulation if the boundary conditions apply to a moving free surface, or to assume continuity if shocks are likely to develop in the flow.

The method of application of the general equations will therefore be illustrated with an example.

4 METHOD OF APPLICATION

The example considered is from the very important class of problems where the material can be taken to have constant density, and where the body force arises from a uniform gravitational field. The assumption of constant density uncouples the momentum equation from the energy equation, which can therefore be neglected in this example. Also, assuming the region R does not extend outside the material of constant density,

$$\int_R f\rho dR = \rho\int_R f dR\,.$$

In a uniform gravitational field such as that which can be assumed at the surface of the earth, the body force f per unit mass may be taken as minus the gradient of a scalar potential gh, where g is the constant acceleration of gravity and h is the vertical height above some arbitrary horizontal datum. Therefore

$$\int_R f\,dR = \int_R -\nabla(gh)\,dR$$

$$= \int_A -n(gh)\,dA$$

using Gauss's Theorem.

The momentum conservation equation then becomes

$$\left[\int_R v\rho\,dR\right]_{t_1}^{t_2} + \int_{t_1}^{t_2}\int_A \rho\{(n.(v-u))v + ngh - n.\sigma/\rho\}\,dA\,Dt = 0 \qquad (4.1)$$

The assumption of constant density therefore transforms the momentum conservation equation into a time difference and a surface integral, the same form as the mass conservation equation.

Equations (3.1) and (4.1) can therefore both be expressed

$$L(t_2) - L(t_1) = \sum_{1A}^{n} I_r \qquad (4.2)$$

where L is the mass or momentum level of the system at any instant and I_r is mass or momentum influx between t_1 and t_2 through the rth of n discrete sections into which the bounding surface A can be divided.

Equation (4.2) is a difference equation which represents (3.1) or (4.1) without approximation. It is therefore extremely attractive for the purposes of numerical simulation, but assuming $L(t_1)$ is given as an initial condition the mass equation and three component momentum equations represented by (4.2) still contain 4n unknown flux components and 4 unknown level components. Considerably more information is therefore needed for closure. This information must be derived from the problem specification.

If the problem is a simple mass balance problem such as a lake with known inflow and outflow, Equation (4.2) alone is sufficient to model the mass level. For a less trivial example, consider a problem of unsteady flow in a river for which the discharge is known at some upstream cross-section and the water level is known at the mouth.

The end boundaries of the control surface are therefore fixed at the upstream cross-section (say section s_0) and at the mouth (say section s_1). The remainder of the control surface is the channel bed, assumed fixed, and the free surface.

First consider a temporary control surface defining a thin slice approximating as closely as desired to an end cross-section. Taking any component of (4.1) parallel to the slice

$$\int_{t_1}^{t_2}\int_A \rho\{ngh - n.\sigma/\rho\}\,dA\,Dt = 0$$

since the component of v parallel to the slice is assumed negligible.

Now, substituting

$$\sigma = -pI + \sigma'$$

where p is the hydrostatic pressure and I is the unit 2-tensor, the above equation becomes

$$\int_{t_1}^{t_2}\int_A \rho\{ngh+np/\rho-n.\sigma'/\rho\}dADt = 0$$

Now, assuming shears parallel to the sides of the slice can be neglected, only the edges of the slice have unit normals with components in the direction considered. Since the edges of the slice can be placed anywhere in the cross-section it is apparent that $p/\rho + gh$ is constant throughout all cross-sections satisfying the above assumptions.

Selecting a frame in which the end sections are fixed means that $u \equiv 0$ at those sections. Using this, the component of (4.1) in the direction of flow is, after division by ρ,

$$[\int_R vdR]_{t_1}^{t_2} + \int_{t_1}^{t_2}\int_{A_{ends}} n.vvdADt + \int_{t_1}^{t_2}\int_{A_{top}} n.(v-u)vdADt$$

$$= \int_{t_1}^{t_2}\int_{A_{ends}} -n(p/\rho+gh)dADt + \int_{t_1}^{t_2}\int_{A_{bed}} n.\sigma'/\rho dADt$$

The mass conservation Equation (3.1) becomes, after division by ρ,

$$[\int_R dR]_{t_1}^{t_2} + \int_{t_1}^{t_2}\int_{A_{ends}} n.vdADt = -\int_{t_1}^{t_2}\int_{A_{top}} n.(v-u)dADt$$

For closure these two equations must be expressed throughout as functions of two unknown dependent variables or known quantities in two difference equations. For instance

$$[\int_R dR]_{t_1}^{t_2}$$

is simply the volume between the free surface at time t_2 and that at time t_1, and this can be related to the difference between the free surface heights at times t_1 and t_2 at either end. Where the free surface is rectangular in plan and of constant width the required volume can be evaluated by a simple trapezoidal rule (method of end areas) to give

$$[\int_R dR]_{t_1}^{t_2} = R(t_2) - R(t_1)$$

$$= \tfrac{1}{2}B\Delta s(y_1(t_2) - y_1(t_1) + y_0(t_2) - y_0(t_1)) \qquad (4.3)$$

where B is the surface width, Δs the surface length and y_1, y_0 are the free surface elevations at the two ends.

The conventional approach produces a term $B\partial y/\partial t$ in place of the above variation in volume - the distance and time mesh widths Δs and Δt have been divided throughout and allowed to tend to zero to produce the required partial differential coefficients. An expression equivalent to (4.3) can therefore readily be obtained by differencing the timewise rate of change of y as an average of the rate of change of y at the two ends.

However where the free surface is irregular in plan and of width varying with depth it is possible to modify expression (4.3) realistically only by proceeding directly from the conservation laws. Then the volume of the region R can be viewed as a function of y_1 and y_0 such that

$$R(t_2) - R(t_1) = R(y_1(t_2), y_0(t_2)) - R(y_1(t_1), y_0(t_1))$$

The change in volume can therefore be related to the change in y_1 and y_0 by any method giving the desired accuracy.

The expression by general functional relationships of the gross characteristics, such as volume, of a system in terms of a few macroscopic coordinates, such as the two surface elevations, is typical of thermodynamic equations of state. The latter are needed for closure of systems of partial differential equations representing energy, momentum and mass conservation, and similarly closure of the exact difference equations (4.2) can be achieved by the specification of enough functional constraints relating the gross characteristics of the system to a limited set of macroscopic coordinates. In the above example the necessary constraint equation is

$$R = R(y_1, y_0) \tag{4.4}$$

As with equations of state, such constraint equations must be matched directly to the properties of the prototype system.

5 BOUNDARY CONDITIONS

As suggested in the above example, part of the formulation of the model involves identifying the surface of the system with surfaces where something is known about conditions. In the example the end surfaces were fixed relative to the reference frame, with additional information available about the mass discharge at one end and the water level (and hence pressure) at the other end. The free surface was a surface through which known mass flows and pressures applied, while below the bed was fixed and impermeable. The channel sides were also fixed and impermeable.

It is of interest to restate these conditions in terms of fluxes - for instance a surface moving at velocity u can be regarded as transmitting a flux of reference space of density unity. The end surfaces therefore have known space flux (zero as they are fixed) together with a known mass flux at one end and a known momentum flux, apart from momentum convected by the mass flux, at the other end. The free surface has a known mass and momentum flux and the bed has a known space and mass flux. The two channel sides also have known space and mass flux.

The unknowns can therefore be paired off at opposite boundaries - a pressure momentum flux and a mass flux at the ends, a space flux and a pressure momentum flux at the top and bottom, and a pressure momentum flux at the sides.

A fuller discussion of the boundary conditions of this particular example is not of general interest and will not be pursued further. However there is a clear indication that if boundary conditions are to be physically relevant to the bounding surface, such conditions must be understood in terms of fluxes of conserved quantities (space is naturally a conserved quantity). It is also clear that the motion of the bounding surface depends entirely on the type of boundary conditions associated with the individual problem, as the free surface can only be modelled with Eulerian elements if the space flux is zero and with Lagrangian elements if the mass flux is zero. It is therefore undesirably restrictive to introduce such specialised elements into the analysis at the conventional early stage.

6 ANALYTICAL SOLUTIONS

Although available only in very restricted circumstances, analytical solutions are still valuable for benchmark tests of numerical solutions and as a means of evaluating the qualitative effects of various parameters.

The usual basis for analytical solutions are partial differential equations which may readily be derived from the general conservation laws (2.1), (3.1), and (3.2) as follows:

(a) The continuum hypothesis is introduced.

This assumes that all the necessary partial derivatives exist at all times everywhere in the solution. Infinitesimal time intervals may then be considered together with infinitesimal volume elements, but the latter depend on the formulation adopted.

(b) Either a Lagrangian or Eulerian volume element is adopted.

(i) Lagrangian - $u \equiv v$

The convective terms drop out of equations (2.1), (3.1) and (3.2) and the mass conservation equation becomes simply

$$\frac{D}{Dt}(\rho dR) = 0$$

However the momentum and energy conservation equations are less simple.

(ii) Eulerian - $u \equiv 0$

Since dR is now fixed it is possible to divide throughout by dR and introduce the definition (Milne-Thomson (1972)).

$$\nabla . X = \lim_{dR \to 0} \frac{1}{dR} \int_A n . X dA$$

The surface integrals then become gradient terms and the mass conservation equation becomes

$$\frac{\partial \rho}{\partial t} + \nabla . (v\rho) = 0$$

The momentum and energy equations follow similarly.

(c) In most cases the Eulerian formulation is presented in Cartesian tensors, implying that the fixed volume element is an infinitesimal rectangular parallelepiped. The mass conservation equation is then further restricted, becoming

$$\frac{\partial \rho}{\partial t} + \frac{\partial}{\partial x_i} (v_i \rho) = 0$$

7 SUMMARY AND CONCLUSIONS

A complete set of macroscopic conservation laws, comprising laws of energy, momentum and mass conservation, results from the application of Galileo's relativity principle to the first law of thermodynamics.

These general laws are not subject to the limitations introduced in the derivation of the conventional partial differential equations such as the assumption of continuity, an Eulerian reference frame and Cartesian coordinates and are therefore more directly related to the requirements of numerical simulation. The application of the laws in general form does not permit the use of the automatic manipulative rules available for partial differential equations but with some thought equivalent manipulations can be devised for the general approach.

An example assuming constant material density and a uniform gravitational field illustrates that the general conservation laws can often be expressed without approximation as simple difference equations although these form a closed numerical model only in the simplest cases. Normally closure must be achieved by the addition of sufficient functional constraint equations relating the gross characteristics of the system to a limited set of macroscopic coordinates. These constraint equations are analagous to thermodynamic equations of state.

The motion of the bounding surface of the system depends entirely on the type of boundary conditions associated with the individual problem, and therefore specialised volume elements such as fixed Eulerian elements should not automatically be introduced into the analysis.

Where analytical solutions are felt to be necessary the conventional partial differential formulations of the conservation laws can readily be obtained from the given general formulation by the introduction of the continuity hypothesis, the required element description and the desired coordinate system.

The approach advocated is a general numerical modelling method, requiring considerable skill and intuition as the standardised frameworks of finite difference or finite element methods are not available. The constraint equations in particular will be quite different in different problem areas. However each final numerical scheme will be directly matched to the requirements of the problem so the accuracy of the simulation can be referred directly to the prototype.

REFERENCES

Landau, L. D. and E. M. Lifshitz (1972) Mechanics and Electrodynamics - A Shorter Course of Theoretical Physics Volume I. 1st Edition, Pergamon Press, Oxford.

Milne-Thomson, L. M. (1972) Theoretical Hydrodynamics 5th Edition, MacMillan Press, London.

Zemansky, M. W. (1968) Heat and Thermodynamics 5th Edition, McGraw-Hill, New York.

Acknowledgement

This paper was prepared in the Systems Laboratory, Ministry of Works and Development, New Zealand, by the kind permission of the Commissioner of Works.

Numerical Simulation of Fluid Motion, ed. John Noye
pp. 261-284

COMPARISON BETWEEN METHODS - A REVIEW

R.A. Pearson

Monash University

Victoria, Australia

SUMMARY

A number of previously published comparisons between the different ways of modelling fluid flows are summarised. Although it seems that no definitive statements can be made, some general conclusions seem justified. Overall, the particular modelling situation will help in the choice of technique. For simple flows within rigid boundaries at high Reynolds numbers, a Galerkin, or pseudo-spectral, technique will give the best results. For steady state calculations, it has been shown that a finite difference technique can be more efficient than the Galerkin technique. If a general open boundary is required, finite difference methods should be used. The comparative advantages of finite element methods has not been examined in sufficient detail. However, where information is required on an irregular grid, or flow at low Reynolds number is considered, a finite element model should have an advantage over other techniques.

INTRODUCTION

Since the early attempts at modelling fluid flows (Richardson 1923) the application of numerical techniques has proceeded concurrently with the development of new methods and the examination of their properties and errors. The continued need for adequately resolving the models has placed more emphasis on completely understanding the properties of the numerical techniques. Many of these properties can be demonstrated using analytical methods (e.g. Lilly 1965). Alternatively errors can be shown by examining the numerical solutions to simple flows.

Both of these approaches yield valuable information. In general the models discussed can be broadly classified into three classes. One class considers the advection of a passive scalar (e.g. Molencamp 1968). Another uses models whose linear analogue has wave type solutions (e.g. Grammeltvedt 1970). In the third class are fluid flows driven by temperature differences. This class includes both elliptic problems (Denny & Clever 1974) and time dependent solutions (Pearson & O'Connor 1976).

In the first class a very common test is to consider a fluid with solid body rotation. A passive scalar is advected where its initial values are zero except in a small region off the axis of rotation. With the advecting field well resolved this test demonstrates the effect of truncation errors and resolution on the scalar field. Using such an experiment Orszag (1971) shows that for a given error, fourth order schemes require less than half the resolution of second order schemes. For a given resolution the Fourier technique will yield much more accurate simulations (see Section A for this test).

For problems with wave-like solution an analytical examination can yield very useful information. In particular it can be shown how the amplitude and phase of the waves are influenced by the numerical technique. Using this approach Kriess & Oliger (1973) suggest that schemes of higher order accuracy should give improved results. Grotjahn & O'Brien (1975) consider the errors in the group velocity introduced by difference schemes. For the short wavelength disturbances and fourth order schemes energy propagates in the opposite direction to that of the continuous system. This error suggests that for some problems a lower order scheme, but with increased resolution, should be used.

In an experimental test of the one dimensional wave equation Orszag & Israeli (1974) give the spatial distribution of errors for both finite difference and Galerkin techniques. Such detailed pointwise evaluations are not common for more complicated models of this class. It is much simpler to show a mean error or examine global conservation properties. Grammeltvedt (1970) examines conservation of energy while Sadourny (1975) considers the distribution of energy between wave-numbers.

The third class of test problems has not been used as often as members of the other two classes. A model of steady state convection between two parallel flat plates has been used to compare Galerkin and finite difference solutions (Denny & Clever 1974). Time dependent numerical models of isolated parcels of warm fluid (or thermals) can show a dramatic instability not apparent in models of the other two classes (Pearson & O'Connor 1976). A very simple finite element model which has been tested shows the instability more drastically than a finite difference technique. A Galerkin expansion in orthogonal functions has not been tested for this form of instability.

A. A TWO DIMENSIONAL ADVECTION TEST OF GALERKIN AND FINITE DIFFERENCE METHODS

A.1 The Model

The test described here is the most common in the first class comparisons. A finite fluid moving with solid body rotation is considered to have a passive scalar property (A). The initial perturbation of this property is zero except in a small region of the fluid. Under the advection the scalar (A) retains its shape. Errors introduced by the numerical technique are shown by the perturbations from the initial conditions. Generally the initial value of A is zero except in a circle. At the centre it has value one and decreases radially to zero (Crowley 1968, Molencamp 1968, Orszag 1971). It should be noted that these initial conditions have a discontinuous slope, with large changes in gradient at the top and sides of the cone (Fig. 1).

In the comparison between finite difference and Fourier techniques the fluid flow is to be confined to a square area with boundary values periodic in both directions. Although this boundary condition was specified by the author to be "for convenience" it does ensure that the Fourier series converges uniformly over the domain.

A.2 The Schemes

Both second and fourth order finite difference analogous to the advectible terms were used. All finite difference schemes were semi-conservative and semi-quadratically conservative.

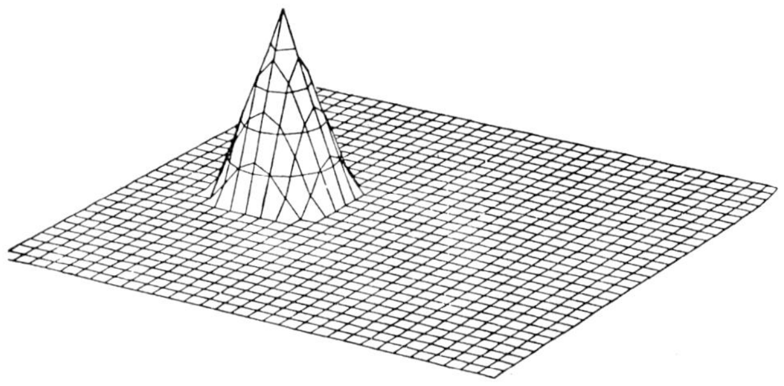

Figure 1: Initial conditions of the passive scalar for the two dimensional test of advection. Basic flow field rotates this "cone" about the centre of the field (from Orszag 1971).

One second order scheme was Arakawa's Jacobian (Arakawa 1966). The other was a staggered second order scheme (Lilly 1961). The fourth order scheme was the fourth order Arakawa Jacobian (1966)*.

The Galerkin (Fourier) approximation considered is a modification of the fast Fourier technique for evaluating convolution sums (Orszag 1971). This modification was introduced to retain the same information on the Fourier space as that on the finite difference grid.

A.3 Results

Only a general summary of the results will be included here. A detailed description is available in the original paper (Orszag 1971). The fields after one revolution are shown in Figure 2. In general the finite difference realisations have a reduced amplitude of the maximum. A trailing wake of spurious non-zero numbers is also observed. The position of the maximum also lags the correct position. These values of bad numbers are consistent with the lagging phase errors in the second order finite difference schemes. The magnitudes of these three errors allow an evaluation of the numerical technique. The values for different schemes and resolutions are given in Table 1.

Numerical method	Order of scheme	Mesh	Number of quarter-revolutions	Maximum at a grid point	Minimum	Lag of maximum (in radians)
Arakawa	Second	96×96	1	0·85	−0·04	0·08
Arakawa	Second	32×32	1	0·79	−0·13	0·19
Arakawa	Second	32×32	4	0·51	−0·23	0·44
Arakawa	Fourth	64×64	1	0·92	−0·03	0·03
Arakawa	Fourth	64×64	4	0·84	−0·04	0
Arakawa	Fourth	32×32	1	0·89	−0·05	0
Arakawa	Fourth	32×32	4	0·83	−0·10	0·08
Arakawa	Fourth	32×32	8	0·74	−0·15	0·08
Fourier	Infinite	32×32 ($K = 16$)	1	0·98	−0·02	0
Fourier	Infinite	32×32 ($K = 16$)	4	0·98	−0·02	0
Fourier	Infinite	16×16 ($K = 8$)	1	0·97	−0·03	0
Fourier	Infinite	16×16 ($K = 8$)	4	0·96	−0·04	0
Fourier	Infinite	16×16 ($K = 8$)	8	0·89	−0·07	0
Fourier	Infinite	16×16 ($K = 8$)	12	0·79	−0·11	0

TABLE 1. Accuracy of numerical simulation of scalar convection

* There is an error in Arakawa 1966. All terms within round brackets in eq. (58) should be minus signs.

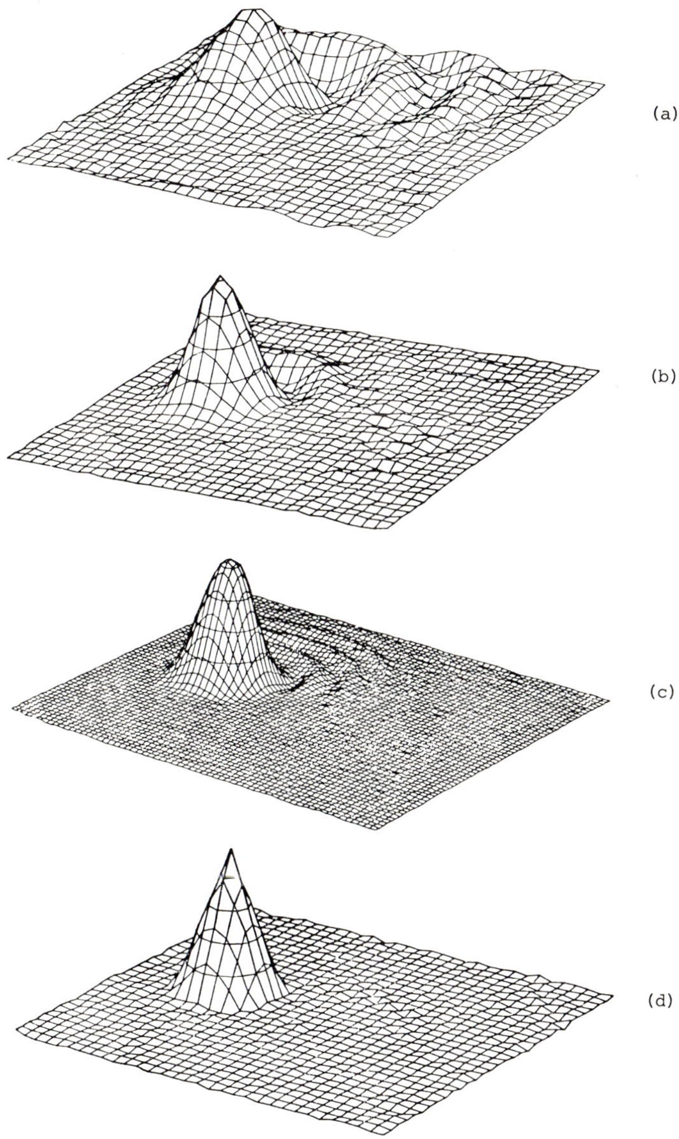

Figure 2: Three-dimensional (x_1, x_2, A) perspective plot of the $A(x,t)$ field obtained after 1 revolution using (a) second-order Arakawa scheme on 32×32 space grid; (b) fourth-order Arakawa scheme on 32×32 space grid; (c) fourth-order Arakawa scheme on 64×64 space grid; (d) cut-off Fourier-expansion scheme on 32×32 space grid.

Overall the following rules seem to apply. A fourth order scheme gives better results than a second order scheme with twice the resolution. A Fourier technique gives results that are comparable to a fourth order scheme with twice the resolution. The amount of computer time required for a Fourier model is not significantly larger than that needed for a finite difference scheme with the same resolution. For highly accurate resolution, in a model where advection dominates, the Galerkin method is preferable.

B. TESTS ON THE ONE DIMENSIONAL WAVE EQUATION

B.1 The Model

Orszag & Israeli (1974) compare Galerkin and finite difference techniques using the one dimensional wave equation. Unfortunately once again no finite element method is considered. The test equation is

$$u_t + u_x = 0, \quad -1 < x < 1 . \qquad \text{B.1.1}$$

B.2 The Finite Difference Solutions

For the second order scheme in the interior a first order scheme at the boundary retains overall second order accuracy. The fourth order internal analog requires a third order boundary condition.

With a boundary condition $u(x,0) = 0$ and forcing $u(-1,t) = \sin m\pi t$ the exact solution is,

$$\begin{aligned} u(x,t) &= 0 & t \leqslant x+1 \\ &= \sin m\pi(t-x-1), & t > x+1 . \end{aligned}$$

Internally a second order analog to B.1.1 is,

$$\frac{\partial u_n}{\partial t} + \overline{u}^x_x = 0, \qquad \text{where} \quad \overline{\alpha}^x_x \equiv D_0\alpha = (\alpha_{n+1} - \alpha_{n-1})/2h ,$$

while fourth order approximation is,

$$\frac{\partial u}{\partial x} = \overline{u}^x_x - \frac{1}{6}\overline{u}^x_{xxx} .$$

At the downstream boundary a condition that allows a wave to propagate through the boundary is required (see e.g. Pearson 1974). To first order this is given by

$$\frac{\partial u_n}{\partial t} + D_- u_n = 0, \qquad \text{where} \quad [D_- = (u_n - u_{n-1})/h] .$$

The asymptotic form of the errors can be calculated analytically. The L_2 errors for each of the finite schemes are given by

$$\varepsilon_2 \sim 5.967\ m^3h^2$$

$$\varepsilon_4 \sim 11.779\ m^5h^4 .$$

These L_2 errors give a measure of the error spread over the whole domain under consideration.

A pointwise evaluation is also of interest. For this particular problem the solution has a discontinuous derivative at x = t-1. Near this discontinuity, errors are large. Maximum error is also evaluated asymptotically. For the finite difference schemes these errors are

$$\varepsilon_2' \approx 0.645\ mh^{2/3}t^{1/3}$$

$$\varepsilon_4' \approx 0.561\ mh^{4/5}t^{1/5}\ .$$

The pointwise error, for schemes of the same resolution but different order are shown in Figure 3. Note that the second order finite difference scheme has an exponential error in front of the discontinuity and in region of trailing wakes. The fourth order scheme has a small negative section before the discontinuity, with trailing errors behind. This suggests that higher order schemes should be used. However this advantage is not clear in other models.

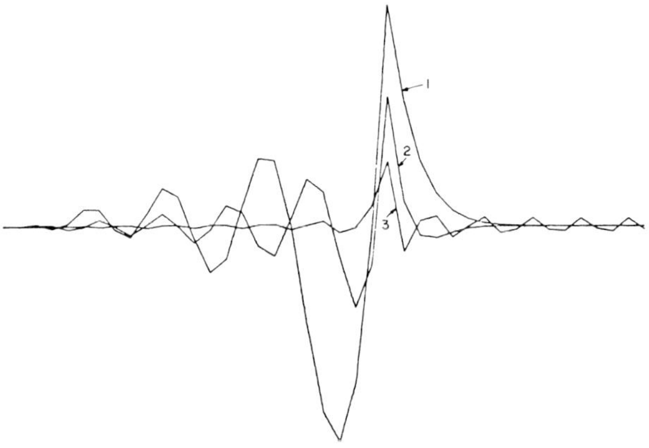

Figure 3: Spatial distribution of error in numerical solution of (8) at t=1.2 when the discontinuity is at x=0.2 (the location of maximum error in all three curves). All curves are drawn for N=40, m=1, and are normalized with respect to the maximum error of the second-order scheme. Curve 1: second order. Curve 2: fourth order. Curve 3: Chebyshev (spectral). (from Orszag & Israeli 1974).

B.3 "Spectral" Methods

There are a number of methods to obtain a series solution to this problem. Care must be taken in choosing the basis functions which should converge uniformly on the closed interval. This condition does not allow Gibbs phenomena. There are three ways of obtaining a solution - these are referred to as Galerkin, Collocation (or pseudo-spectral) and Tau method. For the one dimensional wave equation Chebyshev polynomials are the appropriate basis functions. Each method of solution requires the boundary condition to be explicitly stated. For the Galerkin and Collocation approach the final problem has homogeneous boundary conditions. Unfortunately a non-homogeneous boundary condition, which depends implicitly on the interior flow, no longer gives simple orthogonal conditions on the expansion coefficients. Once again the Fourier transform method for evaluating the convolution sums of a Chebyshev polynomial expansion is used.

The results of the Collocation technique are given in Figures 3 and 4. A 5% error requires no more than 4 polynomials per wavelength. The maximum pointwise error is given asymptotically by $\varepsilon_{\infty}^{!} = \frac{1}{2}mh$. Note however that there is a leading oscillating error ahead of the discontinuity.

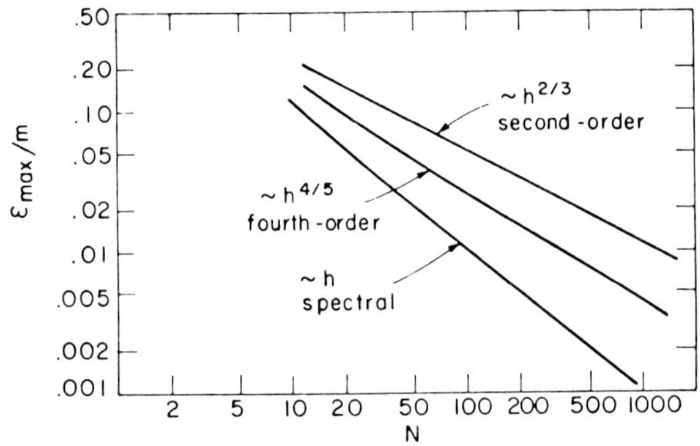

Figure 4: Maximum pointwise error ε_{max} in the numerical solution of (8) at t=1.2 when the discontinuity is at x=0.2. Here N=2/h is the number of grid points, or of Chebyshev polynomials in the spectral method (from Orszag & Israeli 1974).

B.4 Conclusions

The errors are much smaller, for a given resolution, with the Chebyshev polynomial expansion (Collocation technique) than both second and fourth order finite difference methods.

A non-linear modification of this test equation, $u_t = uu_x$ is very commonly used as a test of finite difference methods. This equation is very useful in an examination of non-linear stability. A comparison between Chebyshev, finite difference, and finite element techniques using this equation should give very useful comparison of possible errors.

C. CONVECTION - GALERKIN AND FINITE DIFFERENCE

C.1 The Model

While the previous two comparisons were essentially hyperbolic or parabolic in nature the problem discussed below is elliptic. It falls broadly into the third class of test problem. Once again only the summary will be given here. Details of the model and its implementation are given by Denny & Clever (1974).

For convection between two plates the governing equations are:

$$\nabla^4\psi - \cos\gamma \frac{\partial T}{\partial x} - \sin\gamma \frac{\partial T}{\partial z} = 0 ,$$

$$\frac{\partial\psi}{\partial z}\frac{\partial T}{\partial x} - \frac{\partial\psi}{\partial x}\frac{\partial T}{\partial z} = Ra^{-1}\nabla^2 T ,$$

where $u = \frac{\partial\psi}{\partial z}$, $w = -\frac{\partial\psi}{\partial x}$, γ is the inclination to gravity of the parallel plates, and $Ra = \frac{\alpha g \Delta T d^3}{\nu\kappa}$.

The results of this problem can be characterised by the heat transfer which is given by the Nusselt number,

$$Nu = - \int_0^1 \frac{\partial T}{\partial x}\, dz \Big|_{x=0} .$$

C.2 The Methods

The Alternating Direction Implicit (A.D.I.) scheme is the finite difference technique considered.

For the Galerkin method special trail functions (see Denny & Clever 1974) were used. The set of coefficients that result from this expansion was obtained using Newton-Raphson.

C.3 Results

Comparisons were obtained between both forms of solution at different inclinations and different Rayleigh numbers. For each Rayleigh number the solutions at the previous Rayleigh number were used as a first guess. Both the Nusselt number and the dynamical properties of the flow field are examined at the steady state.

For the heat transfer it was observed that, with increasing resolution, the finite difference technique converged monotonically from above (Figures 5, 6). At low resolution the Galerkin solutions under-estimate the heat transfer. As the resolution increases the Galerkin technique gives a Nusselt number that oscillates slowly about a mean value (Figures 5, 6).

Examining the flow fields shows that the finite difference model represents the boundary layer and the core and localises the error in an over-estimate of the stream function. The Galerkin technique shows errors where the velocities are small and near the boundary layer (Figures 7, 8, 9, 10). This latter error seems consistent with the oscillating error preceding the discontinuity in the previous test.

For a very high accuracy (5% lower) in both flow field and Nusselt number a finite difference technique is preferable. The Galerkin procedure

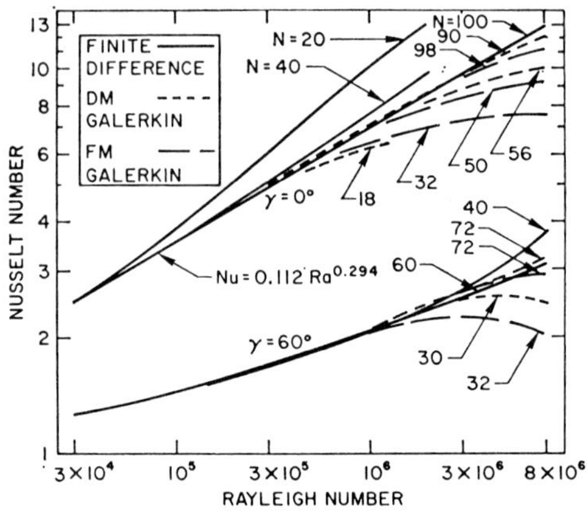

Figure 5: Convergence characteristics of Nusselt number at heated wall with increasing Rayleigh number (from Denny & Clever 1974).

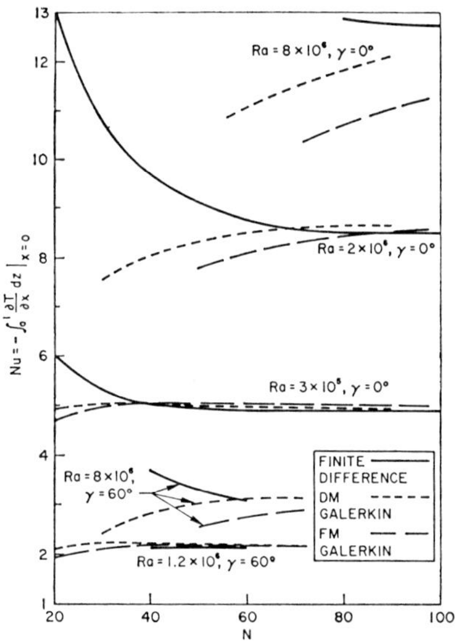

Figure 6: Convergence characteristics of Nusselt number at heated wall with increasing number (N) of strips/terms (from Denny & Clever 1974).

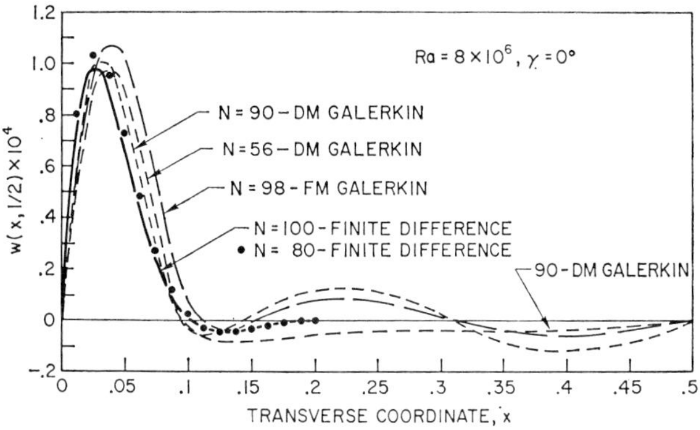

Figure 7: Velocity profile at midheight, $\gamma = 0^\circ$ (from Denny & Clever 1974).

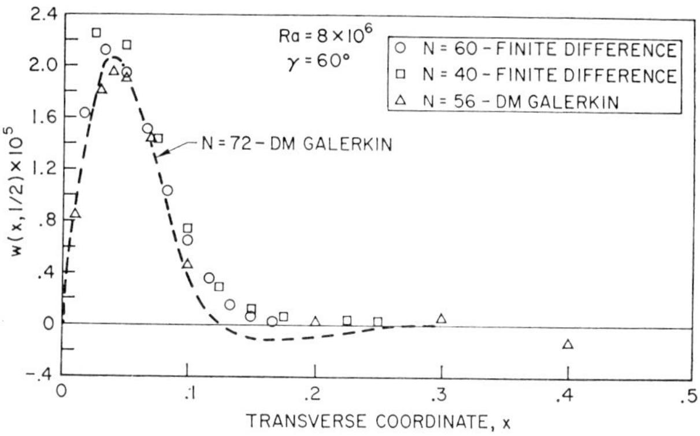

Figure 8: Velocity profile at midheight, $\gamma = 60^\circ$ (from Denny & Clever 1974).

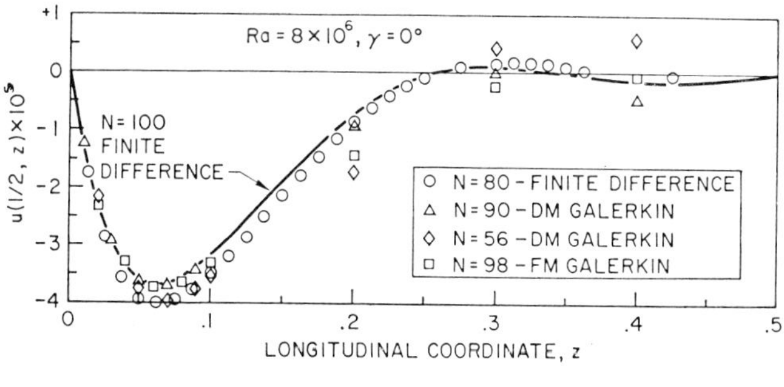

Figure 9: Velocity profile at midplane, $\gamma = 0^\circ$ (from Denny & Clever 1974).

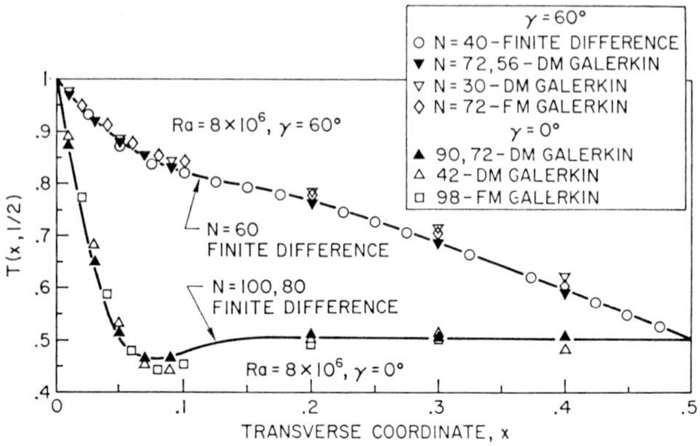

Figure 10: Temperature profile at midheight, $\gamma = 0^\circ$, 60° (from Denny & Clever 1974).

is much more efficient for simulations of less accuracy. For errors of 10% about $^1/_2$ to $^1/_3$ the computer time is required for the Galerkin technique.

D. FINITE ELEMENT - FINITE DIFFERENCE METHODS

Historically the finite element method has been used extensively in engineering, where most applications involve elliptic partial differential equations. These problems have also a large viscous damping. Fluid flows have usually a large Reynolds number and are often hyperbolic or elliptic in nature. The literature has numerous examples of simple test problems for finite difference methods, and a few simple test problems for finite element techniques. Very little has been published on tests similar to fluid flows solved by finite element techniques especially a complete comparison which considers error bounds, or computer time, and space.

The main topic of the paper I consider in this section is an application of a finite element technique to circulation in a lake. The steady state circulations considered here exist with a balance between the input by the wind and the dissipation (drag exerted by the bottom of the lake).

A rectangular lake with the depth a triangular cross-section is considered (Figure 11). To avoid the singularity at zero depth the flow region inside the contour of 5% maximum depth is considered. The finite element method allows a very simple partition of the model domain into non-equal areas (Figure 12).

For the finite difference calculations 1701 points were used. For the finite element method the number of solution parameters were: 627 for the cubic, 209 for the linear field, and 149 for the quadratic field. All methods give similar results (Figures 13, 14). Unfortunately no evaluation of error bounds, either numerical or analytic, is given. No comparable time tests or number of operations required are given in the paper. The author concludes that results show that the higher order element

Finite element analysis

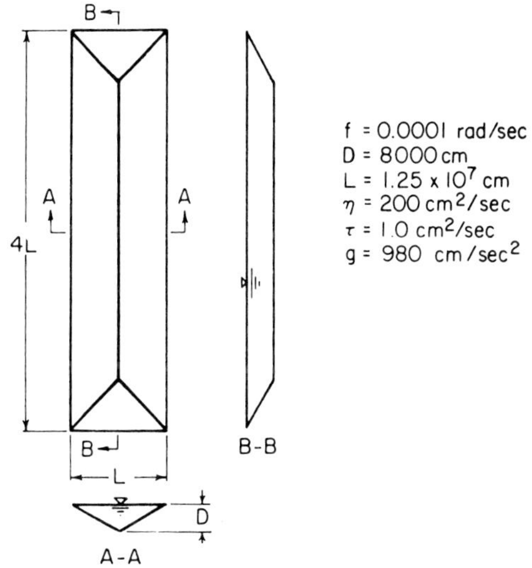

Figure 11: Rectangular lake (from Gallagher & Chan 1973).

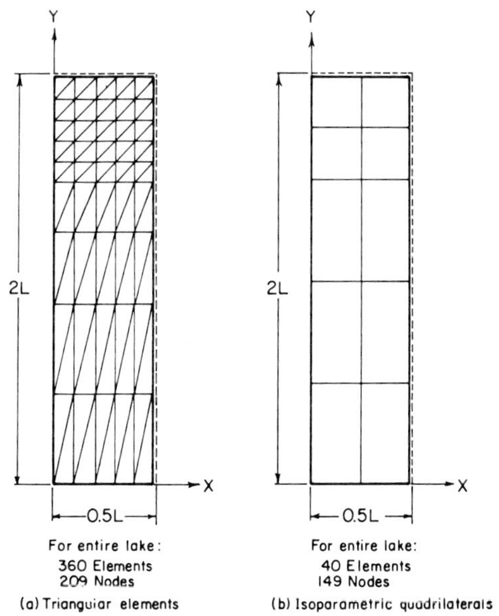

Figure 12: Finite element representation of rectangular lake (from Gallagher & Chan 1973).

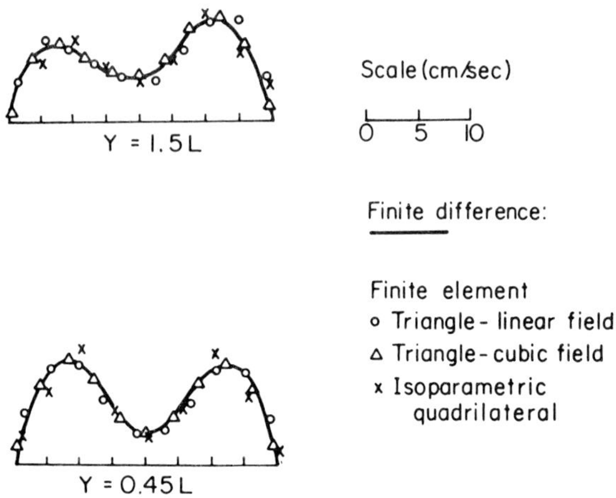

Figure 13: Comparison of total velocity solutions on rectangular lake surface at representative sections (from Gallagher & Chan 1973).

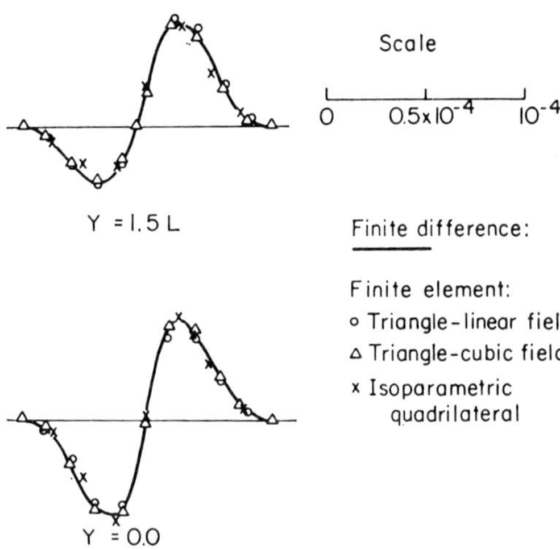

Figure 14: Comparison of stream function solutions for rectangular lake at representative sections (from Gallagher & Chan 1973).

representations give better accuracy and efficiency. However I do not see any hard evidence for this inference.

E. ANALYSIS OF ERRORS

E.1 Amplitude Errors

Although this error did not occur in any of the finite difference schemes used it is included for completeness.

Consider the equation

$$\frac{\partial A}{\partial t} + u\frac{\partial A}{\partial x} = 0$$

with solution $A = \exp\left(ik(x-ut)\right)$.

With a first order upstream difference scheme

$$A_j^{\tau+1} = A_j^{\tau} - u\left(A_{j+1}^{\tau} - A_j^{\tau}\right)$$

the solutions of this equation are of the form

$$A_j^{\tau} = X^{\tau}\, e^{i\theta j}$$

where $\alpha = u\Delta t/\Delta x$,

$$|X|^2 = 1 + 2\alpha(1-\cos\theta)(\alpha - 1)$$
$$\lesssim 1 .$$

As in the continuous system $|X| = 1$ all waves are damped when using this method.

E.2 Phase Errors

These errors occur as the finite difference (and probably finite element method) give the phase of a wave in the discrete system different from that in the continuous system.

The centred second order finite difference analogue to the wave equation considered above is

$$A_j^{n+1} = A_j^{n-1} - \alpha\left(A_{j+1}^{n} - A_{j-1}^{n}\right)$$

$$A_j^{n} = \exp(ikj\Delta x - in\theta)$$

and $$\sin\theta = \alpha\sin(k\Delta x) .$$

It is obvious that $n\theta \neq ku$ (see Table 2 and Figure 15). The actual phase error will vary between schemes. For linear differential equations the error due to the time differencing is in general much smaller than that of the space differencing.

For Galerkin techniques this error only occurs because of the time

$k\Delta x$	Second-order Arakawa scheme	Fourth-order Arakawa scheme	Cut-off Fourier-expansion scheme
15°	0·989	1·000	1·000
30°	0·955	0·998	1·000
45°	0·901	0·989	1·001
60°	0·828	0·966	1·002
90°	0·637	0·851	1·004
120°	0·414	0·622	1·007
150°	0·191	0·310	1·012
180°	0	0	1·017‡

† The values of $\delta = \theta/(k\alpha\Delta x)$ are given for $\alpha = U\Delta t/\Delta x = 0{\cdot}1$.
‡ For $k\Delta x = 180°$, $\delta = 0$. The number listed is $\lim_{k\Delta x \uparrow \pi} \delta$. See text.

TABLE 2. Phase errors (δ)†

(from Orszag 1971).

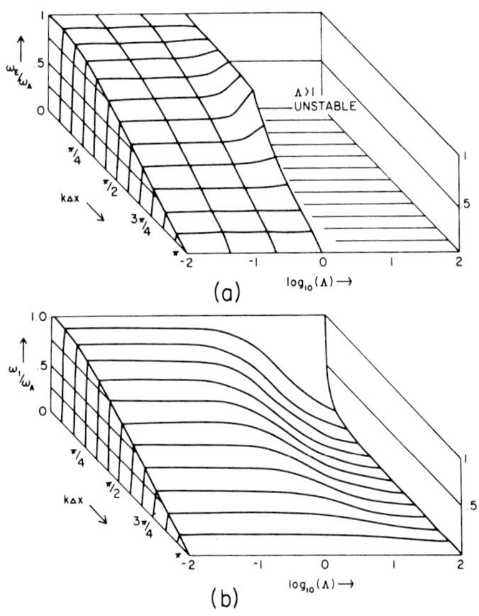

Figure 15: Accuracy of the phase frequencies for the advection equation as a function of the CFL parameter $\Lambda \equiv \alpha$ for (a) an explicit finite difference scheme and (b) for an implicit scheme. The limit for linear stability is $\Lambda = 1$. For $\Lambda < 1$ the two schemes are nearly the same; for $\Lambda > 1$ the implicit scheme maintains stability by slowing down the waves (from Grotjahn & O'Brien 1975).

integration technique.

The large errors in the test of advection are allied to errors in the phase of waves.

E.3 Group Velocity Errors

This error is related to the phase errors discussed above. For our purposes we shall interpret the group velocity as defining the mode of energy propagation of a wave packet (a collection of waves with similar wave number). We shall also limit this discussion to linear systems.

For wavelengths long; compared to the grid distance; the consideration of a quasi continuous group velocity is appropriate in a discrete system. For short wavelengths the actual interaction of two discrete waves would be more appropriate. Examples of this error are given in Figure 16.

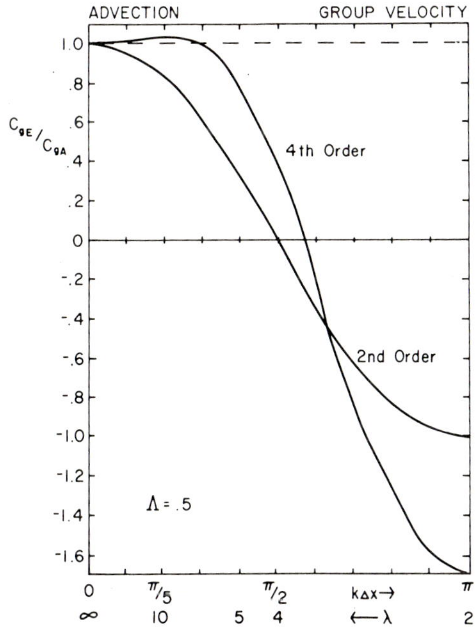

Figure 16: Comparison of the group velocities between a so-called "fourth order" scheme and the "second order", centred-in-space scheme as a function of wavenumber. Long waves are treated quite well by the higher order scheme, however its treatment of the shortest waves is inferior to the second order scheme. (C_{gE} is the group velocity of the finite difference scheme. C_{gA} is the group velocity of the analytical solution) (from Grotjahn and O'Brien 1975).

For coupled systems of equations, especially with dispersive waves, these errors could be very significant. Consider a set of linear equations representing internal waves. The dispersion relation for a wave of the form $A \exp i(kx + mz - wt)$ in the continuous system is

$$w^2 = f^2 + N^2k^2/m^2 .$$

In the finite-difference grid the dispersion relation for an identical wave can be shown, following the method used by Crowley (1968), to be

$$w^2 = a^2f^2 + b^2N^2 \frac{k^2}{m^2} ,$$

where

$$a = 1 - (k^2\Delta x^2/8) - (m^2\Delta z^2/24) ,$$

$$b = 1 - (m^2\Delta z^2/8) - (k^2\Delta x^2/24) .$$

(Pearson 1973).

This dispersion in the finite difference grid would give very significant errors in the group velocity.

E.4 Aliasing

Aliasing, folding and blocking are errors which are inter-related. The actual definition of aliasing is not fixed. The following are some samples.

E.4.1 Definition 1

(Orszag 1971). Consider the arbitrary functions, f(x), g(x) at N equally spaced points.

Then
$$f(x_n) \equiv f_n = \sum_{|k|<K} u(k) \exp(ikx_n) \qquad \text{E.4.1a}$$

$$g(x_n) \equiv g_n = \sum_{|k|<K} v(k) \exp(ikx_n) \qquad \text{E.4.1b}$$

where $N = 2K+1; \quad -K \leqslant k \leqslant K$.

The product $h(x) = f(x)\, g(x)$ has the value $h_n = f_n\, g_n$ with $h_n = \sum_{|k|\leqslant K} w(k) \exp(ikx_n)$,

$$\text{where} \quad w(k) = \sum_{\substack{p+q=k \\ |p|,|q|\leqslant K}} u(p)v(q) + \sum_{\substack{p+q=k+N \\ |p|,|q|\leqslant K}} u(p)v(q) + \sum_{\substack{p+q=k-N \\ |p|,|q|\leqslant K}} u(p)v(q) . \qquad \text{E.4.2}$$

We are using the points as representing the functions, or as samples from the continuous solutions to the differential equations. We are really interested in the Fourier transform representations

$$\left(\bar{f}(x),\, \bar{g}(x),\, \bar{h}(x)\right) = \int_{-\infty}^{\infty} \left(u(k), v(k), w(k)\right) \exp(ikx)dx .$$

One possible error occurs through truncating at finite wave numbers, or spacing. This is considered later in the discussion in the boundary condition errors.

Note that in the transform case above

$$\bar{h}(k) = \int_{-\infty}^{\infty} \bar{u}(k)\ \bar{v}(k-\ell)\ d\ell$$

or for a Fourier series representation

$$\bar{h}(k) = \sum_{p+q=k} \bar{u}(p)\ \bar{v}(q)\ . \qquad \text{E.4.3}$$

The false sums in the expansion (eq. E.4.2) are called, by Orszag, aliasing interactions. The difficulty with this definition is that these aliasing interactions are not necessarily an error.

E.4.2 Definition 2

(Orszag 1972). Consider the best approximation to $h(x) = f(x)\ g(x)$ given that the *exact* values of f and g at the discrete points are $h(x_n) = f(x_n)\ g(x_n)$.

If the values of f and g are not exact then the best approximation to h should be obtained. One definition of the best approximation is that it has the lowest Route mean square error (or L_2 norm). This gives the alias-free convolution sum (eq. E.4.3) as the best approximation to the product. A fully aliased (Definition 2) model of the two dimensional advection (Orszag 1971) shows that this is not a significant error for this test.

E.4.3 Definition 3

The best approximation to h (which is the product of f and g) is that which minimizes the maximum error (or T norm). This definition could lead to a different best approximation than that given previously. For a rectangular domain Chebyshev polynomials would be used for the expansion. (This definition has not been considered in detail).

E.4.4 Definition 4

Aliasing errors are those errors which occur because the finite difference method cannot resolve waves with wavelengths shorter than, or equal to, twice the grid length. This last, less rigorous approach is convenient and will be used throughout the following discussion.

E.5 Folding

Let us consider a general non linear system. The interactions between wave numbers, k, ℓ result in waves with wave numbers $k+\ell$, $k-\ell$. In the finite difference system it is possible that wave number $k+\ell$ is greater than the resolution of the grid, therefore interaction cannot occur as it does in the continuous system. What often happens is that the smallest wavelength wave possible in the system is generated instead. To follow this in more detail I shall follow the discussion of Robert et al (1970).

Consider a finite domain in x of length L, with F(x) having k cycles in L. $F(x) = e^{\frac{2\pi kix}{L}}$. The wave number $= \frac{2\pi k}{L}$, wave length $= \frac{L}{k}$. For the finite difference model divide L into J increments (or J+1) grid points. $L = J\Delta x$, $x_j = j\Delta x$. Then the number of grid intervals per wavelength $\eta = \frac{J}{k}$.

We shall call waves short or long by the following criteria

$$\eta > 4 \quad \text{low frequency}$$

$$2 \leqslant \eta \leqslant 4 \quad \text{high frequency}$$

[Note the non-appearance of $\eta < 2$ is aliasing as defined in definition 4].

As well as the spatially high frequency Robert et al 1970 also show that the theorem below is equally valid when considering temporal frequencies. A folding error is therefore introduced by the time stepping.

Theorem (Folding)

Any high frequency oscillation may be expressed as a low frequency oscillation modulated by the highest frequency ($\eta = 2$).

E.6 False Non-linear Interactions

Definition:- False non-linear interaction. Consider a system of equations representing a non-linear wave-wave interaction. In the continuous system wave numbers k, ℓ, interact forming other wave numbers m, n. The finite difference equations will give an interaction resulting in wave numbers m', n'. Because of aliasing and folding $m' \neq m$ and $n' \neq n$.

This error should be demonstrated in general by considering a broad spectrum force with a spread of wave numbers. Its existence is easily shown by considering a limited component system. Usually a set of Fourier components whose interaction does not introduce further wave lengths is studied. In some numerical finite difference schemes the amplitude of one of the components in a two component (three coefficient) system grows exponentially. This is non-linear instability.

For some quadratically conservative schemes a triad interaction is substituted. In others a four component interaction is possible (see Lilly 1965). For the same equations each finite difference scheme has a completely different closed interaction. These false non-linear interactions vary significantly between schemes.

Sadourny (1975) examines the quasi-stationary distribution of energy between wave numbers for a complete cascade. Two schemes produce very different distributions (Figure 17). The numerical damping (a time averaging) effectively removes short wave length disturbances. It is not clear that exactly equivalent results would be produced with other, more appropriate, forms of time integration (see Lilly 1965 or Grammeltvedt 1970).

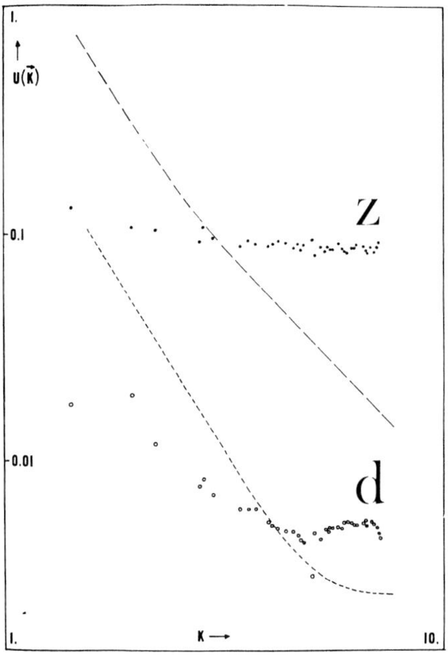

Figure 17: Equilibrium energy spectra in the case of the quasi-inviscid energy-conserving model, showing average energy per vector mode as a function of the pseudo-wavenumber K.Z: rotational energy, d: divergent energy. Dashed lines indicate corresponding spectra in the case of the potential-enstrophy-conserving model, using the same value of dissipativity ($v = 0.005$). Both coordinates are logarithmic. (from Sadourny 1975).

E.7 Blocking

In finite difference methods interactions which would result in waves not resolved by the grid are replaced by spurious interactions. In Galerkin (Fourier) methods no interaction is allowed.

Definition:- Consider the original set of equations where two different wave numbers interact to give other wave numbers. Blocking is the non-existence of this interaction in the Galerkin (Fourier) representation.

The quantitative, and even qualitative effects of blocking are very obscure. In the tests discussed in this review real non-linear wave-wave interactions are not important. Such interactions can be important in geophysical fluid dynamics. In Galerkin models of the ocean, atmosphere (also turbulence) it is convenient to damp the high wave number end of the spectrum by some "eddy viscosity hypothesis" and hope that blocking does not significantly affect the energy cascade to those high wavenumbers.

E.8 Incompressibility Errors

The numerical analog of the continuity equation ($\underset{\sim}{\nabla}.\underset{\sim}{u} = 0$) can be $\underset{\sim}{\nabla}_D \cdot \underset{\sim}{u}_D = \varepsilon_o$. [$\underset{\sim}{\nabla}_D$, $\underset{\sim}{u}_D$ discrete analogous to $\underset{\sim}{\nabla}$, $\underset{\sim}{u}$]. In vorticity forms this error is found in the inversion of vorticity to obtain the stream function ($\zeta = \nabla^2\psi$). With modern algorithms this inversion is possible to machine accuracy. The error is negligible in all the tests considered in this paper.

E.9 Second Differencing Errors (in diffusive terms)

These are defined by the errors in evaluating terms of the form $\frac{\partial^2 A}{\partial x^2}$ by finite differences. For the tests considered this form of error could only occur in the convection example. This error is not considered to be at all significant.

E.10 Boundary Condition Errors

In all these examples flows within simple boundaries have been considered. This eliminates any possible errors. However in practical problems economy, computer space, and available time, forces the modeller to only model a limited region of the whole region. This requires some form of open boundary condition which will vary between models. In finite difference models local conditions give great improvement over closed boundaries (e.g. Pearson, 1974, 1976). The simple forms allow simplicity in programming and require very little extra computer time. To date little work in open boundaries in Galerkin expansions has been published.

One of the main advantages of open boundaries is that they allow scales much larger than the dimensions of the numerical model. In a system with very long waves an open boundary condition allows a mean or linear trend on the domain of the model. In a non precise sense these could be considered as parts of waves whose dimensions are much larger than the domain.

For (non-fluid) examples in engineering the great ease of incorporating irregularly shaped boundaries has been one of the reasons for the great success of finite element methods.

With open boundary conditions a similar simplicity should be possible.

SUMMARY

It is very difficult to give any definite answer about the "best" method for a particular problem. In recent times a number of "tests" have become standard. These are very useful for showing properties of each realisation when applied to simple wave systems, or simple advection. Unfortunately finite element methods have not been used for these standard tests. This makes a comparison very difficult. For more complex systems the occurrence, and influence, of errors is more subtle. Overall a Galerkin expansion seems to give less errors, for a given resolution, where time dependent flows within simple boundaries are considered.

Most derivations, and discussions, of open boundary conditions have used finite difference methods. This is because a local condition is simple to incorporate and gives good results. This preference for a finite difference technique has not been clearly demonstrated.

No tests with complexly shaped boundaries have been published. Practical usage and simplicity indicate a finite element technique should be used for such models at low Reynolds numbers. At high Reynolds numbers insufficient knowledge of errors could cause instabilities and errors.

REFERENCES

Arakawa, A. 1966. Computational design for long-term numerical integration of equations of fluid motion. Two-dimensional incompressible flow. Part 1. J. Comput. Phys. Vol. 1, pp 119-143.

Crowley, W.P. 1968. Numerical advection experiments. Mon. Wea. Rev. Vol. 96, pp 1-11.

Denny, V.E. & Clever, R.M. 1974. Comparisons of Galerkin and finite difference methods for solving highly non-linear thermally driven flows. J. Comput. Phys. Vol. 16, No. 3, pp 271-284.

Dufort, E.C. & Frankel, S.P. 1953. Stability conditions in the numerical treatment of parabolic differential equations. Math. Tables and Other Aids to Computation, Vol. 7, No. 135.

Fromm, J.E. 1969. Practical investigation of convective difference approximations of reduced dispersion. High Speed Computing in Fluid Dynamics. The Physics of Fluids. Supplement II, pp 3-12.

Gallagher, R.H. & Chan, S.T.K. 1973. Higher order finite element analysis of lake circulation. Computers & Fluids, Vol. 1, pp 119-132.

Grammeltvedt, A. 1970. Numerical simulation of the motion of atmospheric fronts using a two-layer model. Tellus, Vol. 12, No. 6, pp 627-637.

Grotjahn, R. & O'Brien, J. Some inaccuracies in finite differencing hyperbolic equations. Submitted to Mon. Wea. Rev.

Kreiss, H.O. & Oliger, J. 1972. Comparison of accurate methods for the integration of hyperbolic equations. Tellus, Vol. 14, No. 3, pp 199-215.

Kreiss, H.O. & Oliger, J. 1973. Methods for the approximate solution of time dependent flows. G.A.R.P. Publication Ser. No. 10. World Met. Org.

Lilly, D.K. 1961. A proposed staggered-grid system for numerical integration of dynamic equations. Mon. Wea. Rev. 1961, Vol. 89, No. 3, pp 59-65.

Lilly, D.K. 1965. On the computational stability of numerical solutions of time dependent non-linear geophysical fluid dynamics problems. Mon. Wea. Rev. Vol. 93, No. 1, pp 11-26.

Molenkamp, C.M. 1968. Accuracy of finite difference methods applied to the advection equation. J. Appl. Meteor., Vol. 7, No. 1, pp 160-167, April 1969.

Orszag, S.A. 1971. Numerical simulation of incompressible flows within simple boundaries. Accuracy. J. Fluid Mech., Vol. 49, Part 1, pp 75-112.

Orszag, S.A. & Israeli, M. 1974. Numerical simulation of viscous incompressible flows. Ann. Rev. Fluid Mech., 1974, pp 281-317.

Pearson, R.A. 1973. Properties of the sea breeze front as shown by a numerical model. J. Atmos. Sci. Vol. 30, No. 6, pp 1050-1060.

Pearson, R.A. 1974. Consistent boundary conditions for numerical models of systems that admit dispersive waves. J. Atmos. Sci. Vol. 31, No. 6, pp 1481-1489.

Pearson, R.A. & McGregor, J.L. 1976. An open boundary condition for models of thermals. Submitted to J. Atmos. Sci.

Pearson, R.A. & O'Connor, S. 1976. A numerical-dynamical instability. To be submitted to Mon. Wea. Rev.

Phillips, N.A. 1959. An example of non-linear computational instability, in The Atmosphere and Sea in Motion. Rossby Memorial Volume. Rockefeller Inst. Press, pp 501-509.

Robert, A.J., Shuman, F.G. & Gerrity Jr., J.P. 1970. On partial differential equations in mathematical physics. Mon. Wea. Rev. Vol. 98, No. 1, pp 1-6.

Sadourny, R. 1975. The dynamics of finite difference models of the shallow water equation. J. Atmos. Sci. Vol. 32, No. 4, pp 680-689.

PART 2

APPLICATIONS

Numerical Simulation of Fluid Motion, ed. John Noye
pp. 285-355

FINITE DIFFERENCE TECHNIQUES APPLIED TO THE SIMULATION OF TIDES AND CURRENTS IN GULFS

B.J. Noye

University of Adelaide, South Australia

K.C.S. Tronson

Bureau of Transport Economics, Canberra, A.C.T.

SUMMARY

The hydrodynamics of gulf systems are formulated in mathematical terms. Four mathematical models, each of which enables a different aspect of the flow to be studied, are described. An outline of the numerical methods used with each is given. Some results from their application to Spencer and St. Vincent Gulfs in South Australia are described.

INTRODUCTION

The description of this work is divided into two different sections, the formulation of the mathematical models with an outline of their method of solution and a description of their application to the South Australian gulf system.

Numerical simulation techniques have been described by which two distinct types of tides can be computed : the astronomically induced tide, which propagates into the gulf from the open boundary as a free wave, and the meteorologically induced tide, produced by weather effects such as the wind stress and the changing atmospheric pressure acting on the sea surface in the gulf. These tides are evident in the form of changing sea-levels and their associated currents. In addition, methods are described to determine the natural oscillations of gulf waters and the equilibrium circulation produced by steady winds.

These methods have been applied to the South Australian gulf system which consists of St. Vincent Gulf, Spencer Gulf, Investigator Strait and Backstairs Passage.

Firstly, the natural frequencies and wave patterns have been calculated for a model sea basin representative of the gulfs. The effect of the Earth's rotation has been evaluated by carrying out the calculations with and without the inclusion of the Coriolis parameter.

Secondly, the tidal motion within the gulfs has been studied. Previously, the nature of the gulf tides was postulated from tidal constituents calculated from records of coastal tides. These have been compared with the numerically predicted tides using both linear and non-linear forms of the equations of motion.

The study of greatest economic significance is the third, on the storm surges of the gulfs. Idealised storm surges caused by wind stress on the surface of the gulf waters and pressure systems moving over the gulfs have been simulated numerically.

Finally, the steady circulation of the waters of the gulfs, caused by various wind systems, has been considered.

1. THE MATHEMATICAL MODELS

A mathematical solution of an environmental problem depends on a set of equations that adequately describe the physical processes which occur. As with most fluid dynamical problems, the relevant equations used to describe the movement of the sea in gulfs are based on the Navier-Stokes equation and the continuity equation. Added to these are the boundary conditions governing the free surface, the boundary layer motion at the sea floor and the complex sea motions near the shoreline.

It is not practical to treat such a problem as a single entity. The topics into which it may be divided are circulation, long wave motion, air-sea interaction, boundary layers on the sea floor, wave motion near the shore, diffusion, turbulence, stratification and internal waves. Long wave motion and circulation are the items of major consideration in this paper, the other topics being considered only with regard to their influence on these two topics.

1.1 The Differential Equations

A formal derivation of the storm surge equations from the Navier-Stokes equation was presented by Fortak (1962). The exact equations include terms arising from lateral friction, boundary friction, atmospheric pressure gradients, wave set-up, horizontal and vertical density gradients, density variations in time, the tide generating force, non-hydrostatic effects and precipitation-minus-evaporation. An alternative derivation is given in Appendix 1, combining the methods used by Stoker (1957) and Welander (1961).

The linearized form of the equations, with forcing terms appropriate to the problem is

$$\frac{\partial U}{\partial t} - fV = -gh\frac{\partial \zeta}{\partial x} - \frac{h}{\rho}\frac{\partial p_a}{\partial x} + \frac{1}{\rho}(F_S - F_B) \quad , \qquad (1.1.1)$$

$$\frac{\partial V}{\partial t} + fU = -gh\frac{\partial \zeta}{\partial y} - \frac{h}{\rho}\frac{\partial p_a}{\partial y} + \frac{1}{\rho}(G_S - G_B) \quad , \qquad (1.1.2)$$

$$\frac{\partial U}{\partial x} + \frac{\partial V}{\partial y} + \frac{\partial \zeta}{\partial t} = 0 \ , \qquad (1.1.3)$$

where the notation used is

t the time,

x,y the space co-ordinates increasing to the east and north respectively,

ζ the elevation of the sea surface above the equilibrium level,

U,V the components of the total stream,

F_S,G_S the components of the surface stress,

F_B,G_B the components of the dissipation forces,

h the undisturbed depth of the water,

f the Coriolis parameter,

p_a the atmospheric pressure at the sea surface,

ρ the density of the water,

g the acceleration of the Earth's gravity.

The assumptions, implicit in the use of Equations (1.1.1), (1.1.2) and (1.1.3), to describe the circulation and long wave motion within Gulf systems, can be listed as

(i) variation of the Coriolis parameter is unimportant,
(ii) non-hydrostatic pressure forces are negligible,
(iii) horizontal diffusion of momentum is not significant,
(iv) convective (non-linear) parts of the acceleration field do not contribute to the motion,
(v) the density is constant in time and space,
(vi) the dissipation of energy may be modelled simply,
(vii) there is no land covered or uncovered with a rise or fall in water level,
(viii) precipitation-minus-evaporation may be neglected,
(ix) the tide generating forces are unimportant.

Assumptions (i) and (ix) are valid as the dimensions of gulf systems are small relative to the length scales associated with a variation in the Coriolis parameter and the tide generating forces. As shown in Appendix 1, the validity of assumption (ii) depends on the condition

$$H^2/L^2 \ll 1, \qquad (1.1.4)$$

where H is a typical depth and L a typical horizontal distance.

Combined with the equation

$$Z/H \ll 1, \qquad (1.1.5)$$

with Z a representative measure of surface elevations encountered, (1.1.4) justifies assumption (iv). In the South Australian gulfs, for example, (1.1.4) is satisfied for all motions in the horizontal plane with a time scale greater than 10 minutes. Condition (1.1.5) is well satisfied for all but the largest displacements of the surface in the shallow areas at the head of each gulf.

Assumption (v) implies that the waters of the gulf are well mixed vertically so that no stratification occurs and that there is little salinity variation between different sections of water. The model is not applicable to estuaries in which large salinity and temperature variations occur between different depths and different regions.

Assumptions (iii) and (vi) imply that certain forms of turbulence are to be neglected, while other forms are modelled by simple terms in the equations of motion. Horizontal diffusion (lateral friction) results from large scale eddies in the horizontal plane. It is normally associated with boundary currents which are thought to have little effect on the height of the sea surface. The usual mathematical description for this diffusion has the form

$$A\left(\frac{\partial^2}{\partial x^2} + \frac{\partial^2}{\partial y^2}\right)(U,V).$$

The coefficient of eddy viscosity, A, is an arbitrary constant which cannot be measured directly. If the interest is on sea levels rather than currents, this term can be omitted.

Although non-dissipative equations can be used to model the fluid motions, a system which dissipates energy is more likely to correspond to reality. The dissipative mechanism is usually equated with vertical turbulent diffusion. Welander (1957) had developed an integro-differential equation which includes vertical turbulent diffusion, but its use is not practical. The quadratic stress as given by Proudman (1953),

$$(F_B, G_B) = \frac{\rho k}{h^2}(U^2 + V^2)^{1/2}(U,V), \tag{1.1.6}$$

where k is the coefficient of quadratic stress, has proved to be a reasonable approximation to the physical processes. The advantages of a linear system of equations is lost if the quadratic stress is used. The usual linear form for the bottom friction is

$$(F_B, G_B) = \frac{\rho\kappa}{h}(U,V), \tag{1.1.7}$$

where κ is the coefficient of linear friction. A derivation of (1.1.7) from physical principles is given by Welander (1961). An alternative linear form for dissipation has been developed by Platzman (1963). For the range of depths within the most gulf systems, the two linear forms are equivalent. To obtain a closed system of equations, the bottom stress terms have been restricted to forms which include no extra dependent variables.

The dissipation of energy in the oceans and continental shelves is very poorly understood. Accordingly, the choice of terms and coefficients to model the dissipation can only be arbitrary. For modelling, the choice of dissipation terms is best made by testing simple mathematical forms for sensible results, rather than by specious arguments about boundary layers.

Although rain is often associated with the weather systems which pass over gulfs and generate storm surges, it is usually small enough to be neglected, except in monsoonal areas. Evaporation, although significant over a year, is unimportant during short periods. If mud flats occur within the gulf system with areas not negligible compared to the area below low tide level then special consideration must be given to allow for shifting land boundaries. If assumption (vii) can be justified, as it can in the South Australian gulfs, it leads to the natural boundary condition that the transport across the fixed shoreline (closed boundary) is zero, or

$$U \sin\psi + V \cos\psi = 0 \tag{1.1.8}$$

at coastal points where ψ is the angle between the normal and the y-axis.

The boundary of a model sea basin consists of sections of shoreline, C_1, and open boundary, C_2, the latter separating the model area from the open sea. The condition imposed at an open boundary is that the surface elevation is known for all time, i.e.

$$\zeta(t,C_2) = F(t,C_2), \tag{1.1.9}$$

where F is a known function. By fixing the open boundary well away from the region of interest, only the major features of F are important in determining the fluid motion. Tidal motion within the basin can be simulated by forcing the open boundary to oscillate at tidal frequencies. For modelling storm surges, the open boundary can be held at zero.

As the system of equations is of the first order in time, a boundary condition for time must be formulated. Either initial values can be given for the dependent variable, or a periodicity condition can be used. If dissipation is included in the equations, the solution is independent of the initial conditions after a time, allowing the use of the simplest initial conditions,

$$U(x,y,t = 0) = V(x,y,t = 0) = \zeta(x,y,t = 0) = 0. \tag{1.1.10a}$$

For oscillatory motion, the periodicity condition is

$$\begin{aligned} U(x,y,t = t_1) &= U(x,y,t = t_1 + T), \\ V(x,y,t = t_1) &= V(x,y,t = t_1 + T), \\ \zeta(x,y,t = t_1) &= \zeta(x,y,t = t_1 + T), \end{aligned} \tag{1.1.10b}$$

where T is the period of the oscillation.

The set of Equations (1.1.1), (1.1.2) and (1.1.3) with boundary conditions (1.1.8), (1.1.9) and (1.1.10a) or (1.1.10b) define a well posed problem if the dissipation has the form of either (1.1.6) or (1.1.7) and the surface stress and air pressure are given. The two forcing terms will be written as a single term.

1.2 Methods of Solution.

Simple analytical techniques can be used to solve the equations only if the bathymetry and forcing functions can be described by elementary functions. The shape and bathymetry of most gulf systems preclude a ready analytic solution of the problem.

Numerical methods applicable to arbitrary gulf systems are described in this study. The following points were considered when determining the approach to be used;

(i) the method should be general so that the equations can be solved for a region of any shape and bathymetry,

(ii) this generalization needs to be included in the method, to enable others to use it directly,

(iii) preferably, the primitive equations should be solved directly, reducing the need to use variables without obvious physical equivalences,

(vi) the use of the primitive equations should minimize the amount of work necessary to modify the method if extra terms are added.

Several numerical methods have been considered. For problems concerning tidal motion, the harmonic method is obvious. The periodicity condition is used and the fourier transforms of (1.1.1), (1.1.2) and (1.1.3) are solved to give tidal amplitudes and phases. This method breaks down when the frequency is close to a natural frequency of the basin. These cases are best studied by solving the associated eigenvalue problem.

The hydrodynamical-numerical (HN-) method, first used by Hansen (1956), is most useful for modelling storm surges. The partial derivatives are approximated by finite differences, and the equations are stepped forward in time.

1.3 Discretization in Space.

A spatial finite-difference form of Equations (1.1.1),(1.1.2) and (1.1.3) is

$$\frac{\partial}{\partial t}\underset{\sim}{U}(t) - F_1\underset{\sim}{V}(t) = -gH_1 D\underset{\sim}{\zeta}(t) + \frac{1}{\rho}(\underset{\sim}{F}_S(t) - F_B(t)), \tag{1.3.1}$$

$$\frac{\partial}{\partial t}\underset{\sim}{V}(t) + F_2\underset{\sim}{U}(t) = -gH_2 E\underset{\sim}{\zeta}(t) + \frac{1}{\rho}(\underset{\sim}{G}_S(t) - G_B(t)), \tag{1.3.2}$$

$$A\,\underset{\sim}{U}(t) + B\,\underset{\sim}{V}(t) + \frac{\partial}{\partial t}\underset{\sim}{\zeta}(t) = 0. \tag{1.3.3}$$

The vectors $\underset{\sim}{U}(t)$, $\underset{\sim}{V}(t)$ and $\underset{\sim}{\zeta}(t)$ are of length k,ℓ and m respectively. Each element of a vector represents the variable in the area around the corresponding grid point.

H_1 and H_2 are diagonal $k\times k$ and $\ell\times\ell$ matrices. The structure of F_1 and F_2 depends on the grid scheme used. Four different grid schemes are in common use. The positioning of the elevation stream points are illustrated in Figure 1.1. They may be described as follows

a) the elevation ζ and the two stream components U,V evaluated at every point: this scheme has been used by Fischer (1959) and Jelesnianski (1965);

b) the elevation ζ and the two stream components U,V evaluated at separate points along alternate lines parallel to the coordinate axes: this has been used by Lauwerier and Damsté (1963) and Heaps (1969);

c) the elevation ζ and flow components (U,V) evaluated at alternate points along lines parallel to the axes, which is b) rotated through 45^0 relative to the axes: this has been used by Sielecki (1968);

d) the elevation ζ and stream component U evaluated at alternate points along a line parallel to the x-axis, and the elevation ζ and stream component V evaluated at alternate points along a line parallel to the y-axis, as used by Platzman (1958), Hansen (1962) and Leendertse (1967).

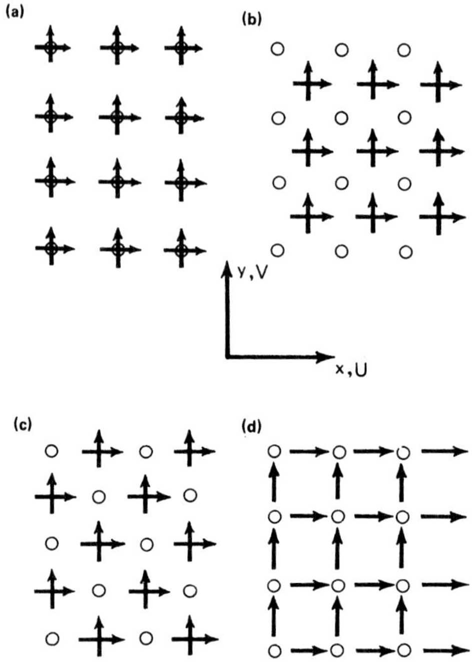

Figure 1.1: *The layout of elevation and stream points in the four common grid schemes: → indicates the position where the x-component U of stream is evaluated; ↑ indicates where the y-component V of stream is evaluated; o is the position where elevation ζ is evaluated.*

The terms $D\underset{\sim}{\zeta}$, $E\underset{\sim}{\zeta}$, $A\underset{\sim}{U}$ and $B\underset{\sim}{V}$ are the finite-difference equivalents in each of these systems of the derivates $\partial\zeta/\partial x$, $\partial\zeta/\partial y$, $\partial U/\partial x$ and $\partial V/\partial y$ respectively.

A grid point within the area of interest, and at which the dependent variable is not defined by a boundary condition, is called an active grid point. By removing all non-active elements from the vectors $\underset{\sim}{U}$, $\underset{\sim}{V}$ and $\underset{\sim}{\zeta}$, the system of Equations (1.3.1), (1.3.2) and (1.3.3) becomes irreducible.

Using the linear form of the bottom friction (1.1.7), the equations can be rewritten as

$$\frac{\partial}{\partial t}\underset{\sim}{z}(t) = G\,\underset{\sim}{z}(t) + \underset{\sim}{s}(t), \qquad (1.3.4)$$

where

$$\underset{\sim}{z}(t)^T = [\underset{\sim}{U}(t)^T, \underset{\sim}{V}(t)^T, \underset{\sim}{\zeta}(t)^T], \qquad (1.3.5)$$

$$\underset{\sim}{s}(t)^T = \frac{1}{\rho}\,[\underset{\sim}{F}_S(t)^T,\ \underset{\sim}{G}_S(t)^T,\ 0] \tag{1.3.6}$$

and

$$G = \begin{bmatrix} -K_1 & F_1 & -gH_1D \\ F_2 & -K_2 & -gH_2E \\ -A & -B & 0 \end{bmatrix}. \tag{1.3.7}$$

The grid scheme used in the application to the South Australian gulfs was that described by Heaps (1969).

The variation of any variable within a distance ε can not be modelled by a finite-difference representation using a grid spacing greater than ε. As well as a minimum distance, there is a minimum time scale, τ_{min}, consistent with a grid scheme. This can be defined as the time taken for a disturbance to travel between adjacent grid points.

1.4 The Harmonic Method.

The transformation of the dependence on time to a dependence on frequency is the basis of the harmonic method. The Fourier transform, denoted by $\hat{}$, of Equation (1.3.4) is

$$i\,\omega\,\hat{\underset{\sim}{z}} - \underset{\sim}{z}(0) = G\,\hat{\underset{\sim}{z}} + \hat{\underset{\sim}{s}}, \tag{1.4.1a}$$

or

$$i\,\omega\,\hat{\underset{\sim}{z}} = G\,\hat{\underset{\sim}{z}} + \hat{\underset{\sim}{s}}, \tag{1.4.1b}$$

depending on which boundary condition, (1.1.10a) or (1.1.10b), is assumed.

Given that dissipation is included, T' is defined as the time taken for the motion to lose all memory of its initial condition. Set

$$\omega_1 = \frac{2\pi}{T'}$$

and

$$\omega_2 = \frac{\pi}{\tau_{min}}.$$

The Fourier transform for an initial value problem with $\underset{\sim}{s}(t) = 0$ for $t < 0$ is

$$\hat{\underset{\sim}{s}}(\omega) = \int_0^\infty e^{-i\omega t}\,\underset{\sim}{s}(t)dt.$$

For $\underset{\sim}{s}(t)$ defined consistently with the spatial grid

$$\hat{\underset{\sim}{s}}(\omega) = 0, \quad |\omega| > \omega_2.$$

If the forcing is periodic with a period less than the dissipation time, T', then

$$\hat{\underset{\sim}{s}}(\omega) = \underset{\sim}{0}, \quad |\omega| < \omega_1.$$

Under these conditions, the time history of the U, V and ζ fields for $T' \le t \le 2T'$ is given by

$$z(t) = \frac{1}{2\pi} \int_{\omega_1}^{\omega_2} \{e^{i\omega t}\,\hat{z}(\omega) + e^{-i\omega t}\,\hat{z}(-\omega)\}d\omega.$$

This method is particularly appropriate for finding the tidal motion resulting from open boundary forcing.

The numerical problem is finding the solution of the matrix equation

$$Q\underset{\sim}{z} = \underset{\sim}{s}, \qquad (1.4.2)$$

where

$$Q = i\omega I - G. \qquad (1.4.3)$$

Equation (1.4.2) has a unique solution if Q is not singular. The conditions under which Q is singular are discussed in Section 1.5.

For the South Australian gulf system, the minimum size of G used was 360×360. To keep the arithmetic real, the size of Q doubles, increasing from $n/2 \times n/2$ to $n \times n$. Halving the grid spacing increases n by a factor of four.

The computing facilities used consisted of CDC 6400 with 50K words of core and a fast access disk store. Direct methods for solving (1.4.2) require $O(\frac{1}{3}n^3)$ operations and extensive use of the fast access storage. The number of operations required by an iterative method is $O(kn^2)$ where k is the number of iterations required. Taking advantage of the sparseness of Q reduces the number to $O(knd)$ where d is the average number of non-zero elements per row, eliminating the need to use the disk store.

With this knowledge and a general lack of software to solve large sparse matrix problems, several iterative schemes were programmed. It was found that the Gauss-Seidel method did not converge. The most successful method was the conjugate gradient method described by Beckman (1960). The number of operations required was $O(7n^2)$.

1.5 The Eigenvalue Problem.

The values of $i\omega$ in Equation (1.4.3) which render Q singular are the eigenvalues of G. They are the finite-difference approximations to the standing wave frequencies of the basin. A knowledge of the standing wave patterns provides an ideal basis for the description of the tides and wind surges of the gulf system. To study the resonances of the basin, the dissipation is set at zero.

Platzman (1972) has shown that, for a Richardson grid and a particular structure of F_1 and F_2, the matrix G has pure imaginary eigenvalues, $i\beta_k$. Resonant iteration was used to calculate an eigenvalue-eigenvector pair requiring $O(n^3)$ operations for each pair.

The procedure followed here is the traditional approach of eliminating U and V between Equations (1.1.1), (1.1.2) and (1.1.3). Ignoring forcing and dissipation terms and assuming a sinusoidal time dependence for ζ of the form

$$\zeta(x,y,t) = \zeta^*(x,y)e^{-i\sigma t},$$

gives the equation

$$\frac{\partial}{\partial x}\left(h\,\frac{\partial\zeta^*}{\partial x}\right) + \frac{\partial}{\partial y}\left(h\,\frac{\partial\zeta^*}{\partial y}\right) + i\tau^*\left(\frac{\partial h}{\partial x}\frac{\partial\zeta^*}{\partial y} - \frac{\partial h}{\partial y}\frac{\partial\zeta^*}{\partial x}\right) + k^2\zeta^* = 0, \qquad (1.5.1)$$

where

$$\tau^* = f/\sigma \qquad (1.5.2)$$

and

$$k^2 = (\sigma^2 - f^2)/g. \qquad (1.5.3)$$

The boundary condition at a closed boundary equivalent to (1.1.8) is

$$\left(\frac{\partial}{\partial n} + i\tau^*\frac{\partial}{\partial s}\right)\zeta^* = 0, \qquad (1.5.4)$$

with n and s taken along the normal and the tangent in a right handed sense. The condition (1.1.9) has the form

$$\zeta^* = 0, \qquad (1.5.5)$$

along an open boundary.

The system of equations has been shown by Platzman (1972) to be self-adjoint. This property is retained if variational methods are used in formulating the finite-difference equations. The variational form of (1.5.1) is

$$\delta\left\{\iint [h|\nabla\zeta|^2 + i\tau h\left(\frac{\partial\bar{\zeta}}{\partial x}\frac{\partial\zeta}{\partial y} - \frac{\partial\zeta}{\partial x}\frac{\partial\bar{\zeta}}{\partial y}\right) - k^2|\zeta|^2]dA\right\} = 0, \qquad (1.5.6)$$

where the superscript * has been dropped, and the over-bar signifies the complex conjugate. Equation (1.5.4) is the natural boundary condition.

This approach is the same as that used by Pnueli and Pekeris (1968), who considered a similar problem, but restricted themselves to closed basins with constant depth. The property that the eigenvalues, k_n^2, are discrete and positive when $\tau < 1$ also holds for a variable depth.

The discretization of Equation (1.5.6) and the choice of the grid scheme is discussed in Appendix 2. The discrete form of the equations can be written as

$$\left(\begin{bmatrix} P & \tau Q \\ -\tau Q & P \end{bmatrix} - k^2\begin{bmatrix} R^2 & 0 \\ 0 & R^2 \end{bmatrix}\right)\begin{bmatrix} \eta \\ \xi \end{bmatrix} = 0 \qquad (1.5.7)$$

where

$$\underset{\sim}{\zeta} = \underset{\sim}{\eta} + i\underset{\sim}{\xi}.$$

The matrices P,Q and R are symmetric, skew-symmetric and diagonal respectively. The eigenvalues, k_n^2, are real and are positive if $\tau < 1$.

Equation (1.5.7) cannot be treated as a fundamental eigenvalue problem; k and τ are not independent, as f is given. Two iterative schemes for finding σ are developed in Appendix 3.

1.6 The Hydrodynamical-Numerical Method.

The HN-method depends on a suitably stable finite-difference approximation for the time dependence. A general time-discretized form of (1.3.1), (1.3.2) and (1.3.3) is

$$\frac{\underset{\sim}{U}(t+\tau)-\underset{\sim}{U}(t)}{\tau} - F_1\underset{\sim}{V}(t+\alpha_1\tau) = -gH_1D\underset{\sim}{\zeta}(t+\beta\tau) + \frac{1}{\rho}\left[\underset{\sim}{F_S}(t+\delta\tau)-F_B(t+\lambda\tau)\right], \tag{1.6.1}$$

$$\frac{\underset{\sim}{V}(t+\tau)-\underset{\sim}{V}(t)}{\tau} + F_2\underset{\sim}{U}(t+\alpha_2\tau) = -gH_2E\underset{\sim}{\zeta}(t+\beta\tau) + \frac{1}{\rho}\left[\underset{\sim}{G_S}(t+\delta\tau)-G_B(t+\lambda\tau)\right], \tag{1.6.2}$$

$$\frac{\underset{\sim}{\zeta}(t+\tau)-\underset{\sim}{\zeta}(t)}{\tau} + A\underset{\sim}{U}(t+\overline{1-\beta\tau}) + B\underset{\sim}{V}(t+\overline{1-\beta\tau}) = 0, \tag{1.6.3}$$

where the dependence on t between the time steps of length τ takes the form

$$\underset{\sim}{\zeta}(t+\beta\tau) = (1-\beta)\underset{\sim}{\zeta}(t) + \beta\underset{\sim}{\zeta}(t+\tau). \tag{1.6.4}$$

For β, α, γ and λ all zero, the time stepping scheme is that used by Heaps (1969). Sielecki (1968) used a system which corresponds to $\beta = 1$, $\lambda = \alpha_1 = 0$ and $\alpha_2 = 1$. Both these schemes may be approximated to what Richardson (1922) referred to as "step-over" differencing, i.e.

$$U(t + \tau/2) = U(t - \tau/2) + \tau f_1(t),$$

$$\zeta(t + \tau) = \zeta(t) + \tau f_2(t + \tau/2).$$

This form of differencing has been used by Harris and Jeslesnianski (1964), Platzman (1963) and many other investigators. The schemes discussed so far are all explicit and the value of τ is limited by stability requirements. An implicit scheme avoids some of the restrictions on τ. A value of 0.5 for β in (1.6.1), (1.6.2) and (1.6.3) gives the implicit scheme used by Uusitalo (1960).

Given $\beta = \alpha_1 = \alpha_2 = \gamma = \lambda = 0$, the system of equations can be written as

$$\underset{\sim}{z}(t+\tau) = G_1\underset{\sim}{z}(t) + \underset{\sim}{s_1}(t) \tag{1.6.5}$$

where

$$\underset{\sim}{s}_1(t)^T = \frac{\tau}{\rho}\left[\underset{\sim}{F}_S(t)^T, \underset{\sim}{G}_S(\underset{\sim}{t})^T - \tau(A\underset{\sim}{F}_S(t)+BG_S(t))^T\right] \tag{1.6.6}$$

and

$$G_1 = \begin{bmatrix} I-\tau K_1 & \tau F_1 & -\tau g H_1 D \\ -\tau F_2 & I-\tau K_2 & -\tau g H_2 E \\ -\tau(A-\tau AK_1-\tau BF_2) & -\tau(\tau AF_1+B-\tau BK_2) & I+g\tau^2(AH_1D+BH_2E) \end{bmatrix} \tag{1.6.7}$$

A program has been written to step $\underset{\sim}{z}$ forward in time, as given in (1.6.5).

1.7 Stability

The stability condition for (1.6.5) is

$$|\lambda_k^{G_1}| \leq 1, \tag{1.7.1}$$

where the $\lambda_k^{G_1}$ are the eigenvalues of G_1. Consider the matrix, G_2, where

$$G_1 = G_2 + O(\tau^2). \tag{1.7.2}$$

The eigenvalues of G_2, $\lambda_k^{G_2}$, can be found from G as

$$G_2 = \tau G + I. \tag{1.7.3}$$

For zero friction, the eigenvalues of G may be assumed to be pure imaginary, $i\beta_k$. The inclusion of friction would change the eigenvalues to the form

$$-\alpha_k + i\gamma_k ,$$

with

$$\alpha_k \geq 0 \quad \text{and} \quad \gamma_k - \beta_k .$$

The eigenvalues of G_2 can now be written as

$$\lambda_k^{G_2} = (1-\tau\alpha_k) + i\tau\gamma_k .$$

Approximating $\lambda_k^{G_1}$ by $\lambda_k^{G_2}$ gives

$$|\lambda_k^{G_1}|^2 \simeq |1-2\tau\alpha_k + \tau^2\alpha_k^2 + \tau^2\gamma_k^2|,$$

which for zero friction $(\alpha_k = 0)$ becomes

$$|\lambda_k^{G_1}| \geq 1.$$

Since the time stepping scheme is stable for some values of τ, the elements in G_1 of the order of τ^2 must be sufficient to reduce $|\lambda_k^{G_1}|$

below 1. This result corresponds to those obtained by Lax and Richtmyer (1956) who used finite fourier series stability analysis.

The boundary conditions and the shape of the model basin are taken into account by (1.7.1). However, this condition cannot be readily transformed into an inequality giving values of τ for which the stepping procedure is stable.

Fischer (1959) carried out a stability analysis, following the methods of von Neumann. No allowance was made for the boundary. The stability conditions found were

$$\tau \le \frac{c_1\delta}{\max(gh)^{1/2}} \tag{1.7.4}$$

and

$$\tau \le \frac{\kappa}{f^2 \max(h)}, \quad f \ne 0, \tag{1.7.5}$$

where δ is the minimum distance between adjacent grid points and c_1 is a constant ($c_1 = 2$ for the grid scheme used in the applications to the South Australian gulfs). The restriction on frequency equivalent to (1.7.4) is

$$\omega \ge \frac{2\pi}{c_1 N\delta} \max(gh)^{1/2} > 0, \tag{1.7.6}$$

as $\omega = 2\pi/N\tau$. There is no restriction on the maximum value of N.

Providing Q is not singular, (1.4.2) may be solved for all values of ω. For large values of ω (i.e. the frequencies of short surface waves), the solution has no physical significance. An upper limit on ω needs to be determined.

If the columns of H are the eigenvalues of G, then (1.4.2) can be rewritten as

$$(i\omega I - H^{-1}GH)H^{-1}\hat{\underset{\sim}{z}} = H^{-1}\hat{\underset{\sim}{s}},$$

giving

$$\hat{z} = -iH\left[\operatorname{diag}\left(\frac{1}{\omega-\lambda_k^G}\right)\right]H^{-1}\hat{\underset{\sim}{s}}.$$

To avoid aliasing, the condition,

$$\omega \le \max(|\lambda_k^G|), \tag{1.7.7}$$

must be satisfied. This can be justified intuitively as the minimum wavelength which can be resolved on a grid is twice the grid spacing. This gives a maximum frequency of

$$\omega \le \frac{\pi \min(gh)^{1/2}}{c_1\delta}. \tag{1.7.8}$$

The value given by (1.7.8) is an upper limit for (1.7.7). For a constant depth model basin, the two conditions would be identical.

For the time stepping regime, (1.7.7) becomes

$$\tau_{min} \geq \frac{2\pi}{N_{max}(|\lambda_k^c|)}, \tag{1.7.9}$$

which is not a stability condition in the sense of (1.7.4) but a measure of the minimum time scale consistent with the spatial discretization.

Problems of instability have been encountered for some combinations of boundary shapes and grid schemes. Using Heaps' scheme with a boundary shape including a corner of 270^0, and assuming zero dissipation, equation (1.6.5) was unstable. Increasing dissipation beyond a certain minimum level was sufficient to overcome the problem. This posed no additional problem as the minimum level of dissipation required for stability was below that which was required to model the observed dissipation. Smith (1973) altered Heaps' (1969) boundary condition at 270^0 corners in order to make the scheme stable in situations where dissipation was small. The Richardson grid described by Platzman (1963) does not appear to induce this form of instability.

The stability of a scheme is sufficient to ensure convergence.

A comparison between the analytic solution and the numerical results obtained using the hydrodynamical-numerical and harmonic methods for a simple model basin is presented by Tronson (1973).

1.8 Steady Motion

For steady motion, (1.1.1), (1.1.2), and (1.1.3) reduce to

$$-fV = -gh\frac{\partial\zeta}{\partial x} + \frac{1}{\rho}(F_S - F_B), \tag{1.8.1}$$

$$fU = -gh\frac{\partial\zeta}{\partial y} + \frac{1}{\rho}(G_S - G_B), \tag{1.8.2}$$

and

$$\frac{\partial U}{\partial x} + \frac{\partial V}{\partial y} = 0. \tag{1.8.3}$$

The form of the bottom stress is assumed to be

$$(F_B, G_B) = \rho r(U,V) \tag{1.8.4}$$

and with r not restricted to a constant value. The stream function, ψ, is defined by

$$\begin{aligned} \partial\psi/\partial y &= -U \\ \partial\psi/\partial x &= V. \end{aligned} \tag{1.8.5}$$

Substituting for U and V and eliminating ζ in Equations (1.8.1) and (1.8.2) gives

$$r\nabla^2\psi + \frac{\partial\psi}{\partial x}\left(-\frac{r}{h}\frac{\partial h}{\partial x} + \frac{\partial r}{\partial x} - \frac{f}{h}\frac{\partial h}{\partial y}\right) + \frac{\partial\psi}{\partial y}\left(-\frac{r}{h}\frac{\partial h}{\partial y} + \frac{\partial r}{\partial y} + \frac{f}{h}\frac{\partial h}{\partial x}\right)$$

$$= \frac{1}{\rho}\left(\frac{\partial G_S}{\partial x} - \frac{\partial F_S}{\partial y} + \frac{F_S}{h}\frac{\partial h}{\partial y} - \frac{G_S}{h}\frac{\partial h}{\partial x}\right). \qquad (1.8.6)$$

Variation in the Coriolis parameter has been neglected. For computing oceanic circulation, r is usually set constant. However, to agree with the form of the linear friction introduced in (1.1.7), the form of r is κ/h. Equation (1.8.6) then becomes

$$\kappa\nabla^2\psi + \frac{\partial\psi}{\partial x}\left(\frac{-2\kappa}{h}\frac{\partial h}{\partial x} - f\frac{\partial h}{\partial y}\right) + \frac{\partial\psi}{\partial y}\left(\frac{-2\kappa}{h}\frac{\partial h}{\partial y} + f\frac{\partial h}{\partial x}\right)$$

$$= \frac{1}{\rho}\left(\frac{\partial G_S}{\partial x} - \frac{\partial F_S}{\partial y} + \frac{F_S}{h}\frac{\partial h}{\partial y} - \frac{G_S}{h}\frac{\partial h}{\partial x}\right). \qquad (1.8.7)$$

The boundary conditions are

$$\psi = \text{constant}$$

on a closed boundary, and

$$\partial\psi/\partial n = 0$$

on an open boundary. The first of these conditions arises from (1.1.8). The second does not correspond to the boundary conditions used earlier. However it is a convenient condition to use, and any errors it produces are insignificant if the open boundary is distant from the region of interest and is in deep water.

The gradient of the surface elevation in terms of ψ is

$$\left(\frac{\partial\zeta}{\partial x},\frac{\partial\zeta}{\partial y}\right) = \frac{-f}{gh}\left(\frac{\partial\psi}{\partial x},\frac{\partial\psi}{\partial y}\right) + \frac{1}{\rho gh}\,(F_S\,,G_S\,) + \frac{\kappa}{gh^2}\left(\frac{\partial\psi}{\partial y},\frac{-\partial\psi}{\partial x}\right).$$

From the open boundary condition on ψ, the rate of change of the surface elevation along the open boundary is

$$\frac{\partial\zeta}{\partial s} = \left\{-\frac{f}{gh}\left(\frac{\partial\psi}{\partial x},\frac{\partial\psi}{\partial y}\right) + \frac{1}{\rho gh}\,(F_S\,,G_S\,)\right\}$$

where s is measured tangentially to the open boundary. The variation of the surface elevation along the open boundary is minimized if h is as large as possible on the boundary. The use of $\partial\psi/\partial n = 0$ in place of $\zeta=0$ as the boundary condition can be justified if the open boundary is chosen to lie in deep water compared with the average depth of the basin.

To model the circulation within the South Australian gulf system, it is necessary to estimate the transport between the mainland and Kangaroo Island. This enables the value of ψ on the two shorelines to be fixed.

Within the South Australian gulf system, the time integrated surfacc forcing does not vary significantly with position, and was assumed constant for this analysis.

The finite-difference scheme used is described in Appendix 4. The numerical problem reduces to solving

$$A\underset{\sim}{\psi} = F_S \underset{\sim}{b_1} + G_S \underset{\sim}{b_2} + \sum_{i=1}^{n} c_i \underset{\sim}{b_{i+2}} . \qquad (1.8.8)$$

Each element of $\underset{\sim}{\psi}$ is the value of the stream function at some grid point and $A\underset{\sim}{\psi}$ is the finite-difference form of the left-hand side of (1.8.7). The circulation driven by a surface wind stress of unit strength to the east and north is $A^{-1}\underset{\sim}{b_1}$ and $A^{-1}\underset{\sim}{b_2}$, respectively. The value of ψ $(= c_i)$ set for each of the n disconnected sections of shore-line is equivalent to assuming the circulation is

$$\sum_{i=1}^{n} c_i A^{-1} \underset{\sim}{b_{i+2}} .$$

By limiting the size of A to 510×510, it was possible to use the program, developed by Curtis and Reid (1971) for the inversion of large sparse matrices, on the CDC 6400 used for the applications to the South Australian gulfs.

1.9 A Non-Linear Model

The linearized equations (1.1.1 to 1.1.3) are correct to the first order in Z/H. This approximation may be improved if non-linear terms to model convection are included in the equations. The equations, accurate to the first order in H^2/L^2, are

$$\frac{\partial U}{\partial t} + \frac{\partial \overline{u^2}}{\partial x} + \frac{\partial \overline{uv}}{\partial y} - fV = -g(h+\zeta)\left(\frac{\partial \zeta}{\partial x} + \frac{\partial p_a}{\partial x}\right) + \frac{1}{\rho}(F_S - F_B), \qquad (1.9.1)$$

$$\frac{\partial V}{\partial t} + \frac{\partial \overline{uv}}{\partial x} + \frac{\partial \overline{v^2}}{\partial y} + fU = -g(h+\zeta)\left(\frac{\partial \zeta}{\partial y} + \frac{\partial p_a}{\partial x}\right) + \frac{1}{\rho}(G_S - G_B), \qquad (1.9.2)$$

$$\frac{\partial U}{\partial x} + \frac{\partial V}{\partial y} + \frac{\partial \zeta}{\partial t} = 0, \qquad (1.9.3)$$

where

$$\overline{uv} = \int_{-h}^{\zeta} uv\,dz .$$

A derivation of these equations is given in Appendix 1.

The HN-method, as described earlier in this chapter, can be used to solve these equations. The spatial finite-difference form for the linearized equation in Section 1.3 must be modified. In matrix notation, the Equations (1.9.1) and (1.9.2) can be written as

$$\frac{\partial}{\partial t}\underset{\sim}{U}(t) + D_1\underset{\sim}{\omega_1}(t) + D_2\underset{\sim}{\omega_3}(t) - F_1\underset{\sim}{V}(t)$$

$$= -g\{H_1 + L_1(t)\}D\underset{\sim}{\zeta}(t) + \frac{1}{\rho}\{\underset{\sim}{F_S}(t) - \underset{\sim}{F_B}(t)\}, \qquad (1.9.4)$$

$$\frac{\partial}{\partial t}\underset{\sim}{V}(t) + D_3\underset{\sim}{\omega}_3(t) + D_4\underset{\sim}{\omega}_2(t) + F_2\underset{\sim}{U}(t)$$

$$= -g\{H_2+L_2(t)\}E\underset{\sim}{\zeta}(t) + \frac{1}{\rho}\{\underset{\sim}{G}_S(t) - \underset{\sim}{G}_B(t)\}. \qquad (1.9.5)$$

The elements of the vectors $\underset{\sim}{\omega}_1(t)$ and $\underset{\sim}{\omega}_2(t)$ and $\underset{\sim}{\omega}_3(t)$ are defined to correspond to the values of $\overline{u^2}$, $\overline{v^2}$ and $\overline{uv}$ at a grid point. An element of $\underset{\sim}{\omega}_1(t)$ can be written as

$$\omega_{1i}(t) = \frac{U_i^2(t)}{h_i + \zeta_i(t)}. \qquad (1.9.6)$$

This enables the finite-difference form of the derivative $\partial\overline{u^2}/\partial x$ to be written as $D_1\underset{\sim}{\omega}_1(t)$. The matrices L_1 and L_2 are diagonal matrices similar to H_1 and H_2 with the i-th position on the diagonal of L_1 being $\zeta_i(t)$. The details of the finite-difference equations are given in Appendix 6.

The non-linear terms do not allow the use of a simple equation of the form of (1.3.4). However, the analysis presented in Section 1.6 on the discretization of the time dependence may be replicated, equation (1.6.1) being rewritten as

$$\frac{\underset{\sim}{U}(t+\tau)-\underset{\sim}{U}(t)}{\tau} + D_1\underset{\sim}{\omega}_1(t+\alpha_1\tau) + D_2\underset{\sim}{\omega}_3(t+\alpha_1\tau) - F_1\underset{\sim}{V}(t+\alpha_1\tau)$$

$$= -g\{H_1+L_1(t+\beta\tau)\}D\underset{\sim}{\zeta}(t+\beta\tau) + \frac{1}{\rho}\{\underset{\sim}{F}_S(t+\delta\tau) - \underset{\sim}{F}_B(t+\lambda\tau)\}. \qquad (1.9.7)$$

A program has been written to step $\underset{\sim}{U}, \underset{\sim}{V}$ and $\underset{\sim}{\zeta}$ forward in time with $\alpha_1 = \beta = \delta = \lambda = 0$ in the non-linear equations.

1.10 Energy

The energy equation corresponding to (1.1.1), (1.1.2) and (1.1.3) is

$$\frac{\partial}{\partial t}\iint_S \frac{1}{2}\left(\frac{U^2+V^2}{h} + g\zeta^2\right)dA$$

$$= -\oint_\ell g\zeta U_n\, d\ell + \iint_S \frac{F_S U+G_S V}{\rho h}\, dA - \iint_S \frac{F_B U+G_B V}{\rho h}\, dA, \qquad (1.10.1)$$

where U_n is the outward normal transport at the boundary.

Harris and Jelesnianski (1964) used this energy equation to check the conservation of energy of the time stepping system described by (1.6.5). They found that the energy was not constant, and concluded that the scheme was not useful. Their system required twice the computer storage that is needed for (1.6.5). Sielecki (1968) demonstrated that the oscillation of the calculated energy could be considerably reduced by accounting for the phase difference between the velocity and elevation fields. The correction was carried out by assuming a phase difference of $\tau/2$.

If the energy equation is derived from the equations

$$\frac{U(t+\tau)-U(t)}{\tau} - fV(t+\alpha_1\tau) = -gh\frac{\partial}{\partial x}\zeta(t+\beta\tau) + \frac{1}{\rho}(F_S(t+\delta\tau) - F_B(t+\lambda\tau)),$$

$$\frac{V(t+\tau)-V(t)}{\tau} + fU(t+\alpha_2\tau) = -gh\frac{\partial}{\partial y}\zeta(t+\beta\tau) + \frac{1}{\rho}(G_S(t+\delta\tau) - G_B(t+\lambda\tau)),$$

and

$$\frac{\zeta(t+\tau)-\zeta(t)}{\tau} + \frac{\partial}{\partial x}U(t+\overline{1-\beta}\tau) + \frac{\partial}{\partial y}V(t+\overline{1-\beta}\tau) = 0,$$

the phase difference between the fields is completely accounted for. For $\beta = \alpha_1 = \alpha_2 = \gamma = \lambda = 0$, one form of the energy equation is

$$\iint_S \left\{\frac{U^2(t+\tau)+V^2(t+\tau)}{2h} + \frac{g}{2}\zeta(t+2\tau)\zeta(t+\tau)\right\} dA$$

$$= \iint_S \left\{\frac{U^2(t)+V^2(t)}{2h} + \frac{g}{2}\zeta(t+\tau)\zeta(t)\right\} dA + \frac{\tau}{2}\iint_S \left\{\frac{f}{h}(U(t)V(t+\tau)-U(t+\tau)V(t)\right\} dA$$

$$+ \frac{\tau}{\rho}\iint_S \frac{1}{h}\left\{(F_S(t)-F_B(t))\left(\frac{U(t+\tau)+U(t)}{2}\right) + (G_S(t)-G_B(t))\left(\frac{V(t+T)+V(t)}{2}\right)\right\} dA$$

$$- g\tau\oint_\ell \zeta(t+\tau)\left\{\frac{U_n(t+\tau)+U_n(t)}{2}\right\} d\ell \,. \qquad (1.10.2)$$

For sinusoidal motion, the correction term for the Coriolis effect always has the same sign.

Numerical experiments have been conducted using (1.10.2) with the time-stepping scheme described by (1.6.5). With zero friction and no rotation, the energy of motion in a closed basin was found to be constant to five significant figures.

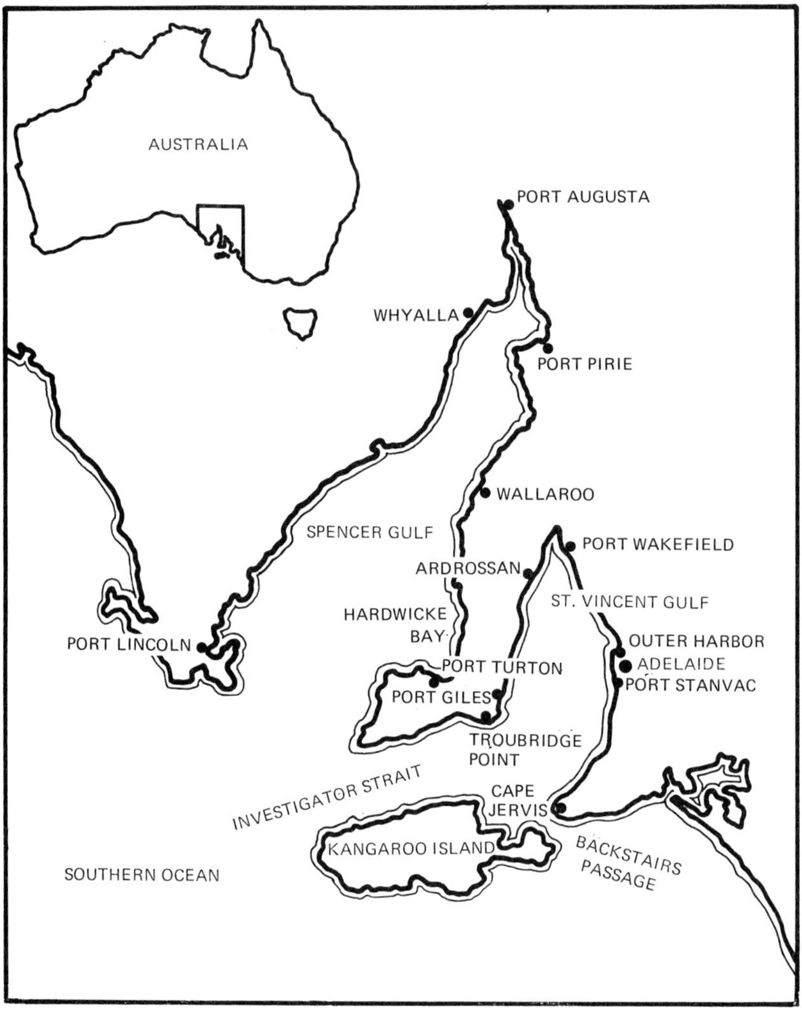

Figure 2.1: The South Australian Gulf System.

2. APPLICATION TO THE SOUTH AUSTRALIAN GULF SYSTEM

The South Australian gulfs lie within a rectangle 400 Km by 160 Km on the eastern side of the Great Australian Bight (Figure 2.1).

Ninety percent of the population of South Australia lives around the shores of the gulfs. Adelaide's ocean terminals, Port Adelaide and Port Stanvac, are situated on the shores of St. Vincent Gulf along with Ardrossan and Port Giles, bulk shipping terminals for the export of cereal crops. Spencer Gulf boasts the ports of Whyalla with its iron and steel and ship building industries, Port Pirie with its lead smelting works and Port Lincoln, the base of a large tuna fishing fleet.

The gulfs are treated as a playground, a thoroughfare, a sewer and a fishery. A knowledge of the hydraulics of the system is essential if these four uses are to be kept from major conflict.

The first investigation of the tides along the coast of South Australia was carried out by Bragg, Inglis, Lyle and Chapman (1892), whose report was presented at the Fourth Annual Session of the Australian Association for the Advancement of Science, held at Hobart in 1892.

Easton (1970) used the information from the lists of tidal constants available for South Australian ports to draw co-tidal and co-range charts for the K_1 and M_2 tides in Spencer and St. Vincent Gulfs. (see Figure 2.2). These components are considered representative of the diurnal and semi-diurnal tide species.

Detailed bathymetric surveys of Spencer and St. Vincent Gulfs and adjacent waters have been conducted by the Royal Australian Navy. The model basis used in this study were set up using the Admiralty Map No. 2389 supplemented by survey charts supplied by the R.A.N. Hydrographic Office.

The continental shelf is of order of 60 Km wide at the western end of Kangaroo Island and considerably wider south of the eastern end of the island. There are rapid changes in depth at the western end of Investigator Strait and at the mouth of Spencer Gulf (see Figure 2.3). The "average" depth of the southern half of Spencer Gulf, Investigator Strait and Backstairs Passage is 40 metres. The depth in St. Vincent Gulf and the northern half of Spencer Gulf decreases northwards from 40 metres down to zero. Much of the water in the two gulfs is less than 10 metres deep.

Along the western side of St. Vincent Gulf there are a series of shoals and sand banks stretching out 10 Km from the shore. Several major shoals exist in the northern half of Spencer Gulf. These run parallel to the axis of the gulf.

There have been few hydrological studies conducted on the waters of the South Australian gulf system. For instance Thomas and Edmonds (1956) published findings on the seasonal variation of chlorinities at Port Lincoln, Port Augusta, Moonta Bay, Port Wakefield, Brighton, Kingscote and Victor Harbor.

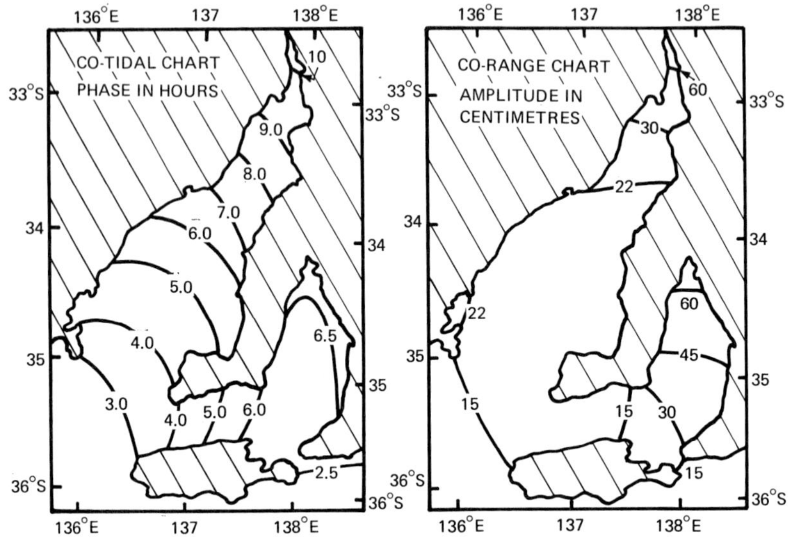

Figure 2.2a: Co-tidal and co-range charts for the M_2 semi-diurnal tidal component in the South Australian gulfs. (after Easton, 1970).

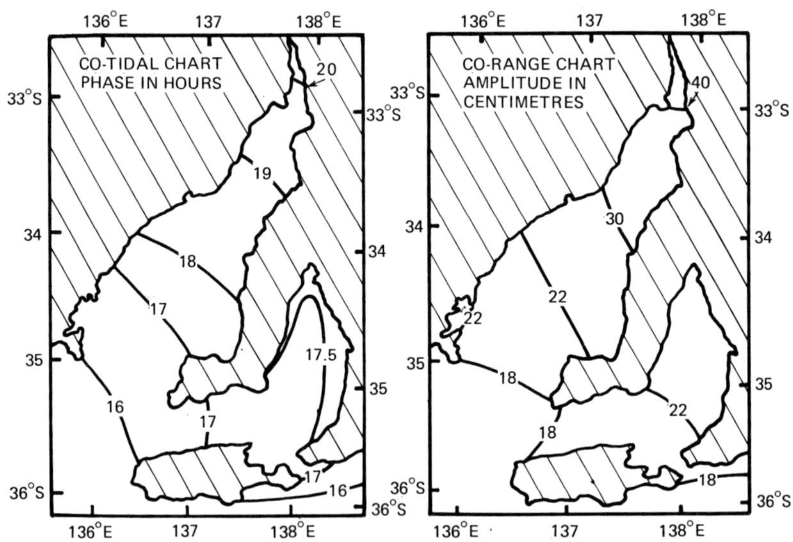

Figure 2.2b: Co-tidal and co-range charts for the K_1 diurnal tidal component in the South Australian gulfs. (after Easton, 1970).

The density was calculated for places at which temperature and salinity measurements were available or could be estimated. There was no more than a 0.5% variation in the density in time and space.

There is no evidence of any significant vertical stratification within the gulf system. To confirm this, vertical profiles of the temperature and salinity were measured in St. Vincent Gulf. Several stations across the gulf at the latitude of 35°S were taken during March 1972. The temperature and salinity were found to be constant at each station.

The overall pattern is of increasing salinity and temperature towards the head of each gulf. Seasonal changes are more marked in areas furthest from the open ocean.

There is a net loss of water from the gulf due to the difference between evaporation and precipitation. As there are no major rivers flowing into either gulf, there must be sufficient water exchange between the gulfs and the Southern Ocean to keep the salinity constant.

Waters (1973) has gathered evidence, in the form of sediment distribution charts, of circulation between the ocean and St. Vincent Gulf.

The hydraulics of the gulf system can be divided into four sections, namely the natural modes of oscillation, tides, storm surges and long term circulation. The first of these provides a basis for the understanding of tidal motion. The significance of the second and third items needs no elucidation. As more wastes are dumped in the gulfs, knowledge of the fourth item, the circulation of the gulf waters, will become more important.

2.1 Resonances of the Gulf System

The standing waves of the gulf system are discussed in this section. The methods used are those outlined in Section 1. A brief analysis of the equations of motion in terms of the eigensolutions of the problem is presented at the end of the Section. Attention is drawn to the significance of certain forms for describing the dissipation and the tidal motion in terms of the eigensolutions.

2.1.1 The Extent of the Analysis

The resonances presented are theoretical approximations. As dissipation and non-linear interactions have been excluded from the model, the frequencies given here are not exactly those which occur in practice. Accordingly, the aim has been merely to identify the significant features of the gulf system which determine the form of the long wave motion.

To keep computing requirements to a minimum, a coarse grid on which model basins are defined has been used. The model basin representing the total gulf system is shown in Figure 2.3. The grid spacing is 12.70 Km. At the bottom right-hand corner, a section with a grid spacing of 6.35 Km was found to be necessary so that Backstairs Passage could be adequately represented.

This model basin was divided into two parts so that the motions in St. Vincent Gulf and Spencer Gulf could be studied separately. The dotted lines at the western end of Investigator Strait and at the mouth of Spencer Gulf were taken as the open boundaries of these models.

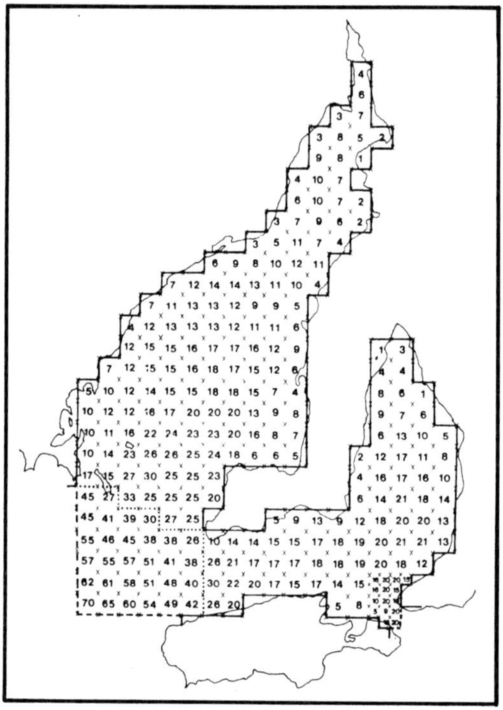

Figure 2.3: The model basin used for analysis of the possible resonances of the gulf system with the depths shown in fathoms (1 fathom = 1.86m); —x— closed boundary; - -x- - open boundary for modelling resonance of connected gulfs; •••x••• open boundary for separate models of Spencer and St. Vincent Gulfs.

The results presented represent the calculation of 27 eigenvalue-eigenvector pairs of 6 matrices. The matrices ranged from a size of 99×99 to 332×332. The eigenvalue and its maximum error are listed with the resonant frequency and period to provide an estimate of the numerical accuracy of the eigenvector.

When rotation has been included, the Coriolis parameter has been set constant, corresponding to a latitude of 35^0S. At this latitude a pendulum day is of length 20.9 hours. As there were no resonant periods greater than 18 hours, no problems have been encountered as a result of τ^* (see Section 1.5) being close to unity.

2.1.2 St. Vincent Gulf

The region referred to as St. Vincent Gulf includes Investigator Strait and Backstairs Passage.

TABLE 1

Eigenvalue, theoretical bound, frequency and period of the lowest modes of oscillation of St. Vincent Gulf, Spencer Gulf and the complete South Australian Gulf System, assuming zero rotation (f=0), or including the effects of the Earth's rotation (f = -0.5363 for St. Vincent Gulf and f = -1.0726 for Spencer Gulf).

	Eigenvalue λ	Bound ε	Frequency σ (rad/hr)	Period T (hrs)
ST. VINCENT GULF	1.3801	9.0×10^{-3}	0.6578	9.55
ZERO ROTATION	4.9275	1.0×10^{-3}	1.2431	5.05
	9.4750	3.0×10^{-3}	1.7237	3.65
ST. VINCENT GULF	1.1313	4.0×10^{-4}	0.667071	9.42
WITH ROTATION	4.66213	2.0×10^{-5}	1.24604	5.04
SPENCER GULF	1.8244	1.4×10^{-6}	0.3819	16.61
ZERO ROTATION	6.95995	6.0×10^{-6}	0.73868	8.51
	17.3914	1.0×10^{-6}	1.16767	5.38
SPENCER GULF	0.8364	7.0×10^{-4}	0.3945	15.93
WITH ROTATION	5.9741	5.0×10^{-4}	0.7480	8.40
	16.2217	2.0×10^{-4}	1.1670	5.38
SOUTH AUSTRALIAN	1.40964	2.0×10^{-4}	0.36416	17.25
	4.55574	2.0×10^{-4}	0.65467	9.60
GULF SYSTEM	5.14435	5.0×10^{-4}	0.69568	9.03
	13.9127	1.0×10^{-2}	1.4406	5.49
ZERO ROTATION	16.0272	1.0×10^{-3}	1.2279	5.12

Assuming zero rotation the three lowest resonant frequencies are listed in Table 2.1. The standing wave patterns associated with each mode of oscillation are shown in Figure 2.4. Although modified by Kangaroo Island partly blocking the opening of the gulf, these standing waves are basically the first three modes of oscillation in a channel open at one end. The distance between the open boundary and the head of St. Vincent Gulf is 230 Km.

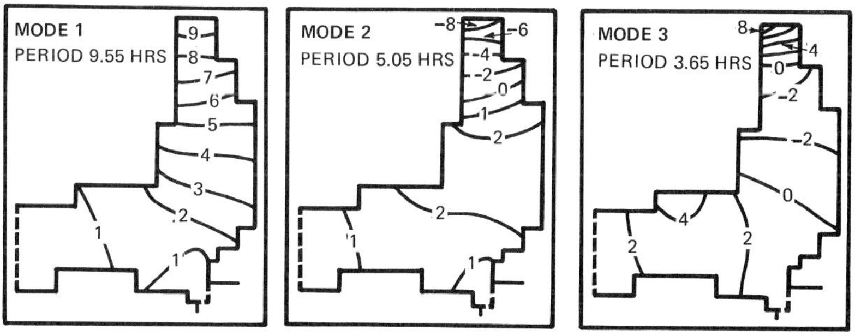

Figure 2.4: Relative amplitude of the surface elevation for the first three modes of oscillation of St. Vincent Gulf neglecting the Earth's rotation.

A discussion of the one-dimensional problem where the depth and breadth are represented as $h_0 x^a$ and $b_0 x^b$ respectively is presented in Appendix 6. The values of a and b used for St. Vincent Gulf can only be approximations. Four possible choices are listed in Table 2. This table includes a comparison between the one-dimensional models and the two-dimensional finite-difference model. If the one-dimensional model is a good approximation to the basin, then the average depth, $\bar{h}$, should be constant for all modes of oscillation.

TABLE 2.

A comparison using Equation (A6.7) of three one-dimensional models for St. Vincent Gulf and Spencer Gulf.

MODEL	MODE	z	Mean depth $\bar{h}$ (metres) St. Vincent Gulf	Spencer Gulf
(i)	1	π	18	11
a=b=0	2	2π	16	10
$\nu=\frac{1}{2}$	3	3π	14	11
(ii)	1	2.405	37	22
a=b=$\frac{1}{2}$	2	5.520	25	15
$\nu=0$	3	8.654	20	16
(iii)	1	2.903	25	15
a=$\frac{1}{2}$,b=0	2	6.033	21	13
$\nu=1/3$	3	9.171	17	14
(iv)	1	3.832	25	14
a=b=1	2	7.016	26	16
$\nu=1$	3	10.173	24	19

Honda *et al* (1908) estimated the fundamental frequency of St. Vincent Gulf by using the one-dimensional model (i) of Table 2. They considered that a line joining Troubridge Point and Cape Jervis defined the southern limit of the gulf, making its length 140 Km. The average depth was assumed to be 21.5 m, giving a fundamental period of 10.8 hours. Oscillations with a period of 10.9 hours, evident in the Port Adelaide tide record, were excited by the tsunami generated by the Krakotoa eruption of 1883.

When compared with the four one-dimensional models introduced earlier (see Table 2) the results are approximately the same for each model, requiring an average depth of 17 metres. However, the eigenvectors of the model basin are two-dimensional in character and cannot be adequately represented by channel modes.

Experiments were conducted with several variations of the grid and bathymetry of the model. These alternatives made no significant changes to the results presented. A model of St. Vincent Gulf with a uniform grid spacing of 6.35 Km was set up to ensure that results from the 12.70 Km grid were correct. The variation between the models was minimal.

Effects of the Earth's rotation on the standing waves in St. Vincent Gulf were also considered. The first two resonant frequencies are given in Table 1 and the standing wave patterns are shown in Figure 2.5.

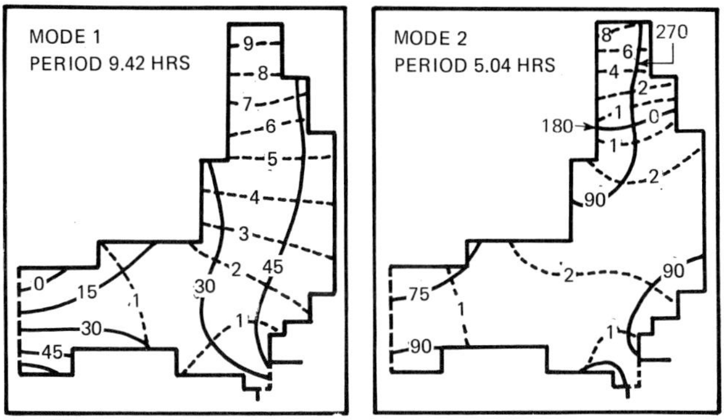

Figure 2.5: The relative amplitude (---) and phase (——) in degrees of the surface elevation for the first two modes of oscillation of St. Vincent Gulf including the effect of the Earth's rotation.

Rotation has only slightly reduced the resonant periods for these two modes and the amplitude pattern has not been significantly altered, except that the node of mode 2 became an amphidrome with rotation. The waves rotated clockwise.

2.1.3 Spencer Gulf

Using the model basin of Spencer Gulf shown in Figure 3 and neglecting the Earth's rotation, the three lowest resonant frequencies of the gulf were calculated. These are listed in Table 1. Figure 2.6 shows the corresponding standing wave patterns.

These modes of oscillation correspond to the first three modes of a channel open at one end. This was expected, as the gulf could be adequately represented by a channel of the form considered in Appendix 6, except for the rapid change in breadth at the southern end of Hardwicke Bay. A comparison of four possible one-dimensional models is given in Table 2. The length of the gulf has been taken as 305 Km.

Models (i) and (iii) appear to be the best one-dimensional representations of the gulf, as the variation of $\bar{h}$ (the average depth) between modes is a minimum. The average depths given in Table 2 of both models are rather small, although model (iii) provides the best estimates, with the depth ranging from 22 m at the open end to zero at the head of the gulf. The breadth is constant for both models, implying that the transport of energy is concentrated along a section of constant breadth in the centre of the gulf.

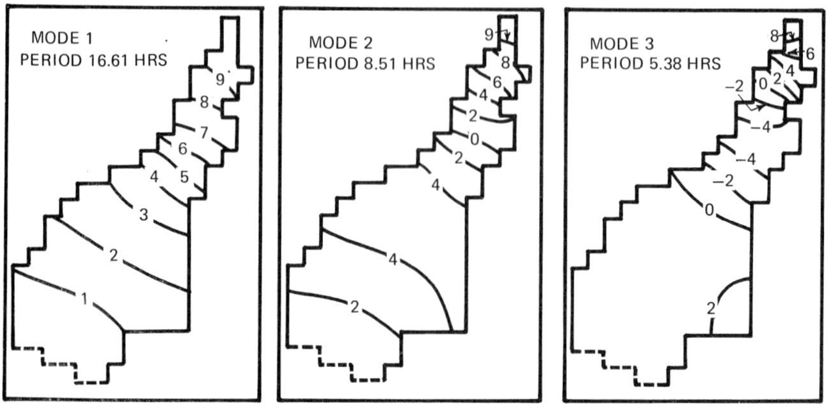

Figure 2.6: The relative amplitude of the surface elevation for the first three modes of oscillation of Spencer Gulf with the Earth's rotation neglected.

When the Earth's rotation was included, the standing wave patterns for the first three modes appeared as shown in Figure 2.7. Table 1 gives the frequencies, periods and eigenvalues for these modes each of which rotated clockwise.

For the lowest mode of both Spencer and St. Vincent Gulfs, inclusion of rotation increased the resonant frequency. This was in agreement with the result, presented by Lauwerier (1960), that "the force of Coriolis has the tendency to increase the lowest eigenvalue" for a rotating rectangular sea of constant depth. He made no reference to the effect of rotation on higher modes.

For a closed basin of constant depth, rotation reduces the frequency of the lowest mode. There is a similar lack of agreement if the results presented by Rao (1966) and Pneuli and Pekeris (1968) for the higher modes are compared with the results presented above.

2.1.4 <u>The Complete Gulf System</u>

The resonance of the total gulf system was studied to determine the interaction between Spencer and St. Vincent Gulfs.

There was found to be no low frequency mode which could be described as a resonance of the total gulf system. The five lowest resonance frequencies, listed in Table 1, are identified with resonances calculated for either Spencer or St. Vincent Gulf. The increase in area of the model basin resulted in a slight decrease in the resonant frequencies, as to be expected. The changes in co-range lines (see Figure 2.8) could also be attributed to this cause. A feature of these modes is the lack of interaction between the two gulfs.

The standing waves for the gulf system, with rotational effects included, have not been described as the results are similar to the corresponding waves calculated for the individual gulfs.

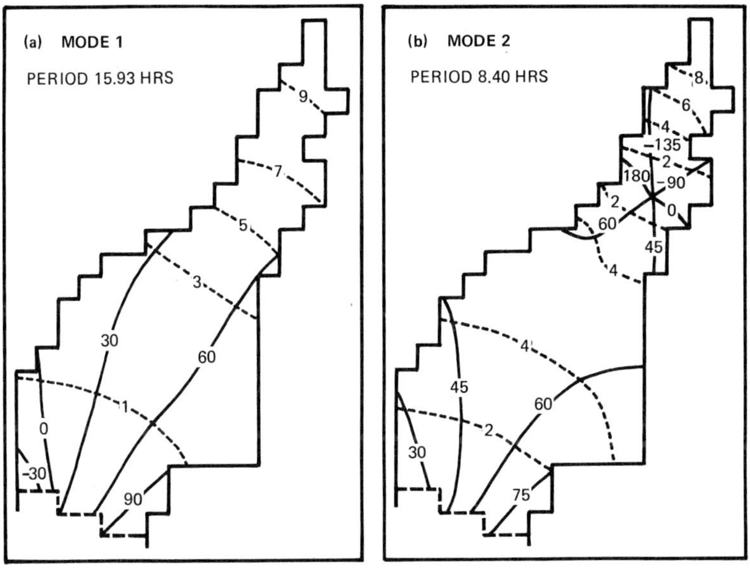

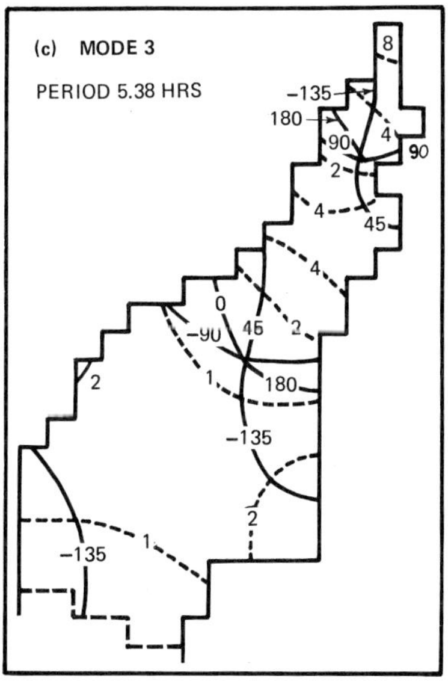

Figure 2.7: The relative amplitude (----) and phase (——) in degrees of the surface elevation for the first three modes of oscillation of Spencer Gulf, including the effect of the Earth's rotation.

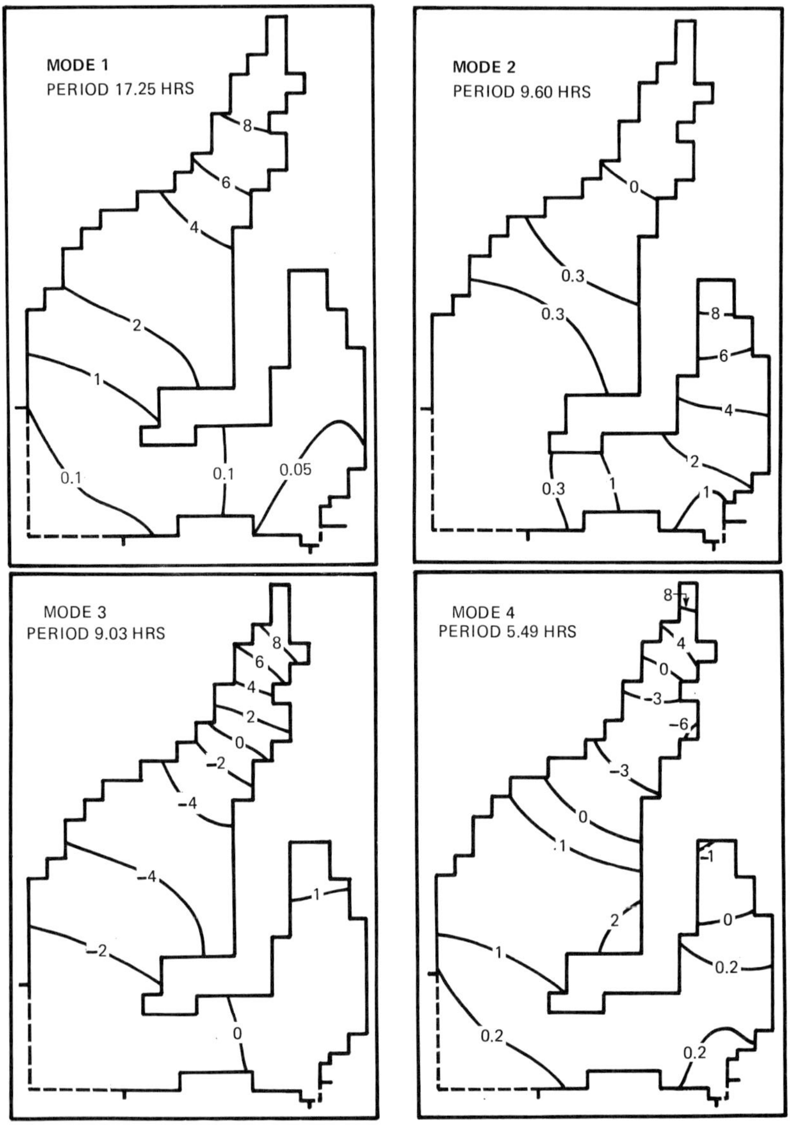

Figure 2.8: The relative amplitude of the surface elevation for the first four modes of resonance of the South Australian gulf system, neglecting the effect of the Earth's rotation.

2.2 Numerical Simulation of the Astronomical Tides

The results presented in this section represent the results obtained from various numerical models for the propagation of tides in the South Australian gulf system. The model results are compared with the tidal measurements presented by Easton (1970) for this region. The primary aim has been to extend the understanding of the factors governing tidal motion in the area rather than to construct a precise predicting model.

2.2.1 The Model

The model basin shown in Figure 2.9a was used in this study. The associated bathymetry is shown in Figure 2.9b. As latitude and longitude were used for the coordinates of position, the distance between similar grid points was expressed as the difference in latitude or longitude of adjacent grid points. This was fixed at 0.0833^0. The distances corresponding to these values are 9.26 Km along a line of longitude, and approximately half this distance along a line of latitude.

On an open boundary, it is necessary to define the surface elevation. Accordingly, the choice of outer boundary reflects a compromise between the knowledge of tidal motion beyond the coastline, the natural position as defined by the bathymetry and the grid-spacing in the model.

The appropriate position for the open boundary in the model was difficult to determine. The co-range and co-tidal lines drawn by Easton (1970) for the South Australian gulfs zone conflicted with those drawn for Southern and South Western Australia south of the gulf system. Accordingly, the boundary was placed as far from the area of interest as possible to minimize the error due to inaccurate specification of the tidal motion on the open boundary.

The loss of energy is important in determining the tidal motion, as the waters of the gulfs are shallow. Experiments were conducted using the linear dissipation term, (1.1.7), and the quadratic friction term, (1.1.6).

2.2.2 Peculiarities of the Numerical Method

The accuracy of any solution to the tidal problem must depend in part on the method of solution. All numerical methods have limitations, and these must be recognized when discussing the solution.

For a rectangular basin, the HN-method provided answers which agreed with analytic solutions to eight significant figures, and no instabilities were encountered with zero friction and no geostrophic forces. With basins of complex shape, it was necessary to include friction to suppress the computational modes. These seemed to be generated at internal corners of 270^0. The frequency of the computational modes was of the order of 0.5 hours for model B, rather than 4.8 minutes (2τ) and it appeared to be independent of the bathymetry. Similar problems have been commented upon by Platzman (1963).

These instabilities were most noticeable around Backstairs Passage. This was probably due to the existence of nine 270^0 corners in that section of the model. Consideration was given to spatial filtering to remove the high frequency waves, but it was felt that further errors would be introduced into areas where the definition provided by the grid was barely sufficient, and where most difficulty was being encountered.

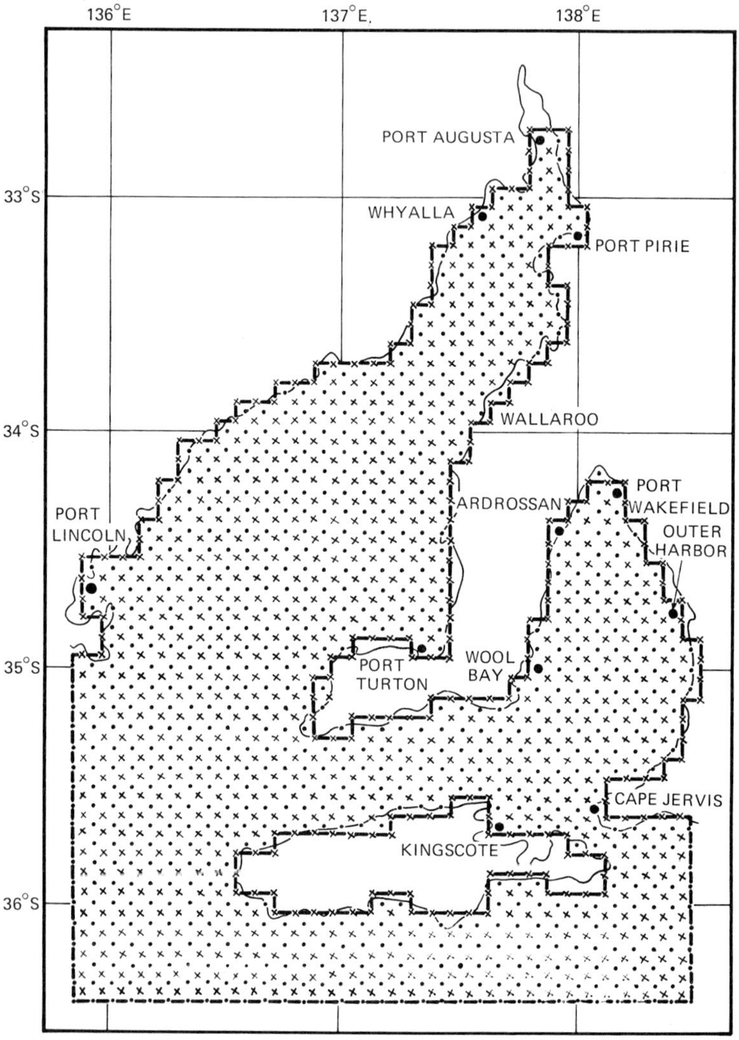

Figure 2.9a: Model of the South Australian gulf system; ·-·-·- open boundary; x-x-x closed boundary; •elevation point; x stream point; o grid point at which computed sea level elevations were compared with observations at a nearby port.

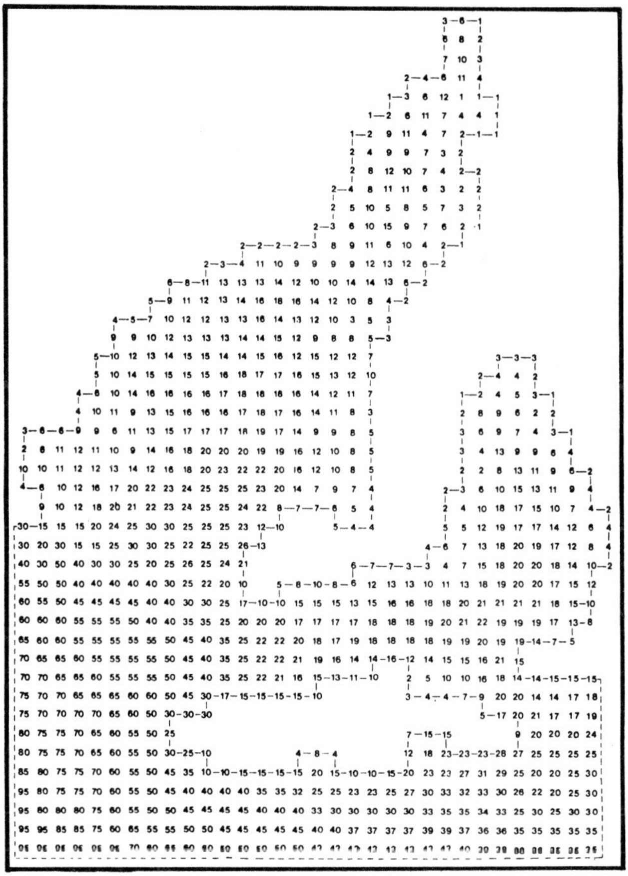

Figure 2.9b: The depth in fathoms at each stream point of the model (1 fathom = 1.86m).

When sufficient friction was included in the model to ensure stability, the free waves induced by the numerical model decayed within several wavelengths of their source. The minimum value of the coefficient of friction necessary to ensure stability was found to increase with a decrease in the frequency of the tidal input.

A numerical experiment showed that the HN-method, as programmed, retained the property of superposition of solutions of linear systems.

To minimize the importance of the numerical errors, the results presented in the following section were calculated during the fourth model day.

2.2.3 The Linear Model

The features used to indicate the response of the gulfs to a tidal input were the amplification and the phase of the tidal component at the head of each gulf (Port Augusta and Port Wakefield) and at Cape Jervis (the northern end of Backstairs Passage). This avoids the need to compare directly the recorded tide with the numerical model results, a task which would have been difficult because the tidal input along the open boundary is unknown, as indicated by Easton's conflicting tidal charts commented upon in section 2.2.1. For each variation discussed below, these features are listed in Table 3.

TABLE 3

The amplification (a) and phase (p hours) of semi-diurnal and diurnal tide components at Port Augusta and Port Wakefield relative to the tidal input for ten numerical experiments and the results of Easton.

MODEL	PERIOD (hours)	PORT AUGUSTA a	PORT AUGUSTA p	PORT WAKEFIELD a	PORT WAKEFIELD p
Linear friction k = 0.003	12	0.97	7.98	1.59	5.30
Linear friction k = 0.002	12	1.68	7.45	1.39	5.00
	24	1.41	6.42	1.39	2.94
Linear friction k = 0.0015	12	2.23	7.18	3.12	4.91
	24	1.73	5.66	1.42	2.37
Linear friction k = 0.001	12	2.94	6.88	4.42	4.80
	24	2.14	4.56	1.14	1.79
Backstairs Passage shallow k = 0.0015	12	2.23	7.18	3.05	4.95
Backstairs Passage closed k = 0.0015	12	2.09	7.15	1.39	6.19
Variable phase-input k = 0.0015	12	2.23	7.21	3.26	5.38
No geostrophic forces	12	2.21	7.36	3.43	5.01
Quadratic friction k = 0.003	12	2.40	7.16	3.12	4.94
	24	1.43	6.17	1.36	2.81
Non-linear model k = 0.003	12	2.41	7.15	3.10	4.91
	24	1.45	6.05	1.36	2.65
Easton (1970)	12	3.0	6.5	4.0	3.5
	24	2.0	3.5	1.5	1.5

From the data analysis presented by Easton (1970), it appears that St. Vincent Gulf amplifies a semi-diurnal component by a factor of 4.0 and a diurnal component by 1.5. For Spencer Gulf the amplifications are 3.0 and 2.0 respectively. Using the amplification values presented in Table 3 for the four values of the linear friction coefficient tested, the coefficient of friction was found to be approximately 0.0012 m/sec. Clearly it is dependent on the velocity of the motion being modelled. The value of the coefficient of friction used for the remainder of the study is 0.0015 m/sec.

A comparison of the measured phase lags with the predicted phase lags provided further information about the performance of the model. Easton's (1970) measured phase lags between the outer boundary and the heads of Spencer and St. Vincent Gulfs were, respectively, 3.5 and 1.5 hours for a diurnal wave and 6.5 and 3.5 hours for a semi-diurnal wave. These values varied somewhat from the lags predicted by the numerical models using each of the four values of friction tested.

To provide a more detailed picture of the predicted tidal motion, the co-range and co-tidal charts for the diurnal and semi-diurnal components have been drawn in Figure 2.10. These may be compared with the charts drawn by Easton (1970) (Figure 2.2).

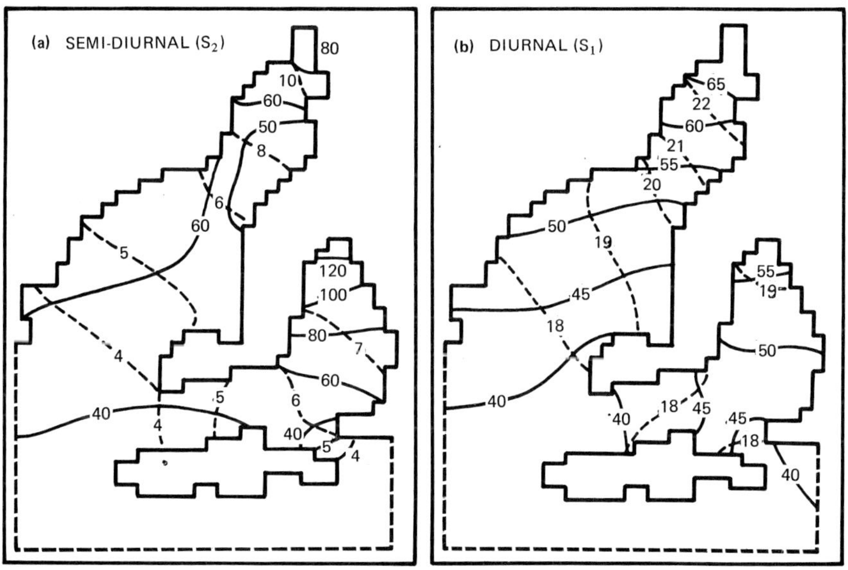

Figure 2.10: Predicted co-tidal and co-range charts for (a)S_2 semidiurnal tidal component, (b)S_1 diurnal tidal component, using the model with linear friction; ---- co-tidal lines (hours), —— co-range lines (cm).

The co-tidal chart for the M_2 component suggests a near resonant situation in St. Vincent Gulf as evidenced by the rapid phase change through Investigator Strait and almost constant phase in St. Vincent Gulf. This would account for the large amplification of the semi-diurnal component compared with the diurnal component. The fundamental period of St. Vincent Gulf of 9.55 hours calculated in Section 2.1 is sufficiently close to 12 hours to support this observation. The phase change of 2 hours in the semi-diurnal tide through Investigator Strait was reproduced by the numerical model. However, in St. Vincent Gulf, the model predicts a progressive wave slowed by friction.

The co-range charts for the semi-diurnal tides in St. Vincent Gulf have similar patterns. The major differences occur at the southern end of the gulf and appear to be due to Backstairs Passage being made too wide in the model. Results for the other major semidiurnal tides for the region, viz. μ_2, S_2 and K_2, were similar to that for the M_2 tide.

In St. Vincent Gulf, the computed diurnal tide only differs from the K_1 component by Easton in that the phase between the head of the gulf and the ocean is greater. Similar results were obtained for the other major diurnal tidal components in the gulfs, the Q_1, O_1 and P_1.

The eight tidal components referred to above account for over 98 percent of the tidal range at any point in the gulf system. The response of the system to the SO_3, MN_4, $2MS_6$ and $2SM_6$ components were also calculated. Good agreement was obtained for all but the SO_3 component, for which the difference in phases is particularly large. [The main reason for this is that the SO_3 component is generated locally by non-linear interactions in the relatively shallow waters of the gulf system. Similar interactions would be expected to contribute to the M_4, MN_4, $2MS_6$ and $2SM_6$ components. From the work of Munk and Cartwright (1966), it would seem that it is incorrect to model the higher order tidal components using a progressive wave model.

The differences between the model tide and the measured tide in Spencer Gulf have the same characteristics as those in St. Vincent Gulf. By choosing a suitable value for the coefficient of friction, the amplification of the model tide can be made to agree with the observed amplification. However, the model tide takes too long to travel the length of the gulf. The inability to align both the amplification and phase of the model tide with the measured tide is probably due to inaccurate modelling of the dissipation mechanisms. An alternative form for the dissipation term is tested in the following Section.

The co-range lines for the diurnal component in the southern half of Spencer Gulf predicted by the numerical model are different from those drawn by Easton. This can be accounted for by the differences in the tidal input.

The co-tidal charts for the semi-diurnal component in Spencer Gulf are remarkably similar. However, this is not the case for the co-range chart. From the lack of information in Figure 2.2, it must be concluded that it was difficult for Easton (1970) to determine the co-range pattern. This would be the case if Figure 2.10a were correct. The tidal pattern explained by Taylor (1921) as due to oppositely travelling Kelvin waves is observed in Spencer Gulf. Hendershott and Speranza (1971) have shown that such a pattern exists in a rectangular gulf due to dissipation of energy at the head of the gulf. This is the situation in Spencer Gulf, as the dissipation occurs in the shallow water at the head of the gulf.

Several further experiments were conducted on the tidal motion to determine the significance of certain features of the linear model.

Two alternative forms of modelling Backstairs Passage were tried. As the width of the passage in the model was fixed by the grid, the only way to reduce its importance was to reduce the depth of the passage by one third. This had very little effect on the semi-diurnal tide as shown in Table 3. As this result supported the theory of near resonance, in a further experiment the passage was then closed off completely. The results from Section 2.1 indicated that this would change the resonant frequencies of the model and hence the result for the semi-diurnal wave. Such was indeed the case, with the amplification being reduced from 3.12 to 1.39 at the head of St. Vincent Gulf. The phase lag was not altered significantly, but the increase in phase from the ocean to Port Wakefield was more uniform than for other models.

An experiment with an input of variable phase on the open boundary was conducted. The phase of the forced semi-diurnal input varied from 3.0 hours on the western end of the southern section of the open boundary to 4.0 hours at the eastern end. The only change in results was an increase of the order of 0.35 hours in the phase of the tide in St. Vincent Gulf. This reinforces the observation that Backstairs Passage is too large in the model, and hence its effect is over-emphasized.

When geostrophic forces were omitted from the model, the tidal waves took the form expected for waves in a channel of variable depth and breadth. The Kelvin wave pattern evident in Spencer Gulf disappeared. As seen from the information presented in Table 3, the amplification and phase at the head of each gulf were not changed.

The harmonic method as described in Section 1.4 was also tried. The results obtained were the same as those calculated by the HN-method with linear friction, except for small differences in phase.

2.2.4 Non-linear Models

The discrepancy between the model results and the observed tide is usually claimed to be due to the use of linear models rather than a more correct non-linear model.

To evaluate the importance of the non-linear terms in the equations of motion, the numerical system was extended to include quadratic friction and the convective terms. The value of the coefficient of quadratic friction used in this analysis was 0.003 as has been used successfully by Marchuk, Kagan' and Tamsaly (1969) and Dunn-Christensen (1971).

Rather than attempt to model the total tide, it was decided to run a model to provide order of magnitude estimates. The major components of the tide in the South Australian gulf system are diurnal and semi-diurnal. Accordingly, the input tide was modelled by the sum of two waves with periods of 24 and 12 hours, both with an amplitude of 0.4 metres, and with a phase of 17 and 3 hours respectively. The amplitude and phase of the waves within the model basin with periods of 24, 12, 8 and 6 hours were calculated from the predicted surface elevation using least squares techniques.

The co-range and co-tidal charts for the diurnal and semi-diurnal waves are drawn in Figures 2.11, for a model incorporating

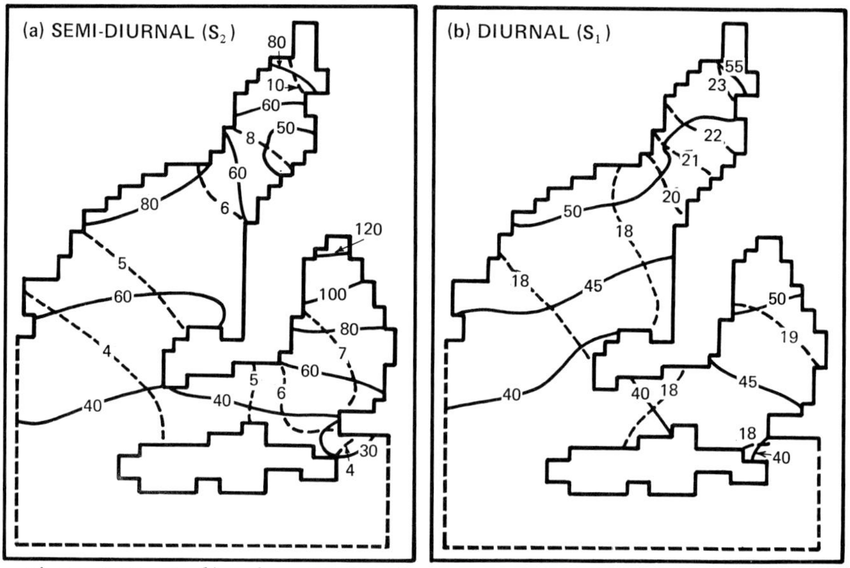

Figure 2.11: Predicted co-tidal and co-range charts for (a) S_2 semidiurnal, (b) S_1 diurnal tidal component, using the non-linear model with quadratic friction only; ---- co-tidal lines (hours), —— co-range lines (cm).

quadratic friction. A comparison, using the results included in Table 3 of the amplification and phase lag at the three grid points, demonstrated the similarity of the quadratic and linear friction models. The most significant differences occurred for the diurnal wave. This wave took longer to reach the head of either gulf than that in the linear model. Unfortunately, this change in phase further reduced the agreement between observed and computed tides.

An important factor in this model was the generation of higher order harmonics. At Port Augusta, the amplitude of the third harmonic was 0.1 dm. This reduced rapidly to 0.05 dm at the latitude of Port Pirie and from there decreased steadily to 0.01 dm at the mouth of Spencer Gulf. The co-tidal chart indicated that the wave was propagating away from the mouth of the gulf, both towards the open boundary and northwards. In St. Vincent Gulf, this harmonic had a maximum amplitude of 0.04 m at the head of the gulf and propagated away from that point.

The fourth order harmonic was similar in character, but had maximum amplitudes of 0.04 and 0.006 dm in Spencer and St. Vincent Gulfs respectively.

These values were not due to the numerics, as similar calculations were carried out using the linear model and the amplitudes of the third and fourth harmonics produced numerically were both less than 0.0005 dm.

Figure 2.12 shows the water level for Port Augusta as predicted by the linear friction and quadratic friction models. The distortion from a purely sinusoidal wave due to non-linearity can be clearly seen.

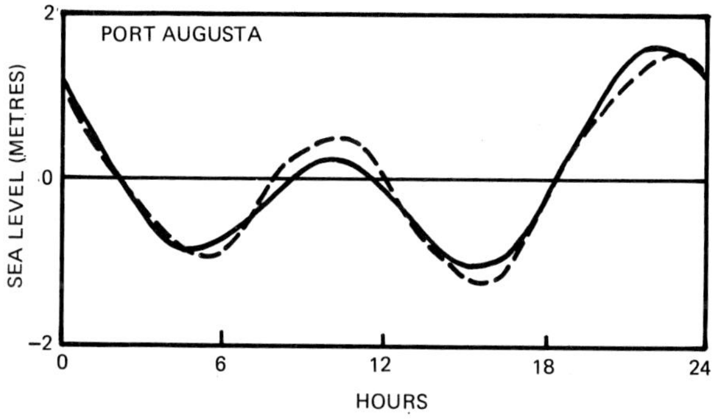

Figure 2.12: The tidal records for Port Augusta as computed using linear (——) and quadratic (----) bottom stress laws.

When the convective and quadratic friction terms were included, the co-range and co-tidal charts for the diurnal and semi-diurnal waves were identical with those calculated for the quadratic friction model. However, the maximum amplitudes of the third and fourth harmonics were 0.83 and 0.69 dm at Port Augusta and 0.76 and 0.10 dm at Port Wakefield. These waves were propagated in the same directions in Spencer Gulf as those generated solely by quadratic friction, but they appeared to travel faster (see Fig. 2.13). In St. Vincent Gulf, the waves originated along the western shoreline rather than at the head of the gulf.

The formation and propagation of these waves depended on the interaction of the forced waves, so only the main features have been outlined. From the results, it would appear that turbulent dissipation, which the quadratic stress law attempts to model, is more important than convection when modelling non-linear interactions. Equally, the results indicated that a quadratic stress law did not give much better results than could be obtained using a simple linear dissipation term, and the advantages of linearity were lost.

2.2.5 Tidal Energy

The amplitude of the two waves used to model the total tide for the non-linear model is approximately equal to the sum of the amplitudes of each constituent of the tidal species which each wave represents. This corresponds to a spring tide. Accordingly the energies calculated within the model are for spring tides, but the orders of magnitude are correct for all phases of the tide.

For the model, using a linear dissipation law with $k = 0.0015$ m/sec, the predicted energy during the fourth model day is drawn in Figure 2.14a.

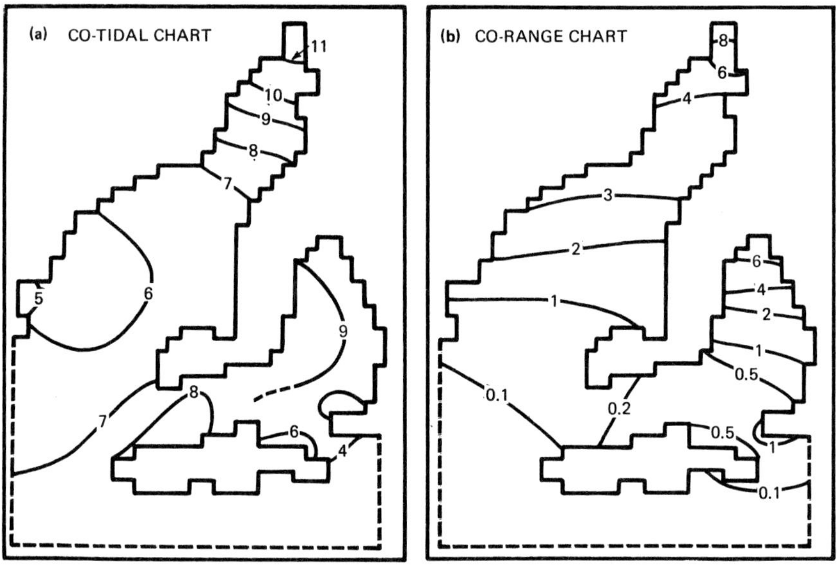

Figure 2.13: Computed (a) co-tidal lines in hours, (b) co-range lines in cm. for the S_3 tri-diurnal tide component produced by the interaction of S_1 diurnal and S_2 semi-diurnal tide components propagating into the gulfs from the open boundary.

The "kinetic" energy (calculated from the transport) and "potential" energy (calculated from the surface elevation) are also drawn. For comparison, the total energy predicted using quadratic dissipation is included in Figure 2.14a. The difference between the two results is only small.

More surprising is the similarity between the dissipation rate predicted using the two dissipation laws (see Figure 2.14b). The energy dissipated in 24 hours is 5.2×10^{14} joules, which corresponds to an average rate of dissipation rate of 6.5×10^9 joules/sec. These values are of the same order of magnitude as the daily variation in energy ranging from 5.5×10^{14} to 0.72×10^{14} joules.

With the numerical methods used, the net energy input into the basin during each time step can be calculated. This enables lower bounds of 6.8×10^{14} and 1.6×10^{14} joules for the energy input and output over 24 hours to be fixed. The ratio of the reflected energy, E_r, to the incident energy, E_i, is $(1.6 \times 10^{14} + E_a)/(6.8 \times 10^{14} + E_a)$ where E_a is included to account for the difference between the gross and net energy input during each time step. Since the maximum energy of the basin, the dissipation over 24 hours and the measurable energy input are approximately equal, the energy input would not be greater than twice that calculated. This maximum value of E_a provides an upper bound of the reflected energy so that

$$0.24 \leq \frac{E_r}{E_i} \leq 0.60.$$

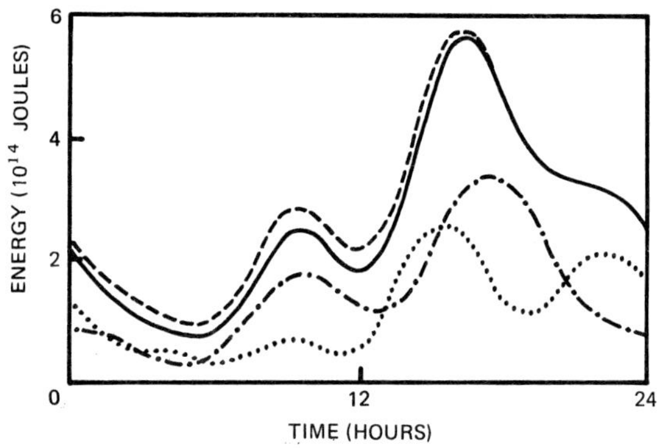

Figure 2.14a: The energy of the tidal motion predicted for the South Australian gulfs during spring tides, ——— total energy, -·-·-·- potential energy, ···· kinetic energy, all calculated assuming linear dissipation of energy; ---- total energy, assuming quadratic energy dissipation.

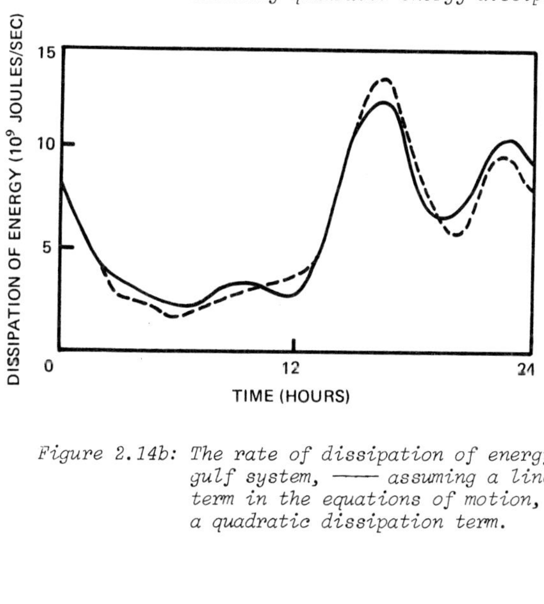

Figure 2.14b: The rate of dissipation of energy within the gulf system, ——— assuming a linear dissipation term in the equations of motion, ---- assuming a quadratic dissipation term.

An alternative statement is that between 76% and 40% of the tidal energy incident on the open boundary of the gulf system is dissipated within the gulfs.

These results may be compared with similar results calculated by Hendershott and Speranza (1971) and Miller (1966) for the M_2 tide in the Gulf of California with an area approximately four times that of the

South Australian gulf system. They obtain a dissipation of the order of 2.0×10^{10} joules/sec, and an absorption of at least 70% of the incident energy.

2.3 Simulations of Idealized Surges

The HN-method has been used extensively to calculate storm surges in the North Sea, the Great Lakes of North America and the Baltic Sea. The papers on this subject in the scientific literature usually present a comparison of the recorded water levels with those calculated after the event. From the results presented by Welander (1961), Platzman (1963), Heaps (1969) and Dunn-Christensen (1971), it must be concluded that the HN-method can be used to provide reliable predictions of the water level if the meteorological conditions are known.

No attempt was made to prove the numerical model in the storm surge situation. From the experiments conducted, it was felt that the only serious difficulty in evaluating the performance of the numerical model in predicting the surges of the gulf system would be the collection of adequate meteorological data and the adequate incorporation of the part produced externally which propagates in as a free wave.

The driving mechanisms for the storm surges which occur within the gulf system are wind stress, air pressure variations and propagation of long waves generated in the Great Australian Bight and beyond. The result of modelling the first two of these is described in this section.

2.3.1. The Response to Wind Stress

The performance of the numerical model was tested by calculating the sea level change due to a SW wind of 10 m/sec which was constant in space and started at time zero, with initial surface elevation and stream velocities set at zero. To determine the importance of the wind direction, the experiment was repeated using a NW wind.

Figure 2.15 shows the predicted sea levels for Port Augusta, Port Lincoln and Outer Harbor. As seen from the predicted curves for Port Augusta and Outer Harbor, both of which were representative of other ports at the head of each gulf, the fundamental modes of oscillation were not excited. However, cross channel oscillations were set up in several areas. The best example is the sea level response at Port Lincoln.

From the observations of the numerical simulations, the impression gained is that the axis of an oscillation must be normal to the surface forcing if that oscillation is to be excited. If the axis of the oscillation is not a straight line, it cannot satisfy the above condition. The cross channel oscillation as recorded at Port Lincoln approximately satisfies the condition for a SW wind but not for a NW wind. This is a comment on the numerical method rather than on the physical processes which occur. Rao (1962) has studied the possible excitation of the free modes of oscillation, in a one-dimensional lake model, due to a moving stress band. He concluded that oscillations were only excited for a very limited set of conditions.

The time scale of the motions in response to a step change in the surface forcing is of the order of 10 hours. This value was found to be approximately

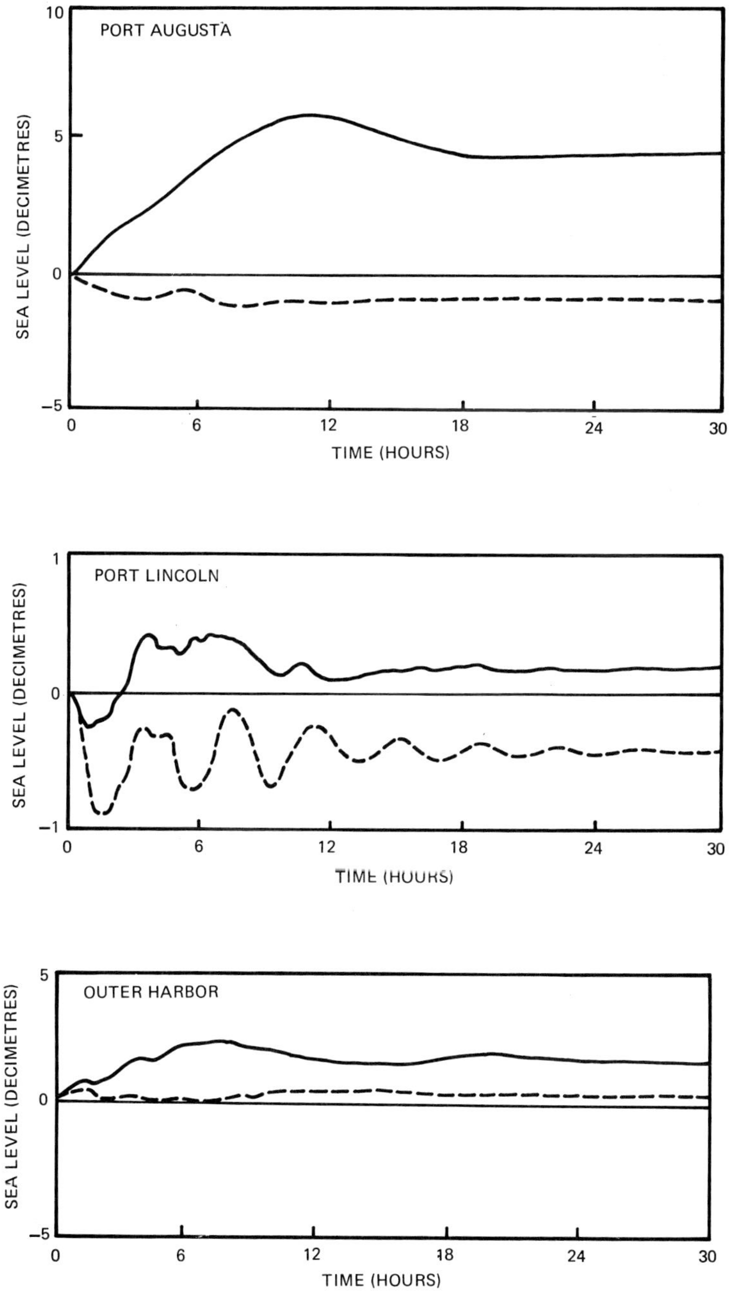

Figure 2.15: *The predicted sea level for Port Augusta, Port Lincoln and Outer Harbor in response to wind of 10m/sec. which was constant in space and started at time zero; —— SW wind; ---- NW wind.*

independent of friction. However, the rate of decay of free oscillations depends on the value set for the coefficient of friction.

2.3.2. The Response to a Moving Pressure System

The strong winds which partially generate the storm surges are associated with low pressure systems. As these approach, the air pressure can drop by 40 mb or more in 12 hours. The static response of the sea level to this change is a rise of 0.4 m. However, the results presented by Tronson and Noye (1973) suggest that this response is amplified.

To examine the response of the gulf system to a moving pressure field, the numerical model was run with the pressure field shown in Figure 2.16 being moved from west to east across the model at constant speed. This simplified field was used as it incorporates the major features of the weather systems which pass across the gulf, while requiring a minimum number of parameters to define it. The significant features can be listed as: the maximum pressure drop, the distance between the line of minimum pressure and the leading and trailing edges (marked as d_1 and d_2 in Figure 2.16a) and the angle, θ_p, between the pressure gradient and a line of longitude.

The standard pressure field for this experiment moved at a speed of 10 Km/hr with $\theta_p = 45^0$, d_1 = 100 Km and d_2 = 300 Km. The pressure drop used was equivalent to 25 cm of water, but as the model is linear, it is the amplification of the static response which is of interest.

A curve, typical of the sea level response, is shown in Figure 2.16b. The static response is included for comparison. The features of the dynamic response are a drop in the water level as the system approaches but before the leading edge arrives, a rapid increase in the sea level as the pressure falls, reaching a maximum approximately one hour after the pressure minimum has past, and a relatively rapid return to the undisturbed level.

This pattern is disturbed in such areas as Investigator Strait and Backstairs Passage because of the flow induced by the pressure system.

For the standard case, the amplification at Outer Harbor was 1.9, a result which is in accord with those presented by Tronson and Noye (1973). This value was representative of the amplification obtained in St. Vincent Gulf with values of approximately 1.0 in Investigator Strait and Backstairs Passage. The maximum amplification of 2.3 was obtained at Port Augusta. At the mouth of Spencer Gulf, the amplification calculated was 0.5 (Port Turton) and 0.34 (Port Lincoln).

Increasing the rate of change of pressure with respect to time increased the amplification everywhere except in areas where the flow is constricted. This increase in $\partial p/\partial t$ at a point can be accomplished by either reducing d_1 and d_2 or by increasing the speed of the pressure field.

The changes in the amplification resulting from changing θ_p depend on the shape of the boundaries. Increasing θ_p from 45^0 to 60^0 decreased the amplification in Spencer Gulf but increased the amplification in St. Vincent Gulf, giving a value of 2.1 at Outer Harbour.

Except for a few local oscillations, the resonances of the system were not excited by the pressure changes modelled.

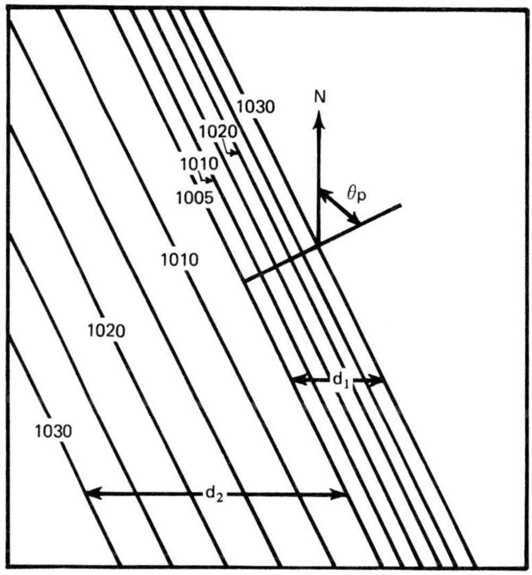

Figure 2.16a: The pressure system, showing a pressure drop of 25mb, which was moved across the model of the gulf system at a constant velocity.

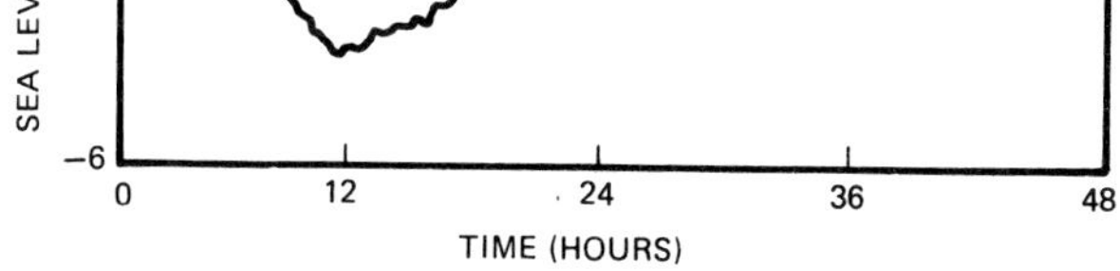

Figure 2.16b: The response of the surface elevation at Wallaroo to the moving pressure field; —— computed dynamic response, ---- static response.

Although it is possible to extend this investigation in its present form, a better understanding may be obtained by studying equivalent analytical models.

2.4 Circulation in the Gulf System

As mentioned previously, some form of long term circulation exists within the South Australian gulf system, although the strength and pattern are unknown. It appears that there is a net southward flow through Backstairs Passage, the strength of which can only be approximately estimated.

Two forms of long term circulation must be considered. The first is an average of the circulations with a time scale of hours due to weather conditions at the air-sea interface, and to non-linear interactions. The second is a "true long term circulation" and is due to relatively constant forcing over a long time, such as wind stress due to trade winds.

In the deep ocean, the second form of circulation is usually dominant, enabling measurements to be made relatively simply. Within the gulf system, the variations in the weather are such that the first form of circulation masks all other modes.

The distribution of bottom sediments gives a weighted average of the bottom currents over geological time. The weighting must describe the conditions under which bottom sediments move and hence must be significant in determining their distribution. Apart from this, there are no measurements which would be likely to give the long term circulation within the gulfs. A possible method of determining the circulation is suggested at the end of this section.

2.4.1. Driving Mechanisms and Models

The circulation of gulf waters results from four major driving mechanisms which may be listed as: non-linear interactions, wind stress at the air-sea interface, external forcing by ocean currents and horizontal density variations. The last of these is not considered here.

Non-linear interactions, depending on turbulent and convective motions, cannot be modelled adequately at this stage. This applies to the theoretical considerations concerning the form the model equations should take and the numerical solution of these equations. Attempts were made at calculating the steady state circulation due to non-linear interactions between tidal components when the results for Section 2.2 were being prepared. However, it appeared that numerical errors dominated the results. This problem was not investigated further as it was felt that turbulence, the major non-linear interaction, cannot be more accurately represented on a 10 Km grid than by a linear dissipation term.

The possibility of steady state circulations existing in the gulf depends on the time scale of the driving forces and resulting motion. The time constant for the gulf system is of the order of 10 hours. As the time scale of the variations in weather conditions is of the order of 2 days, it is possible for the meteorologically induced circulation to be constant for several days. The approximations made when modelling steady ocean circulation in the horizontal plane hold for the depth and length scales of the gulf system. Accordingly, the steady state circulation models used to describe ocean circulations constant over a long period of time can be used to describe circulation patterns which occur in the gulfs.

Equations (1.8.1) to (1.8.3) describe a linear circulation model independent of time. These may be solved in three different ways.

By assuming a time dependence of the form used for the tidal equations, the HN-method may be used with constant surface forcing. The solution of the equations is given by the U, V and ζ fields, which are constant after the decay of the starting transients. A second approach is to eliminate U and V from the equations, giving a second order partial differential in ζ which may then be solved. Alternatively, ζ may be eliminated and U and V defined in terms of a stream function, ψ. This results in a second order partial differential equation in ψ, the solution of which can be used to determine U and V.

It was found that each method only gave reliable results for the variables in which it was formulated. This was due to the build-up of numerical errors when integrals along a line of grid points were calculated. To provide a numerical comparison with earlier work, the first and last methods were used. No attempt at a detailed comparison is made here, although the results from each method are used contiguously.

As seen from (1.8.7), the circulation and hence the surface elevation is dependent on the value assumed for linear friction. As the friction coefficient is only an estimate, it is necessary to have some understanding of how the dependent variables are affected by friction.

Consider the one-dimensional model described by the steady state equations,

$$0 = \frac{\partial U}{\partial x}, \tag{2.4.1}$$

$$0 = -gh\frac{\partial \zeta}{\partial x} - rU + F, \tag{2.4.2}$$

with $U=0$ at the closed end of the channel, $x = x_1$, and $\zeta = 0$ at the open end of the channel, $x = x_2$. For a constant depth channel, the surface elevation and transport within the channel are

$$\zeta(x) = \frac{F}{gh}(x - x_2), \tag{2.4.3}$$

and

$$U(x) = 0, \tag{2.4.4}$$

respectively. The solution for a channel open at both ends, with boundary conditions $\zeta(x_1) = \zeta_1$ and $\zeta(x_2) = \zeta_2$ is

$$\zeta(x) = \zeta_1 + \frac{(x - x_1)}{(x_2 - x_1)}(\zeta_2 - \zeta_1), \tag{2.4.5}$$

$$U(x) = c = \frac{F}{r} - \frac{gh}{r}\left(\frac{\zeta_2 - \zeta_1}{x_2 - x_1}\right). \tag{2.4.6}$$

The inverse relationship between transport and friction is clearly demonstrated in this example.

A comparison of the two models shows the conditions under which the surface elevation obeys the static relationship, (2.4.3). When the flow is unrestricted except by friction, the surface elevation depends only upon the surface elevation at the boundary. As the flow is restricted, the surface elevation will depend directly on the surface forcing term.

The features of these solutions will dominate the two-dimensional solutions obtained for the gulf system although the effects of the Earth's rotation, variable depth and complex shape have been included.

2.4.2. The HN-Model

The HN-method, using the model basin shown in Figure 2.9, was used to obtain estimates of the velocity and surface elevation fields caused by a constant wind stress.

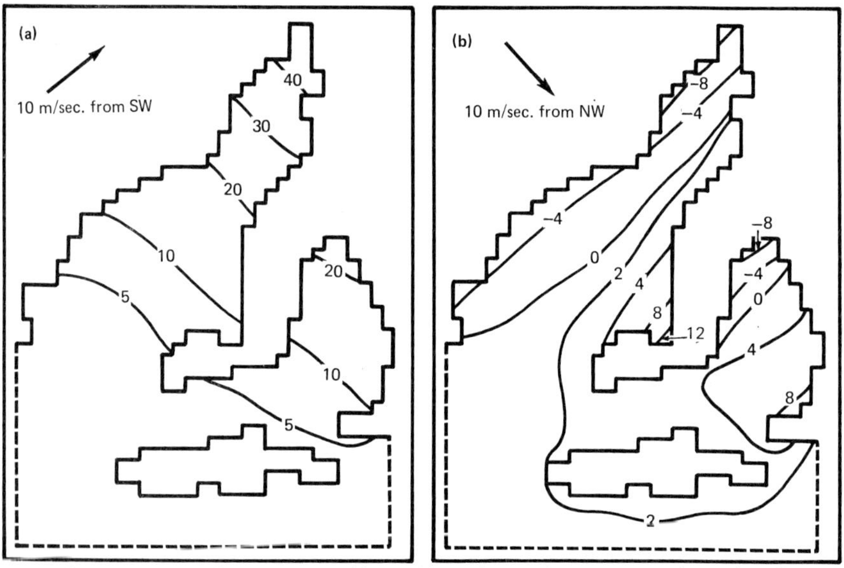

Figure 2.17a: The co-range lines (in cm) of the surface elevation in equilibrium with a steady wind of 10m/sec. (a) from the south-west (SW), (b) from the north-west (NW).

Figure 2.17a shows the surface elevation over the gulfs due to a SW wind of 10 m/sec. As the model is linear, the surface elevation increases linearly with wind stress, or as the square of the wind speed, since the boundary conditions hold the surface elevation at zero along an open boundary.

The response to the SW wind could be estimated by a one-dimensional model as the contours of constant surface elevation are normal to the direction of the wind stress. The maximum elevation of 46 cm occurs at the head of Spencer Gulf. The surface elevation south of Kangaroo Island has a maximum height of 3 cm, but at most grid points it is less than 1 cm.

The response of the surface elevation to a NW wind is considerably less than for a SW wind. This is to be expected since the fetch for the NW wind is the width of the gulf, compared to the length of the gulfs for the SW wind. The lines of constant elevation are normal to the wind stress except where the boundary conditions and shoreline make this impossible, as happens at the mouth of St. Vincent Gulf.

The depth-mean velocity at each grid point due to a SW wind is shown as Figure 2.17b.

A characteristic of the velocity fields is strong boundary currents. The velocity, u, is calculated using the formula $u = U/h$, so shallow water near the shore can result in large velocities if the transport is constant. This is not necessarily a numerical error, as large velocities can occur in practice. At the northern end of Spencer Gulf, the velocities are large at some grid points. However, the spacing of the grid points is too coarse to properly define the field.

Both SW and NW winds induce a considerable movement of water in the southern part of the gulf system. Several features of the circulation induced by a SW wind may be clearly seen in Figure 2.17b. These may be listed as: two circulation cells in St. Vincent Gulf with water moving northwards along the eastern and western shores and southwards in the centre, a similar circulation pattern in the southern half of Spencer Gulf and an exchange of water between the northern and southern sections, and nett southward transport of water through Backstairs Passage.

The circulation due to the NW wind retains the circulation cells in each gulf but the flow is in the opposite direction. A strong circulation through Investigator Strait and Backstairs Passage is induced by a NW wind.

The flow through Backstairs Passage may be calculated from these models. From continuity, this must be equal to the flow through Investigator Strait. For a SW wind, the flux is 2.7×10^4 m^3/sec and for a NW wind it is 1.2×10^5 m^3/sec.

2.4.3 Circulation Model.

To provide estimates of the circulation, Equation (1.8.5) was solved for two model basins representing St. Vincent Gulf and Spencer Gulf. The grid is the same as that used for Section 2.1 but with the grid spacing reduced by a factor of 2 to 6.35 Km. An averaged form of the bathymetry used for the following study is shown in Figure 2.3.

The conditions along an open boundary for this model are different from those used in the HN-method. The velocity was forced to be normal to the open boundary rather than the elevation being held at some constant level. This change in boundary conditions had little effect on the solution since the flow indicated by the HN-method (see Figure 2.17b) is approximately normal to the open boundaries defined in the models used. To further avoid any changes in the solution due to the boundary conditions, the open boundaries were sited as far from the area of interest as possible while keeping the computing requirements to a minimum.

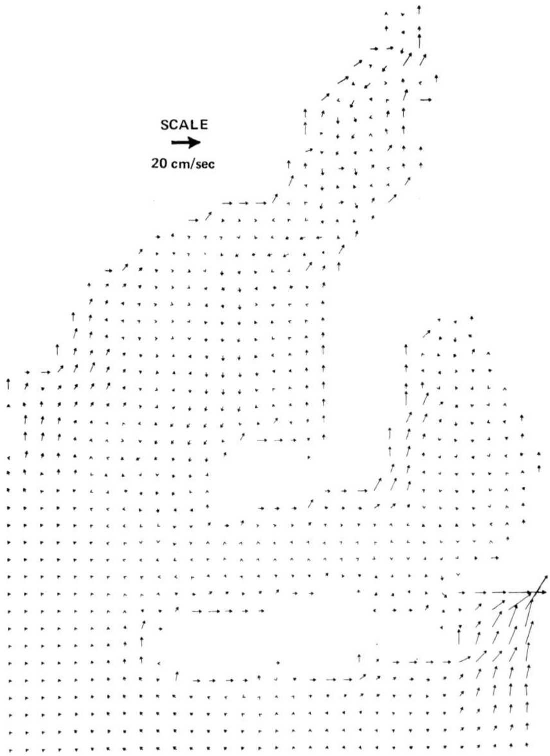

Figure 2.17b: The equilibrium velocity field driven by a steady SW wind of 10m/sec. computed for the South Australian gulf system using the HN-method. The arrows indicate direction of the depth averaged velocity and a length of 1cm ≡ 0.4 m/sec.

At a closed boundary, the boundary condition, ψ constant, is equivalent to zero flow across the boundary. For a model basin with disconnected shorelines, the difference between the value of ψ prescribed on separate boundaries is the nett flow between them. Hence, for a model basin representing St. Vincent Gulf, an estimate, ψ_K, of the flow through Investigator Strait, Backstairs Passage is required.

Assuming Investigator Strait and the continental shelf south of Backstairs Passage can be considered as infinite sources of sinks (i.e. $\zeta_2 = \zeta_1 = 0$ in Equation (1.2.6)), then the velocity predicted, using Equation (1.2.6), in response to a wind of 10 m/sec parallel to the passage is 0.20 m/sec if the friction, r, is taken as $(0.0015/h)\text{sec}^{-1}$. This is an order of magnitude larger than the velocity predicted from the HN-model. However, velocities of 0.4 m/sec (1 knot) are not uncommon in the area.

For an increase in surface elevation of ζ, the drop in velocity predicted from (1.2.6) is $(h\zeta/3)$m/sec assuming the passage is 20 Km long. A rise in sea level of 3.0 cm is sufficient to reduce the current to zero. As changes in the surface elevation of this order of magnitude are common in the area, estimation of the currents through the passage is almost impossible.

The circulation was calculated for a range of values of ψ_K. Large values of ψ_K correspond to the flow being driven by the ocean circulation on Flinders current (Bye, 1972).

Figures 2.18a,b and c show the circulation in St. Vincent Gulf due to a south wind, a west wind and a non-zero transport through Backstairs Passage. The coefficient of friction used was 0.001 m/sec. As the equations are linear, these solutions may be combined to give the circulation arising from a given wind and transport.

Increasing the coefficient of friction, κ, reduced the transport accordingly to the simple inverse relation, $\psi \propto 1/\kappa$, and did not alter the circulation patterns. From (1.8.7), increasing κ by an order of magnitude is equivalent to decreasing the Coriolis parameter, f, and the surface forcing, F_S and G_S, by an order of magnitude. The magnitude of the transport will change proportionally with changes in F_S and G_S. If geostrophic forces are significant, the circulation pattern will change if f is altered. As this did not occur, it can be stated that the Earth's rotation does not affect the circulation within the gulf system.

Figure 2.19a shows the calculated circulation in response to a 10 m/sec NW wind, assuming a southward flow through Backstairs Passage of $10^5\text{m}^3/\text{sec}$.

For comparison, the circulation was recomputed assuming a southward flow through Backstairs Passage of $10^4\text{m}^3/\text{sec}$. The flow in the northern section of the Gulf is the same in both cases. The flow in the southern section of the model is dominated by the eastward flow induced by the boundary conditions. With the lower value of ψ_K, the circulation cells in the central section of the gulf are not absorbed into the total flow but as ψ_K is increased these disappear, with the circulation pattern of Figure 2.18c becoming dominant. It is interesting to note that the higher values of ψ_K increase the rate of renewal of water in the central areas of the gulf. This is not obvious from the three components shown in Figure 2.18.

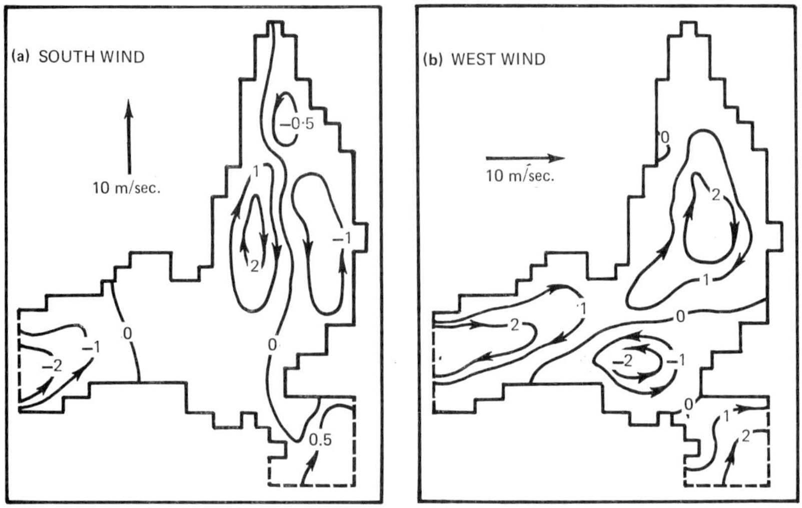

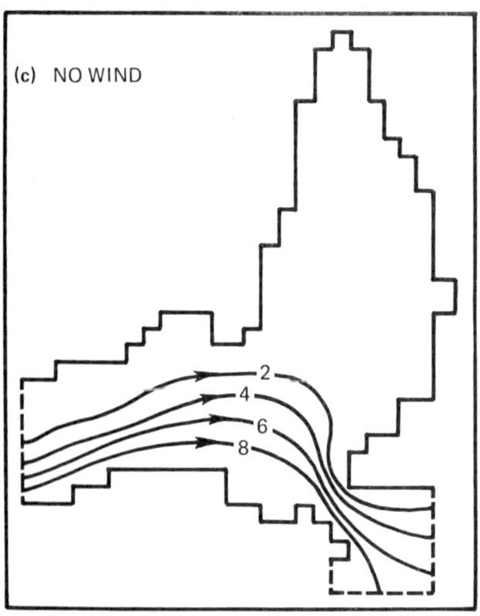

Figure 2.18: Streamlines for the equilibrium circulation in St. Vincent Gulf driven by (a) south wind of 10 m/sec. and no transport in Backstairs Passage (units $10^2 m^3/sec$); (b) west wind of 10 m/sec. and no transport in Backstairs Passage (units $10^2 m^3/sec$); (c) no wind and a southward flow of $10 m^3/sec$. through Backstairs Passage (units $1 m^3/sec$).

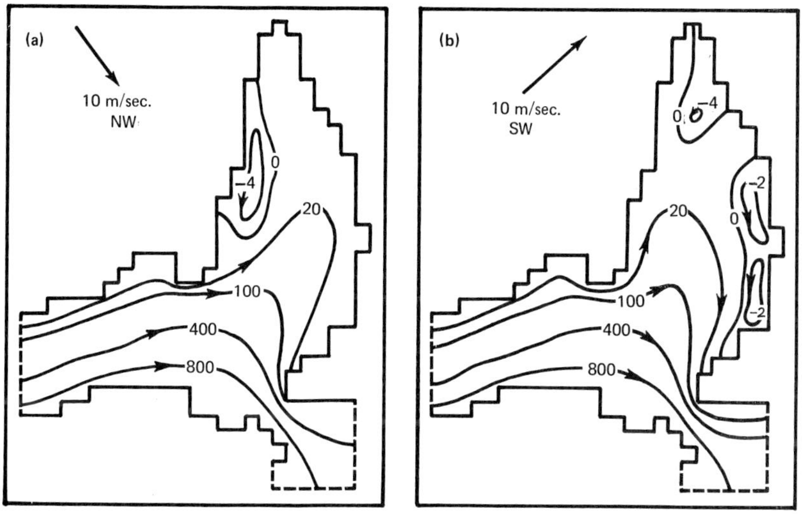

Figure 2.19: Streamlines (values in $10^2 m^3/sec$) for the circulation in St. Vincent Gulf driven by a 10m/sec. wind and a southward flow of $10^5 m^3/sec$. through Backstairs Passage, the wind being from (a) the north-west (NW) (b) the south-west (SW).

For a 10 m/sec SW wind, the circulation patterns given by Figure 2.19b corresponding to a southward flow of $10^5 m^3/sec$ through Backstairs Passage.

The distribution of bottom sediments (Waters, 1973) suggests that there exists a northward transport from Troubridge Shoals along the western shore of St. Vincent Gulf and a circulation cell at the head of the gulf. It also seems that the circulation along the eastern shore is separated from the remainder of the gulf and that this circulation may be divided into cells with a null point south of Adelaide. These features of the circulation are apparent in Figure 2.19b. South-west winds are more frequent and stronger than other winds and are usually associated with weather conditions which may persist for two days. These winds generate considerable surface wave activity. In the shallow waters of the gulfs, the waves increase the turbulent motion on the sea floor, enabling the bottom sediment to be more easily moved. With a constant circulation existing for several days, large amounts of material could be moved compared with that moved during "quiet" periods. Accordingly, it is perhaps not surprising that the bottom sediment distribution partly reflects the circulation driven by SW winds.

Figures 2.20a and b show the circulation in Spencer Gulf due to a wind from the south and west respectively. The strength of the circulation is for a friction coefficient of 0.001 m/sec. Increasing this coefficient reduces the circulation with a small reduction in the relative strength of each cell. The circulation induced by a south wind consists of two cells in the southern section and two in the northern half of the gulf. There is a little exchange

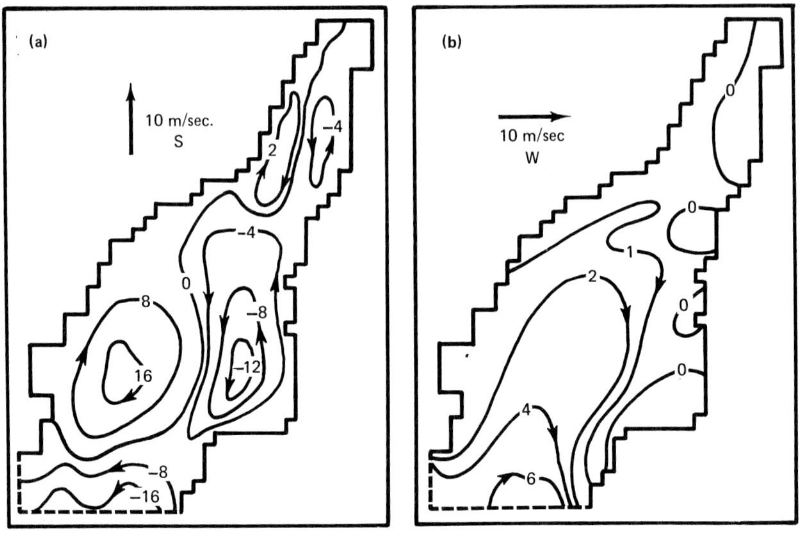

Figure 2.20: Streamlines (units of $10^3 m^3/sec$) for the equilibrium circulation in Spencer Gulf driven by a wind of 10m/sec. from (a) the south (S) (b) the west (W).

between the northern and southern halves of the gulf. With a west wind the cell structure of the circulation evidenced earlier does not exist, but the rate of replacement of water in the southern section is increased. There is very little circulation in the northern half of the gulf. The flow of water into and out of the gulf is concentrated near the shoreline for both wind systems.

The results for Spencer Gulf were combined to provide estimates of the circulation drawn in Figures 2.21a and b due to SW and NW winds. The major feature of the circulation due to a SW wind is the transport towards the head of the gulf on both sides with a balancing outward flow down the centre. This is similar to the results shown in Figure 2.17b. The transport across the line AB on Figure 2.21a is $10^3 m^3/sec$. This is an estimate of the interchange of water between the upper reaches of Spencer Gulf and the ocean due to a SW wind.

The major features of the circulation induced by a NW wind predicted by the HN- and circulation models are the same. These features consist of a strong outward flow near the eastern shoreline, a broad poorly defined inward flow on the western shoreline with circulation cells in the area, and a weak circulation in the northern half of the gulf. The flow across the line AB is 600 m^3/sec, approximately half the flow induced by a SW wind.

The volume of the northern half of Spencer Gulf is of the order of $10^{10} m^3$. Theoretically, it would take two years to replace this water at a replacement rate of $10^3 m^3/sec$. As SW winds do not blow continuously, the value of $10^3 m^3/sec$ represents a maximum value of the circulation between the

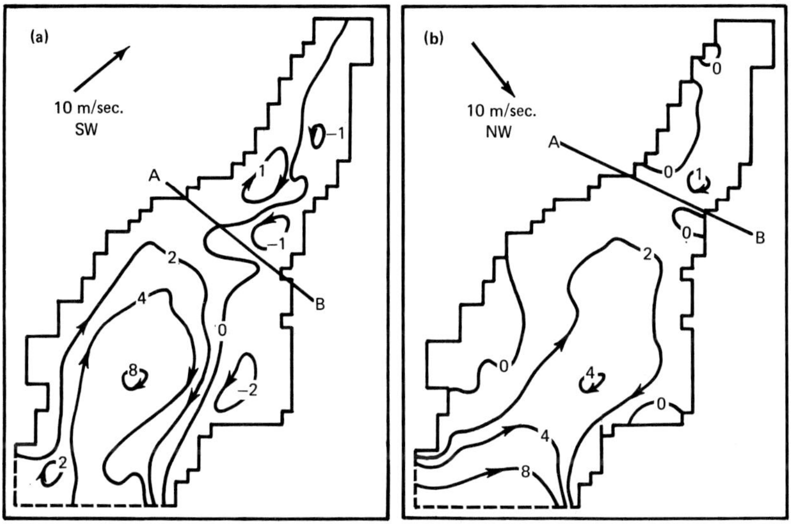

Figure 2.21: Streamlines (units of $10^3 m^3/sec$) for the equilibrium circulation in Spencer Gulf driven by a wind of strength 10 m/sec. from the (a) south-west (SW), (b) north-west (NW).

two halves of the gulf. The circulation is such that the water in certain areas would be replaced many times before the water in other areas was replaced once. This depends on local conditions. The estimate of two years for the time required to circulate all the water of the northern half of Spencer Gulf once is probably an order of magnitude too small.

The time scale for the circulation of water through St. Vincent Gulf depends on the value used for the flow through Backstairs Passage. As this may vary over two orders of magnitude, no estimate can be attempted.

REFERENCES

Beckman, F. (1960): The Solution of Linear Equations by the Conjugate Gradient Method, Mathematical Methods for Digital Computers (A. Ralston and H.S. Wilf, eds.), John Wiley and Sons, Inc., New York.

Bragg, W.H., Inglis, A., Lyle, T.R. and Chapman, R.W. (1892): The Tides of the Coast of South Australia, Austr. Assoc. Adv. Science, 4, 230-232.

Bye, J.A.T. (1972): Oceanic Circulation South of Australia, Antarctic Res. Series 19, 95-99.

Curtis, A.R. and Reid, J.K. (1971): Fortran Subroutines for the solution of Sparse Sets of Linear Equations. Atomic Energy Research Establishment, (Harwell), Report No. R6844.

Dunn-Christensen, J.T. (1971): Investigations on the Practical Use of a Hydrodynamic Numeric Method. Deutsche Hydrographische Z 24, 210-226.

Easton, A.K. (1970): The Tides of the Continent of Australia, Ph.D. Thesis, The Flinders University of South Australia.

Fischer, G. (1959): Ein numerisches Verfahren zur Errechnung von Windstau und Gezeiten in Randmeeren. Tellus, 11, 60-76.

Fortak, H.G. (1962): Hydrodynamic Equations with respect to Basic Storm Surge Equations, National Hurrican Research Project, Report No. 51

Hansen, W. (1956): Theorie zur Errechnung des Wasserstandes und der Strömungen in Randmeeren nebst Anwendungen, Tellus 8, 287-300.

Hansen, W. (1962): Hydrodynamical methods applied to oceanographic problems, Proc. Symp. math. hydro. methods of phys. oceanography, Sept. 1961, Mitt. Inst. Meeresk. Univ. Hamburg No.1, 25-34.

Harris, D.L. and Jelesnianski, C.P. (1964): Some Problems Involved in the Numerical Solutions of Tidal Hydraulics Equations, Mon. Wea. Rev. 92, 409-422.

Heaps, N.S. (1969): A Two-dimensional Numerical Sea Model, Proc. Roy. Soc. (A) 265, 93-137.

Henderschott, M.C. and Speranza, A. (1971): Co-oscillating Tides in Long, Narrow Bays. Deep Sea Research, 18, 959-980.

Honda, K., Terada, T., Yoshida, Y. and Isitani, D. (1908): An Investigation on the Secondary Undulations of Oceanic Tides. J. College Sci., Tokyo, 24, 1-110.

Jelesnianski, C.P. (1965): A numerical calculation of storm tides induced by a tropical storm impinging on a continental shelf. Mon. Weath. Rev. 93, 343-358.

Lamb, H. (1932): Hydrodynamics, Cambridge University Press, Cambridge.

Lauwerier, H.A. (1960): Free Motions of a Rotating Rectangular Bay. Pcok. H. ned. Akad. Wet. A 63, 423-438.

Lauwerier, H.A. and Damste, B.R. (1963): The North Sea problem VIII: A numerical treatment. Proc. K. Ned. Akad. Wet. A66, 167-184.

Lax, P.D. and Richtmyer, R.D. (1956): The Stability of Finite Difference Equations. Comm. Pure. Appl. Math. 9, 267-293.

Leendertse, J.J. (1967): Aspects of a computational model for long-period water-wave propagation. Memo. Rand Corp., Santa Monica, Calif. RM-5294-PR.

Marchuk, G.I., Kagan, B.A. and Tamsaly, R.E. (1969): Numerical Method of Calculating Tidal Motion in Border Seas. Atm. and Ocean Physics, 5, 694-703.

Miller, G.R. (1966): The Flux of Tidal Energy out of the Deep Oceans. J. Geophys. Res. 71, 2485-2489.

Munk, W.H. and Cartwright, D.E. (1966): Tidal Spectroscopy and Prediction. Proc. Roy. Soc. (A) 259, 533-581.

Platzman, G.W. (1958): A numerical computation of the surfe of 26 June 1954 on Lake Michigan. Geophysica 6, 407-438.

Platzman, G.W. (1963): The Dynamical Prediction of Wind Tides on Lake Erie. Met. Mons., 4, 1-44.

Platzman, G.W. (1972): Two-dimensional Free Oscillations in Natural Basins. J. Phys. Oceanog. 2, 117-138.

Pneuli, A. and Pekeris, C.L. (1968): Free Tidal Oscillations in Rotating Flat Basins. Proc. Roy. Soc. 263 (A), 149-171.

Proudman, J. (1953): Dynamic Oceanography. Methuen, London.

Rao, D.B. (1962): The Response of a Lake to a Time-dependent Wind Stress. Univ. of Chicago, Department of the Geophysical Sciences, M.S. Thesis.

Rao, D.B. (1966): Free Gravitational Oscillations in Rotating Basins. J. Fluid Mech. 25, 523-555.

Richardson, L.F. (1922): Weather Prediction by Numerical Process, Cambridge University Press, Cambridge.

Sielecki, A. (1968): An Energy-conserving Difference Scheme for the Storm Surge Equation. Mon. Wea. Rev. 96, 150-156.

Smith, E.B. (1973): Wind driven and seiche forced water motion in Grand Traverse Bay, Michigan. Ph.D. dissertation (Engineering Mechanics), Univ. of Michigan, 130 pp.

Stoker, J.J. (1957): Water Waves, Interscience, New York.

Taylor, G.I. (1921): Tidal Oscillations in Gulfs and Rectangular Basins. Proc. London Math. Soc. 20, 148-151.

Thomas, I.M. and Edmonds, S.J. (1956): Chlorinities of Coastal Waters in South Australia. Trans. Roy. Soc. S.A. 79 152-166.

Tronson, K. (1973): The Limits of Time Stepping when used to solve the Long Wave Equations Numerically. Bull. Austr. Math. Soc. 8, 321-332.

Tronson, K. and Noye, B.J. (1973): A Statistical Model for prediction of Storm Surges on the Adelaide Foreshore. Proc. Conf. Dynamics of Coastal Zone, Sydney, 1973, 190-196.

Uusitalo, S. (1960): The Numerical Calculation of Wind Effects on Sea Level Elevations, Tellus 12, 427-435.

Waters, C. (1973): Unpublished data.

Wilkinson, J.H. (1965): The Algebraic Eigenvalue Problem, Clarendon Press, Oxford.

Wilson, B.W. (1960): Note on Surface Wind Stress Over Water at Low and High Wind Speeds. J. Geophys. Res. 65, 3377-3381.

Welander, P. (1957): Wind Action on a Shallow Sea. Tellus, 9, 45-52.

Welander, P. (1961): Numerical Predictions of Storm Surges. Adv. Geophys. 8, 316-379.

APPENDIX 1

Derivation of Long Wave Equations.

The equations of motion and continuity for a fluid, in terms of Eulerian variables with the axes fixed relative to the Earth are

$$\frac{\partial u}{\partial t} + u\frac{\partial u}{\partial x} + v\frac{\partial u}{\partial y} + w\frac{\partial u}{\partial z} - fv = -\frac{1}{\rho_0}\frac{\partial p}{\partial x} + \frac{1}{\rho_0}\frac{\partial \tau_x}{\partial z}, \tag{A1.1}$$

$$\frac{\partial v}{\partial t} + u\frac{\partial v}{\partial x} + v\frac{\partial v}{\partial y} + w\frac{\partial v}{\partial z} + fu = -\frac{1}{\rho_0}\frac{\partial p}{\partial y} + \frac{1}{\rho_0}\frac{\partial \tau_y}{\partial z}, \tag{A1.2}$$

$$\frac{\partial w}{\partial t} + u\frac{\partial w}{\partial x} + v\frac{\partial w}{\partial y} + w\frac{\partial w}{\partial z} = -\frac{1}{\rho_0}\frac{\partial p}{\partial z} - g, \tag{A1.3}$$

and

$$\frac{\partial u}{\partial x} + \frac{\partial v}{\partial y} + \frac{\partial w}{\partial z} = 0. \tag{A1.4}$$

The rectangular coordinates x,y and z are defined with the xy plane horizontal, the z-axis vertical and positive upwards. The remainder of the notation is as follows:

- (u,v,w) the velocity components
- p the pressure
- ρ_0 the density (assumed constant)
- g the acceleration due to gravity
- (τ_x,τ_y) the horizontal components of the frictional stress $\underset{\sim}{\tau}$
- f the Coriolis parameter.

Molecular viscosity has been neglected in these equations. The terms in τ_x and τ_y are included to model vertical turbulent diffusion.

By setting the origin on the undisturbed sea surface, the free surface and the bottom may be represented by $z = \zeta(x,y,t)$ and $z = -h(x,y)$, respectively. Denoting the wind stress and bottom stress components as (F_S,G_S) and (F_B,G_B) respectively and the surface pressure as p_a, the boundary conditions become

$$\left.\begin{aligned} u = v = w &= 0 \qquad &(A1.5)\\ (\tau_x,\tau_y) &= (F_B,G_B) \qquad &(A1.6)\end{aligned}\right\} \text{ at } z = -h,$$

$$\left.\begin{aligned} (\tau_x,\tau_y) &= (F_S,G_S) \qquad &(A1.7)\\ p &= p_a \qquad &(A1.8)\\ \frac{\partial \zeta}{\partial t} + u\frac{\partial \zeta}{\partial x} + v\frac{\partial \zeta}{\partial y} &= w \qquad &(A1.9)\end{aligned}\right\} \text{ at } z = \zeta,$$

Condition (A1.9) states that the free surface follows the fluid.

Dimensionless variables may be introduced through the two length scales, H and L, which are characteristic of the depth and horizontal distance (wavelength) respectively. A compatible set of non-dimensionalized units, denoted by $\sim$, is:

$$\left.\begin{aligned}
(\tilde{x},\tilde{y},\tilde{z}) &= \left(\frac{x}{L},\frac{y}{L},\frac{z}{H}\right),\\
\tilde{t} &= t\sqrt{gH}/L\ ,\\
(\tilde{u},\tilde{v},\tilde{w}) &= (u,v,Hw/L)/\sqrt{gH}\\
\tilde{f} &= fL/\sqrt{gH},\\
(\tilde{p},\tilde{\tau}) &= (\rho,L\tau/H)/\rho gH,\\
\text{and}\qquad&\\
(\tilde{h},\tilde{\zeta}) &= (h,\zeta)/H.
\end{aligned}\right\} \qquad \text{(A1.10)}$$

Following Stoker (1957), the dimensionless parameter,

$$\sigma = H^2/L^2, \qquad \text{(A1.11)}$$

is introduced. When this parameter is small, the water is considered shallow.

The non-dimensionalized forms of the Equations (A1.1), to (A1.9) are

$$\sigma\left(\frac{\partial u}{\partial t} + u\,\frac{\partial u}{\partial x} - v\,\frac{\partial u}{\partial y} - fv\right) + w\,\frac{\partial u}{\partial z} = \sigma\left(-\,\frac{\partial p}{\partial x} + \frac{\partial \tau_x}{\partial z}\right), \qquad (\overline{\text{A1.1}})$$

$$\sigma\left(\frac{\partial v}{\partial t} + w\,\frac{\partial v}{\partial x} + v\,\frac{\partial v}{\partial y} + fu\right) + w\,\frac{\partial v}{\partial z} = \sigma\left(-\,\frac{\partial p}{\partial y} + \frac{\partial t_y}{\partial z}\right), \qquad (\overline{\text{A1.2}})$$

$$\sigma\left(\frac{\partial w}{\partial t} + w\,\frac{\partial w}{\partial x} + v\,\frac{\partial w}{\partial y}\right) + w\,\frac{\partial w}{\partial z} = \sigma\left(-\,\frac{\partial p}{\partial z} - 1\right), \qquad (\overline{\text{A1.3}})$$

$$\sigma\left(\frac{\partial u}{\partial x} + \frac{\partial v}{\partial y}\right) + \frac{\partial w}{\partial z} = 0, \qquad (\overline{\text{A1.4}})$$

$$\left.\begin{aligned}
u = v = w &= 0, & (\overline{\text{A1.5}})\\
(\tau_x,\tau_y) &= (\tilde{F}_B,\tilde{G}_B) & (\overline{\text{A1.6}})
\end{aligned}\right\}\ \text{at}\quad z = -h$$

$$\left.\begin{aligned}
(\tau_x,\tau_y) &= (\tilde{F}_S,\tilde{G}_S) & (\overline{\text{A1.7}})\\
p &= p_a & (\overline{\text{A1.8}})\\
\sigma\left(\frac{\partial \zeta}{\partial t} + u\,\frac{\partial \zeta}{\partial x} + v\,\frac{\partial \zeta}{\partial y}\right) &= w & (\overline{\text{A1.9}})
\end{aligned}\right\}\ \text{at}\quad z = \zeta,$$

It is assumed that u, v, w, ζ and p can be written as perturbation expansions in the form,

$$u = u_0 + \sigma u_1 + \sigma^2 u_2 + \ldots \quad . \tag{A1.12}$$

Substituting these in Equations $(\overline{A1.1})$ to $(\overline{A1.9})$, the terms of zero order in σ give

$$w_0 = 0. \tag{A1.13}$$

The theory presented by Stoker for an irrotational fluid has u_0 and v_0 independent of z.

The equations of the first order in σ from $(\overline{A1.1})$ to $(\overline{A1.9})$ are

$$\frac{\partial u_0}{\partial t} + u_0 \frac{\partial u_0}{\partial x} + v_0 \frac{\partial u_0}{\partial y} + w_1 \frac{\partial u_0}{\partial z} - f v_0 = - \frac{\partial p_0}{\partial x} + \frac{\partial \tau_x}{\partial z}, \tag{A1.1)$_1$}$$

$$\frac{\partial v_0}{t} + u_0 \frac{\partial w_0}{x} + v_0 \frac{\partial v_0}{y} + w_1 \frac{\partial v_0}{z} + f u_0 = - \frac{\partial p_0}{y} + \frac{\partial \tau_y}{z}, \tag{A1.2)$_1$}$$

$$0 = \frac{\partial p_0}{\partial z} + 1, \tag{A1.3)$_1$}$$

$$\frac{\partial u_0}{\partial x} + \frac{\partial v_0}{\partial y} + \frac{\partial w_1}{\partial z} = 0, \tag{A1.4)$_1$}$$

$$\frac{\partial \zeta_0}{\partial t} + u_0 \frac{\partial \zeta_0}{\partial x} + v_0 \frac{\partial \zeta_0}{\partial y} = w_1 \quad \text{at} \quad z = \zeta_0. \tag{A1.9)$_1$}$$

Integration of Equation $(A1.3)_1$ form $z = \zeta_0$ to z yields the hydrostatic pressure relation,

$$p_0(x,y,z,t) = p_a(x,y,\zeta_0,t) + \zeta_0(x,y,t) - z. \tag{A1.14}$$

This is not a proof of (A1.14), but shows that it is the first order approximation in terms of the parameter σ in the formal perturbation analysis.

For numerical work, the transport equations are more useful. Integrating Equations $(A1.1)_1$, $(A1.2)_1$ and $(A1.4)$ from $z = -h$ to $z = \zeta_0$ gives

$$\frac{\partial U}{\partial t} + \frac{\partial}{\partial x} \overline{u^2} + \frac{\partial}{\partial y} \overline{uv} - fV = -(h+\zeta_0)\left(\frac{\partial \zeta_0}{\partial x} + \frac{\partial p_a}{\partial x}\right) + (F_S - F_B), \tag{A1.15}$$

$$\frac{\partial V}{\partial t} + \frac{\partial}{\partial x} \overline{uv} + \frac{\partial}{\partial y} \overline{v^2} + fU = -(h+\zeta_0)\left(\frac{\partial \zeta_0}{\partial y} + \frac{\partial p_a}{y}\right) + (G_S - G_B), \tag{A1.16}$$

and

$$\frac{\partial \zeta_0}{\partial t} + \frac{\partial U}{\partial x} + \frac{\partial V}{\partial y} = 0, \tag{A1.17}$$

where

$$(U,V) = \int_{-h}^{\zeta_0} (u_0, v_0)\,dz,$$

$$(\overline{u^2}, \overline{v^2}) = \int_{-h}^{\zeta_0} (u_0^2, v_0^2)\,dz,$$

and

$$\overline{uv} = \int_{-h}^{\zeta_0} u_0 v_0\,dz.$$

There have been no extra approximations made in deriving (A1.15), (A1.16) and (A1.17).

If it is assumed that the particle velocity and free surface elevation and their derivatives are small quantities, their squares and products can be neglected in comparison with their linear terms, and Equations (A1.15), (A1.16) and (A1.17) can be linearized and written as

$$\frac{\partial U}{\partial t} - fV = -h\left(\frac{\partial \zeta_0}{\partial x} + \frac{\partial p_a}{\partial x}\right) + (F_S - F_B), \tag{A1.18}$$

$$\frac{\partial V}{\partial t} + fU = -h\left(\frac{\partial \zeta_0}{\partial y} + \frac{\partial p_a}{\partial y}\right) + (G_S - G_B), \tag{A1.19}$$

$$\frac{\partial \zeta_0}{\partial t} + \frac{\partial U}{\partial x} + \frac{\partial V}{\partial y} = 0. \tag{A1.20}$$

These assumptions are equivalent to assuming that

$$Z/H \ll 1,$$

where Z is the characteristic surface elevation.

APPENDIX 2

Discretization of Variational Formulation of the Long Wave Equation.

A convenient form of the variational formulation of the long wave equation given as (1.5.6) is

$$\delta\{I(\eta,\xi)\} = \delta\left\{\iint_S \left[h\left(\left(\frac{\partial\eta}{\partial x}\right)^2 + \left(\frac{\partial\eta}{\partial y}\right)^2 + \left(\frac{\partial\xi}{\partial x}\right)^2 + \left(\frac{\partial\xi}{\partial y}\right)^2\right)\right.\right.$$

$$\left.\left. -2\tau h\left(\frac{\partial\eta}{\partial x}\frac{\partial\xi}{\partial y} - \frac{\partial\eta}{\partial y}\frac{\partial\xi}{\partial x}\right) - k^2(\eta^2+\xi^2)\right]dA\right\}, \tag{A2.1}$$

or simply

$$\delta\{I(\eta,\xi)\} = \delta\left\{\iint_S F\left(x,y,\eta,\xi,\frac{\partial\eta}{\partial x},\frac{\partial\eta}{\partial y},\frac{\partial\xi}{\partial x},\frac{\partial\xi}{\partial y}\right)dA\right\}, \tag{A2.2}$$

where ζ has been written as

$$\zeta = \eta + i\xi. \tag{A2.3}$$

The grid used is shown in Figure A2.1. This figure also shows the numbering system for labelling grid points. The discrete form of F is

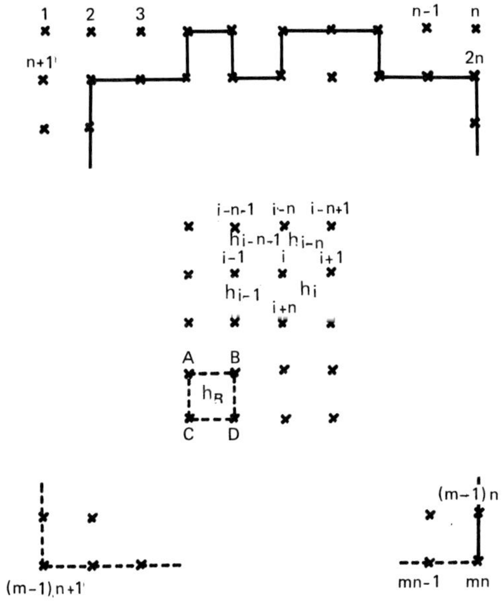

Figure A2.1: The gridscheme on which the finite-difference form of Equation (A2.1) is defined. The closed and open boundaries are marked as —x— and --x-- respectively.

$$F = \frac{h}{2}\left[\frac{(\eta_B-\eta_A)^2}{\varepsilon_x^2} + \frac{(\eta_C-\eta_D)^2}{\varepsilon_x^2} + \frac{(\eta_B-\eta_C)^2}{\varepsilon_y^2} + \frac{(\eta_A-\eta_D)^2}{\varepsilon_y^2} + \frac{(\xi_B-\xi_A)^2}{\varepsilon_x^2}\right.$$

$$\left. + \frac{(\xi_C-\xi_D)^2}{\varepsilon_x^2} + \frac{(\xi_B-\xi_C)^2}{\varepsilon_y^2} + \frac{(\xi_A-\xi_D)^2}{\varepsilon_y^2}\right]$$

$$- \frac{\tau\, h_R}{2\varepsilon_x\,\varepsilon_y}\left[(\eta_B-\eta_A+\eta_C-\eta_D)(\xi_A-\xi_D+\xi_B-\xi_C)\right.$$

$$\left. -(\xi_B-\xi_A+\xi_C+\xi_D)(\eta_A-\eta_D+\eta_B-\eta_C)\right]$$

$$- \frac{k^2}{4}\left[\eta_A^2+\eta_B^2+\eta_C^2+\eta_D^2+\xi_A^2+\xi_B^2+\xi_C^2+\xi_D^2\right], \tag{A2.4}$$

where h_R is a representative value of the depth within the rectangle ABCD and η_A is an average value of η around the point A.

By restricting the boundary of S to consist of sections parallel to the x or y-axis and connecting adjacent grid points, as shown in Figure A2.1, the finite-difference equivalent of (A2.2) is

$$\delta\{I(\eta,\xi)\} = \delta\left\{\sum_{i=1}^{n} w_i \quad F_i\,(\eta_j\ \xi_j\ )\right\}, \tag{A2.5}$$

for $j = 1,\dots,n$. The summation is over the n points within S and on the boundary. The weight function w_i is $\varepsilon_x\,\varepsilon_y$ for all grid points, as the grid is uniform over S. Equation (A2.5) may be written as the linear set of equations,

$$\frac{\partial}{\partial\eta_j}\sum_{i=1}^{n} w_i\,F_i\,(\eta_j\ ,\xi_j\ ) = 0,$$

$$\frac{\partial}{\partial\xi_j}\sum_{i=1}^{n} w_i \quad F_i\,(\eta_j\ ,\xi_j\ ) = 0,$$

for $j = 1,\dots,n$. The derivative $\partial/\partial\eta_j$ is the sum of terms of the form $\partial/\partial\eta_A$, arising from the four quadrants around the j-th grid point. In matrix notation, these equations are

$$\left(\begin{bmatrix} P & \tau Q \\ -\tau Q & P\end{bmatrix} - k^2\begin{bmatrix} R^2 & 0 \\ 0 & R^2\end{bmatrix}\right)\begin{bmatrix}\eta \\ \xi\end{bmatrix} = 0, \tag{A2.6}$$

The matrix, R^2 is a diagonal matrix defined as

$$R_{ii}^2 = \frac{m_i}{2},$$

where m_i is the number of quadrants within S at the ith grid point. From the construction of F, it can be readily shown that P is real and symmetric and Q is real and skew-symmetric.

APPENDIX 3

A Cubic Eigenvalue Problem, Numerical Methods.

Equation (1.5.7) can be written as a cubic equation in σ with

$$(A_0 + \sigma A_1 - \sigma^3 I)\underset{\sim}{x} = 0, \tag{A3.1}$$

where

$$A_0 = f\begin{bmatrix} 0 & R^{-1}QR^{-1} \\ -R^{-1}QR^{-1} & 0 \end{bmatrix}, \tag{A3.2}$$

$$A_1 = \begin{bmatrix} R^{-1}PR^{-1}+f^2 I & 0 \\ 0 & R^{-1}PR^{-1}+f^2 I \end{bmatrix} \tag{A3.3}$$

and

$$\underset{\sim}{x} = \begin{bmatrix} \eta \\ \xi \end{bmatrix}. \tag{A3.4}$$

Following Wilkinson (1965), the eigenvalue problem defined by (A3.1) can be solved by inverse iteration. The sequence of vectors,

$$\underset{\sim}{y} = \sigma\, \underset{\sim}{x},$$

$$\underset{\sim}{z} = \sigma\, \underset{\sim}{y},$$

is defined, enabling (A3.1) to be written as a fundamental eigenvalue problem,

$$\begin{bmatrix} 0 & I & 0 \\ 0 & 0 & I \\ A_0 & A_1 & 0 \end{bmatrix}\begin{bmatrix} x \\ y \\ z \end{bmatrix} = \sigma \begin{bmatrix} x \\ y \\ z \end{bmatrix}. \tag{A3.5}$$

Inverse iteration is defined by the equations,

$$\begin{bmatrix} -pI & I & 0 \\ 0 & -pI & 0 \\ A_0 & A_1 & -pI \end{bmatrix}\begin{bmatrix} x_{s+1} \\ y_{s+1} \\ z_{s+1} \end{bmatrix} = \begin{bmatrix} u_s \\ v_s \\ w_s \end{bmatrix} \tag{A3.6}$$

and

$$\left[u_{s+1}^T ,\ v_{s+1}^T ,\ w_{s+1}^T \right] = \frac{[x_{s+1}^T ,\ y_{s+1}^T ,\ z_{s+1}^T]}{t_{s+1}}, \tag{A3.7}$$

where t_{s+1} is the maximum element of $\left[x_{s+1}^T ,\ y_{s+1}^T ,\ z_{s+1}^T\right]$ and p is some estimate of the eigenvalue being calculated. An alternative form of (A3.6) is

$$\left.\begin{aligned} y_{s+1} &= u_s + p\, x_{s+1} , \\ z_{s+1} &= v_s + p\, y_{s+1} , \\ (A_0+pA_1-p^3)x_{s+1} &= w_s + pv_s + p^2 v_s + p^2 u_s - A_1 u_s . \end{aligned}\right\} \tag{A3.8}$$

The system of Equations (A3.8) and (A3.7) can be iterated to obtain an estimate of the eigenvector. The eigenvalue is a solution of the cubic equation,

$$\sigma^3 - (f^2 + a)\sigma - fb = 0, \tag{A3.9}$$

where

$$a = \eta^T R^{-1} P R^{-1} \eta + \xi^T R^{-1} P R^{-1} \xi$$

and

$$b = \xi^T R^{-1} Q R^{-1} \eta - \eta^T R^{-1} Q R^{-1} \xi.$$

These equations arise from considering the Rayleigh quotient for (1.5.7).

This approach, although efficient with regard to the number of computing operations required, is rather awkward to program.

An alternative method has been used because of its simplicity. Two separate iterations are carried out simultaneously. They may be defined as

$$T_i = \begin{bmatrix} R^{-1}PR^{-1} & \tau_i R^{-1}PR^{-1} \\ -\tau_i R^{-1}QR^{-1} & R^{-1}PR^{-1} \end{bmatrix} \tag{A3.10}$$

$$(T_i - pI)x_{s+1} = u_s , \tag{A3.11a}$$

$$(T_i - b_s I)x_{s+1} = v_s , \tag{A3.11b}$$

$$u_{s+1} = \frac{x_{s+1}}{\max(x_{s+1})} , \tag{A3.12a}$$

$$v_{s+1} = \frac{x_{s+1}}{(x_{s+1}^T \; x_{s+1})^{1/2}} , \tag{A3.12b}$$

$$\mu_{s+1} = \frac{x_{s+1}^T \; T_i \; x_{s+1}}{x_{s+1}^T \; x_{s+1}} , \tag{A3.13}$$

$$\varepsilon_{s+1} = [v_{s+1}^T (A' - \mu_{s+1} I)(A - \mu_{s+1} I) v_{s+1}]^{1/2}, \tag{A3.14}$$

$$\sigma_{i+1}^2 = \frac{\mu_{i+1}^2}{1 - \tau_i^2} , \tag{A3.15}$$

$$f_{i+1} = \tau_i \;\; \sigma_{i+1} , \tag{A3.16}$$

and

$$\tau_{i+1} = f/\sigma_{i+1} . \tag{A3.17}$$

Equations denoted as a and b are alternatives.

For a constant value of τ_i, Equations (A3.10) to (A3.14) define the sequence used for inverse iteration. Equation (A3.14) is included to provide a measure of the convergence of σ_i to an eigenvalue of T_i as there is at least one eigenvalue of T_i in the interval $(\mu_s - \varepsilon_s, \mu_s + \varepsilon_s)$. Equations (A3.10), (A3.13) and (A3.15) to (A3.17) form the second sequence.

An initial estimate of σ (i=0) is made and an eigenvalue of T_0 is calculated using inverse iteration. The second sequence is used to update the value of τ, given the better estimate of σ. The two iterations are repeated with the updated matrix T_1. The procedure is terminated when f_i has converged to f and the bound, ε_i, on σ_i are less than some predetermined value.

The number of iterations used to obtain a final result depends on the initial choice of σ, the accuracy to which the eigenvalue of T_i is calculated and how the two alternative forms, a and b, of inverse iteration are used.

APPENDIX 4

The Finite-Difference Formulation of the Time Independent Circulation.

The same grid scheme as defined in Appendix 3 has been used. Partial derivatives are approximated by first order central difference. The grid spacing is restricted to forms with $\varepsilon_x = \varepsilon_y = \varepsilon$. Equation (1.8.7) in finite-difference form with F_S and G_S constant is

$$\frac{K}{\varepsilon^2}(\psi_{i-n} + \psi_{i+n} + \psi_{i+1} + \psi_{i-1} - 4\psi_i)$$

$$+ \frac{1}{\varepsilon^2}(\psi_{i+1} - \psi_{i-1})\left(\frac{2\kappa}{h}(h_{i-n-1} + h_{i-1} - h_{i-n} - h_i) + f(h_{i-1} + h_i - h_{i-n} - h_{i-n-1})\right)$$

$$+ \frac{1}{\varepsilon^2}(\psi_{i-n} - \psi_{i+n})\left(\frac{2\kappa}{h}(h_{i-1} + h_i - h_{i-n} - h_{i-n-1}) + f(h_{i-n} + h_i - h_{i-n-1} - h_{i-1})\right)$$

$$= \frac{1}{\varepsilon\rho}(F_s(h_{i-n} + h_{i-n-1} - h_{i-1} - h_i) + G_s(h_{i-n-1} + h_{i-1} - h_{i-n} - h_i)), \tag{A4.1}$$

where

$$h = (h_i + h_{i+1} + h_{i-n} + h_{i-n-1})/2.$$

This equation is set up for each grid point within the model sea basin. In matrix notation, (A4.1) becomes

$$A\underset{\sim}{\psi} = F_s \underset{\sim}{b}_1 + G_s \underset{\sim}{b}_2 + \sum_{i=1}^{n} c_i \underset{\sim}{b}_{i+2} \tag{A4.2}$$

where coefficients, c_i, are the values of the streamfunction on the n disconnected lengths of shoreline.

APPENDIX 5

The Finite-Difference Form of the Convection Terms.

The finite-difference forms used for the convection terms in Equation (1.9.1) are

$$\frac{\partial \overline{u^2}}{\partial x} + \frac{\partial \overline{uv}}{\partial y} = \frac{(U_{i+1}^2/h_{i+1} - U_{i-1}^2/h_{i-1})}{2\varepsilon_x}$$

$$+ \frac{(U_{i+n}V_{i+n}/h_{i+n} - U_{i-n}V_{i-n}/h_{i-n})}{2\varepsilon_y}, \qquad \text{(A5.1)}$$

where the same numbering system as shown in Figure A2.1 is used, and ε_x and ε_y are the distances between adjacent stream points along the x- and y-axes respectively. A similar approximation was used for the convection terms in (1.9.2).

For the stream points at which (A5.1) cannot be used, such as those points on the closed boundary and those adjacent to the open boundary, the non-linear terms have been neglected. As the number of grid points where this approximation has been made is less than 5% of the total, the resulting error is not serious.

APPENDIX 6

Long Waves in Channels.

The linearized equation describing the motion of the surface elevation for flow in a channel with a variable cross section is

$$\frac{\partial^2 \zeta}{\partial t^2} = \frac{g}{b}\frac{\partial}{\partial x}\left(hb\,\frac{\partial \zeta}{\partial x}\right), \tag{A6.1}$$

(see Lamb, 1932). If the depth, h, and breadth, b, are assumed to take the form,

$$h = h_0 x^a , \tag{A6.2}$$

$$b = b_0 x^b , \tag{A6.3}$$

with $0 \leq a < 2$ and $b \geq 0$, where h_0 and b_0 are constants, and

$$\zeta(x,t) = \zeta^*(x)e^{-i\sigma t} ,$$

then (A6.1) may be written as

$$\frac{d^2\zeta^*}{dx^2} + \frac{a+b}{x}\frac{d\zeta^*}{dx} + \frac{\sigma^2}{gh_0 x^a}\,\zeta^* = 0. \tag{A6.4}$$

The boundary conditions are that ζ^* is finite at $x=0$ and $\zeta^*=0$ at $x=\ell$.

This Sturm-Liouville problem is satisfied by the functions

$$\zeta = x^{\alpha} J_{\nu}(\beta x^{\gamma}), \tag{A6.5}$$

where

$$\alpha = \frac{1-a-b}{2} ,$$

$$\gamma = 1 - \frac{a}{2},$$

$$\nu^2 = \left(\frac{1-a-b}{2-a}\right)^2 ,$$

$$\beta = \frac{\sigma}{\sqrt{gh_0}}\,\frac{2}{2-a} ,$$

and σ satisfies the equation

$$J_{\nu}(\beta \ell^{\gamma}) = 0.$$

If z denotes the zero of J_{ν} corresponding to σ and h_e $(=h_0\,\ell^a)$ denotes the depth at the open end of the channel, it may be shown that

$$\frac{gh_e}{\ell^2} = \left(\frac{2}{2-a}\;\frac{\sigma}{z}\right)^2 . \tag{A6.6}$$

The left-hand side of this equation is constant for all modes of oscillation. This provides a method of comparing various one-dimensional models with results obtained from two-dimensional models.

The average depth $(\bar{h})$ of a basin is defined as

$$\bar{h} = \frac{1}{\ell} \int_0^\ell h(x)dx.$$

For a basin with a depth described by (A6.2),

$$\bar{h} = \frac{h_\ell}{1+a} .$$

Equation (A6.6) may be rewritten as

$$\frac{g\bar{h}}{\ell^2} = \frac{1}{1+a} \left\{\frac{2}{2-a} \frac{\sigma}{z}\right\}^2, \qquad \text{(A6.7)}$$

As $\bar{h}$ and ℓ^2 are approximately constant for different models describing the same basin, Equation (A6.7) can be used for comparing possible models.

Numerical Simulation of Fluid Motion, ed. John Noye
pp. 357-369

A TIDAL MODEL OF CORIO BAY, VICTORIA

A. K. Easton

Caulfield Institute of Technology,

Victoria, Australia

SUMMARY

A finite difference approach has been adopted to approximate the depth-average equations governing the movement of tides. This paper discusses the development of the model and its application to simulate the waver movement in Corio Bay. Predicted values of the tide height and currents are presented.

1. INTRODUCTION

Corio Bay is the harbour of the major port of Geelong situated at the end of the western arm of Port Phillip Bay, Victoria (Figure 1.1). As well as being used for fishing, sailing and swimming it serves as a sink for waste discharges from industries such as Shell, Phosphate Cooperative and Ford which flank the Bay.

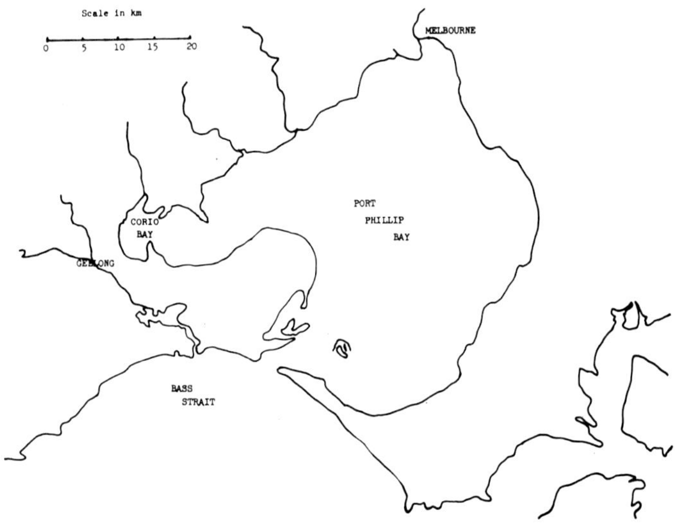

Figure 1.1 Port Phillip Bay, Victoria.

Pollution of Corio Bay has become an important issue. In November 1974, the Public Health Department prohibited the fishing for mussels due to the high concentration of the metal cadmium.

The aim of this project is to develop a water quality model which will simulate the transport within the Bay. This information will be very valuable to the Environment Protection Authority of Victoria whose task it is to set levels of discharge for the various industries. In this paper, the model chosen for the water movement and the problems involved will be discussed.

2. CORIO BAY

The movement of water (and ships) into and out of Corio Bay is severely restricted by the sand bar extending northwards from Point Henry (Figure 2.1). Apart from this constriction, the area has an almost uniform depth of approximately 8 m. Small ships may use the channels to the north of the sand bar but most shipping passes along the narrow Hopetoun Channel which is maintained at 11 m depth. The current has been traced by the Geelong Harbour Trust along this channel from Point Henry, where a current of 1.8 knots was observed, into Corio Bay. However, from a comparison of the cross sectional areas it appears that this deep channel carries only about 35% of the water into the Bay compared with 55-60% to the north of the sandbar.

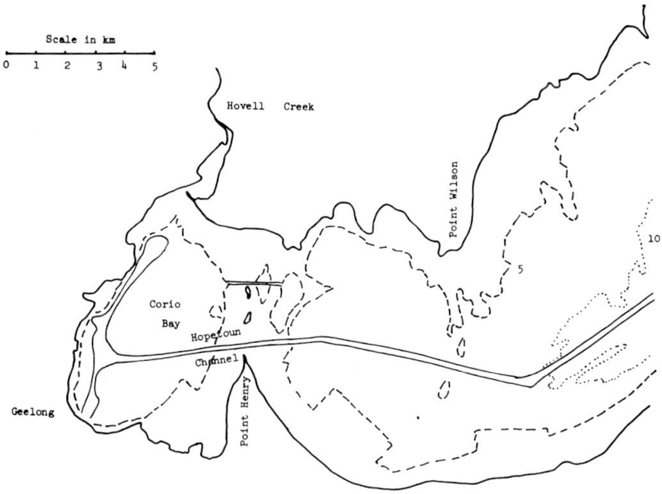

Figure 2.1 Shipping channels and depth contours (in m) in Corio Bay.

Tide records from Geelong show a marked effect of wind and seiches but the main water movement in Corio Bay is tidal with a mean range of less than a metre. Freshwater flow into the Bay is negligible.

3 WATER MOVEMENT MODEL

3.1 Equations

When the equations of continuity and motion are integrated over the depth, they become

$$\frac{\partial Z}{\partial t} = - \frac{\partial HU}{\partial x} - \frac{\partial HV}{\partial y} \tag{3.1}$$

$$\frac{\partial U}{\partial t} = - U\frac{\partial U}{\partial x} - V\frac{\partial U}{\partial y} + fV - g\frac{\partial Z}{\partial x} - r\frac{U\sqrt{U^2 + V^2}}{H} + \alpha\frac{U_a\sqrt{U_a^2 + V_a^2}}{H} \tag{3.2}$$

$$\frac{\partial V}{\partial t} = - U\frac{\partial V}{\partial x} - V\frac{\partial V}{\partial y} - fU - g\frac{\partial Z}{\partial y} - r\frac{V\sqrt{U^2 + V^2}}{H} + \alpha\frac{V_a\sqrt{U_a^2 + V_a^2}}{H} \tag{3.3}$$

where

$Z(x,y,t)$ is the height of the water level above datum,

$U(x,y,t)$ is the x-component of the velocity of the water,

$V(x,y,t)$ is the y-component of the velocity of the water,

$H(x,y,t)$ is the total depth of water,

f is the coriolis parameter and is assumed constant ($f = -0.00009\ \sec^{-1}$),

g is the gravitational acceleration ($g = 9.80\ \mathrm{m\ sec^{-2}}$),

r is the frictional coefficient due to drag at the lower water surface (0.002),

α is the frictional coefficient due to drag at the upper water surface (0),

$U_a(x,y,t)$ is the x-component of the velocity of the wind,

$V_a(x,y,t)$ is the y-component of the velocity of the wind.

These equations of motion include the non-linear advection and friction terms as well as those due to the gravitation and rotation.

3.2 Difference equations

Numerous researchers, perhaps the most well known being Hansen (1966), Leendertse (1967) and Miyazaki (1965), have solved these equations numerically using finite difference techniques. Here, the method adopted follows Hansen who used central difference approximations for derivatives where ever possible and forward difference approximations elsewhere. If Δx, Δy and Δt represent incremental changes in x,y,t and the subscripts j,k,n represent discrete values of x,y and t respectively,

$$\frac{\partial Z}{\partial t} \simeq \frac{Z(x,y,t+\Delta t) - Z(x,y,t-\Delta t)}{2\Delta t}$$

$$= \frac{Z_{j,k,n+1} - Z_{j,k,n-1}}{2\Delta t}$$

Then, equation (3.1) becomes

$$Z_{j,k,n+1} = Z_{j,k,n-1} - \frac{\Delta t}{\Delta x}[(H\ U)_{j+1,k,n} - (H\ U)_{j-1,k,n}]$$
$$- \frac{\Delta t}{\Delta y}[(H\ V)_{j,k+1,n} - (H\ V)_{j,k-1,n}]$$

This equation requires the U velocity at some points and the V velocity at others and the depth D below datum at both. Hence, we choose the following grid representation

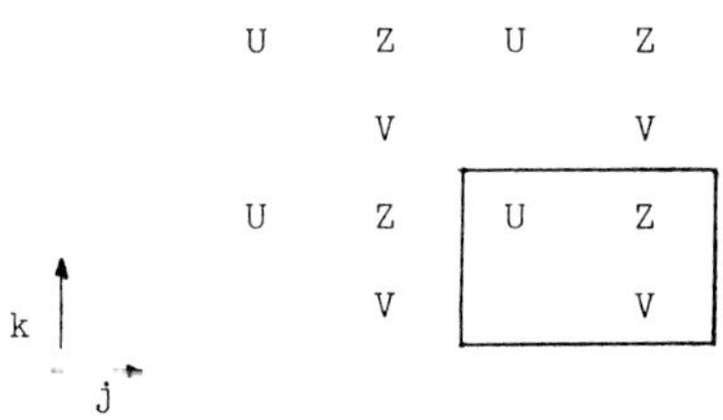

Then, in order to save computer space, these are grouped so that a Z, U and V point have the same coordinates j and k, and we write

$$Z_{j,k,n+1} = Z_{j,k,n-1} - \frac{\Delta t}{\Delta x}[(Hu\ U)_{j+1,k,n} - (Hu\ U)_{j,k,n}]$$
$$- \frac{\Delta t}{\Delta y}[(Hv\ V)_{j,k+1,n} - (Hv\ V)_{j,k,n}] \qquad (3.4)$$

(for convenience, the depths at the U and V velocity points have been stored in separate arrays Hu and Hv respectively).

In equation (3.2), forward differences in time are used and

$$
\begin{aligned}
U_{j,k,n+2} = U_{j,k,n} &- \frac{\Delta t}{\Delta x} U_{j,k,n} \left(\frac{U_{j+1,k,n} - U_{j-1,k,n}}{2}\right) \\
&- \frac{\Delta t}{\Delta y} V_{BAR} \left(\frac{U_{j,k+1,n} - U_{j,k-1,n}}{2}\right) + \Delta t\ f\ V_{BAR} \\
&- \frac{\Delta t}{\Delta x} g\ (Z_{j,k,n} - Z_{j-1,k,n}) \\
&- \Delta t\ r\ \frac{U_{j,k,n}}{Hu_{j,k,n}} \sqrt{U^2_{j,k,n} + V^2_{BAR}} \qquad (3.5) \\
&+ \Delta t\ \alpha\ \frac{Ua_{j,k,n}}{Hu_{j,k,n}} \sqrt{Ua^2_{j,k,n} + Va^2_{j,k,n}}
\end{aligned}
$$

In this equation it is necessary to use the following approximations

$$Hu_{j,k,n} = Du_{j,k} + \tfrac{1}{2}\ (Z_{j+1,k,n+1} + Z_{j,k,n+1})$$

where $Du_{j,k}$ is the depth of water at a U point below the mean level

and $$V_{BAR} = \tfrac{1}{4}[V_{j,k,n} + V_{j,k+1,n} + V_{j-1,k,n} + V_{j-1,k+1,n}]$$

since V velocity and total depth are not calculated directly at the U points.

Similarly,

$$
\begin{aligned}
V_{j,k,n+2} = V_{j,k,n} &- \frac{\Delta t}{\Delta x} U_{BAR} \left(\frac{V_{j+1,k,n} - V_{j-1,k,n}}{2}\right) \\
&- \frac{\Delta t}{\Delta y} V_{j,k,n} \left(\frac{V_{j,k+1,n} - V_{j,k-1,n}}{2}\right) - \Delta t\ f\ U_{BAR} \\
&- \frac{\Delta t}{\Delta y} g\ (Z_{j,k,n} - Z_{j,k-1,n}) \qquad (3.6) \\
&- \Delta t\ r\ \frac{V_{j,k,n}}{Hv_{j,k,n}} \sqrt{U^2_{BAR} + U^2_{j,k,n}} \\
&+ \Delta t\ \alpha\ \frac{Va_{j,k,n}}{Hv_{j,k,n}} \sqrt{Ua^2_{j,k,n} + Va^2_{j,k,n}}
\end{aligned}
$$

where $Hv_{j,k,n} = Dv_{j,k} + \tfrac{1}{2}(Z_{j,k,n+1} + Z_{j,k-1,n})$

and

$$U_{BAR} = \tfrac{1}{4}[U_{j,k,n} + U_{j,k-1,n} + U_{j+1,k,n} + U_{j+1,k-1,n}]$$

4. MODELLING CORIO BAY

4.1 Grid

Several criteria were considered in deciding on a grid for Corio Bay. Firstly, a good representation of the small channel width and the abrupt changes in depth particularly near Point Henry can only be achieved by a small grid length. On the other hand, this smallness is limited by the storage requirement and the calculation time for running on the computer. As a compromise, a grid length of 375 m was chosen for both the x and y directions i.e. $\Delta x = \Delta y = 375$ m.

This finite difference model requires that the boundary be approximated by straight line segments parallel to the x and y axes, another factor which requires small grid length in order to model such features as Point Henry. The open boundary was taken opposite Point Wilson on the assumption that any errors in the conditions imposed there would not affect the flow in the Bay proper; this was considered to be a very desirable feature despite the increase in dimension to 37 x 31. The sand bar of Point Henry was modelled by a thin barrier.

The grid outline chosen to represent the Bay and a contour of the depths at the grid points is shown in Figure 4.1.

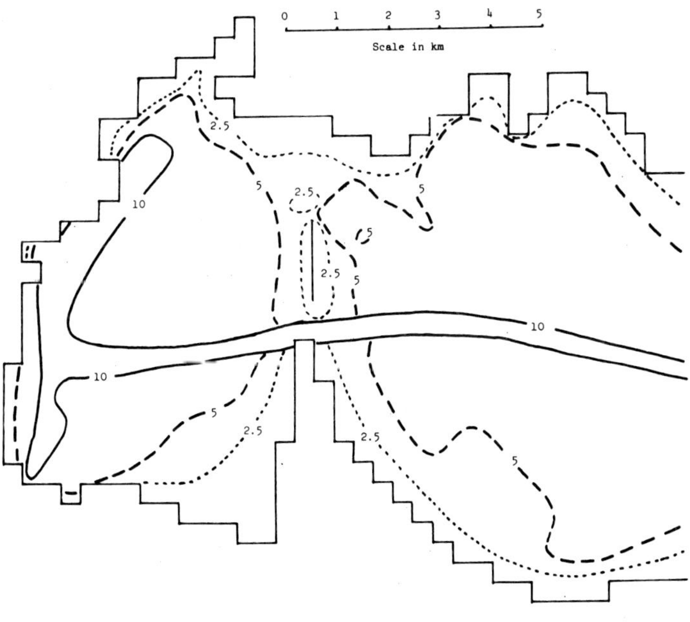

Figure 4.1 Outline of grid chosen for Corio Bay showing depth in metres below mean water level.

4.2 Boundary Conditions

Since there is no flow across a closed boundary, V = 0 on those straight line boundary segments parallel to the x axis and U = 0 on those parallel to the y axis. No attempt has been made to have a moveable boundary.

On the open boundary, the height Z is specified in the form

$$Z(t) = a \sin \omega t$$

where a = 0.5 m, $\omega = 0.00014 \text{ sec}^{-1}$ corresponding to the M_2 tide.

4.3 Initial Conditions

Initially, it was assumed that all heights Z and velocities U and V were zero. After incrementing the height values at the open boundary, the velocities were computed using equations (3.5) and (3.6) and the height using equation (3.4). Then, the new boundary values are specified and the cycle of computing U, V and Z in turn is repeated. The calculations are made using the equations specified at all interior grid points except those adjacent to the boundaries where forward or backward differences are used in place of central ones. For convenience of calculation, all points have been assigned an integer number which indicates the required form of the equations to be used.

These finite difference equations will be stable if the time step Δt is limited by

$$\Delta t < \frac{\Delta x}{\sqrt{2g\,H}}$$

In reality, slightly larger values than the maximum of 25 seconds can be used; a step of 30 seconds has been used here.

4.4 Filtering

Experience with similar models [Easton (1972)] has shown that waves with wavelengths of the order of the grid size are generated and hence they were expected to be found here too. Their effect has been minimised in this model by the use of the averaging filter

$$Z_{j,k,n} = \frac{1}{8}[Z_{j+1,k,n} + Z_{j-1,k,n} + Z_{j,k-1,n} + Z_{j,k+1,n} + 4Z_{j,k,n}]$$

at every interior point at every time step [see Hamming (1962)].

4.5 Bottom and Surface Friction

Friction with the bottom of the Bay has been allowed for through a quadratic velocity term with a friction coefficient of 0.002. This term is small compared with some of the other terms and hence the value of the coefficient should not be critical.

Although the model has been developed to include the influence of wind on the water surface, this has been omitted from the present investigation by choosing the drag coefficient α = 0.

5. RESULTS

The computations were carried out on the ICL 1904A computer at the Caulfield Institute of Technology. Before the model was used on Corio Bay, it was checked on a rectangular bay. For the Corio Bay model, the arrays for height, velocity, depth and identification have dimension 37 × 31 and using a 30 second time step, 30 minutes calculation time were used for every 120 minutes real time.

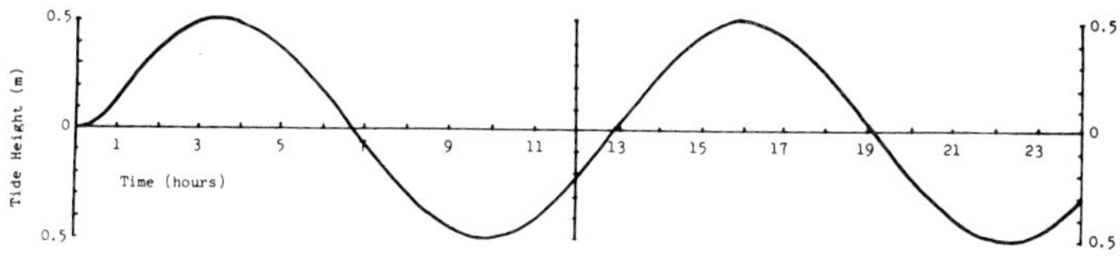

Figure 5.1 Tide height (m) at grid point (17, 16) just west of Point Henry.

As the initial conditions of mean height and zero velocity are obviously in error, there was at first an adjustment period. This is well shown for the grid point (17, 16) near Point Henry in Figure 5.1. However, the transient time was short and after the first 3 hours (real time) the tide flow had become completely regular. As expected, the U and V velocity components show a more marked initial effect but these effects are also minimal after the first 3 hours (Figure 5.2). That these velocities are much more sensitive to change than the height is shown by the persistence of some irregularity throughout the first day.

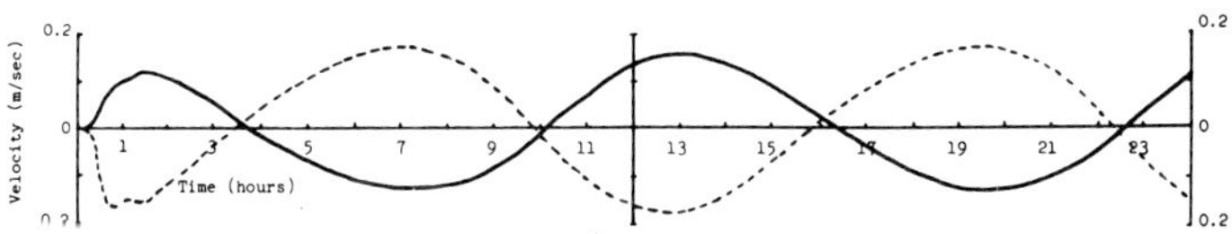

Figure 5.2 Tide velocity components U and V (m/sec) at grid point (17, 16) just west of Point Henry.

Closer examination shows that local oscillations have been initiated in Corio Bay. Figure 5.3 shows the very small but oscillatory motion with a period of 6 minutes at grid point (11, 27). These may be continually excited near grid point (7,25) where there is a bay and an abrupt change in depth. Simple calculations show that the natural period for Corio Bay would be of the order of 20 minutes.

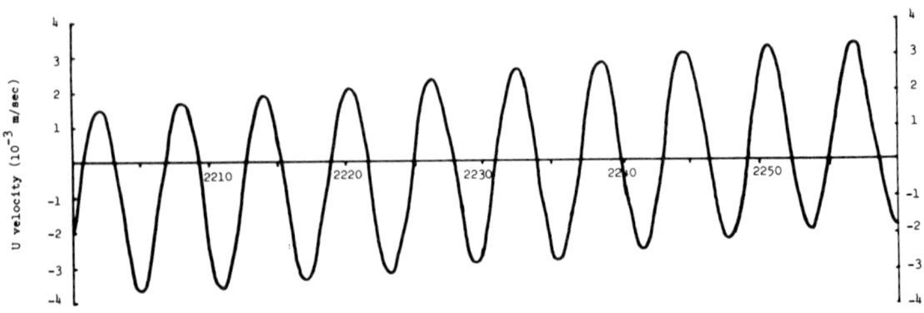

Figure 5.3 Local oscillations in Corio Bay (at grid point (11,27)).

The output from the computer program is in the form of tide heights and velocities at a given time or as maximum and minimum values over a specified time interval. The maximum heights computed for the third tide cycle displayed several characteristic features. Although there was a slight amplitude increase into the Bay from the boundary, the maximum change was only 5% and hence has not been graphed. The phases (Figure 5.4) show a lag of over 40 minutes in Hovell Creek. There is virtually no lag between the open boundary and Point Henry and also within Corio Bay itself which has a remarkably uniform lag of 30 - 35 minutes. The very rapid change in phase in the channel off Point Henry is a strong indication that most of the water which enters the Bay passes through the passage to the north of the sandbar.

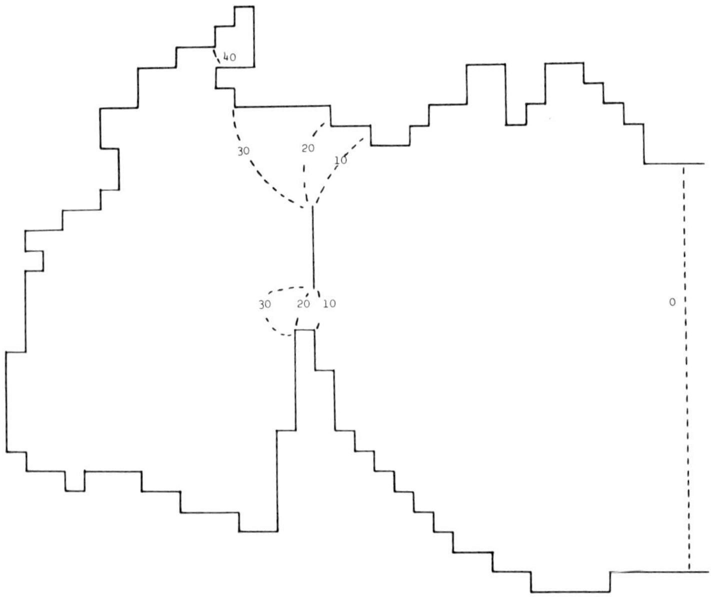

Figure 5.4 Phase lag of maximum tide in minutes compared with Point Wilson.

A comparison of Figures 5.1 and 5.2 shows quite clearly that the tide height and velocities are 90° out of phase as is exhibited by a standing wave. Thus, the velocities at mean tide in Corio Bay shown in Figures 5.5 and 5.6 must be maximum in that area.

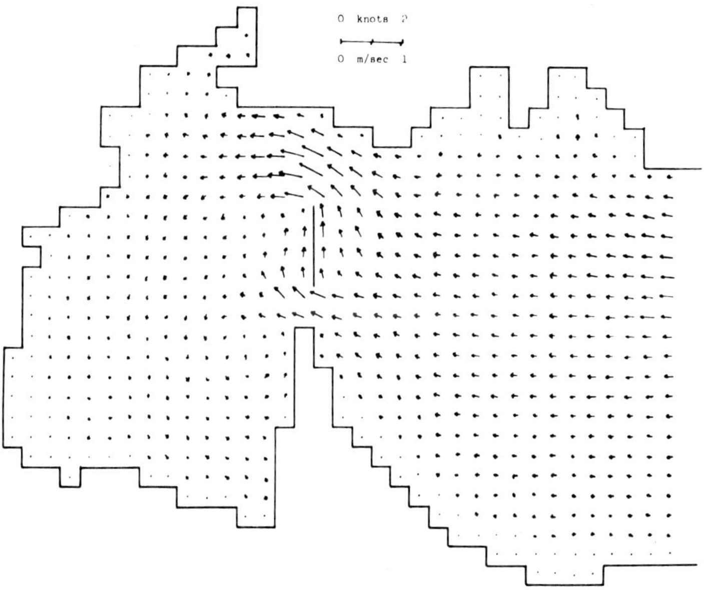

Figure 5.5 Flood tidal velocity at mean height in Corio Bay.

The main features can be summarised as follows:

(1) The ebb and flood tides are almost identical except for their opposite directions.

(2) The tides along the coastline including at Geelong are virtually zero.

(3) The tide floods into the Bay on the north side of the sandbar, then toward the west side of the Bay, due to the water flowing through the main shipping channel, and finally to the south. The tide ebbs in the reverse directions.

(4) The maximum tidal streams on the north side of the bar are 0.5 m/sec compared with 0.4 m/sec in the main shipping channel.

(5) Maximum tidal velocities within Corio Bay are of the order of 0.1 m/sec which would result in a horizontal movement of about 1 km.

(6) There are weak eddys near Point Henry and near the sandbar.

(7) Flow near the open boundary exhibits the characteristic shape with maximum values of 0.2 m/sec in mid stream.

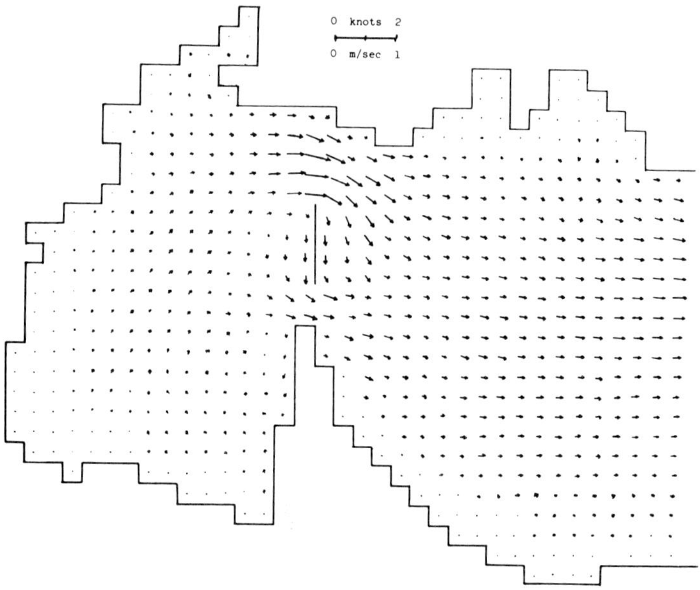

Figure 5.6 Ebb tidal velocity at mean height in Corio Bay.

The regular pattern of the flow to the east of Point Henry shows that the choice of the position of the open boundary is a very good one.

The only weaknesses in the model might be that the channel has a width of only one grid length and that the model imposes that there is no flow across the sandbar whereas in reality, the sandbar dries at low tide. This modification should only affect the flow in the immediate vicinity due to the small flow across the bar. It is noted that the maximum tidal flows are only half of the reported 1.8 knots near Point Henry.

6. CONCLUSIONS

1. The maximum tidal flow occurs to the north of the sandbar.
2. Corio Bay is virtually motionless. With a maximum horizontal tidal movement of less than 2 km, only the water at its northern end ever leaves the Bay. Even the water north of the sandbar travels less than 10 km i.e. it remains within the area modelled. Normal tidal flow conditions, as investigated in this model, are certainly insufficient to flush the Bay.
3. It is expedient that the major industries are located on the north side of the Bay since any effluent is carried by the ebb tide directly toward the deeper water of Port Phillip Bay by these currents.
4. The flood tide carries water from the north of the Bay toward the south but under normal tidal conditions effluent from the north shore would not reach the south.
5. Tidal flow near Point Henry is partly across the channel, NW - SE, and suggests that sand must be continually eroded from the sandbar and deposited in the channel.

7. RECOMMENDATIONS

1. An experiment using floats and current meters and an analysis of tide data from the 3 existing tide gauges should be carried out to check the above conclusions.
2. This investigation should be extended to determine the effect of diffusion processes and wind transport.
3. Consideration should be given to the possible relocation of the shipping channel to the north of the sandbar.

8. ACKNOWLEDGEMENTS

I wish to thank the Environment Protection Authority and the Geelong Harbour Trust for their interest in this work, and the Caulfield Institute of Technology for the use of their facilities including the ICL 1904A computer.

9. REFERENCES

Easton, A.K. (1972) Tides of Hudson Strait, Bedford Institute of Oceanography, Report Series BI-R-72-6

Hamming, R.W. (1962) Numerical Methods for Scientists and Engineers McGraw-Hill

Hansen, W. (1966) Mitt. Inst. f. Meereskunde, Univ. Hamburg

Leendertse, J.J. (1967) Aspects of a Computational Model for Long-period Water Wave Propagation Rand Cooperation

Miyazaki, M. (1965) A Numerical Calculation of the Storm Surge in the Gulf of Mexico Oceanogr. Mag., 17, 109 - 140

Numerical Simulation of Fluid Motion, ed. John Noye
pp. 371-381

SENSITIVITY ANALYSIS OF A TROPICAL CYCLONE SURGE MODEL

B.A. Harper, R.J. Sobey and K.P. Stark

James Cook University of North Queensland

Australia

SUMMARY

Various numerical aspects of an explicit finite-difference model for tropical cyclone storm surge generation and propagation are investigated numerically. The open boundary conditions pose a particular problem as the geophysical scale of a cyclone normally exceeds the area that can feasibly be modelled on a digital computer. Four alternative open boundary conditions are discussed. The relationship between the geophysical scale of the cyclone and the model grid spacing is also considered. The paper concludes with a discussion of the interaction among the various terms of the governing equations.

1. INTRODUCTION

The numerical hydrodynamic model SURGE has been developed on the James Cook University DEC System-10 to study tropical cyclone storm surge generation and propagation along the Queensland coast. Numerical modelling of long wave propagation is not a new problem; satisfactory computations of astronomical tide propagation in seas or wide estuaries were first undertaken in the 1950's. There are however several unique aspects of the present problem that warranted detailed investigation. These problems stem from the complex character and geophysical extent of the meteorological forcing and include the specification of suitable open boundary conditions and the relationship between the length scale of the tropical cyclone and the model grid spacing.

2. THE NUMERICAL HYDRODYNAMIC MODEL

Long Wave Equations

Storm surge propagation is governed by the Long Wave Equations (Welander, 1964), expressing respectively the conservation of mass and the conservation of momentum in spatial directions x and y:

$$\frac{\partial \eta}{\partial t} + \frac{\partial U}{\partial x} + \frac{\partial V}{\partial y} = 0 \tag{2.1a}$$

$$\frac{\partial U}{\partial t} + \frac{\partial}{\partial x}\left(\frac{U^2}{h+\eta}\right) + \frac{\partial}{\partial y}\left(\frac{UV}{h+\eta}\right) - fV$$

$$= -g(h+\eta)\frac{\partial \eta}{\partial x} - \frac{h+\eta}{\rho_w}\frac{\partial p_s}{\partial x} + \frac{1}{\rho_w}(\tau_{xs} - \tau_{xb}) \tag{2.1b}$$

$$\frac{\partial V}{\partial t} + \frac{\partial}{\partial x}\left(\frac{UV}{h+\eta}\right) + \frac{\partial}{\partial y}\left(\frac{V^2}{h+\eta}\right) + fU$$

$$= -g(h+\eta)\frac{\partial \eta}{\partial y} - \frac{h+\eta}{\rho_w}\frac{\partial p_s}{\partial y} + \frac{1}{\rho_w}(\tau_{ys} - \tau_{yb}) \tag{2.1c}$$

The xy datum plane is located at the mean water level with the z axis directed vertically upwards. The water surface elevation with respect to datum is $\eta(x,y,t)$, the sea bed is $h(x,y)$ below datum, U and V are depth-integrated flow velocities or discharges per unit width, f is the Coriolis parameter, and τ_b is bed friction.

The hydrodynamic forcing influence of the tropical cyclone is introduced into the governing equations by the specification of suitable values for the wind stress τ_s and atmospheric pressure p_s at the water surface. For this purpose it is necessary to adopt a suitable model of the atmospheric boundary layer that will allow the estimation of spatial and temporal variations in the shear stress and pressure at the water surface. Suitable models have been proposed by the National Hurricane Research Project of the U.S. Weather Bureau, (Graham and Nunn (1959), Jelesnianski (1965), and U.S. Weather Bureau (1968)). The simplified model of Jelesnianski (1965) was adopted for the present purposes, although the more realistic model of the U.S. Weather Bureau (1968) will be used in production runs.

Typical radial profiles of M.S.L. pressure and sustained azimuthal (or tangential) wind speed at height 10 m are shown in Figure 1. The principal features of these profiles are the sharp gradients around the radius of maximum wind R and the much more gradual changes beyond this position; in fact the spatial extent of the organised motion within the atmospheric boundary layer can extend some 1000 km from the eye.

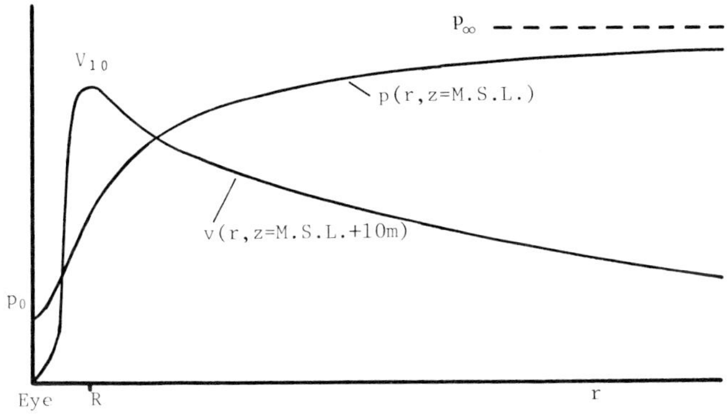

FIGURE 1. TYPICAL WIND VELOCITY AND PRESSURE PROFILES WITHIN A TROPICAL CYCLONE

Finite Difference Equations

Discrete values of the variables are specified on a space (x,y) and time (t) staggered grid, whose node points are defined as $(i\Delta x, j\Delta y, n\Delta t)$. Depth h or D below M.S.L. and water surface elevation η or H are located at (i,j,n), depth-integrated flow U at $(i+\frac{1}{2}, j, n+\frac{1}{2})$ and depth-integrated flow V at $(i, j+\frac{1}{2}, n+\frac{1}{2})$. The finite difference equations, which constitute an explicit leap-frog operator; are:

(a) x momentum equation centred at $(i+\frac{1}{2}, j, n)$

$$\frac{U^{n+\frac{1}{2}}_{i+\frac{1}{2},j} - U^{n-\frac{1}{2}}_{i+\frac{1}{2},j}}{\Delta t} - f\,\overline{\overline{V}}^{n-\frac{1}{2}}_{i+\frac{1}{2},j}$$

$$= -g\,\overline{DX}^{n}_{i+\frac{1}{2},j}\,\frac{H^{n}_{i+1,j} - H^{n}_{i,j}}{\Delta x} - g\,\overline{DX}^{n}_{i+\frac{1}{2},j}\,\frac{B^{n}_{i+1,j} - B^{n}_{i,j}}{\Delta x}$$

$$+ WX^{n}_{i+\frac{1}{2},j} - \frac{\lambda}{8}\sqrt{(U^{n-\frac{1}{2}}_{i+\frac{1}{2},j})^2 + (\overline{\overline{V}}^{n-\frac{1}{2}}_{i+\frac{1}{2},j})^2}\;\frac{U^{n+\frac{1}{2}}_{i+\frac{1}{2},j}}{(\overline{DX}^{n}_{i+\frac{1}{2},j})^2} \qquad (2.2a)$$

(b) y momentum equation centred at $(i,j+\frac{1}{2},n)$

$$\frac{V^{n+\frac{1}{2}}_{i,j+\frac{1}{2}} - V^{n-\frac{1}{2}}_{i,j+\frac{1}{2}}}{\Delta t} + f\,\bar{\bar{U}}^{n-\frac{1}{2}}_{i,j+\frac{1}{2}}$$

$$= -g\,\overline{DY}^{n}_{i,j+\frac{1}{2}}\,\frac{H^{n}_{i,j+1} - H^{n}_{i,j}}{\Delta y} - g\,\overline{DY}^{n}_{i,j+\frac{1}{2}}\,\frac{B^{n}_{i,j+1} - B^{n}_{i,j}}{\Delta y}$$

$$+ WY^{n}_{i,j+\frac{1}{2}} - \frac{\lambda}{8}\sqrt{(\bar{\bar{U}}^{n-\frac{1}{2}}_{i,j+\frac{1}{2}})^2 + (V^{n-\frac{1}{2}}_{i,j+\frac{1}{2}})^2}\;\frac{V^{n+\frac{1}{2}}_{i,j+\frac{1}{2}}}{(\overline{DY}^{n}_{i,j+\frac{1}{2}})^2} \qquad (2.2b)$$

(c) continuity equation centred at $(i,j,n+\frac{1}{2})$

$$\frac{H^{n+1}_{i,j} - H^{n}_{i,j}}{\Delta t} + \frac{U^{n+\frac{1}{2}}_{i+\frac{1}{2},j} - U^{n+\frac{1}{2}}_{i-\frac{1}{2},j}}{\Delta x} + \frac{V^{n+\frac{1}{2}}_{i,j+\frac{1}{2}} - V^{n+\frac{1}{2}}_{i,j-\frac{1}{2}}}{\Delta y} = 0 \qquad (2.2c)$$

where $H^{n}_{i,j} = H(i\Delta x, j\Delta y, n\Delta t)$

$U^{n+\frac{1}{2}}_{i+\frac{1}{2},j}$, $V^{n+\frac{1}{2}}_{i,j+\frac{1}{2}}$ similarly

$B^{n}_{i,j} = \frac{1}{\rho_w g}\,p_s(i\Delta x, j\Delta y, n\Delta t)$

$WX^{n}_{i+\frac{1}{2},j} = \frac{1}{\rho_w}\,\tau_s\,((i+\frac{1}{2})\Delta x, j\Delta y, n\Delta t)$

$WY^{n}_{i,j+\frac{1}{2}}$ similarly

$\Delta x = \Delta y = \Delta s$ = spatial grid dimension

Δt = time step, λ = friction factor

$\overline{DX}^{n}_{i+\frac{1}{2},j} = \frac{1}{2}(D_{i,j} + H^{n}_{i,j} + D_{i+1,j} + H^{n}_{i+1,j})$

$\overline{DY}^{n}_{i,j+\frac{1}{2}}$ similarly

$\bar{\bar{U}}^{n-\frac{1}{2}}_{i,j+\frac{1}{2}} = \frac{1}{4}(U^{n-\frac{1}{2}}_{i-\frac{1}{2},j} + U^{n-\frac{1}{2}}_{i+\frac{1}{2},j} + U^{n-\frac{1}{2}}_{i+\frac{1}{2},j+1} + U^{n-\frac{1}{2}}_{i-\frac{1}{2},j+1})$

$\bar{\bar{V}}^{n-\frac{1}{2}}_{i+\frac{1}{2},j}$ similarly

The finite difference approximations to the convective terms are consistent with the above equations and have been omitted for simplicity. Reduced forms of these equations are necessary in the neighbourhood of open (H) and closed (U,V) boundaries, and the Great Barrier Reef.

Stability and Wave Deformation Characteristics of the Linearised Equations

Formal analytical evaluations of stability and wave deformation can only be attempted for the linearised equations. Terms such as bed friction, convective accelerations and meteorological forcing cannot be included, and it must be assumed that the computational segment considered is far away from the disturbing influence of any boundary.

The numerical stability requirement for the explicit leap-frog scheme was first discussed by Platzman (1956) who gave the following upper limit for the incremental time step:

$$\Delta t \not> \Delta s \; / \; (2gh_{max})^{\frac{1}{2}} \qquad (2.3)$$

In addition to stability, numerical distortion of the physical surge wave needs investigation. Some measure of this distortion can be had from a comparison of the numerical and analytical solutions of the linearised equations. An investigation of various numerical schemes by Sobey (1970), using the propagation factor concept of Leendertse (1967) showed that the present scheme has negligible amplitude distortion when the above stability condition is satisfied.

The numerical deformation characteristics of the full long-wave equations are beyond analytical evaluation and can only be investigated by numerical experiment. A number of such experiments under various conditions have been completed. The most important objectives were to establish the applicability of the Courant condition as an estimate of the maximum incremental time step, the suitability of different open boundary conditions, the resolution capability of the grid spacing and the interaction of the various terms of the conservation equations. The former objective was adequately demonstrated and will not be further considered.

3. OPEN BOUNDARY CONDITIONS

In terms of meteorological tides, the spatial extent of a tropical cyclone would approach 1000 km, although the region of peak positive and negative surges would have a spatial scale of the order of the radius of maximum winds (typically 25 km). Ideally, the open boundaries of the computation field should be sufficiently distant from the cyclone that boundary water levels can be taken as mean sea level (or tide level). However, the storage limitations of modern time-sharing computer systems preclude the adoption of a computational field that has linear dimensions of the order of 1000 km, and simultaneously reproduces details on the scale of the radius of maximum winds. A practical compromise to this conflict of scales must be adopted and the forcing influence of the cyclone outside the computational field must be included in the open boundary conditions; the problems associated with computer storage limitations are discussed in the following section.

Four possible open boundary conditions are briefly discussed and an evaluation of their respective performances is made. Open boundary conditions can be either a water level or flow condition. The alternatives considered were:

(a) Open boundary water levels set equal to the pressure surge, the head of water equivalent to the pressure deficit, i.e. -

$$H^n_{i,j} = \frac{p_\infty - p^n_{s\;i,j}}{\rho_w g} \qquad (3.1)$$

(b) Open boundary water levels set to M.S.L. (Heaps, 1969), i.e. -

$$H^n_{i,j} = 0 \qquad (3.2)$$

(c) Flow gradient perpendicular to open boundary set to zero (Jelesnianski, 1965), e.g. -

$$V^{n+\frac{1}{2}}_{i,j+\frac{1}{2}} = V^{n+\frac{1}{2}}_{i,j-\frac{1}{2}} \tag{3.3}$$

at the northern open boundary of a grid.

(d) Open boundary water levels set equal to the local bathystrophic storm tide, i.e., the quasi-static profile described, for example by -

$$0 = -g(h+\eta)\frac{\partial\eta}{\partial x} - \frac{h+\eta}{\rho_w}\frac{\partial p_s}{\partial x} + \frac{\tau_{xs}}{\rho_w} \tag{3.4}$$

when the open boundary is in the x direction.

For comparison purposes, a hypothetical continental shelf region extending 195 n mile (361 km) along a length of straight coast was considered. The sea bed sloped uniformly at a slope of 0.00055 away from a depth of 2.0 m at the coast. At the outer open boundary, 137.5 n mile (255 km) from the coast, the depth was 145 m. The shelf region was assumed centred about 18°S latitude. A test cyclone, similar in many respects to tropical cyclone Althea (Townsville, 1971), was adopted. The cyclone was initially placed in the centre of the flow field and the model allowed a period of 10 iterations for initial build-up from still water before the cyclone was moved perpendicular to the coast at the nominated speed. Total simulation time was 5 hours, with landfall after 3¼ hours. Grid spacing was 5 n mile and the time step was 150 secs.

The first of the four boundary conditions was selected as a basis for comparison. Using this condition, a peak surge of 3.3 m occurred approximately 30 minutes after cyclone landfall and 20 km south of the landfall position. Encouragingly, the phase and amplitude of the peak surge in all cases was almost identical. Such was not the case however away from the peak surge location and close to the open boundaries. Ideally, realistic simulation as close as possible to the open boundaries is required. Detailed comparisons of the relative performance of the boundary conditions can be found in Sobey et al (1975). Figure 2 reproduces some typical coastal water level profiles (observer facing seawards) and the overall behaviour is summarised below.

<u>Mean Sea Level Condition</u> (Figure 2a)

The numerical simulations with boundary conditions (a) and (b) gave broadly equivalent flow patterns; phase agreement was generally good but water level differences averaged 20 cm with a maximum of some 35 cm over the field (Figure 2a). It appeared that the times immediately before and after landfall of the cyclone were most sensitive to these water level deviations, with recovery taking place approximately an hour later. This level decrease could in part be due to a widespread drawdown around the eye for the M.S.L. boundary condition.

<u>Asymptotic Flow Condition</u> (Figure 2b)

Although the peak amplitude and phase matching was good throughout, significant water level differences occurred near the open boundaries. Figure 2b is a typical result, drawdown north of the landfall position and build-up to the south both being accentuated.

The original appeal of this condition (c) was its ability to allow variations in water level at the open boundaries, hopefully producing a more realistic flow pattern in these regions. However, spurious effects were observed in the flow patterns, particularly at intersections of open boundaries. Physically, it is difficult to see what is particularly significant about the flow gradient normal to an open boundary, as the position and especially the orientation of this boundary are essentially arbitrary.

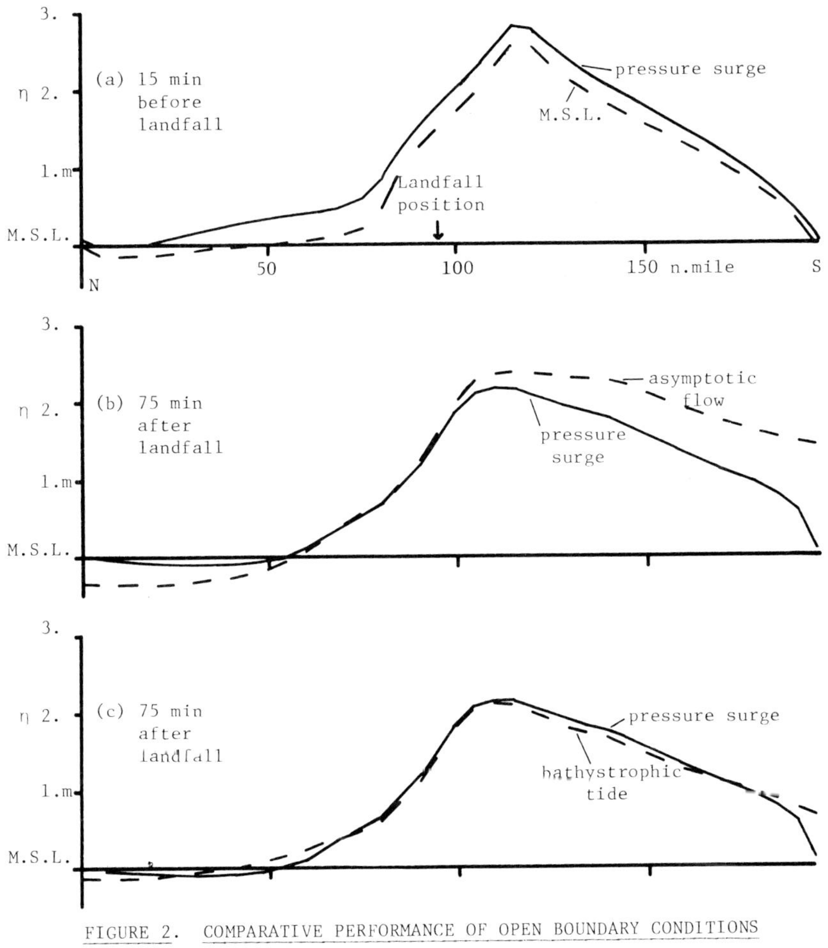

FIGURE 2. COMPARATIVE PERFORMANCE OF OPEN BOUNDARY CONDITIONS

Bathystrophic Storm Tide Approximation (Figure 2c)

Again from a physical viewpoint the bathystrophic tide condition (d) should provide the most realistic boundary condition as it involves a lowest order momentum balance along the open boundary; in practice it has proved very effective. The bathystrophic tide profile would only be used along those open boundaries which encounter land at some point and it would be used in conjunction with the pressure surge condition (a) in deep water.

Basic flow patterns and peak surge phase and amplitude remained unchanged throughout the simulation and resulting water surface contour patterns were well behaved and appeared the most plausible of the four results. Figure 2c, showing the resultant surge profile along the coast 75 minutes after cyclone landfall typifies the result for condition (d). It can be comapred to the corresponding profile for the asymptotic flow condition in Figure 2b, where water levels to the right of the eye seem unnaturally high.

4. RESOLUTION OF GRID SPACING

In the previous section the inclusion of the large scale forcing of a tropical cyclone in a computational field of smaller dimensions was discussed. It remains advantageous, however, to model as large an area as is physically possible, subject of course to computer storage limitations. A tropical cyclone has a definite spatial wind and pressure structure, and this information can only be delivered to the model at the discrete grid points within the field. If then the grid spacing is too large, possible loss in resolution of the true cyclone behaviour and consequent wave deformation may result. The adopted grid spacing also directly determines the computer storage requirements and the maximum incremental time step (explicit scheme) for a particular situation. The large time step, coupled with a small number of grid points, results in substantial advantages in computation time for a 10 n mile grid over a 5 n mile grid, so that it would be preferable to use as large a grid spacing as is hydrodynamically justified.

These problems are directly related to the ratio of the fundamental length scale of the cyclone, the radius of maximum winds R, to the grid spacing of the finite difference field, Δs. The effect of this R/Δs ratio on the computed wave was evaluated using grid spacings of 10 n mile, 5 n mile and 3.3 n mile respectively; the models were identical in all other respects, in particular R was 19 n mile. The pressure surge boundary condition was used in all three cases.

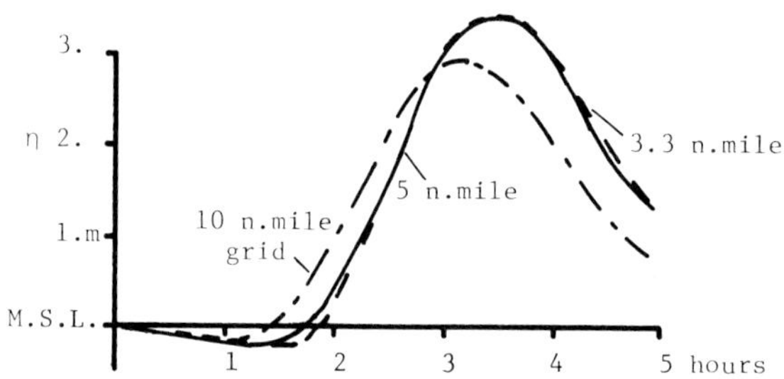

FIGURE 3. TYPICAL GRID RESOLUTION RESULT

Figure 3 is a typical result, and shows the water level histories at the position of the peak surge. It appears that decreasing the R/Δs ratio results in a lowering of the peak surge level, with an associated phase advance of the computed wave. Because of the large variations in peak levels and phase away from the other two cases, the 10 n mile grid was considered unsatisfactory for the prediction of the peak surge at the coast. However the results show that it is capable of adequately modelling the pressure surge in deep water, which may be useful in certain cases. Additionally there appears little advantage in using the smaller 3.3 n mile grid in place of the 5 n mile grid, especially when interest centres on the phase and amplitude of the peak surge. However, if finer resolution of coastal detail is required, then a smaller grid must be used.

5. INTERACTION OF TERMS

A useful and instructive part of a comprehensive sensitivity analysis is the observation of the relative importance of the terms of the conservation equations at particular locations within the computational field. The largest terms in the momentum equations, as expected, are the temporal accelerations, the water surface gradients, and the surface wind stress. The atmospheric pressure gradient can also be a major term, particularly in deep water, with bed shear, Coriolis, and convective terms being of lesser importance. However, at certain locations these latter terms can become significant.

Figure 4 shows the interactions which occurred between the dependent variables and the terms of the y momentum equation during the course of a normal simulation. The location considered is in shallow water, one grid point off the coast and in the path of the eye of the cyclone. The open boundary condition was condition (d), the local bathystrophic tide.

At this particular location the passage of the cyclone can be observed in the behaviour of the forcing terms, in particular the fluctuation in the surface pressure gradient and the reversal of the wind stress term. The water surface gradient term is seen to steadily increase in magnitude, resulting in the characteristic set-up to the south of the eye and set-down to the north. Of particular interest here is the increasing influence of the bed friction term in this shallow water location, and the importance of the Coriolis and convective accelerations especially as the eye passes.

An examination of the interactions between the dependent variables shows that in the initial stages flow was basically north-east and away from the coast, driven by the strong circular wind field about the cyclone. This resulted in a slight drop of the water surface below mean level. Approximately one hour before the eye passed, however, the x component of flow (U) was directed towards the coast and the surge wave (η) build-up was initiated. An important point is that the peak surge lags the passing of the eye by some twenty minutes, followed by a strong return flow away from the coast in the latter stages. Mechanical energy (E) throughout the simulation can be seen to closely follow the behaviour of the water surface elevation.

6. CONCLUSIONS

A numerical hydrodynamic model that adequately describes the generation and propagation of tropical cyclone storm surge has been developed. Variations in cyclone intensity (deepening and filling) and in speed and direction of the eye can be varied continuously (after each time step if necessary).

A detailed sensitivity analysis has resolved the open boundary condition problem, demonstrating the suitability of the bathystrophic tide condition in representing the forcing influence of the tropical cyclone beyond the boundaries of the computational field. Realistic flow patterns can now be obtained in the vicinity of the open boundaries.

Additionally, the sensitivity analysis has shown the importance of the scale relationship between the model grid spacing and the tropical cyclone dimensions. It appears that an $R/\Delta s$ ratio of at least 4 is necessary to adequately represent the meteorological forcing.

7. ACKNOWLEDGEMENTS

This study was conducted as part of a sponsored research project on tropical cyclone storm surge generation and propagation along the Queensland coast. Financial support from the Beach Protection Authority, Brisbane, and the

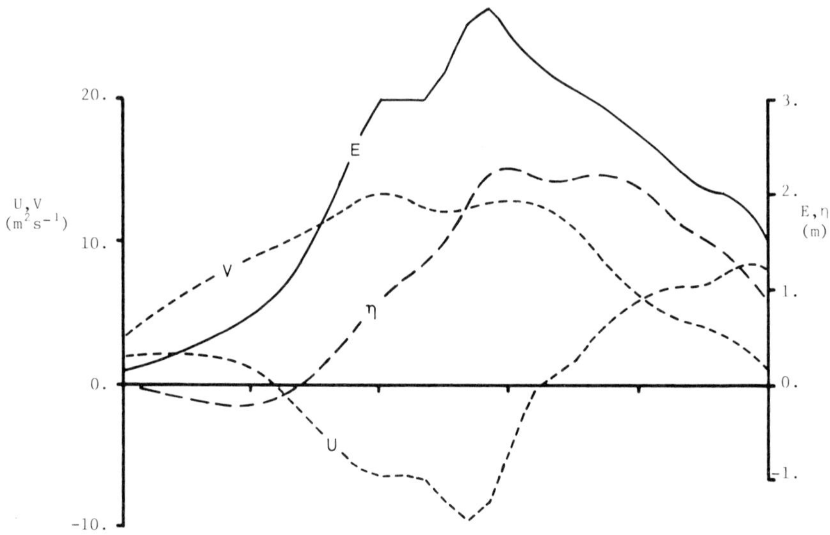

(a) Dependent Variables and Mechanical Energy

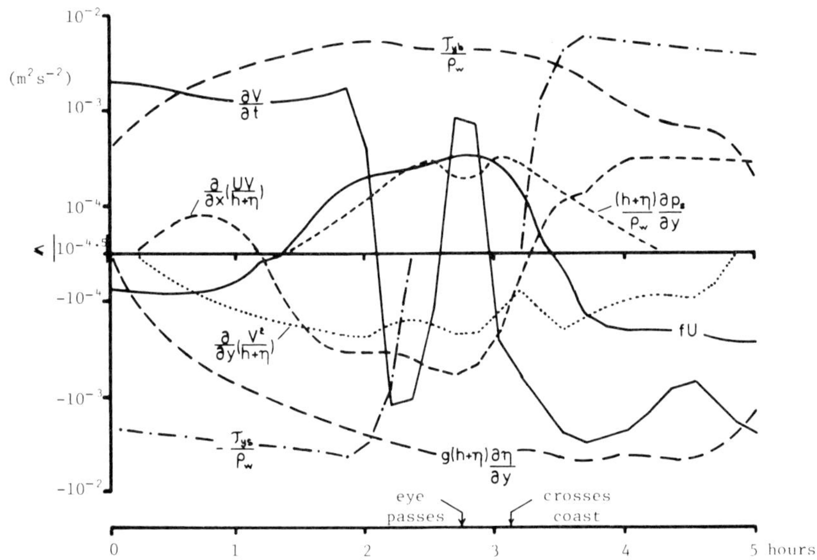

(b) Terms of y-Momentum Equation

FIGURE 4. SHALLOW WATER MOMENTUM BALANCE DURING PASSAGE OF CYCLONE

co-operation of the James Cook University Computer Centre are gratefully acknowledged.

8. REFERENCES

GRAHAM, H.E. and NUNN, D.E. (1959) - "Meteorological Considerations Pertinent to Standard Project Hurricane, Atlantic and Gulf Coasts of the United States", National Hurricane Research Report No.33, U.S. Weather Bureau.

HEAPS, N.S. (1969) - "A Two-Dimensional Numerical Sea Model", Proc. Royal Society, London, A265, pp. 93-137.

JELESNIANSKI, C.P. (1965) - "A Numerical Calculation of Storm Tides Induced by a Tropical Storm Impinging on a Continental Shelf", Monthly Weather Review, 93, pp. 343-358.

LEENDERTSE, J.J. (1967) - "Aspects of a Computational Model for Long-Period Water-Wave Propagation", Rand Corporation, RM-5294-PR.

PLATZMAN, G.W. (1956) - "A Numerical Computation of the Surge of 26 June 1954 on Lake Michigan", Geophysica, 6, pp. 407-438.

SOBEY, R.J. (1970) - "Finite-Difference Schemes Compared for Wave Deformation Characteristics in Mathematical Modelling of Two-Dimensional Long-Wave Propagation", U.S. Army Corps of Engineers, Coastal Engineering Research Center, Tech. Memo. No. 32, October.

SOBEY, R.J., HARPER, B.A. and STARK, K.P. (1975) - "Progress Report on Simulation Model for Cyclonic Storm Surges on the Queensland Coast", Department of Engineering, James Cook University, July.

SOBEY, R.J. (1976) - "The Generation and Propagation of Cyclonic Storm Surges", Annual Engineering Conference, Institution of Engineers, Australia, Townsville, May, pp.284-289.

U.S. WEATHER BUREAU (1968) - "Meteorological Characteristics of the Probable Maximum Hurricane, Atlantic and Gulf Coast of the United States", Hurricane Research Interim Report HUR 7-97.

WELANDER, P. (1964) "Numerical Prediction of Storm Surges", Advances in Geophysics, 8, pp. 315-379.

Numerical Simulation of Fluid Motion, ed. John Noye
pp. 383-405

TIDAL AND THERMAL PROPAGATION IN A TIDAL ESTUARY

M.D. Teubner and B.J. Noye

University of Adelaide

South Australia

SUMMARY

The problem of tidal and thermal propagation in the waters within a network of shallow tidal inlets is described, with specific reference to the Port River Estuary in South Australia. A numerical model is developed from a set of one-dimensional equations using explicit finite differences to approximate the tidal flow equations and implicit finite differences to model the heat flow equation. The model is designed to include a number of interconnecting channels with appropriate boundary conditions to determine transports, elevations and temperatures along each of the channels.

The numerical model is applied to the Port River Estuary, where the Torrens Island Power Station pumps heated water into one of the channels within the estuary. Tidal elevations obtained from the model compare favourably with data collected at various points in the estuary. The temperature results suggest that recirculation of heated water from the outflow region to the point where water is drawn into the Power Station does occur at most stages of the tidal cycle at spring tide which must lower the efficiency of the Power Station.

1. THE MATHEMATICAL MODEL

One-dimensional mathematical models of channel flow have a number of attributes which make their use worthwhile. They are relatively simple, both mathematically and physically, which makes them reasonably easy to work with; and yet they can throw a great deal of light on important aspects of the flow being modelled. If necessary, more complicated two- and three-dimensional models can then be used to look at special features in fine detail.

In this section a numerical model is developed from the one-dimensional tidal and thermal flow equations for channels of different types of cross-section.

1.1 The Governing Equations

The one-dimensional equations describing tidal and thermal propagation in a narrow, shallow channel have been derived by Teubner (1976). These governing equations are:

The equation of continuity:

$$b\frac{\partial\zeta}{\partial t}+\frac{\partial Q}{\partial x}=Q_S\ ; \qquad (1.1.1)$$

The momentum equation:

$$\frac{\partial Q}{\partial t}+\frac{\partial Q^2/A}{\partial x}+gA\frac{\partial\zeta}{\partial x}+g\frac{Q|Q|}{AC^2R_h}=Q_S(U_{in}-U_f)\ ; \qquad (1.1.2)$$

The temperature equation:

$$\frac{\partial AT}{\partial t}+\frac{\partial QT}{\partial x}=\frac{\partial}{\partial x}(AE_L\frac{\partial T}{\partial x})-\frac{AK_e T}{\rho ch}+AH_S\ , \qquad (1.1.3)$$

where x,t are the longitudinal and time coordinates respectively,

$b(x,t)$ is the width of the channel,
$\zeta(x,t)$ is the water elevation relative to the mean sea level,
$Q(x,t)$ is the volumetric transport,
$Q_S(x,t)$ is the lateral volumetric inflow per unit length,
$A(x,t)$ is the cross-sectional area of fluid flow,
$C(x,t)$ is the Chezy coefficient of friction, defined to be

$$C(x,t)=\frac{1.003}{n(x)}(R_h(x,t))^{\frac{1}{6}}\ [m^{\frac{1}{6}}\ sec^{-1}]$$

(see Harleman and Lee, 1969) where $n(x)$ is Manning's roughness coefficient,
$R_h(x,t)$ is the hydraulic radius, defined by

$$R_h(x,t)=A(x,t)/P_W(x,t),$$

where $P_W(x,t)$ is the wetted perimeter,

$U_{in}(x,t)$ is the speed of the inflow,
$U_f(x,t)$ is the speed of the water in the vicinity of the inflow,
$T(x,t)$ is the increase in water temperature above ambient,
$E_L(x,t)$ is the longitudinal dispersion coefficient,
$h(x,t)$ is the mean depth of water across the channel,
$H_S(x,t)$ is the heat sink/source term,
g is the acceleration due to gravity,

K_e is the heat exchange coefficient,
ρ is the density of sea water,
c is the specific heat of water.

The last four parameters are assumed to be constant. The wind stress term has been omitted from the momentum equation under the assumption that wind effects are not important in the regions to be considered.

1.2 Boundary Conditions

To solve the tidal equations within a network of shallow inlets the following boundary conditions are used:

(a) For a channel closed at both ends, the transports $Q(0,t)$ and $Q(L,t)$ are assigned, where $x = 0$ and $x = L$ are the two ends of the channel.

(b) For a channel closed only at one end, say $x = 0$, the transport $Q(0,t)$ and the elevation $\zeta(L,t)$ are assigned.

(c) For a channel open at both ends, the elevations $\zeta(0,t)$ and $\zeta(L,t)$ are assigned.

Where $Q(x,t)$ is assigned, it is given either the value zero, when there is no transport through the closed end, or some value $Q_{inf}(t)$, where $Q_{inf}(t)$ is some function representing a known inflow into the channel. Where $\zeta(x,t)$ is assigned, it is given the value $\zeta_0(t)$, representing either a periodic tidal input to the channel or some known elevation determined in a connecting channel.

For the temperature equation, the boundary conditions are:

(a) At the head of each channel $(x = 0)$, $T(0,t)$ is assigned the value $T_0(t)$, representing a time-dependent input to the channel either from an external source or from a connecting channel.

(b) (i) For a closed-ended channel $(x = L)$,

$$(QT)_{x=L} = 0 \quad \text{and} \quad \left(E_L \frac{\partial T}{\partial x}\right)_{x=L} = 0 ,$$

the first condition arising from the fact that $Q(L,t)$ is zero at a closed end, and the second condition following the assumption that no heat is lost to the lateral boundaries.

(ii) For an open-ended channel,

$$\left(\frac{\partial QT}{\partial x}\right)_{x=L} = 0 ,$$

assuming that the water temperature approaches ambient beyond the point $x = L$, and

$$\left(E_L \frac{\partial T}{\partial x}\right)_{x=L} = 0 .$$

1.3 Solutions to the Equations

Because of the non-linear nature of Equations (1.1.1) - (1.1.3), analytic solutions cannot be found except in the simplest of cases, for instance when the equations are applied to a rectangular channel and all non-linear terms are ignored or simplified. For this reason, finite difference techniques have been used to solve the equations.

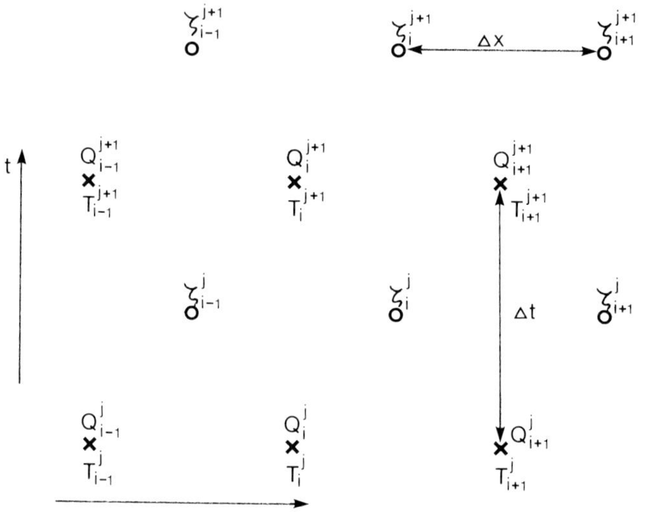

Figure 1.1 The finite difference grid.

1.3.1. The Finite Difference Grid

Figure 1.1 shows the finite difference grid in (x,t) space used to solve the general equations. Volumetric transports and temperatures are calculated at points denoted by $\times$ and tidal elevations at points denoted by O. The grid system is staggered in both time and space, with the time derivatives being calculated in the vertical direction, the space derivatives in the horizontal direction. The finite difference scheme is based on the principle that, knowing the values of the dependent variables at the jth time level, the values at the $(j+1)^{th}$ level can be calculated.

Using the grid scheme shown in Figure 1.1, the general one-dimensional equations can be approximated by the following finite difference equations

$$b_i^j \left(\frac{\zeta_i^{j+1}-\zeta_i^j}{\Delta t}\right) + \frac{Q_{i+1}^j - Q_i^j}{\Delta x} - (Q_s)_i^j = 0 \qquad (1.3.1)$$

$$\frac{Q_i^{j+1}-Q_i^j}{\Delta t} + \frac{D_{i+1}^j - D_{i-1}^j}{2\Delta x} + g\, A_i^{j+1}\left(\frac{\zeta_i^{j+1}-\zeta_{i-1}^{j+1}}{\Delta x}\right) + g\,\frac{Q_i^j\,|Q_i^j|}{A_i^{j+1}F_i^j}$$

$$= (Q_S)_i^j \left[(U_{in})_i^j - (U_f)_i^j\right] \qquad (1.3.2)$$

$$\frac{(AT)_i^{j+1}-(AT)_i^j}{\Delta t} + \frac{(QT)_{i+1}^{j+1}+(QT)_{i+1}^j-((QT)_{i-1}^{j+1}+(QT)_{i-1}^j)}{4\Delta x}$$

$$= \frac{(AE_L)_i^{j+1}}{2(\Delta x)^2}\left[T_{i+1}^{j+1} + T_{i+1}^j - (T_i^{j+1} + T_i^j)\right]$$

$$- \frac{(AE_L)_{i-1}^{j+1}}{2(\Delta x)^2}\left[T_i^{j+1} + T_i^j - (T_{i-1}^{j+1} + T_{i-1}^j)\right] - \frac{\kappa}{2}\left\{\left(\frac{A}{h}T\right)_i^{j+1} + \left(\frac{A}{h}T\right)_i^j\right\}$$

$$- (AH_S)_i^j \;, \qquad (1.3.3)$$

where Δx and Δt are the grid spacings for the horizontal and time coordinates respectively, ϕ_i^j represents the value of each parameter $(\zeta, Q, T, \text{etc.})$, evaluated at the grid point (i,j), and

$$D_i^j = \left(\frac{Q^2}{A}\right)_i^j \;,\; F_i^j = (C^2R_h)_i^j \;,\; \kappa = \frac{K_e}{\rho c} \;.$$

The finite difference scheme used in approximating the temperature equation (1.3.3) has been adapted from the system first used by Crank and Nicolson (1947) in which values of each parameter are calculated as means of their values at time levels j and $j+1$.

Equations (1.3.1) - (1.3.3) can be rearranged so that all the unknown terms appear on the left-hand side of each equation as follows:

$$\zeta_i^{j+1} = \zeta_i^j - \frac{\Delta t}{b_i^j\,\Delta x}\left[Q_{i+1}^j - Q_i^j - \Delta x\,(Q_S)_i^j\right] \;, \qquad (1.3.4)$$

$$Q_i^{j+1} = Q_i^j - \Delta t\left\{\frac{D_{i+1}^j - D_{i-1}^j}{2\Delta x} + g\,A_i^{j+1}\left(\frac{\zeta_i^{j+1}-\zeta_{i-1}^{j+1}}{\Delta x}\right) + g\,\frac{Q_i^j\,|Q_i^j|}{A_i^{j+1}F_i^j}\right.$$

$$\left. - (Q_S)_i^j\left[(U_{in})_i^j - (U_f)_i^j\right]\right\} \;, \qquad (1.3.5)$$

$$\left[\frac{Q_{i+1}^{j+1}}{4\Delta x} - \frac{(AE_L)_i^{j+1}}{2(\Delta x)^2}\right] T_{i+1}^{j+1} + \left[A_i^{j+1}\left(\frac{1}{\Delta t} + \frac{K}{2h_i^{j+1}}\right) + \frac{(AE_L)_i^{j+1} + (AE_L)_{i-1}^{j+1}}{2(\Delta x)^2}\right] T_i^{j+1}$$

$$- \left[\frac{Q_{i-1}^{j+1}}{4\Delta x} + \frac{(AE_L)_{i-1}^{j+1}}{2(\Delta x)^2}\right] T_{i-1}^{j+1} = \left[\frac{(AE_L)_i^{j+1}}{2(\Delta x)^2} - \frac{Q_{i+1}^j}{4\Delta x}\right] T_{i+1}^j$$

$$+ \left[A_i^j\left(\frac{1}{\Delta t} - \frac{K}{2h_i^j}\right) - \frac{(AE_L)_i^{j+1} + (AE_L)_{i-1}^{j+1}}{2(\Delta x)^2}\right] T_i^j + \left[\frac{Q_{i-1}^j}{4\Delta x} + \frac{(AE_L)_{i-1}^{j+1}}{2(\Delta x)^2}\right] T_{i-1}^j$$

$$+ (AH_S)_i^j \tag{1.3.6}$$

It will be noticed that the terms ζ_i^{j+1} and A_i^{j+1} are included in the right hand side of Equation (1.3.5). It is assumed that Equation (1.3.4) is solved before Equation 1.3.5., so that ζ_i^{j+1}, and hence A_i^{j+1}, are known for all i. This is also true in Equation (1.3.6) for Q_i^{j+1} and $(AE_L)_i^{j+1}$.

Equations (1.3.4) and (1.3.5) explicitly define one unknown, ζ_i^{j+1} and Q_i^{j+1} respectively, so that the values of ζ and Q at each x point on the grid can be determined at time level $(j+1)$ by substituting the known values at a lower time level into the right hand sides of these equations. Equation (1.3.6), however, involves three unknowns, T_{i-1}^{j+1}, T_i^{j+1} and T_{i+1}^{j+1}. To enable this equation to be solved, the tridiagonal matrix inversion algorithm described by Carnahan, Luther and Wilkes (1969) is used.

1.3.2 Initial and Boundary Conditions

To begin the computational procedure, values of ζ, Q and T need to be defined for the time level $j = 1$. These initial conditions are:

$$\begin{aligned} \zeta_i^1 &= a_0 && \text{for all } i, \\ Q_i^1 &= 0 && \text{for all } i, \\ T_i^1 &= T_0 && \text{for } i = 1, \\ T_i^1 &= 0 && \text{for } i > 1, \end{aligned}$$

where a_0 is the amplitude of the tidal wave at the open end of the main channel at time $t = 0$ and T_0 is the initial temperature at the head of each channel.

The boundary conditions for the tidal equations follow from the general conditions mentioned in Section 1.2 for all types of channels. These are discussed further in Section 1.5 when side channels are considered in more detail.

Two boundary conditions are applied to the temperature equation. The first, namely that the value of the temperature is assigned at the head of each channel, is discussed in Section 1.5. The second set of conditions for the end of each channel can be incorporated directly into the finite difference equations for the x grid point $i = n$, where n is the last horizontal grid point along each channel. The finite difference equation for an open-ended channel then becomes:

$$\frac{(AT)_n^{j+1} - (AT)_n^j}{\Delta t} = - \frac{(AE_L)_{n-1}^{j+1}}{2(\Delta x)^2} \left[T_n^{j+1} + T_n^j - (T_{n-1}^{j+1} + T_{n-1}^j) \right]$$

$$- \frac{K}{2} \left\{ \left(\frac{A}{h}T\right)_n^{j+1} + \left(\frac{A}{h}T\right)_n^j \right\} + (AH_S)_n^j \quad . \qquad (1.3.7)$$

1.3.3 Stability

In the previous section, the complete set of finite difference equations for the numerical model have been developed. These equations need to be analysed to establish criteria which define the conditions under which the numerical schemes can be considered stable. A thorough examination of stability is impossible for the complete non-linear equations and can be attempted only for linearized equations. The criteria found for these equations can then be used as guidelines for the criteria required for the non-linear case.

The stability conditions required for the explicit finite difference approximations to the linearized tidal equations have been found by Teubner (1976) to be

$$\Delta t \leqslant \Delta x/\sqrt{gh} \quad . \qquad (1.3.8)$$

The finite difference approximations to the linearized temperature equation are unconditionally stable but for large values of $\Delta x/E_L$, they are not necessarily convergent (see Teubner, 1976). The conditions required for convergence of the temperature equation are found to be

$$u \; \Delta t \leqslant \Delta x \quad , \qquad (1.3.9)$$

$$E_L \; \Delta t \leqslant (\Delta x)^2 \quad , \qquad (1.3.10)$$

$$\Delta x \leqslant \frac{x}{3.25} \left(\frac{E_L}{ux}\right)^{0.453} \quad , \qquad (1.3.11)$$

where x in the last condition represents the extent of the tidal excursion, and u in Equation (1.3.9) is the maximum speed of the tidal stream.

1.4 Cross-sectional Schematizations

Three different cross-sectional schematizations are used in the numerical model to enable the various types of naturally occurring channels to be modelled as accurately as possible.

(i) Rectangular configuration.

This type of cross-section assumes that the width of the channel remains approximately constant with the rise and fall of the tide, and is represented diagramatically in Figure 1.2(a). The rectangular cross-section can be used to model deeper rivers and shipping channels in estuaries in which very little shoreline becomes exposed at low water. The depth at any time is defined by

$$h(x,t) = d(x) + \zeta(x,t) \qquad (1.4.1)$$

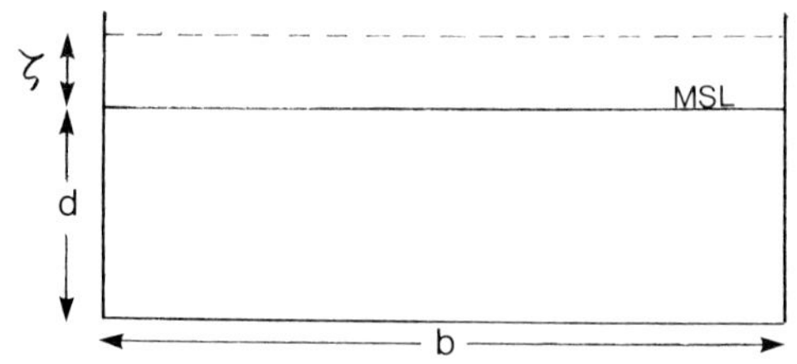

Figure 1.2(a) Rectangular configuration.

where $h(x,t)$ = average depth of water across the channel at time t,
$d(x)$ = average depth of water at mean sea level,
and $\zeta(x,t)$ = width-averaged elevation about mean sea level.

The hydraulic radius $R_h(x,t)$ is found to be

$$R_h(x,t) = \frac{b(x).h(x,t)}{b(x)+2h(x,t)} , \qquad (1.4.2)$$

where $b(x)$ is the width.

(ii) Triangular configuration.

In dealing with channels in which large areas of mud flats or beach become exposed as the tide falls, a triangular cross-section can be used to account for the variation in width across the channel. Figure 1.2(b) shows

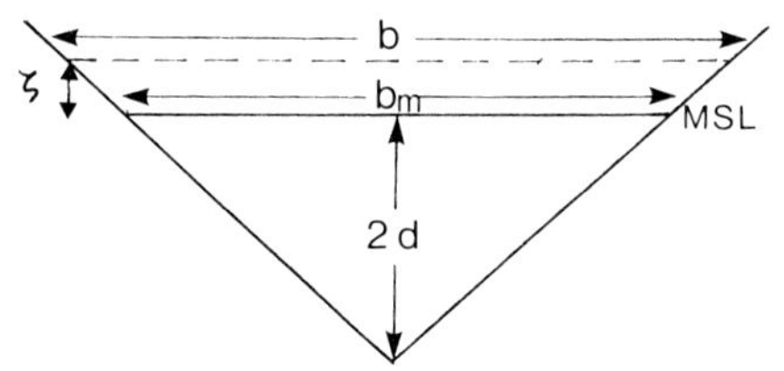

Figure 1.2(b) Triangular configuration.

this type of cross-section diagramatically. The relationships involved in this schematization are

$$h(x,t) = d(x) + \tfrac{1}{2}\zeta(x,t) , \qquad (1.4.3)$$

$$b(x,t) = b_m(x)\ (1 + \zeta(x,t)/2d(x)) , \qquad (1.4.4)$$

and

$$R_h(x,t) = \frac{h(x,t)}{[1+(4d(x)/b_m(x))^2]^{1/2}} , \qquad (1.4.5)$$

where $b_m(x)$ = width of the channel at mean sea level.

(iii) Storage configuration.

To account for a more general type of channel, a third type of cross-sectional schematization has been developed for use in the model. This configuration considers a core area in the middle of the channel, which transports most of the water, with sloping storage areas on either side of the main core area. This schematization may be used to model most natural estuaries, where a falling tide will often leave exposed mud banks on both sides of a main narrow channel. Figure 1.2(c) shows this type of channel diagramatically.

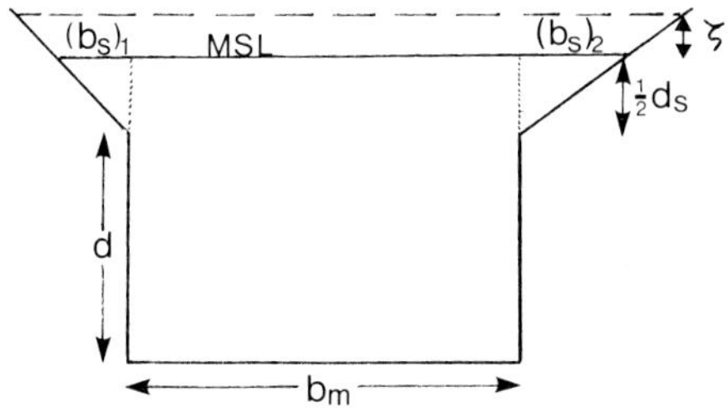

Figure 1.2(c) Storage configuration.

From the diagram, the following relationships are seen to hold:

$$h(x,t) = d(x) + \zeta(x,t) + \tfrac{1}{2}d_S(x), \qquad (1.4.6)$$

$$A_c(x,t) = b_m(x).h(x,t), \qquad (1.4.7)$$

$$b(x,t) = b_m(x) + [\zeta(x,t) + \tfrac{1}{2}d_S(x)][b_S(x)/d_S(x)], \qquad (1.4.8)$$

$$A(x,t) = A_c(x,t) + \tfrac{1}{2}[\zeta(x,t) + \tfrac{1}{2}d_S(x)]^2[b_S(x)/d_S(x)], \qquad (1.4.9)$$

$$R_h(x,t) = \frac{A_c(x,t)}{b_m(x)+2h(x,t)}, \qquad (1.4.10)$$

where $d_S(x)$ = tidal range,
$A_c(x,t)$ = cross-sectional area of core region,
$b_S(x) = (b_S)_1 + (b_S)_2$ = sum of widths of storage areas,
and $A(x,t)$ = total cross-sectional area.

A_c is the area associated with all terms involving the volumetric transport, as the horizontal velocity in the storage regions is assumed to be zero.

1.5 Schematization of the Estuary

The model has been developed to solve the difference equations along each channel of an interconnecting estuarial network as follows: the estuary is subdivided into a main channel, M, along which most of the water is assumed to be transported, and a series of side channels, $S_1,\ldots,S_N$ running off either M or each other. The finite difference approximations to the tidal and temperature equations are solved firstly for channel M by establishing grid points at equal spaces Δx_M along the channel. (The equal spacing is for convenience in solving the difference equations, but it is not necessary). The values of ζ and T determined at grid points along M are used as boundary values for the side channels. Figure 1.3 shows how this is done. The values of Q_i^m, ζ_i^m, T_i^m, $i = 1,\ldots,n$ are determined for the main channel at time $t = m\Delta t$ from the finite difference equations. Then the side channels are considered in turn with the already-calculated values of ζ_i^m and $(T_i^m+T_{i+1}^m)/2$ being used as inputs to channel S_j. Thus the following equations are used as boundary conditions for each S_j at the point A:

$$(\zeta_j)_1^m = \zeta_i^m, \qquad (1.5.1)$$

$$(T_j)_0^m = T_i^m + T_{i+1}^m - (T_j)_1^{m-1}, \qquad (1.5.2)$$

where $(\zeta_j)_1^m$ and $(T_j)_0^m$ represent the elevation and temperature at the start of channel S_j, $(T_j)_0^m$ being the point shown on Figure 1.3; its value is

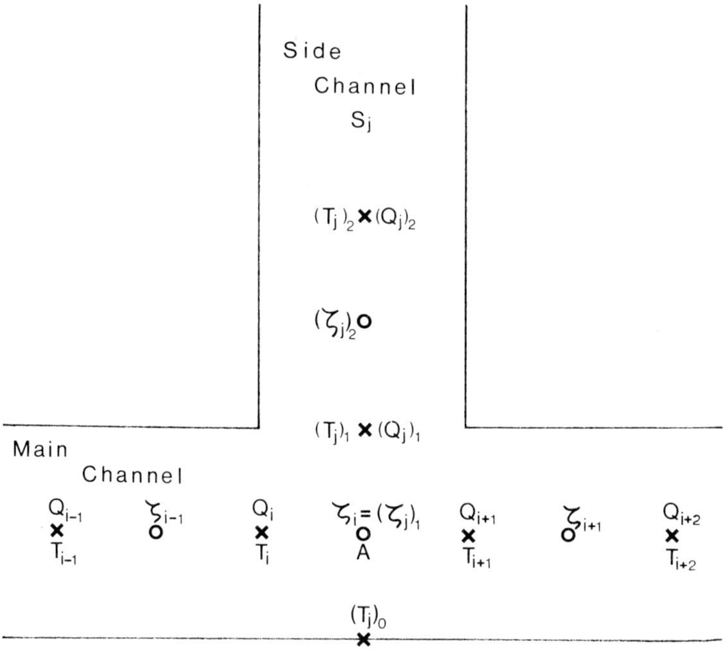

Figure 1.3 Side channel schematization.

determined from linear extrapolation of the temperature at A (assumed to be $(T_i^m + T_{i+1}^m)/2$) and $(T_j)_1^{m-1}$ along channel S_j.

Using boundary conditions (1.5.1) and (1.5.2), together with the appropriate conditions defined in Sections 1.2 and 1.3.2, the values of $(\zeta_j)_i^m, (Q_j)_i^m$ and $(T_j)_i^m$, $i = 1,\ldots,n_j$, can be calculated for each side channel S_j, where n_j is the number of horizontal grid points along S_j. Having calculated these values, the flows and temperatures along the side channels can now be considered as inflows and outflows to the main channel at time $t = (m+1)\Delta t$. The volumetric inflow per unit length into channel M at the point A is defined as

$$(Q_S)_i^{m+1} = (Q_j)_1^m/\Delta x_M \quad , \qquad (1.5.3)$$

and is included in the equation of continuity.

The mathematical formulation for the temperature inflow into channel M is determined by considering the convection across the mouth of channel S_j. At the temperature point T_i in the main channel, the convection can be approximated by

$$\frac{[Q_{i+1}\ T_{i+1} - (Q_j)_1 (T_j)_1] - Q_{i-1}\ T_{i-1}}{2\Delta x} \quad . \qquad (1.5.4a)$$

At the point T_{i+1}, the convection becomes

$$\frac{Q_{i+2}\ T_{i+2} - [Q_i T_i + (Q_j)_1 (T_j)_1]}{2\Delta x} \quad . \qquad (1.5.4b)$$

Since the heat source term, H_S, is on the right hand side of the temperature equation, while the convection term is on the left hand side, the temperature inflow can be determined from Equations (1.5.4a,b) to be

$$(AH_S)_i^{m+1} = (AH_S)_{i+1}^{m+1} = (Q_j)_1^m (T_j)_1^m / 2\Delta x_M \quad . \qquad (1.5.5)$$

Because the convection term in the temperature equation is not modelled explicitly as suggested by Equations (1.5.4a,b), the inflow terms $(AH_S)_i^{m+1}$ and $(AH_S)_{i+1}^{m+1}$ can be replaced by the mean of the inflows at the time levels m and $m+1$.

1.6 The Computer Program

The finite difference equations which form the basis of the numerical model have been solved on a CDC 6400 computer. The model runs for a certain number of time steps m_T, usually a multiple of the tidal period. In the first instance the tidal equations are solved for the complete estuary and required values of Q_i and ζ_i for each channel are stored on a permanent file. These values are then used to calculate temperatures T_i throughout the whole estuary .

Since the convergence conditions required by the temperature equation are far less restrictive than the stability criterion required by the tidal equations, a larger time step can be used for the temperature equation, requiring fewer values of Q_i and ζ_i to be written on the permanent file. However, the temperature equation requires a larger "model" time to converge, while the tidal equations converge very quickly. For instance, considering a tidal input of period T_p, the tidal equations are found to converge within $(\frac{1}{2})T_p$, but the temperature equation requires $3T_p$ or longer. Therefore, the tidal equations can be solved over a period of $(3/2)T_p$ with values of Q_i and ζ_i being stored on permanent file from time $(\frac{1}{2})T_p$ at every k^{th} time interval, where $k = \Delta t_T/\Delta t_H$, with Δt_T being the time step required for convergence for the temperature equation, and Δt_H for the tidal equations. These values are then read off the permanent file and used to solve the finite difference approximations to the temperature equation.

2. APPLICATION OF THE MODEL TO THE PORT RIVER ESTUARY

2.1 Geography of the Region

The numerical model developed in Section 1 has been applied to the Port River Estuary, shown in Figure 2.1. This diagram shows the main bodies of water within the Estuary, including the Port River, Barker's Inlet, North Arm and Angas Inlet.

2.1.1 Angas Inlet

Angas Inlet is a narrow, shallow channel which was originally connected to North Arm separating Torrens and Garden Islands. With the construction of the Torrens Island Power Station, a solid causeway was built across the south-western end of the Inlet to prevent heated water discharged from the Power Station from mixing with the intake water in the North Arm.

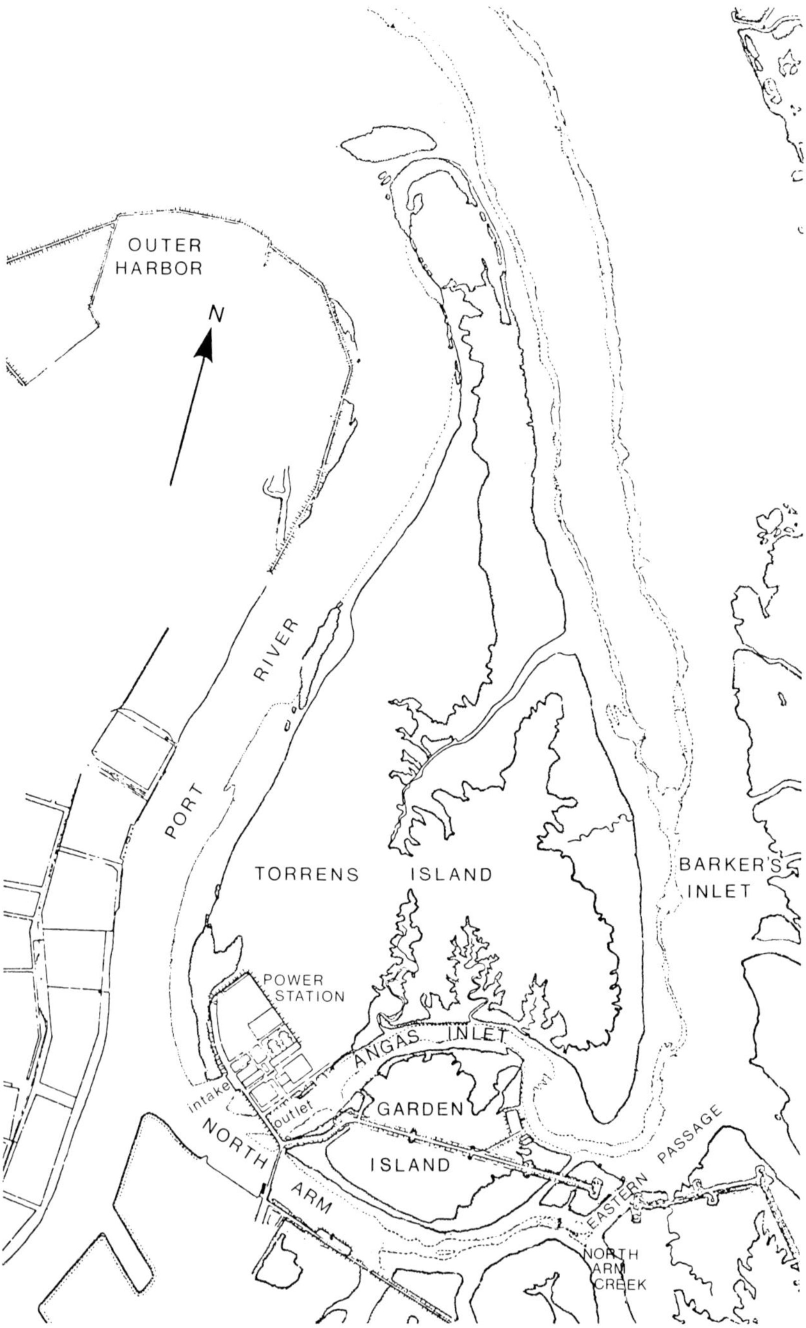

Figure 2.1. The Port River Estuary. Dotted lines represent low water mark.

The Inlet is approximately 2.5 km in length, with width-mean depths ranging from 4 m. at mean sea level (MSL) near the Power Station to 2 m. at the end of the Inlet, near Eastern Passage.

Figure 2.1 shows the rather large variations in width which occur near the Angas Inlet - Eastern Passage junction. At MSL the width of the Inlet 2.4 km from the causeway is 250 m. Considering a tidal range of 2.8 m., which is the maximum spring tide range, the width at low water is approximately 150 m. This large variation in width along the Inlet, as at other points throughout the estuary, has been carefully considered when applying the numerical models.

2.1.2 The Torrens Island Power Station

The Torrens Island Power Station is situated on the south-western corner of Torrens Island. It consists of four 120 Megawatt generating units. Each unit is served by two water pumps which extract water from the North Arm, west of the causeway, at the rate of 1,000 $m^3.hr^{-1}$ This water is passed through the Power Station to cool the condensers. 8,000 $m^3.hr^{-1}$ of heated water is thus discharged from the Power Station into Angas Inlet just beyond the causeway.

2.2 Schematization of the Estuary

To enable the numerical models to be applied to the Port River Estuary, a geometrical schematization of the whole area has been developed. Figure 2.2 shows diagramatically how this is achieved. The main channel is assumed to be made up of Angas Inlet, Eastern Passage, North Arm and the Lower Port River. The radius of curvature of this channel is large so that, to a first approximation, the channel can be extended in one direction. Because of the variations of width with tide height, the cross-section of the main channel is modelled using the storage configuration described in Section 1.4, where a core area of primary flow is assumed to transport most of the water, with variable-width storage areas on both sides to account for the covering or uncovering of tidal mudbanks.

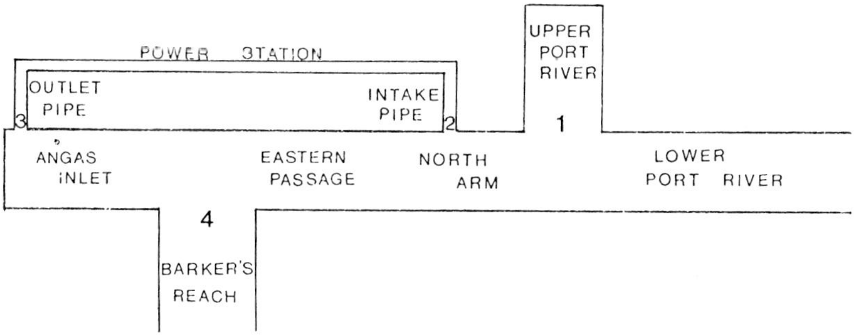

Figure 2.2. Schematization of the Port River Estuary.

Various other important features of the estuary are incorporated into the model:

1. The Upper Port River is assumed to be a closed-ended side channel running off the main channel. As the width of the river varies only slightly with time along this branch of the River, this side channel is modelled using the rectangular configuration, where width is assumed constant with time.

2. The intake pipe withdraws water from the North Arm at a constant rate Q_I m^3. sec^{-1}.

3. The outlet pipe pumps water from the Power Station into Angas Inlet at the same rate, Q_I m^3. sec^{-1}. Associated with this outflow is an increase in water temperature, T_I $^\circ C$, which becomes the boundary condition for the temperature equation.

4. Barker's Reach is assumed to be an open-ended side channel, with the tidal elevation defined at the seaward end, taken from available tidal data. The cross-sectional area is approximated by the triangular configuration as the Reach is very wide at high tide (almost 1 km. at some points) but reduces to less than 200 m. at low tide, with the flow being constant over the whole width.

Provision can also be made to include the channels running off the North Arm and Barker's Reach, but depth and breadth measurements are not at present available for these regions.

2.3 Inputs for the Numerical Model

The breadths and depths used in the numerical model have been taken from the Department of Marine and Harbors Sounding Plans. Figure 2.3 shows the depths at MSL used for the main channel. All measurements are averaged across the channel, and each point represents the average depth 200m. before and after the given point (to provide for a grid step of 400m.). Because of the difficulty in obtaining accurate measurements, the width-mean depth of water in Barker's Reach is assumed to be constant at MSL, taking a value of 1.4m. The width-mean depth of water in the Upper Port River varies linearly along the channel, being 6 m. at the open end and 7m. at the closed end at MSL.

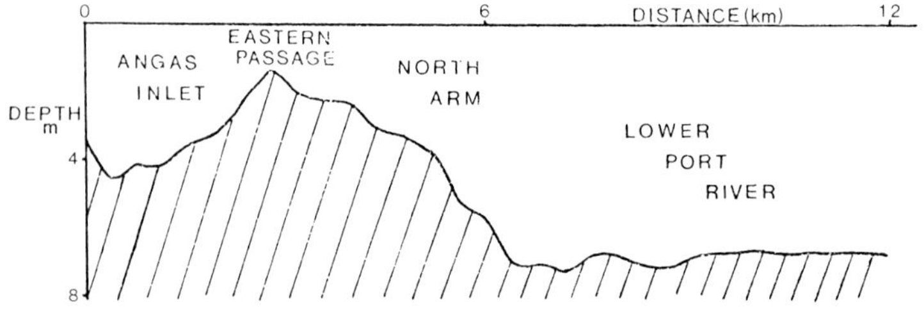

Figure 2.3. Width-averaged depths at mean sea level in the main channel.

TABLE 2.1

INPUT VALUES FOR THE NUMERICAL MODEL

Parameter	Main Channel	Barker's Reach	Port River
Δx (m.)	400	800	800
Time step for hydraulics Δt_H (secs.)	←	50	→
Time step for temperature Δt_T (secs.)	←	300	→
Number of grid points	31	10	7
Starting point (in relation to main channel)	1	8	17
E_L (m^2/sec.)	80	100	200
$K_e/\rho c$ (m/sec.)	←	9.256×10^{-6}	→
Manning's coefficient	←	0.03	→
Tidal range (m)	←	2.84	→
Tidal period (hrs.)	←	24.5	→

Table 2.1 defines further values used in the numerical model. These have been chosen to satisfy the stability and convergence conditions defined in Section 1.3.3. A value of 800m. for Δx for each of the side channels, compared with 400m. for the main channel, has been chosen because accurate values of T, ζ and Q are not required for the side channels. It is their effect on the main channel which is important and this is, to a certain extent, independent of Δx .

The number of grid points for each channel defines the number of horizontal space steps into which each channel is divided, while the starting points refer to the grid point in the main channel associated with the beginning of each side channel. Figure 2.4 shows the exact location of each grid point along each channel throughout the whole estuary. Values of depth and breadth defined at each elevation point, and calculated values of ζ, T and Q at elevation and temperature/transport points are assumed to be averaged over the region $(i-\frac{1}{2})\Delta x$ and $(i+\frac{1}{2})\Delta x$ for the grid point i .

The value of $K_e/\rho c$ in Table 2.1 has been derived using Harleman and Stolzenbach's (1972) value for K_e , and taking ρ and c, the density and specific heat of water, to be constants at 1,025 kgm. m^{-3} and 1,000 cal./kgm.$^{\circ}$C respectively.

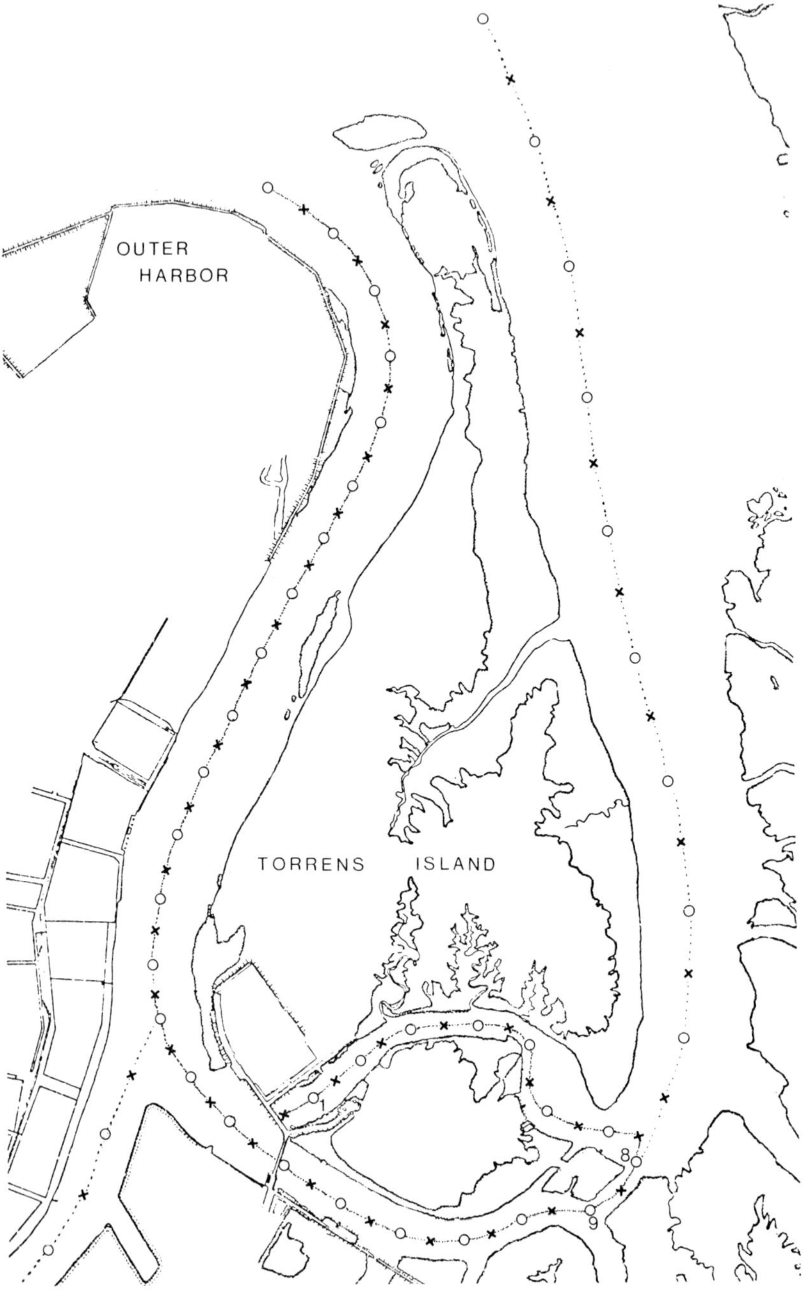

Figure 2.4. Geometric schematization of the Estuary, showing elevation points (o) and transport/temperature points (x).

Since no measurements were available to accurately determine the dispersion coefficient, E_L, for the Port River Estuary, and since its value is generally quite arbitrary, values ranging from 5 to 1,000 $m^2 sec^{-1}$ being given by authors such as Freeman (1970), Bansal (1971), Hansen and Rattray (1972), it was decided to make its value depend on the convergence condition (1.3.11), with x being taken as the tidal excursion defined by $u T_p/2\pi$, where T_p is the tidal period. This leads to the values for E_L presented in Table 2.1 for each channel.

The only other inputs to the numerical model involve boundary conditions at the closed end of the main channel, i.e. at the causeway. These are defined as:

$$Q_1 = 22.9 \ m^3 sec^{-1}$$

$$T_1 = (5.4 + T_k)^{\circ}C ,$$

where T_k is the increase in temperature above ambient at the intake pipe in North Arm. Both values have been supplied by the Electricity Trust of South Australia.

2.4 Comparison of Tidal Results.

2.4.1 Tidal Elevations

Tidal elevation measurements have been supplied by the Electricity Trust for the spring tide of 14 February 1974 at the following points:

(1) Outer Harbor, representing the open end of the main channel (data to be used as the elevation input for that channel).

(2) At the seaward end of Barker's Reach (to be used as the elevation input for the first side channel);

(3) The Electricity Trust jetty, 200m. beyond the causeway along Angas Inlet, the approximate position of the first elevation grid point (to be used comparatively);

(4) Pole 29 Black, near the Angas Inlet - Barker's Reach - Eastern Passage junction, the approximate position of elevation grid point 8 (also to be used comparatively); and

(5) North Arm Creek, very near the position where the North Arm Creek flows into Eastern Passage, close to elevation grid point 9 (to be used as a third comparison with the numerical model).

Each set of tidal data has been graphed for the complete tidal cycle of 24.5 hours: the inputs to the system from positions (1) and (2) are shown in Figure 2.5(a), the Barker's Reach record having been shifted by 30 minutes in time as the two records are almost identical for most of the tidal period, and the elevation records for positions (3), (4) and (5) are shown in Figures 2.5(b), (c) and (d) respectively.

Using the same tidal inputs taken from positions (1) and (2) above, the numerical model was run for 1¼ tidal cycles, i.e. for a period of $5T_p/4$ hours, where T_p is the tidal period of 24.5 hours. Experience has shown that, because of the small time step of 50 secs. used, a relative convergence of better than 99% seems to occur within ¼ of a tidal cycle (i.e. the difference in elevations at times $T_p/4$ and $5T_p/4$ at corresponding positions is less than 1%),

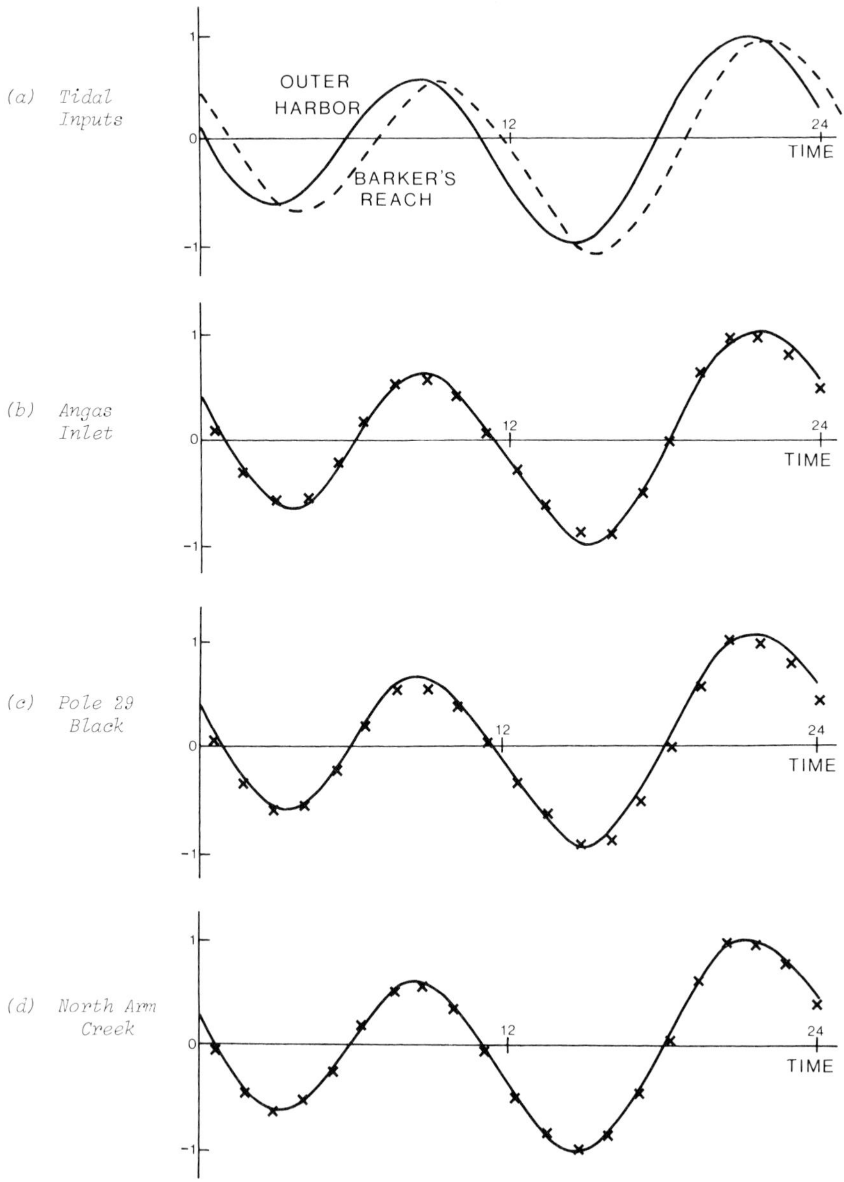

FIGURE 2.5 Inputs for the spring tide (Barker's Reach data shifted forward 30 mins.) and comparisons of outputs at (b) Angas Inlet, (c) Pole 29 Black and (d) North Arm Creek (**×** represents the numerical model and continuous line the measured data).

suggesting that all initial transients have been effectively removed by the time $T_p/4$.

The numerical results have been plotted against the measured data in Figures 2.5(b), (c) and (d). These plots show good agreement with the measured elevations. The discrepancies for the Power Station region (Figure 2.5(b)) are quite likely due to the method of approximating the breadths and depths throughout Angas Inlet, particularly around the junction area where exposed mud flats on the south-eastern corner of Torrens Island have been poorly represented in the model. Even so, for the greater part of the tidal cycle, the numerical model agrees remarkably well with the measured results, considering the one-dimensional nature of the determination of the elevation. Since no measurements are available for either the transports or temperatures for the estuary over a similar tidal cycle, this agreement of calculated values with measured elevations must form the basis for the verification of the numerical model until further data is available.

2.5 Volumetric Flow Charts

For convenience, the volumetric flows at any given time are diagramatically represented on the rectangular estuary model given in Figure 2.2. This has been reproduced in Figure 2.6(a) showing the exact location of each feature used in the model.

Figures 2.6(c) - (j) represent the volumetric flows within the Port River Estuary at various stages of the tidal cycle. These stages have been superimposed on the graph representing the tidal elevations at Outer Harbor in Figure 2.6(b). The relevant times for the volumetric flows are shown on the diagrams (c) - (j).

In most of these diagrams, stagnation points are seen to occur at some point along Barker's Reach, a phenomenon which is well known by users of the area. The apparent existence of stagnation points in Eastern Passage and the North Arm as well is due to the different flow forces acting within this region. The only unexpected stagnation point occurs in Figure 2.6(f), which represents the flow just after high tide at Outer Harbor (cf. Figure 2.6(b)), when the transports are very small, of the order of $5m^3.sec^{-1}$.

Another feature which is clearly shown in the charts is the positive nature of the flow along Angas Inlet up to 1 km. from the causeway at all stages of the tidal cycle (i.e. the flow is always away from the outlet pipe) resulting in stagnation points in Figures 2.6(e) and (i), caused by an incoming tide meeting the outflowing water from the Power Station. This feature has been borne out by qualitative observation within the Inlet itself.

2.6 Temperature Results

Temperature increases have been computed throughout the Port River Estuary for the spring tide of 14 February 1974 by using the calculated values of volumetric transports and elevations for the numerical model. The temperature models were run for a model time of $3T_p$, where T_p = 24.5 hours, temperatures being printed out for the last complete tidal cycle from $2T_p$ to $3T_p$. These results showed that most of the initial transients had been damped out after $2T_p$ hours, as variations of less than 5% occurred at corresponding times within the tidal cycle.

The solutions obtained from the model are presented in Figures 2.7(a),(b) and (c), which show the temperatures at three points along the main channel for a complete tidal cycle. The points shown are:

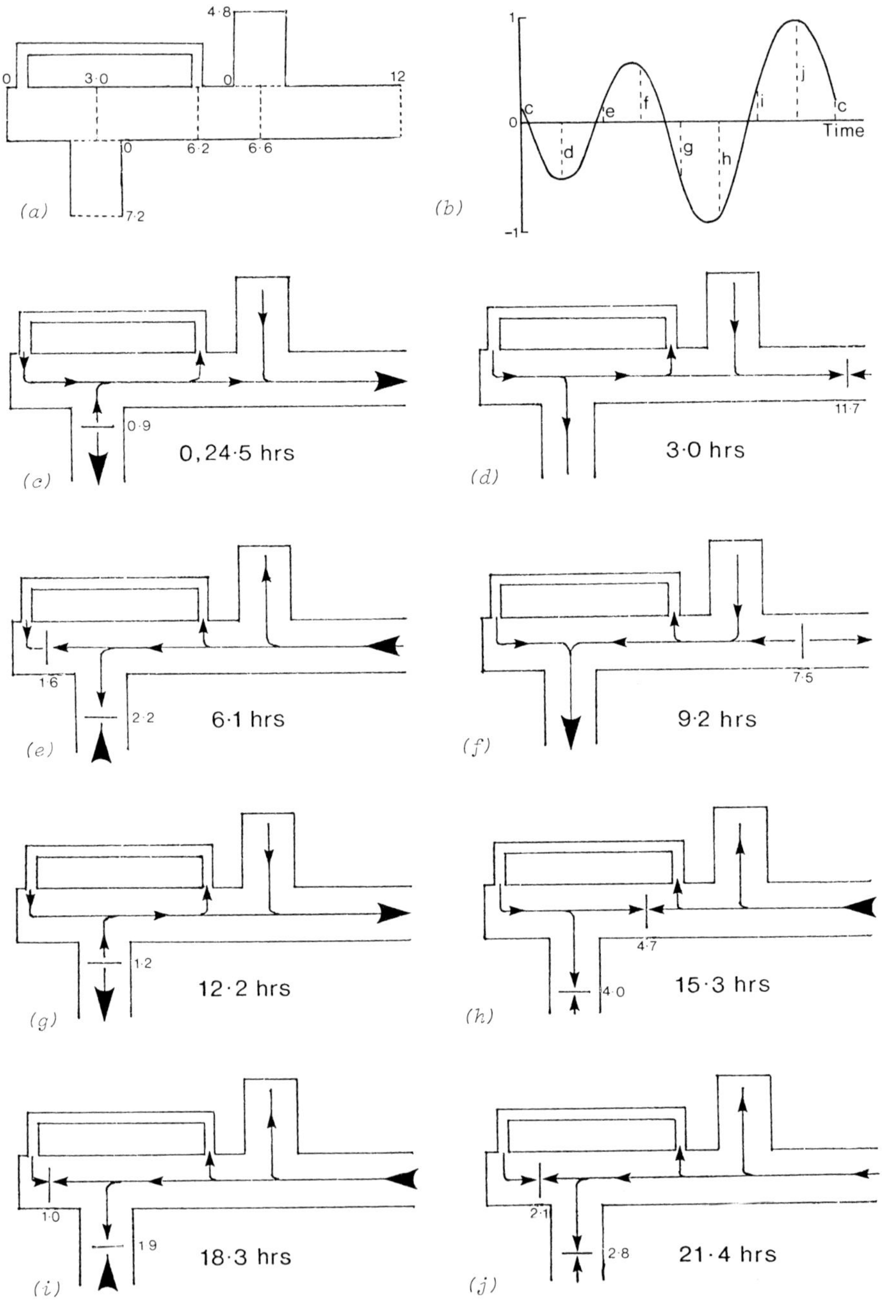

FIGURE 2.6 Volumetric flow charts. All distances are in kilometres and larger arrows indicate greater volumetric transport.

(a) T_1, the temperature increase at the first grid point, representing the outlet pipe in Angas Inlet;

(b) T_8, the temperature increase at the Angas Inlet - Barker's Reach junction;

(c) T_{15}, the temperature increase in the North Arm, approximately 200m. before the intake pipe.

The graphs show definite turning points, with maxima at approximately 3 and 15 hours, times of low water, and minima at 9 and 21 hours, when high tide occurs at Outer Harbor. The graph of T_8 has turning points which differ slightly from the other two graphs. This is undoubtedly due to the interaction of the two tidal systems at the Barker's Reach - Angas Inlet - Eastern Passage junction.

The T_1 graph gives an idea of the amount of recirculation occurring for this spring tide. With a temperature increase of 5.4°C through the Power Station, the temperature at the intake pipe is the difference between the T_1 graph and the line $T = 5.4^{\circ}$. These results indicate that recirculation occurs over the complete tidal cycle.

An interesting comparison can be made between Figures 2.7(b) and (c), which show the temperature increases throughout the whole estuary at the times 9.2 and 15.3 hours respectively. This interval of time covers a complete ebb tide, where flows are predominantly seawards along each channel (c.f. Figure 2.6(g)). Heat is carried from the outflow pipe by advection and dispersion, thus leading to large increases in temperature throughout the whole estuary.

The range of temperatures in Angas Inlet can be approximately determined from Figures 2.7(b) and (c), as (b) represents a tidal minimum and (c) a maximum. Figure 2.7(b) suggests that the minimum temperature increase at the Barker's Reach end of Angas Inlet, 2.8 km. from the outflow pipe, is 3°C, while from Figure 2.7(c) the maximum is 5.5°C. This is supported by Figure 2.7(a), which shows that this maximum increase occurs only over a short period of time, while increases of less than 3°C occur more frequently. These results do suggest however, that the temperature throughout the Inlet remains more than 3°C above ambient for the greater part of the tidal cycle.

3. CONCLUSION

The numerical model developed in Section 1 has been applied to the Port River Estuary in South Australia, and its accuracy verified by comparison with available tidal elevation data, which shows that the model is very accurate in its prediction of tidal elevations. As previously mentioned, no worthwhile temperature or transport data is at present available to verify the solutions obtained for these parameters. The computed temperature increases suggest that recirculation of heated water occurs during spring tide, and that this problem will undoubtedly be magnified if the Power Station is increased in size, unless some other form of heat disposal is used.

As well as having a detrimental effect on the Power Station's efficiency, an increase in the volumetric rate of water being discharged from an expanded Power Station on Torrens Island could also have a disastrous effect on the estuarine environment, particularly that of Angas Inlet.

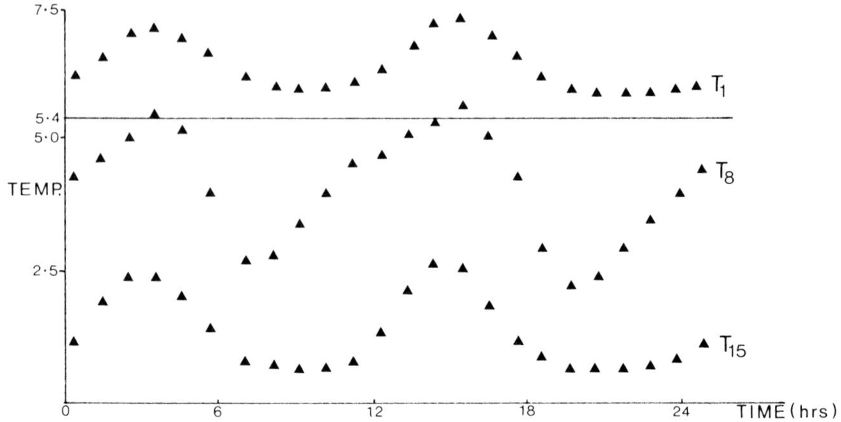

(a) Temperatures (°C) at the points T_1, T_8 *and* T_{15}.

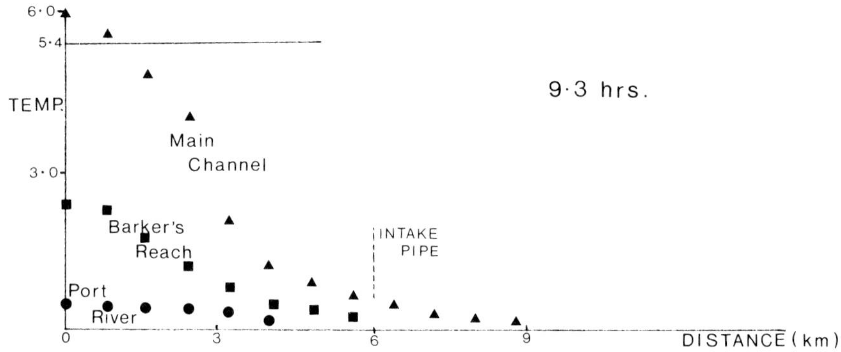

(b) Temperatures (°C) throughout the estuary at time 9.3 hours.

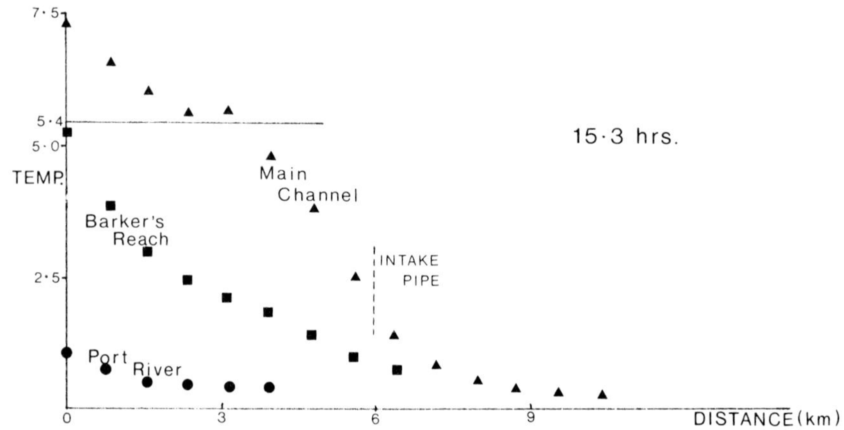

(c) Temperatures (°C) throughout the estuary at time 15.3 hours.

FIGURE 2.7 Computed temperatures for the numerical model.

REFERENCES

Bansal, M.K. (1971), "Dispersion in Natural Streams", J. Hyd. Div., Proc. Am. Soc. Civil Engrs., HY11.

Carnahan, B., Luther, B.H. and Wilkes, J. (1969), "Applied Numerical Methods", Wiley, New York.

Crank, J. and Nicolson, P. (1947),"A Practical Method for Numerical Evaluation of Solutions of Partial Differential Equations of the Heat-Conduction Type", Proc. Camb. Phil. Soc., 43(50).

Freeman, N.G.S. (1970), "The Application of Sigma Coordinates to the Numerical Modelling of Great Lakes Hydrodynamics", M.Sc. Thesis, University of Waterloo.

Hansen, D.V. and Rathay, M. (1972), "Estuarine Circulation Induced by Diffusion," J. Marine Res., Vol. 30 (3).

Harleman, D.R.F. and Lee, C.H. (1969), "The Computation of Tides and Currents in Estuaries and Canals," U.S. Army Corps of Engrs., Comm. on Tidal Hydraulics, Tec. Bull. No.16.

Harleman, D.R.F. and Stolzenbach, K.D. (1972), "Fluid Mechanics of Heat Disposal from Power Generation", Ann. Rev. of Fluid Mechanics, Vol.4.

Teubner, M.D. (1976), "Tidal and Thermal Propagation in the Port River Estuary", Ph.D. Thesis, University of Adelaide.

Numerical Simulation of Fluid Motion, ed. John Noye
pp. 407-434

A PRACTICAL APPLICATION OF A NETWORK MODEL

W.R.B. Morrison and P. Arnot Smith

Oceanics Australia Pty. Ltd.

Brisbane, Queensland

SUMMARY

This paper discusses a typical application of a network hydraulic model, including the information required to describe the model and the results which can be obtained.

1. INTRODUCTION

Many hydraulic problems involve two-dimensional flows such as flows over broad flood plains or flow in estuaries where there is complex channelisation. Flows of this kind can be adequately represented by network models, from which satisfactory results can be obtained at a modest cost.

This paper gives a brief description of one such hydraulic modelling system, and describes its application to a particular problem involving tidal hydraulics and flooding.

2. HYDRAULIC MODELLING SYSTEM

The model described here has been set up to simulate flood and tidal flows in streams, estuaries, lakes and bays. The body of water under consideration is separated into a series of nodal areas connected by equivalent channels.

The waterways may be braided and branched, and large bodies of water may be represented by a network of nodes and channels. Channels and storage areas may be specified to come into operation above certain water levels only, to allow representation of floodways and tidal flats.

Boundary conditions may be specified in a variety of ways. At any node an inflow may be specified either as a constant flow rate, as a flow rate related to stage height (e.g. as a restricted flow from a large reservoir), or as a flow hydrograph (inflow rate vs. time) either directly or by a unit hydrograph/hyetograph convolution. Alternatively for any node, the water level may be specified either as a combination of sunusoidal tides, as a stage hydrograph or surge (height vs. time), or as height as a function of total inflow rate (e.g. upstream of a weir).

Channel resistance is normally specified by Mannings "n", but may be described in terms of a resistance coefficient which can be a function of water depth. Bridges and weirs are modelled according to the procedures recommended by the U.S. Department of Transportation (1973).

The continuity equation for flow in a uniform channel may be written:

$$\frac{\partial}{\partial x}(UA) + B\frac{\partial H}{\partial t} = 0 ,$$

and the momentum equation:

$$\frac{\partial U}{\partial t} + \frac{\partial U}{\partial x} + g\frac{\partial H}{\partial x} + K|U|U = 0,$$

where
- U = average channel velocity (=Q/A)
- A = cross-sectional area of channel
- H = height of water surface above datum
- B = channel width $(= \frac{\partial A}{\partial H})$
- L = channel length
- g = acceleration due to gravity
- K = friction coefficient (= K("n",A,B))
- s = nodal surface area
- x = distance along channel
- t = time

At node j the continuity equation becomes

$$\frac{\partial H_j}{\partial t} = \frac{1}{s_j} \Sigma\, U_i A_i ,$$

where the "i" is summed over all the channels connected to node j.

For a constant-section channel the momentum equation may be written, for channel i connecting nodes j1 and j2:

$$\frac{\partial U_i}{\partial t} - \frac{1}{2} \frac{U_i B_i}{A_i} \frac{\partial (H_{j1} + H_{j2})}{\partial t} + \frac{g}{L_i} (H_{j2} - H_{j1}) + K_i |U_i| U_i = 0$$

These equations are strongly non-linear since A, B, K and s are all functions of H.

The momentum equation is solved for velocities at the end of a time step, using the properties calculated at the end of the previous step. The water levels at the nodes are then obtained from continuity, and the new channel properties calculated.

These equations are solved either by an explicit Runge-Kutta method or by an implicit matrix inversion procedure. In general the matrix method is more accurate while the explicit method is faster. Both procedures are available and the optimum method for a particular problem is found by trial.

The output consists of channel velocities, volume flows, integral volume flows and nodal heights, all at a series of specified times. In addition, the maximum and minimum values of the above quantities are given, along with their times of occurrence.

3. THE PROBLEM

On the "Sunshine Coast", a resort area approximately 100 km north of Brisbane, a group of companies has an interest in developing 1600 hectares of land, extending in an almost unbroken strip from Mount Coolum in the north to the Maroochy River in the south (Figure 1) and representing in all, about one half of the total area between the river and the coast.

Except for a small section of land near Mount Coolum and a strip adjacent to the coastal dunes, this entire region forms part of the flood plain of the Maroochy River, being all below an RL of 3 metres (above A.H.D.) and quite flat. In major floods, large areas are inundated.

One of the alternatives being considered by the Suncoast Group is to develop these properties as a system of canals or lakes and this investigation was commissioned to examine some of the hydraulic and hydrological consequencies of such action; particularly the effect of the development on flooding and on tidal hydraulics of the river.

The concern about flooding is that development on the flood plain would involve the raising of these properties to be above flood level, and since this inevitably means that some existing flood flow paths will be blocked and a certain amount of existing flood storage would be lost, there is clearly a potential problem in that flood levels may be increased.

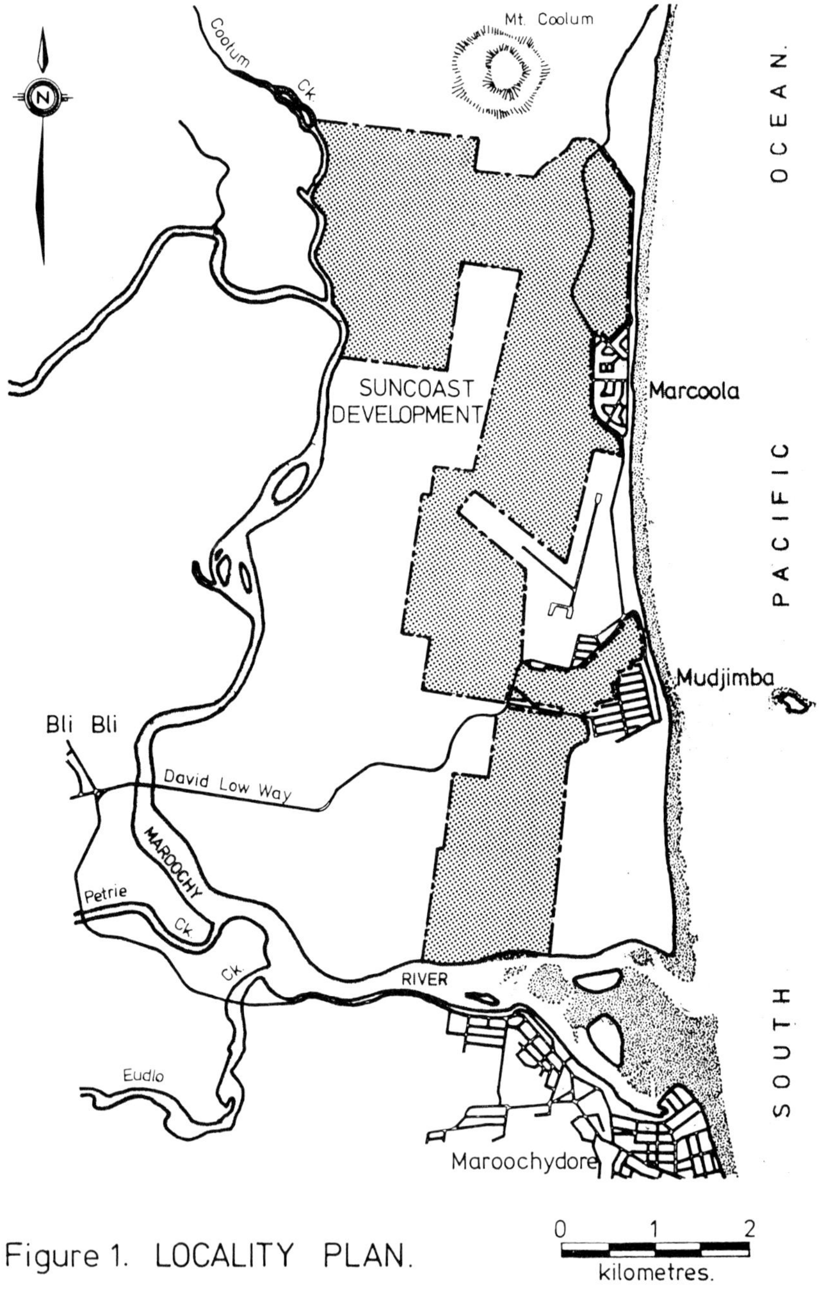

Figure 1. LOCALITY PLAN.

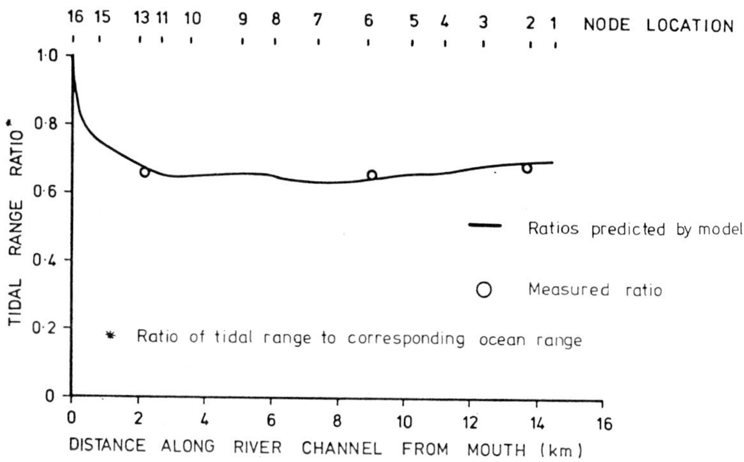

Figure 2. COMPARISON BETWEEN MEASURED AND PREDICTED TIDAL RANGE RATIOS

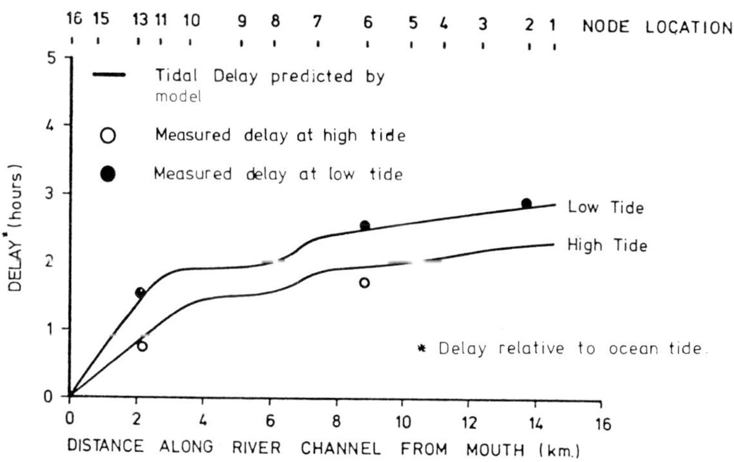

Figure 3. COMPARISON BETWEEN MEASURED AND PREDICTED TIDAL DELAY

In this study, several development alternatives were assessed with the express purpose of demonstrating not only that the flooding problem can be minimised but also that a degree of flood mitigation can be achieved by suitable arrangement of the canals.

The question of the effect of the development on the tidal hydraulics of the Maroochy River is particularly important because the waterway systems contemplated are large in comparison with the tidal compartment of the river itself. This investigation was designed to examine the effect of various alternative layouts on the tidal hydraulics of the combined canal-river system and it was hoped that this would provide guidelines to minimise potential problems by appropriate design of the system.

The investigation was pursued only far enough to demonstrate the general feasibility of the scheme, the primary objective being to find at least one arrangement of canals which would provide satisfactory performance. There was no attempt at refinement or optimisation, the layout recommended being just one workable solution, not necessarily the best possible design.

4. DEVELOPMENT ALTERNATIVES

The various development alternatives examined in this investigation were all based on one or other of the two basic configurations shown in Figures 4a and 4b, cases T2 and T3; the initial development proposal and an alternative scheme incorporating a flood bypass respectively.

The waterway system depicted in case T2 was the favoured scheme at the time this investigation was initiated.

The alternative scheme shown in case T3 was examined in this study since the general arrangement seemed to offer potential advantages, primarily from the point of view of flood mitigation but also based on considerations of the suitability of canal entrance locations, and of ecological factors.

The central feature of this scheme is a flood bypass canal with branch canals diverging from this central canal, arranged generally so that they follow existing major drainage paths. The exact geometry of these branch canals is not critical to the hydraulic performance of the whole system, the geometry shown being merely one possible layout that has about the required tidal compartment.

An essential feature of this scheme is the provision of a floodway which blocks tidal flow but allows a certain amount of flood flow through the bypass when levels exceed the base of the floodway.

This kind of system will clearly reduce flood levels upstream but may also increase levels near the downstream end. Whether or not a satisfactory result can be achieved may depend in this situation on the relative phasing of flood peaks from the Maroochy River, Petrie Creek and Eudlo Creek catchments.

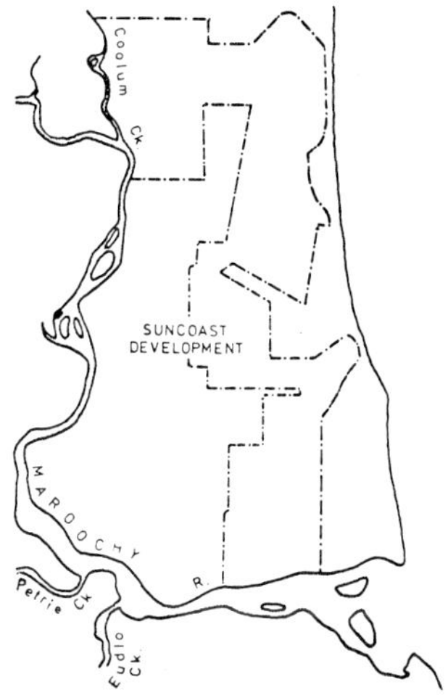

Case T1 : EXISTING SITUATION

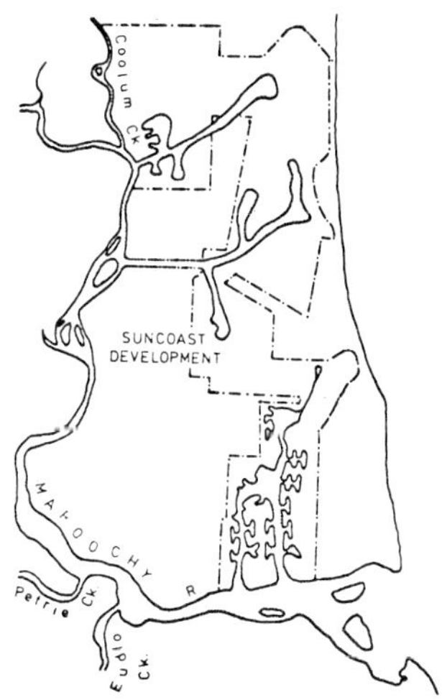

Case T2 : INITIAL DEVELOPMENT PROPOSAL

Figure 4a. CANAL SYSTEM ALTERNATIVES SIMULATED FOR TIDAL FLOW

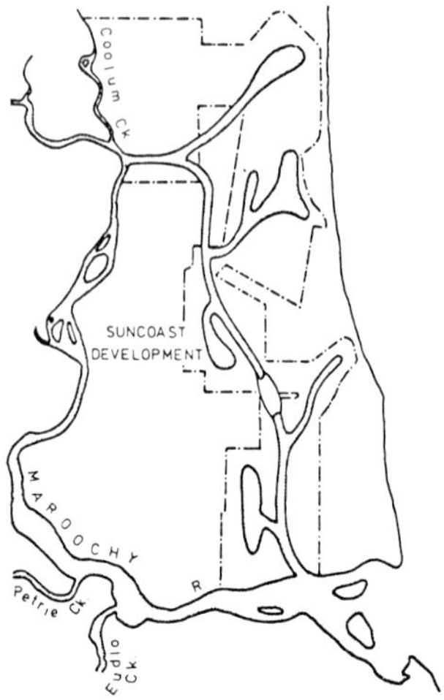

Case T3 : DUAL CANAL SYSTEMS SEPARATED BY A FLOODWAY

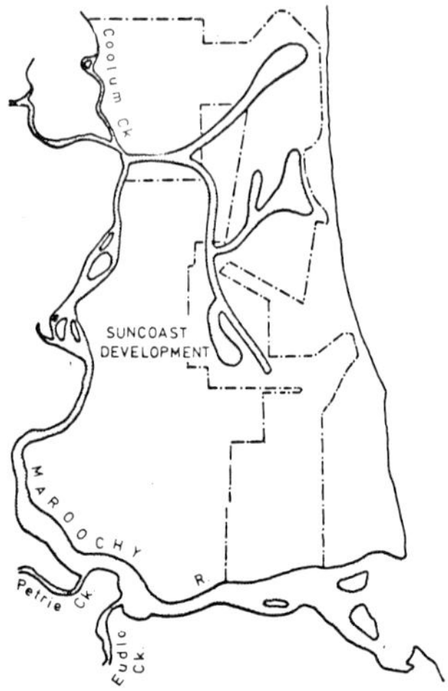

Case T4 : CANAL SYSTEM WITH CONNECTION NEAR COOLUM CREEK

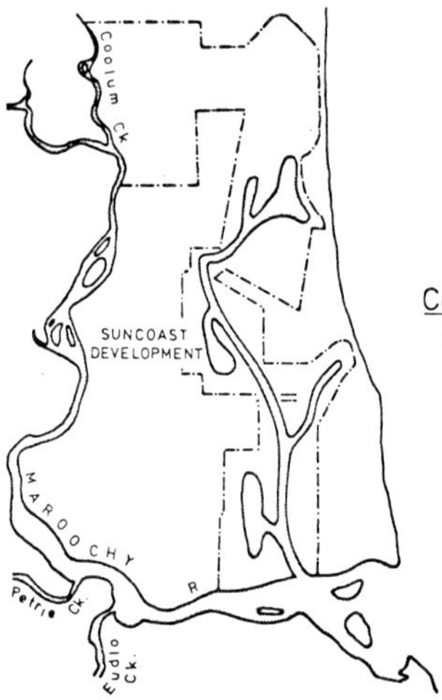

Case T5 : CANAL SYSTEM WITH CONNECTION NEAR MAROOCHY RIVER MOUTH

Fig.4b CANAL SYSTEM ALTERNATIVES SIMULATED FOR TIDAL FLOW

5. TIDAL HYDRAULICS

Connecting a large waterway system to a tidal river will amost certainly cause some alteration in the tidal hydraulics of the river and this inevitably will lead to an adjustment in the natural regime of the river and possibly some subsequent modification to the ecology of the area.

For example, if the adjustment in the tidal hydraulics is such that the tidal range is decreased in a part of the river where there is a fisheries habitat, then the productivity of this area could be expected to decrease. On the other hand, if the changes are such as to cause the tidal flow into the river to increase, then the river mouth will scour so that navigation would be improved and flood levels in the immediate vicinity would be reduced.

The approach adopted here was to examine various alternatives in an attempt to find an arrangement of canals which would minimise the problems associated with altering the tidal hydraulics while at the same time retaining the attendant advantages. In particular, for schemes incorporating a flood bypass, the performance of the system in providing flood mitigation is greatly enhanced if the river channel downstream scoured by increased tidal flow. A system which would provide this scouring of the lower reaches of the river was actively sought for this reason.

The technique used here for investigating the effects of various canal schemes on the tidal hydraulics of the canal-river system was to simulate the flow using a network model as described below. The model was first adjusted to reproduce existing behaviour and then used to predict anticipated flows for each development alternative.

5.1 Tidal Flow Model

The region modelled includes the Maroochy River from its mouth upstream to Coolum Creek, and a portion of Eudlo and Petrie Creeks, together with the canal system appropriate for each case. Each system was simulated as a number of equivalent channels and nodes (for example Figure 5) where the network representation for tidal flow for case T5 is given.

Each equivalent channel was assigned, as functions of depth, information on cross-sectional area, wetted perimeter and surface width, based on cross-sections measured in the river and creeks, and on proposed cross-sections for the canals. Channels were also assigned friction factors in the form of Manning coefficients, tables in Chow (1959) being used as a guide.

To account for the storage effect in each section of river or canal, nodes were assigned an appropriate surface area to depth relation, again based on the river geometry and the measured cross-sections.

The input to the model was the tidal level outside the mouth of the Maroochy River, and for all cases sinusoidal tides of the same amplitude (1.4 metres) were assumed. Simulations were started at mid-tide with all levels at mean water level (assumed equal to A.H.D.) and all velocities set to zero. Consequently there were small starting transients, but these generally decayed to an insignificant level at the end of the first tidal cycle. For this reason, velocities, flows and levels were read out during the second cycle.

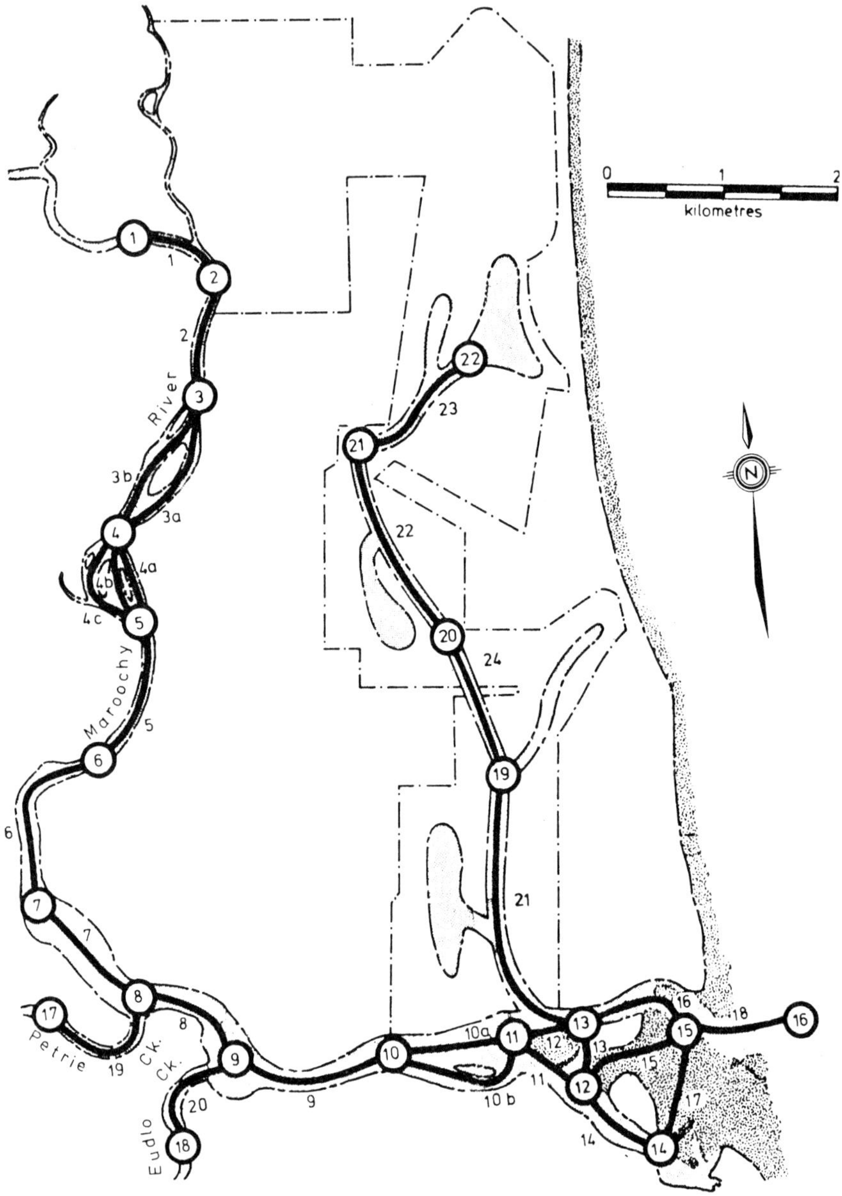

Figure 5. NETWORK APPROXIMATION FOR TIDAL FLOW-CASE T5

The model was calibrated by first simulating the existing tidal situation, adjusting the friction in various channels until existing behaviour was reproduced.

5.2 Simulation of Existing Tidal Behaviour

Tidal flow in the river as it exists was simulated using a network representation similar to that shown in Figure 5.

Since the region modelled did not include a detailed representation of the upper reaches of the Maroochy River and Eudlo and Petrie Creeks it was necessary to augment the storage associated with the nodes at the upstream boundaries (nodes 1, 17 and 18) so that tidal flow to and from the model across these boundaries was faithfully represented. The technique employed here to estimate the appropriate magnitudes for these storages was to measure the tidal prisms passing these boundaries for a particular tidal range, and then to calculate an equivalent storage area. Use of areas calculated in this way for other tidal ranges implies that tidal prism will vary linearly with tidal range, and since the measurements were taken during tides of relatively large range this assumption was thought to be adequate.

At its mouth, the Maroochy River forms a small delta with two major islands and a number of sand banks which are submerged at various stages of the tide. All of the major flow paths were represented by separate equivalent channels with their geometry specified to represent the closure of some of the channels below certain levels.

At a number of other places the river divides to form smaller channels around islands and, in general, where they would all have similar frictional characteristics parallel paths were modelled as one equivalent channel.

The procedure used for calibrating the model was to adjust channel friction to reproduce the measured tidal gradients along the river, the actual parameter used for comparison being the ratio of tidal range at a point to the corresponding tidal range at the ocean. A comparison between measured and predicted tidal ranges is given in Figure 2.

The three points shown in Figure 2 as representing measured ranges are in fact average ratios for a number of tides measured at these positions. At two of the locations the results were based on tide records extending over periods of several days and at the other location on tide staff readings taken over one tide cycle. Corresponding ocean tides were taken as the predicted levels at Caloundra which are available from the Department of Harbours and Marine.

At the beginning of the calibration procedure, values of channel friction were selected on the basis of tabulations given by Chow(1959) In the process of calibration the only adjustments necessary were in the immediate vicinity of the river mouth, where an increase in channel friction was necessary to obtain the reduction in tidal range that does occur in this part of the river. Without any other tailoring of channel friction, the model faithfully reproduced tidal range variation over the remainder of the river including the slight amplification in range which was observed from Bli Bli to Coolum Creek.

With the model adjusted to reproduce tidal gradient, the variation in tidal delay along the river provided an indication of the success of the simulation and a comparison of predicted and measured delays is given in Figure 3. Delays were measured at either high or low tide relative to the ocean tide, and again points representing measured delays were the averages for a number of tides.

Clearly the measured and predicted delays agree quite closely and in particular the model reproduces the observed difference between the delays at high and low tides. For this river the delay at low tide is always significantly larger, because of the geometry of the entrance which becomes significantly more restricted at low tide.

5.3 Tidal Flow Simulation for Various Alternative Canal System Layouts

5.3.1 Alternative Schemes Simulated

Tidal flow was simulated for the river with the four different canal system arrangements shown in Figure 4a and 4b as Cases T2, T3, T4 and T5. A typical example of the network representation used for these cases is shown in Figure 5.

The tidal simulation of the river in its existing condition was used as a basis for comparison with these various alternative systems and is referred to here as Case T1.

In general terms, when a large canal system is connected to a river causing the tidal hydraulics to be altered, there are two possible major effects - a change in tidal range along the river, and a change in tidal velocities or equivalently a change in tidal prisms along the river. Because of this, the results of the various simulations have been summarised in terms of the predicted changes in tidal range and tidal prism, the latter being defined as the volume of water flowing past any particular location during one tidal cycle.

5.3.2 Alteration in Tidal Range

All of the canal systems examined here will cause some alteration in the tidal range along the river and the predicted changes are shown in Figure 6.In every case, tidal range would be reduced over the entire length of the river, and in some instances the reduction would be substantial.

It is evident that systems which have major canal volumes connected to the upper reaches of the river will cause the greatest reductions. For example, for Case T3 it could be expected that the tidal range in the vicinity of Coolum Creek would be reduced by about 30% while for case T5,a system of canals with only one connection near the river mouth, the range at Coolum Creek would be reduced by about 12%.

A significant reduction in tidal range would probably be considered undesirable from an ecological point of view, since reduced tidal range would mean a reduced supply of saline water to mangrove areas. It is important, however, in the context of assessing the effect of any tidal range reduction, to consider also that the river in its existing condition is subject to major variations in tidal range as a result of naturally occurring changes in the river mouth. There is of course an important difference between a reduction in tidal range caused by siltation at the mouth and a reduction caused by connecting a canal system to the river. In the latter case the change will be permanent and probably additional to any naturally occurring alteration.

There is one other aspect of this simulation which should be emphasised. In every case, no allowance has been made for possible readjustment of the river channel geometry as a result of any modification in the tidal hydraulics caused by canals. The predicted reduction in tidal range may be conservatively high in the sense that there may be less reduction in range when the river has had time to adjust, particularly for cases where the river mouth may be enlarged by increased tidal flow, since tidal range in the river is largely controlled by the restriction of the river mouth.

On the basis of these comparisons it would seem that if reductions of tidal range are to be minimised then the canals should be connected as close to the river mouth as possible, and a layout based on case T5 would be the most appropriate choice.

5.3.3 Alteration in Tidal Prism

The simulation results indicate that all of the canal schemes examined will also produce significant changes in tidal velocities and flows.

The various alternative developments are compared here in terms of how the tidal prism along the river is altered, and these results are summarised in Figure 7. Tidal prism is a convenient measure of change in tidal flow but is also the most appropriate parameter for assessing the extent of siltation or scour that would follow any alteration in tidal condition. The points shown in Figure 7 represent tidal prisms for each equivalent channel, calculated from the simulation results by integrating tidal flows over one tide (from low water to high water). All of the results were for a fixed tidal range at the ocean of 1.4 metres and can therefore be compared directly.

In general the schemes represented by cases T2, T3 and T4 will all result in increased tidal flow from about Petrie Creek upstream to the canal entrance near Coolum Creek, although beyond Coolum Creek the flow would be substantially reduced. At the river mouth however, in all of these cases, there will be virtually no changes whatever in tidal flow.

On the other hand the scheme represented by case T5 would increase tidal flow at the mouth by about 12%, at the same time causing flows upstream of the canal entrance to decrease by about the same percentage.

The flows modelled here represent the conditions that would apply immediately after the canals are connected to the river, and these flows would undoubtedly change as the river channels adjusted to the altered flow conditions. For cases T2, T3 and T4 the changes in flow during this adjustment period would probably be small because of

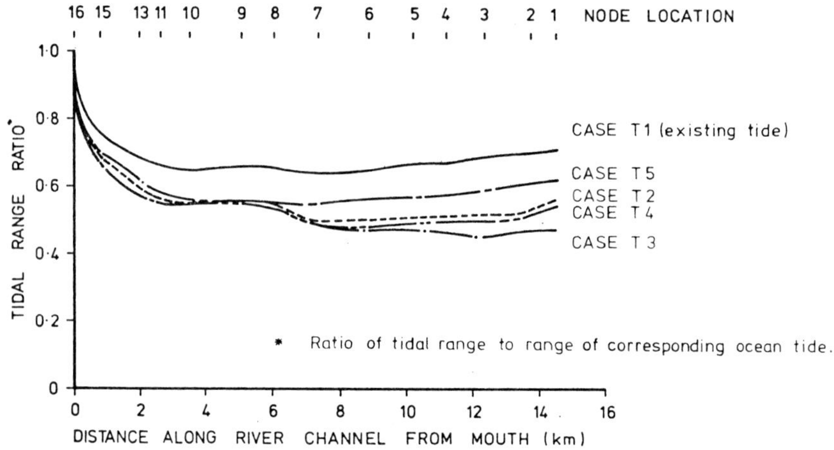

Figure 6. PREDICTED VARIATION IN TIDAL RANGE FOR VARIOUS DEVELOPMENT ALTERNATIVES

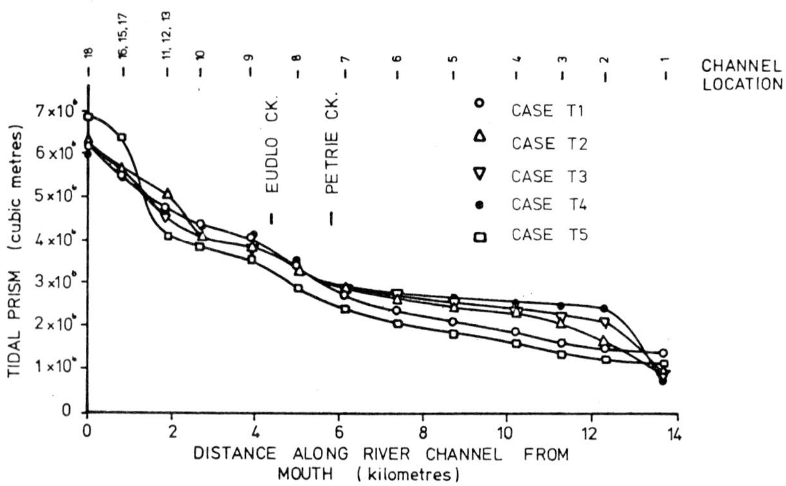

Figure 7 PREDICTED VARIATION IN TIDAL PRISM FOR VARIOUS DEVELOPMENT ALTERNATIVES

the controlling effect of the river mouth, which would not be altered. Significant readjustment, however, could be anticipated for case T5 because of scouring at the river mouth and the most probable result is that tidal flows over the majority of the river would tend to return to existing levels while flows near the river mouth would increase.

Estimates have been made of the possible changes in the regime of the river, based on the well known relationship between stable cross-section area and tidal prism, and the results of this are presented in Figure 8. On the basis of these, it could be expected that for case T5 the river mouth would probably scour to increase its cross-section area by about 10% while in all other cases anticipated readjustments would be minor.

In summary, the layout based on case T5 would, after a period of readjustment at the river mouth, probably result in a situation where both tidal range and tidal flows over the majority of the river would be essentially the same as currently exist and this would be an acceptable result. The other three cases examined would lead to significant changes that may be considered unacceptable.

6. FLOODING

The objective of this investigation was to assess how flooding will be affected by the various development alternatives being considered and in particular to assess to what extent flood levels will be modified by reclamation on the flood plain. The technique used was to model flood flows using a network representation basically similar to the one used for modelling tidal hydraulics. In assessing the flooding problems, the existing situation was modelled first and the resulting flood gradients were then used as a basis for comparison with the various development proposals.

It is always desirable when modelling floods to have records available from actual floods against which the model can be calibrated. For the Maroochy River however there is a relative scarcity of recorded information on flood flows, rainfall and flood levels, and in the case of this model, calibration was based on a simulation of the July 1973 flood, the largest flood in recent years and one for which there were at least a few recorded flood levels.

Although the available calibration information could not be considered entirely adequate for every purpose it was probably sufficient in this instance where the investigation was essentially comparative.

6.1 Flood Flow Paths

As the first step in setting up a model of flood flows in the region of interest it is necessary to know :

(a) The paths through which flood flows occur, their cross-sectional areas of flow at various heights, and the resistance to flow offered by the vegetation in these paths.

(b) The areas in which flood flows are stored, and the volume of water which will be stored at various flood heights.

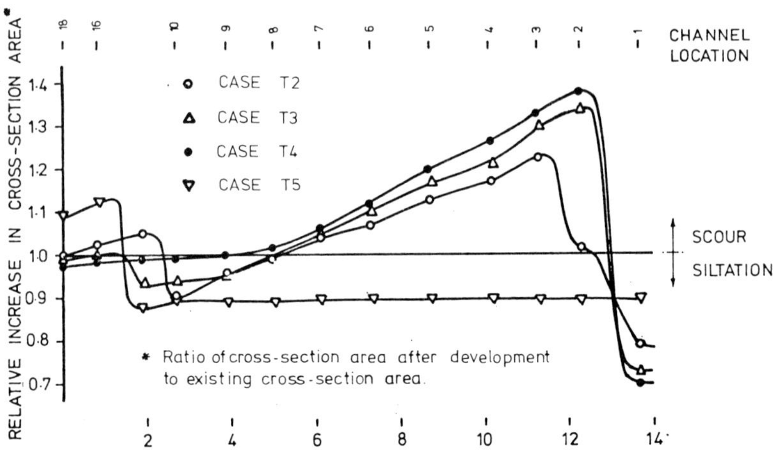

Figure 8. ANTICIPATED READJUSTMENT OF RIVER CHANNEL CAUSED BY VARIOUS CHANNEL DEVELOPMENTS

This information was obtained from a combination of spot level surveys, site examination, discussion with local residents, interpretation of aerial photographs, and the vegetation classification for the site. This classification was followed closely in drawing contours, since different species of swamp and flood-plain vegetation tend to be segregated quite sharply according to the height of the ground above water.

The resulting contours and the flood flow paths are shown in Figure 9. It is apparent that large areas of the land proposed for development are currently inundated by major floods, providing substantial flood storage which would act to reduce the magnitude of flood peaks downstream. There are minor flood channels which pass through the vegetation on both banks of the river along most reaches, and there is a major flood bypass over the region immediately to the east of the David Low Bridge. The David Low Way effectively forms a dike across the flood plain, preventing direct flood flow across it.

6.2 Flood Flow Model

Flood flows were simulated using a model that is basically the same as the tidal flow model outlined in section 5.1 above. In essence, each network representing a particular canal-river configuration was expanded by the addition of further equivalent channels and nodes to represent the flows over the flood plain and the effect of increased storage capacity. Equivalent channels over the flood plain were specified so that they became operative only when flood heights exceeded appropriate levels.

Inputs to the model include the water level at the ocean - in this case a sinusoidal tide approximating the tidal conditions applicable during the July 1973 flood - and flow hydrographs from each of the major tributaries, these again being the estimated flows for the July 1973 flood.

Calibration of the model was based partly on the tidal flow calibration, and partly on adjustment to reproduce known flood levels. The calibration procedure is as follows. Firstly, all of the main river channels were assumed to have the same frictional resistance as developed in the process of the tidal flow calibration described in 5.2. Since in the flooding simulation these channels are limited in width strictly to the main river course*, and since depth of flow does not increase greatly during a flood the frictional resistance should be unchanged. Having fixed the friction of these channels, the remainder of the model is adjusted to reproduce recorded flood levels when simulating the July 1973 flood. Because the tidal calibration is a relatively accurate procedure its inclusion in the process should improve the reliability of the overall flood calibration.

The appropriate choice of Manning's n for channels on the flood plain depends mainly on the kind of vegetation encountered and for this reason a brief vegetation survey was undertaken of existing vegetation types.

---

* This is done because, in general, the overbank flows pass through relatively densely vegetated areas where the frictional resistance is considerably higher than for the main river channel. Overbank flows are all represented by separate channels.

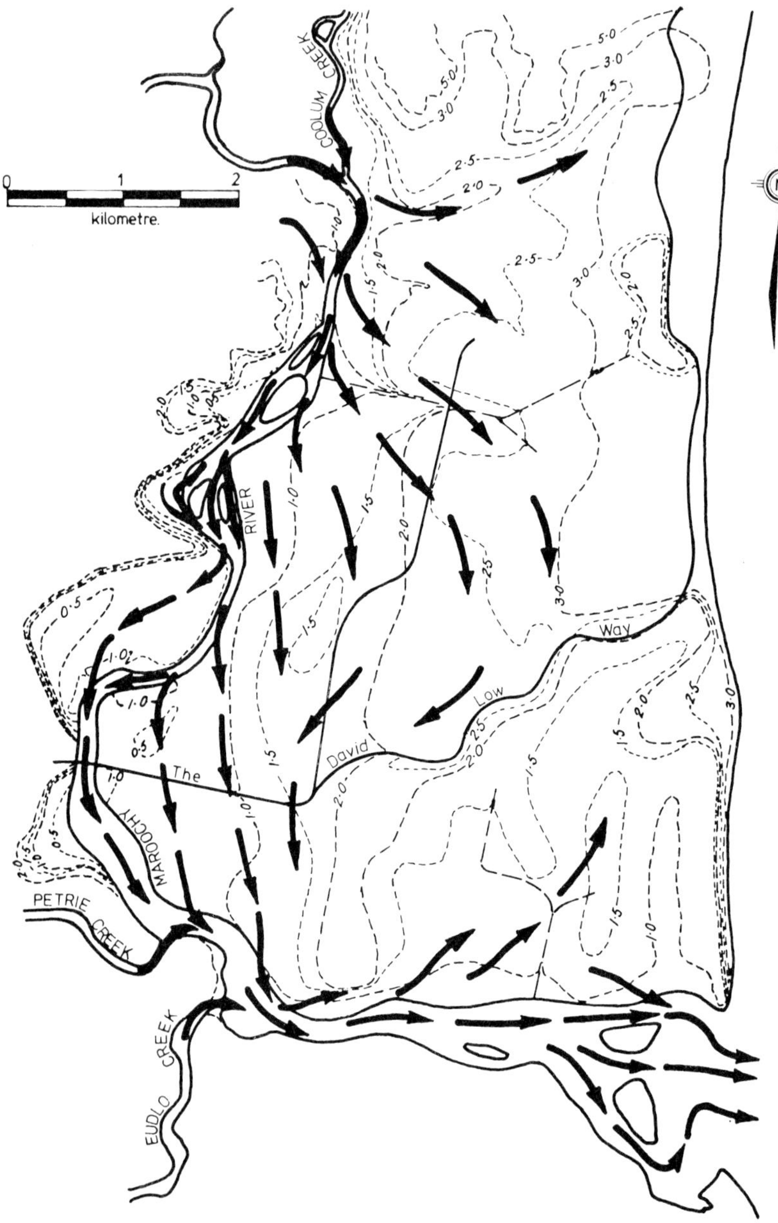

Figure 9 FLOOD FLOW PATHS.

The majority of the flood plain is covered by sugar cane, and melaleuca and sedge swamps, with a narrow mangrove fringe along either side of the river.

Equivalent friction coefficients for each channel were estimated, based on a comparison of existing vegetation with the pictorial catalogue given by Chow (1959), and other experience in simulating flood flows through cane fields in the Tweed River.

Flood flows arriving at the boundaries of the region modelled in this study originate from one or other of three sources (see Figure 10), the Maroochy River - Coolum Creek catchment, Petrie Creek catchment and Eudlo Creek catchment.

Flow hydrographs from each of these catchments were developed for the July 1973 flood and these were applied as boundary conditions in all of the flood simulations.

6.3 Simulation of the Existing Flooding Behaviour

Existing flooding in the lower reaches of the river was simulated using a network representation similar to that shown in Figure 13.

After calibration the model reproduced the flood levels recorded during the July 1973 flood except for some minor discrepancies which probably reflect the fact that some of the levels would have been influenced by local runoff.

Results of the simulation, in terms of typical variations in level during the flood for several locations along the river, are given in Figure 11.

A significant feature of the results is that north of Petrie Creek the flood plain carries a large proportion of the total flow, in parts accounting for as much as 40 percent of the total. Clearly, any development which blocks a significant amount of this flow could be expected to increase flood levels substantially.

Downstream of Petrie Creek the amount of overland flow (of the order of only 2% of the total flow) and flood plain storage are relatively small. Development on the flood plain in this area should therefore have an insignificant effect on flood levels.

6.4 Flood Simulation for Various Alternative Canal System Layouts

6.4.1 Alternative schemes simulated

Flood flow simulations were undertaken for four different canal system arrangements, shown in Figures 12a and 12b as cases F2, F3, F4 and F5.

Case F2 is the initial development proposal of Figure 2 while cases F3 and F4 were based on the canal layout of case T3, the only difference between these two cases being the base level of the floodway. For case F3 the base level was assumed to be at an RL of 0.8 metres (above AHD) whereas for case F4 an RL of 1.5 metres

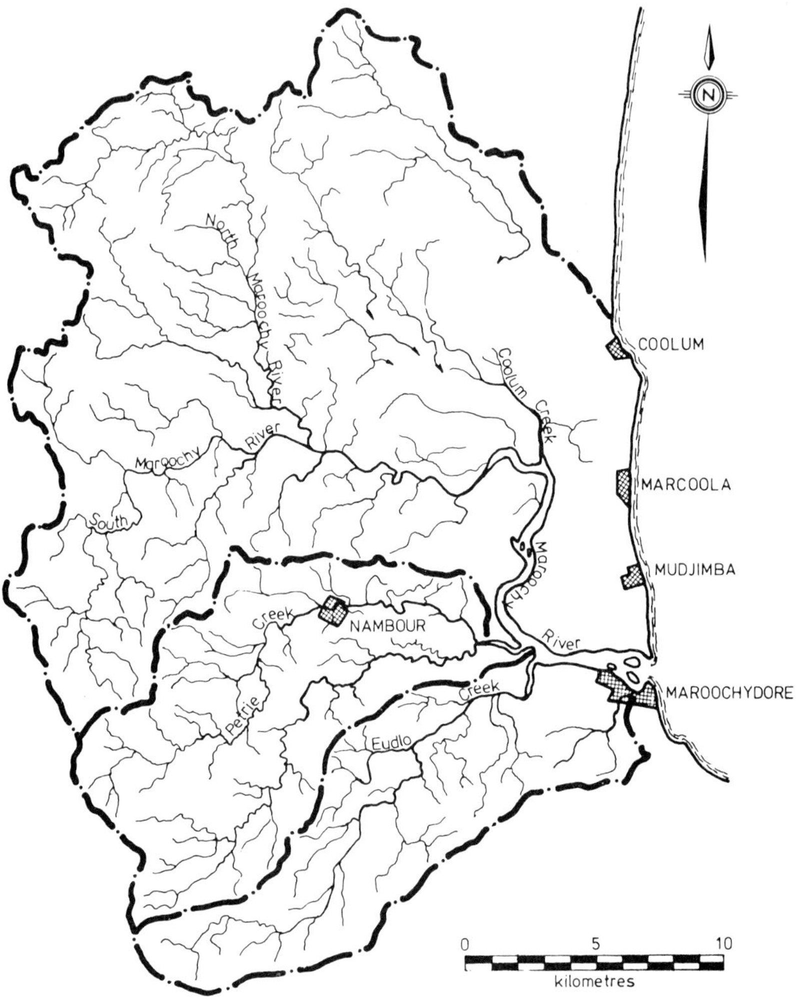

Figure 10. MAROOCHY RIVER CATCHMENT

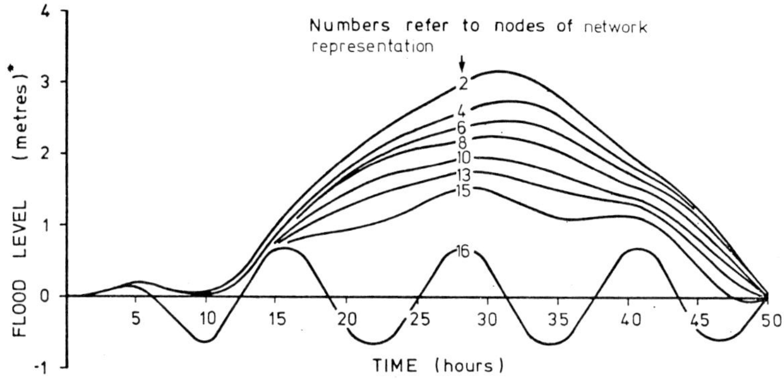

* RELATIVE TO A.H.D.

Figure 11
CALCULATED FLOOD LEVELS ALONG MAROOCHY RIVER FOR JULY, 1973 FLOOD

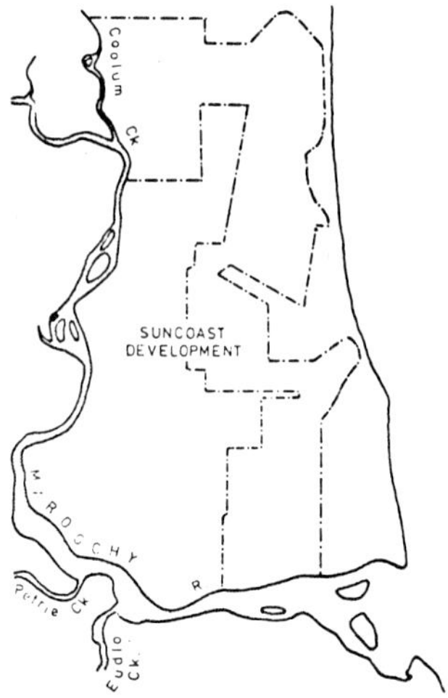

Case F1 : EXISTING SITUATION

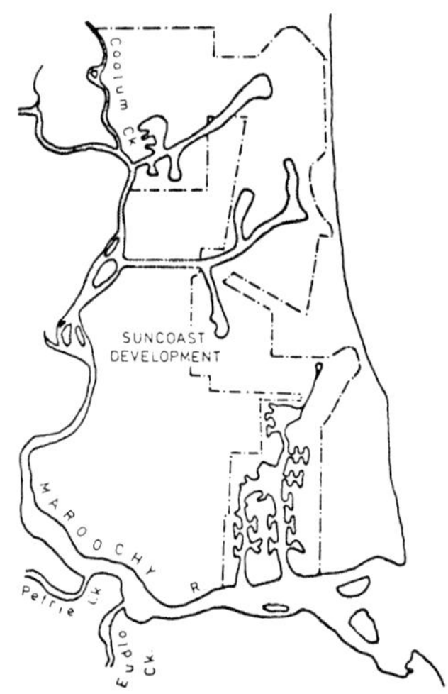

Case F2 : INITIAL DEVELOPMENT PROPOSAL

Figure 12a CANAL SYSTEM ALTERNATIVES SIMULATED FOR FLOODING

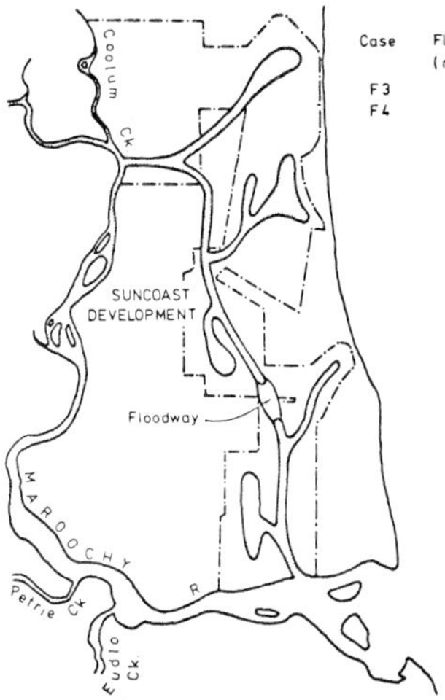

Case	Floodway Base Level (metres above A.H.D.)
F3	0·8
F4	1·5

Cases F3 & F4 : CANAL SYSTEM INCORPORATING FLOOD BYPASS WITH FLOODWAY SOUTH OF MUDJIMBA AIRPORT

Case F5 : CANAL SYSTEM INCORPORATING FLOOD BYPASS WITH FLOODWAY NORTH OF MUDJIMBA AIRPORT

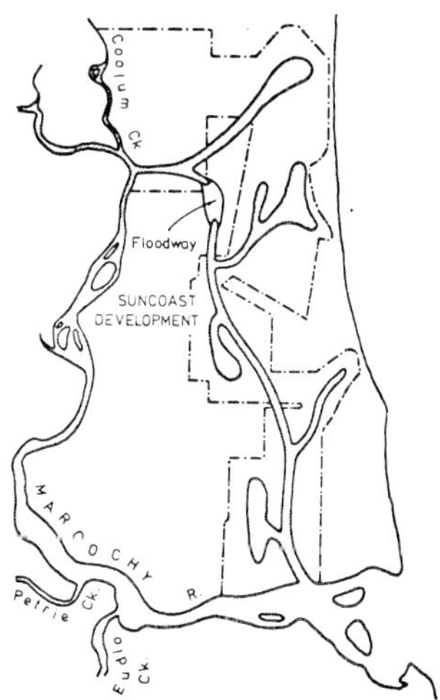

Figure 12b CANAL SYSTEM ALTERNATIVES SIMULATED FOR FLOODING

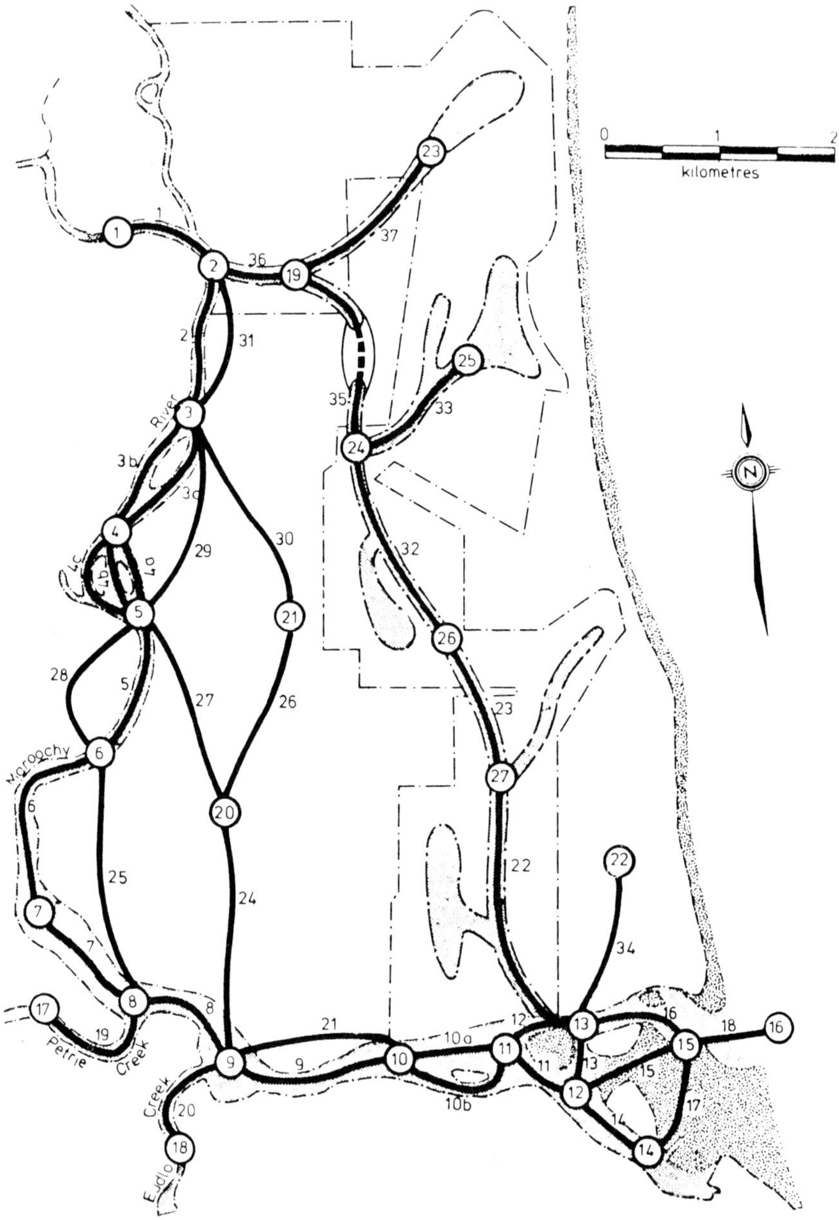

Figure 13. NETWORK APPROXIMATION FOR FLOODING — CASE F5.

was used. Two different floodway levels were examined to establish the effect of varying the proportion of flow carried by the bypass.

The final canal arrangement modelled, case F5, was basically similar to cases F3 and F4 except that the floodway was positioned toward the northern end of the flood bypass. With this arrangement, the canals to the south of the floodway, which were connected close to the river mouth, have a very much larger tidal compartment than for the system represented by cases F3 and F4. As far as tidal flow is concerned case F5 should be very similar in performance to the system modelled as case T5, the only difference being the addition of a relatively small canal at the northern end of the system. In particular, it could reasonably be expected that this development would result in additional scour of the river entrance as discussed in 5.3.3 above, so in modelling this case for flooding, it was assumed that the mouth has been enlarged by scour by the amount indicated in Figure 8. The estimate of scour should be conservatively low for this case. The reason for including river geometry changes in this case alone and not in any of the other flood simulations is that flooding is very sensitive to changes in the geometry of the river mouth and this was the only case where any change could be expected at the mouth.

A typical finite element approximation used to represent these cases is given in Figure 13.

The simulation of existing flooding as described in 6.3 above was used as a basis for comparison with these various alternative systems and is referred to here as case F1.

6.4.2 Comparison of the Effect of the Various Development Alternatives on Flooding

Developments of the kind examined here alter the existing flood situation in a number of ways, but there are only two aspects which have any significant bearing on the acceptability of any particular scheme.

Of primary concern is the effect of the development on flood levels, and a comparison of relative increase in flood levels caused by each of the cases examined is given in Figure 14.

The second aspect of the results which needs to be considered is the degree to which any development will modify flow along the main river channel. This is an important consideration since flood flow plays at least some part in maintaining the river in its existing stable condition. It is especially relevant in the case of a flood bypass scheme where flows along the main river channel could be reduced, leading to siltation in the river which could partially offset the advantage provided by the bypass.

A development along the lines of the initial proposal could be expected to lead to a general increase in flood levels along the lower reaches of the river from Coolum Creek to the mouth. For a flood of the magnitude of the July 1973 flood, an increase in peak level of the order of 5 to 7 centimetres could be expected over the majority of the river in the region modelled.

The predominant effect of including a flood bypass canal is to reduce flood levels immediately downstream of the bypass entrance, a result of the fact that the bypass diverts flow from the main river channel. It is also clear that this reduction is achieved at the expense of an increase in levels at the river mouth because the bypass augments the maximum flood flow downstream of its outlet.

Results for cases F3 and F4 demonstrate that it is possible to limit the increase in flood levels near the mouth by reducing the amount of flow through the bypass, but this is offset by increased levels upstream. For example, with the floodway base at an RL of 0.8 metres (case F3), existing flood levels would be reduced upstream (for the particular flood examined) by as much as 30 centimetres, but at the same time flood levels at Maroochydore would be increased by some 8 centimetres. By raising the floodway base to an RL of 1.5 metres and thereby reducing the flow through the bypass (from 25% of the total flood flow to about 14% - refer Table 1), levels near the river mouth can be lowered slightly (by about 2 centimetres), but this is accompanied by a rise in levels upstream by about 20 centimetres.

TABLE 1

Proportion of Total Flood Flow Carried by Flood Bypass

Development Alternative	Floodway RL (metres above AHD)	Fraction of Total Flow carried by Flood Bypass
Case F3	0.8 m	25%
Case F4	1.5 m	14%
Case F5	0.8 m	27%

Both of the schemes examined here would lead to reduced flow along the main river course as illustrated in Figure 14. According to these predictions, peak flood flows would be reduced by as much as 20% for case F3 and 10% for case F4, the extent of the reduction depending on the amount of flow allowed through the flood bypass.

It would appear that the best possible result with this kind of canal layout would be a situation where the bypass canal carries about the same flow as would otherwise have passed over the flood plain in the areas blocked by the development. If this could be achieved, flood levels should then be about the same as existing levels at all points along the river, except perhaps near the mouth where levels would be slightly increased. Flow in the river channel would be changed slightly, reflecting only the effects of reduced flood plain storage.

On the basis of the simulation results it could be expected that for the canal layout modelled as case T5, flood levels would be reduced over the entire length of the river, by as much as 33 centimetres near Coolum Creek (for this particular flood) and by a few centimetres at Maroochydore.

The essential difference between this scheme and the ones represented by cases F3 and F4, at least as far as flooding is concerned, is the

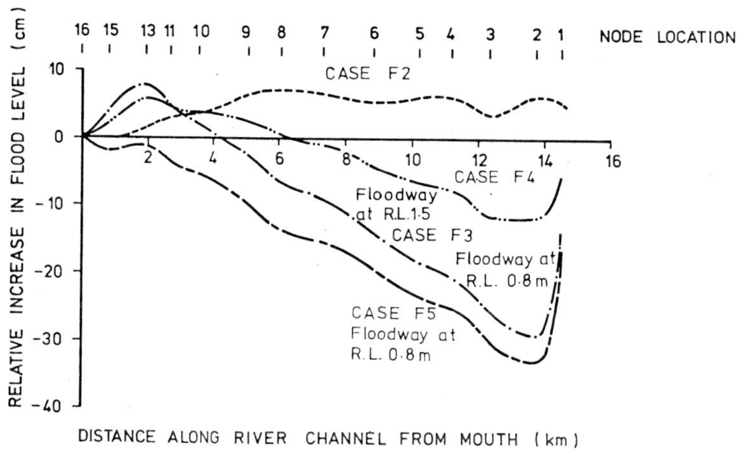

Figure 14
INCREASE IN FLOOD LEVEL RELATIVE TO EXISTING FLOOD

slightly enlarged river entrance. While the maximum flow rates in the channels near the mouth have been increased by the bypass flow, in this case, the increased cross-section area of the channels has more than compensated for this, so that gradients and levels are lower.

As in the case of the other bypass arrangements examined, this scheme results in a significant change in flow along the main river channel, the greatest reduction being of the order of 20%. The extent of this reduction could be controlled by adjusting bypass flow, as demonstrated by the comparison between cases F3 and F4 above, and there is little doubt that a satisfactory level could be achieved. The extent of change in flood flow along the main river channel that could be tolerated would depend largely on the exact stable tidal condition achieved in the river. Because tidal flow has a greater influence on the river geometry it may be possible to accept slightly reduced flood flows if tidal flows are larger.

Any reduction in bypass flow would of course alter flood levels, tending to increase levels upstream and reduce those near the mouth.

Overall it is clear that this kind of scheme could be tailored to produce a very satisfactory result as far as flooding is concerned.

7. CONCLUSION

This simple network model has given a clear understanding of the hydrologic implications of various development options at a reasonable cost. The model will also be valuable for refining design details and evaluating changes as the project is put into practice.

In all, this model has now been used in some fifteen different studies and in every case it has proven to be a very cost effective tool.

8. ACKNOWLEDGEMENT

The authors wish to thank the Suncoast Group of Companies for permission to publish this information.

9. REFERENCES

CHOW, V.T. (1959). *Open Channel Hydraulics.* McGraw-Hill.

U.S. Dept. of Transportation/Federal Highway Administration (1973). Hydraulics of Bridge Waterways. Hydraulic Design Series No.1.

Numerical Simulation of Fluid Motion, ed. John Noye
pp. 435-451

FINITE DIFFERENCE SCHEMES FOR THE SOLUTION OF LINEARISED WIND EFFECT EQUATIONS

B.J. Noye

University of Adelaide

South Australia

SUMMARY

A numerical method is described for obtaining the frequency response of shallow lakes, acted on by moderate winds, by solving the linearised depth integrated equations of motions and continuity. Irrespective of the nature of the grid-scheme used the one-dimensional models were stable at low frequencies; the two dimensional models were unstable at all frequencies. This has not restricted the application of the method because error build up is so slow that relatively large grid-schemes can still be accommodated.

1. INTRODUCTION

The action of wind stress is the main cause of variation of water levels and circulation of water masses within closed lakes. Recent studies by Walsh and Noye (1974,1976) have been concerned with the development of mathematical models, using a frequency response concept, for predicting steady state effects of moderate strength winds on shallow closed lakes. The schemes developed have been applied successfully to predict wind induced effects in the Murray Mouth lake system in South Australia. The finite difference numerical approach used avoids the usual problem of inverting a large sparse matrix, replacing it with a very much smaller, dense matrix.

One dimensional models applicable to elongated lakes were found to be numerically stable over a definite frequency range; the two dimensional models developed were found to be unstable at all frequencies, although this did not restrict their usefulness because of the very slow error build up produced in the frequency response functions on marching through the spatial grid. For two dimensional grids of large size, double precision is needed for all calculations. Different algorithms and other grid schemes are being examined in order to produce a numerically stable system for very large two dimensional grids.

2. THE WIND EFFECT EQUATIONS

The linearised depth-integrated equations of motion and the continuity equation, referred to a set of cartesian coordinate axes with the z axis directed vertically upwards and the x and y axes in the plane of the undisturbed lake surface, are well known (Groen and Groves, 1960) and may be written as

$$\partial U/\partial t = -rh^{-1}U - gh\partial\zeta/\partial x + K\tau_x \qquad (2.1a)$$

$$\partial V/\partial t = -rh^{-1}V - gh\partial\zeta/\partial y + K\tau_y \qquad (2.1b)$$

$$\partial U/\partial x + \partial V/\partial y = -\partial\zeta/\partial t \qquad (2.1c)$$

where

(a) $\tau_x = \tau_x(x,y,t)$, $\tau_y = \tau_y(x,y,t)$ are respectively x and y components of tangential wind stress at time t;

(b) $\zeta = \zeta(x,y,t)$ is the displacement of the surface (from the undisturbed level) due to wind stress;

(c) $U = U(x,y,t)$, $V = V(x,y,t)$ are respectively x and y components of volume transport per unit width due to wind stress;

(d) $h = h(x,y)$ is the depth of the lake;

(e) $K = (1+m)/\rho$, where m is a dimensionless bottom friction parameter with value about .05 (Francis, 1954) and ρ is the density of water;

(f) g is the acceleration due to gravity;

(g) r is a damping parameter which accounts for internal dissipation of energy - for shallow lakes $r \sim 6 \times 10^{-4} m.s^{-1}$ (Noye and Walsh, 1975).

Assumptions made in the derivation of (2.1a-c) include (i) the surface displacement is much smaller than the depth, a condition which is violated close to the lake shore; (ii) the horizontal dimensions of the lake are much greater than the depth; (iii) the Coriolis force has negligible effect, a condition brought about by the small size of the lakes and the large frictional effects. Since only wind effects on the lakes are being considered, the above equations are solved subject to the condition that no flow occurs across the horizontal lake boundary, thus neglecting the effect of inflow and outflow of water.

Consider the steady state response of this system to a spatially homogeneous wind stress, oscillating in time with a given circular frequency ω, viz.

$$\tau_x = Xe^{j\omega t}, \quad \tau_y = Ye^{j\omega t} \tag{2.2}$$

where X,Y are complex constants and $j = \sqrt{-1}$. Since Equations (2.1) are linear, the response of the system takes the form

$$\zeta = Z(x,y)e^{j\omega t}, \quad U = P(x,y)e^{j\omega t}, \quad V = Q(x,y)e^{j\omega t}. \tag{2.3}$$

Substituting (2.2) and (2.3) into (2.1) yields

$$(j\omega + rh^{-1})\, P = -gh\partial Z/\partial x + KX, \tag{2.4a}$$

$$(j\omega + rh^{-1})\, Q = -gh\partial Z/\partial y + KY, \tag{2.4b}$$

$$\partial P/\partial x + \partial Q/\partial y = -j\omega Z, \tag{2.4c}$$

subject to the condition that the component of the vector (P,Q) normal to the boundary contour of the lake is zero everywhere along the contour.

3. ONE-DIMENSIONAL FINITE-DIFFERENCE MODELS

For elongated lakes only one dimension need be considered. Measuring the x-coordinate along the longitudinal axis of the lake, so that $x=0$ and $x=L$ are the coordinates at the ends of the lake, the Equations (2.4) become

$$(j\omega+rh^{-1})\, P = -gh\partial Z/\partial x + KX \tag{3.1a}$$

$$\partial P/\partial x = -j\omega Z. \tag{3.1b}$$

subject to the boundary conditions

$$P=0 \quad \text{at} \quad x=0, x=L. \tag{3.1c}$$

Two grid-schemes with associated algorithms have been used to solve numerically the system (3.1).

Forward and Backward Finite Differences.

For a given frequency ω, the frequency response functions $P(x)$ and $Z(x)$ are calculated at equispaced intervals $x=id$, $i=0,1,2,\ldots,N$ where $d=L/N$, in the following way:
Using a forward difference approximation for the derivative in (3.1b) at $x=id$ we can use $P_i = P(id)$ and $Z_i = Z(id)$ to compute P_{i+1} by means of the relation

$$P_{i+1} = P_i - j\omega dZ_i \tag{3.2a}.$$

Using a backward difference approximation for the derivative in

(3.1a) at $x = (i+1)d$ we use Z_i, P_{i+1} to determine Z_{i+1} from

$$Z_{i+1} = Z_i + d\{KX - (j\omega + rh_{i+1}^{-1})P_{i+1}\}/gh_{i+1}. \quad (3.2b)$$

A solution for this system, with boundary conditions $P_0 = P_N = 0$ can be obtained by the inversion of a large, complex, square matrix. This matrix, though sparse, has dimensions which tax the capacity of most computers. An alternative approach which requires the inversion of only a small dense matrix is described below.

Commencing with the boundary condition $P_0 = 0$ and assuming a value for Z_0 we may march through the grid, calculating P_1 from (3.2a), Z_1 from (3.2b), then P_2 from (3.2a), Z_2 from (3.2b), and so on. Then P_N will be zero only if Z_0 has been correctly chosen.

In order to find the correct value for Z_0, we note that P_N is a linear function of Z_0, i.e.

$$P_N = AZ_0 + B \quad (3.3)$$

where A and B are complex constants. Clearly, the correct choice for Z_0 is $Z_0^{(s)} = -B/A$. A and B may be determined by calculating P_N for any two different choices of Z_0. For example, choosing $Z_0 = 0$ and marching through the grid as described will yield $P_N = P_N^{(o)}$; substitution in (3.3) gives $B = P_N^{(o)}$. Choosing $Z_0 = 1$ yields $P_N = P_N^{(1)}$ whence $A + B = P_N^{(1)}$. Thus

$$Z_o^{(s)} = P_N^{(o)}/(P_N^{(o)} - P_N^{(1)}). \quad (3.4)$$

Using this starting value for Z_0, a final run may be made through the grid to determine the correct values of P and Z at each gridpoint.

A stability criterion for this scheme will be determined using a discrete perturbation stability analysis. A small round off error is introduced at one step of the computation of P (or Z) and an analysis is made of the manner in which the error propagates as further values of P(or Z) are calculated.

Suppose that an error ΔZ_i is introduced in the computation of Z_i, so that we obtain instead

$$Z_i^* = Z_i + \Delta Z_i$$

and further

$$P_{i+1}^* = P_{i+1} - j\omega d\Delta Z_i ,$$

$$Z_{i+1}^* = Z_{i+1} + \left\{1 - \frac{d^2\omega^2}{gh_{i+1}} + j\frac{d^2 r\omega}{gh_{i+1}^2}\right\}\Delta Z_i .$$

The error, ΔZ_i, propagates with diminished magnitude if

$$\left|1 - \frac{d^2\omega^2}{gh_{i+1}} + j\frac{d^2 r\omega}{gh_{i+1}^2}\right| < 1,$$

which upon rearrangement gives

$$d < \left[\frac{2gh_{i+1}}{\omega^2 + r^2/h_{i+1}^2}\right]^{\frac{1}{2}} \quad (3.5a).$$

The introduction of an error ΔP_i introduced in the computation of P_i leads to a similar condition for the diminished propagation of the error. It is clear then that by choosing a grid length d such that

$$d < d_o = \left[\frac{2gh_m}{\omega^2 + r^2/h_m^2}\right]^{\frac{1}{2}}, \qquad (3.5b)$$

where $h_m = \min\{h_i\ ;\ i=1,2,\ldots,N-1\}$, the stability of the iterative scheme will be guaranteed. Condition (3.5b) is a sufficient stability condition, but is not a necessary condition. This is because the scheme may, according to (3.5a), increase the introduced error at one step, but because of changing depths, decrease the error at the next step. Experience indicates that it is better to suggest that a necessary stability condition for the scheme is

$$d < d_o = \left[\frac{2gH}{\omega^2 + r^2/H^2}\right]^{\frac{1}{2}}, \qquad (3.5c)$$

where H is the average depth of the lake. Note that with $r=0$, conditions (3.5) are equivalent to the Courant-Friedrichs condition for stability of time-stepping schemes given by Fischer (1959).

Centred Finite Differences

For a given frequency ω, the frequency response functions $P(x)$ and $Z(x)$ are calculated alternately at equispaced intervals $x=id/2$, $i=0,1,2,\ldots,N$ where $d=L/N$. Using the notation $P_i = P[id]$ and $Z_i = Z[(i+\frac{1}{2})d]$, and approximating spatial derivatives by centred finite differences in (3.1a) and (3.1b), we obtain the same scheme as before described by relations (3.2a) and (3.2b), where h_i is the depth at gridpoint $x=id$.

Commencing with $P_0=0$ and the appropriate starting value $Z_0 = Z_0^{(s)}$ we can march through the grid calculating P_1, Z_1, P_2, Z_2, and so on, finally obtaining $P_N = 0$. The starting value $Z_o^{(s)}$ is given by (3.4).

The stability criterion for this scheme is that given by (3.5). For small values of ω, such as those associated with diurnal and semi-diurnal wind regimes, the scheme was stable for $d < 0.5$ km for the Murray Mouth lakes which have depths of the order of 2m.

4. TWO-DIMENSIONAL FINITE-DIFFERENCE MODELS

The numerical solution of (2.4) by finite-difference methods requires the placement of a two dimensional array of points over the surface of the given lake, and hence an approximation of the boundary contours by horizontal and vertical straight line segments. The schemes described here are staggered ones, in which the current response functions (P,Q) are computed at different points in space to those where surface elevation response functions (Z) are computed.

P,Q calculated at different positions

In this system the gridpoints are at the points of intersection of lines drawn parallel to the x and y axes. Along the lines parallel to the x axis Z and P are evaluated at alternate gridpoints a distance Δx apart, and along the lines parallel to the y axis Z and Q are evaluated

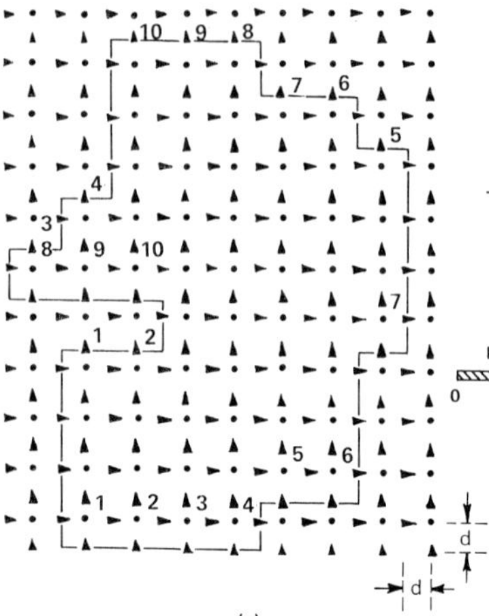

(a)

4	4	8	8	8	4	4	4	4
4	4	5	6	6	7	8	4	4
4	4	5	6	6	6	6	7	4
4	8	5	6	6	6	6	6	5
8	5	6	6	6	6	6	6	3
1	2	2	6	6	6	6	6	3
4	8	8	5	6	6	6	2	3
4	5	6	6	6	6	6	3	4
4	5	6	6	6	6	6	3	4
4	5	6	6	6	2	2	3	4
4	1	2	2	2	3	4	4	4

(b)

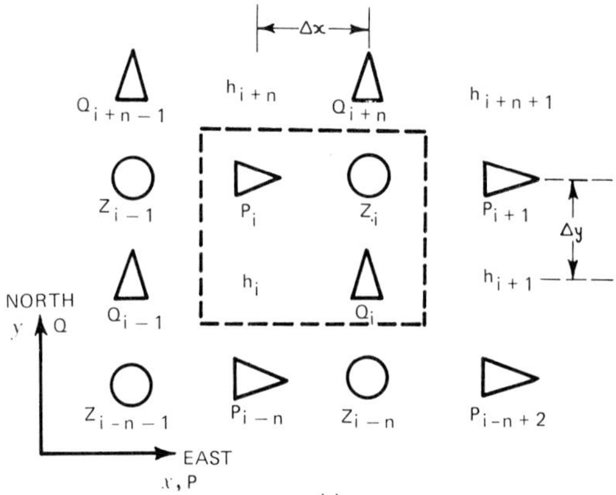

(c)

Fig.1a. Two dimensional array (11×9) for modelling behaviour in Lake Albert using the system of Equations (4.2). "Starting" values of Z are defined at the 10 numbered grid-points O; "end" values of Q occur at the numbered grid-points Δ .

1b. Corresponding class labels for each element of the model.

1c. Indices assigned to elevation frequency response Z and transport responses P,Q and depth h . The i-th element is shown surrounded by the pecked lines.

at alternate grid points a distance Δy apart (cf. Leendertse, 1967). The numbers of each kind of point across the grid in the x and y directions respectively are denoted by n and m respectively.

Shown in Fig.1a is a model of Lake Albert, South Australia, constructed from this scheme with $m = 11$, $n = 9$. For a given ω, $Z(x,y)$ is evaluated only at points marked o, $P(x,y)$ only at point marked ▷ and $Q(x,y)$ only at points marked Δ. The horizontal and vertical array axes (x and y axes respectively) coincide with the west-east and south-north alignments. For the Lake Albert model $\Delta x = \Delta y = d = 850$m.

In order to simplify the notation used it is assumed that the grid is composed of "elements", the depth within each element being assigned at the gridpoint (x,y), say. Within the element there is also one Z-point at $(x + \Delta x, y + \Delta y)$, one P-point at $(x, y + \Delta y)$ and one Q-point at $(x + \Delta x, y)$, as in Fig. 1c. The elements are then numbered 1,2,3,...,mn starting in the south-west corner and counting across the grid from west to east, proceeding one row at a time from south to north. Each "element" belongs to a particular class; there are eight such classes listed in Table 1. In Fig.1b, each element of the Lake Albert model of Fig.1a has been designated by its class label.

The values of Z,P,Q and h at their respective points in the ith element are denoted Z_i, P_i, Q_i and h_i, and the class label of the element by L_i.

Consider the element and surrounding gridpoints shown in Fig.1c. At the gridpoint where P_i is evaluated, we can replace the derivative $\partial Z/\partial x$ by the centred finite-difference form $(Z_i - Z_{i-1})/2\Delta x$ and the depth h by the mean value $(h_{i+n} + h_i)/2$. Equation (2.4a) therefore becomes

$$\left(j\omega + \frac{2r}{h_{i+n} + h_i}\right) P_i = -g\left(\frac{h_{i+n} + h_i}{2}\right)\left(\frac{Z_i - Z_{i-1}}{2\Delta x}\right) + KX. \qquad (4.1a)$$

Similarly, at the gridpoint where Q_i is evaluated, Equation (2.4b) becomes

$$\left(j\omega + \frac{2r}{h_{i+1} + h_i}\right) Q_i = -g\left(\frac{h_{i+1} + h_i}{2}\right)\left(\frac{Z_i - Z_{i-n}}{2\Delta y}\right) + KY, \qquad (4.1b)$$

and at the gridpoint where Z_i is evaluated, Equation (2.4c) becomes

$$\left(\frac{P_{i+1} - P_i}{2\Delta x}\right) + \left(\frac{Q_{i+n} - Q_i}{2\Delta y}\right) = -j\omega Z_i . \qquad (4.1c)$$

Equations (4.1), together with the boundary condition that $P_i = 0$ if a vertical segment of the boundary passes through the ith element and $Q_i = 0$ if a horizontal segment of the boundary passes through the ith element, form a closed set in the complex unknowns Z,P and Q evaluated for all non-exterior elements. A solution for this system can be obtained by the inversion of a large, complex, square matrix. This matrix, though sparse, has dimensions about (250 × 250) and its inversion is beyond the capacity of most computers. An alternative approach which requires the inversion

TABLE 1

LABELS ASSOCIATED WITH CLASSES OF ELEMENTS AND MODIFICATIONS TO EQ. (4.2)

The ith element is shown by the pecked line, the land-water boundary is indicated by the solid line, the land region is shaded.

LABEL	ELEMENT	MODIFIED SCHEME
1	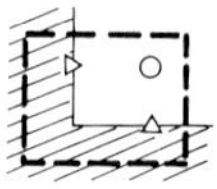	$Q_i = 0$; Z_i suitably chosen as described in text; $P_i = 0$
2	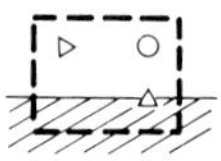	$Q_i = 0$; Z_i suitably chosen as described in text; P_i as in Eq.(4.2c)
3	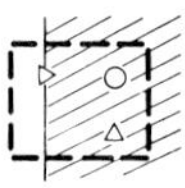	$Q_i = 9999$; $Z_i = 9999$; $P_i = 0$
4	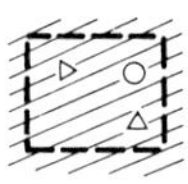	$Q_i = 9999$; $Z_i = 9999$; $P_i = 9999$
5	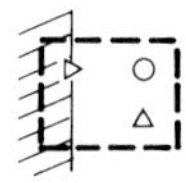	Q_i as in Eq.(4.2a); Z_i as in Eq.(4.2b); $P_i = 0$
6	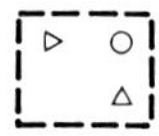	Q_i as in Eq.(4.2a); Z_i as in Eq.(4.2b); P_i as in Eq.(4.2c)
7	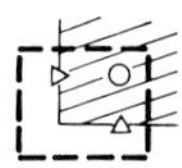	Q_i as in Eq.(4.2a); $Z_i = 9999$; $P_i = 0$
8	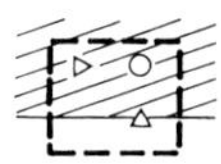	Q_i as in Eq.(4.2a); $Z_i = 9999$; $P_i = 9999$

TABLE 2

CLASSIFICATION OF ELEVATION AND STREAM POINTS FOR USE WITH EQ. (4.18)

STREAM POINTS

Group no.	TYPE	Group no.	TYPE
1	Exterior pt	2	Interior pt
3	North boundary pt	4	South boundary pt
5	West boundary pt	6	East boundary pt
7	North-west 90° corner pt	8	North-west 270° corner pt
9	South-west 90° corner pt	10	South-west 270° corner pt
11	South-east 90° corner pt	12	South-east 270° corner pt
13	North-east 90° corner pt	14	North-east 270° corner pt

ELEVATION POINTS

Group no.	TYPE	Group no.	TYPE
15	Exterior pts	16	Pts immediately (i) south of North boundary
17	Interior pts other than group 16 pts		(ii) east of West boundary
			(iii) south-east of group 8 pt

of only a dense (10 × 10) matrix, is described in the following.

With a square grid scheme in which $\Delta x = \Delta y = d$, we may rewrite Equation (4.1c) with i-n for i, in the form

$$Q_i = P_{i-n} + Q_{i-n} - 2jd\omega Z_{i-n} - P_{i-n+1}, \tag{4.2a}$$

Equation (4.1b) in the form

$$Z_i = Z_{i-n} + \frac{4d}{g(h_{i+1} + h_i)} \left\{KY - \left(j\omega + \frac{2r}{h_{i+1} + h_i}\right) Q_i\right\}, \tag{4.2b}$$

and equation (4.1a) in the form

$$P_i = \left\{KX - \frac{g(h_{i+n} + h_i)(Z_i - Z_{i-1})}{4d}\right\} \Big/ \left(j\omega + \frac{2r}{h_{i+n} + h_i}\right). \tag{4.2c}$$

Then, knowing values of P_{i-n}, Q_{i-n}, Z_{i-n}, P_{i-n+1}, Z_{i-1} values of Q_i, Z_i, P_i can be calculated in turn.

In order to initiate a marching scheme which produces values of Q, Z and P at all the gridpoints of the array, we may simply choose a "starting value" for Z in each element labelled 1 or 2. Then values for Q, Z and P for all other elements in the array are either defined as in Table 1 or calculated using the equations (4.2). For instance, for an ith element labelled $h_i = 6$, Q_i is calculated using Equation (4.2c). All other classes of elements have one or more of these equations replaced by defined values for the variable since the iterative scheme involves values of Q, Z and P either within the lake region (e.g. elements labelled 6) or on the lake boundary (e.g. elements labelled 1,2,3,5,7,8). In the latter case at least one of the values P,Q is zero as they must satisfy the condition of no flow across the boundary. Values allocated to Z,P or Q at points outside the model lake may be chosen arbitrarily since they will not be involved in the calculations. They can therefore be chosen to control the output format. For instance, for elements labelled 4 which are exterior to the model lake, we may set Q = Z = P = 9999 (say) which are unrealistically large values easily identifiable as being associated with exterior points in the output of elevation and flow component fields. The class labels for each element of this Lake Albert model are shown in Fig. 2b.

Proceeding west to east along each row and treating each row in order from south to north, we can use the marching scheme defined for the various classes of element as in Table 1. In this system all boundary conditions are satisfied regardless of starting values, except on east-west boundaries in the elements of class 7 and 8. The boundary conditions Q=0 in these elements will be satisfied only if the correct starting values for Z have been chosen.

We must therefore, develop a method which provides the appropriate "starting values", viz. those values of Z in elements of class 1 and 2 which produce the required "end values", Q=0, in the elements of class 7 and 8.

For a consistent system, the number of starting values and the number of end values are the same, say k. For any set of starting values $\{s_p, p=1,k\}$ we can run through the iterative scheme to determine the corresponding set of end values $\{\varepsilon_p, p=1,k\}$, the members of the sets being numb-

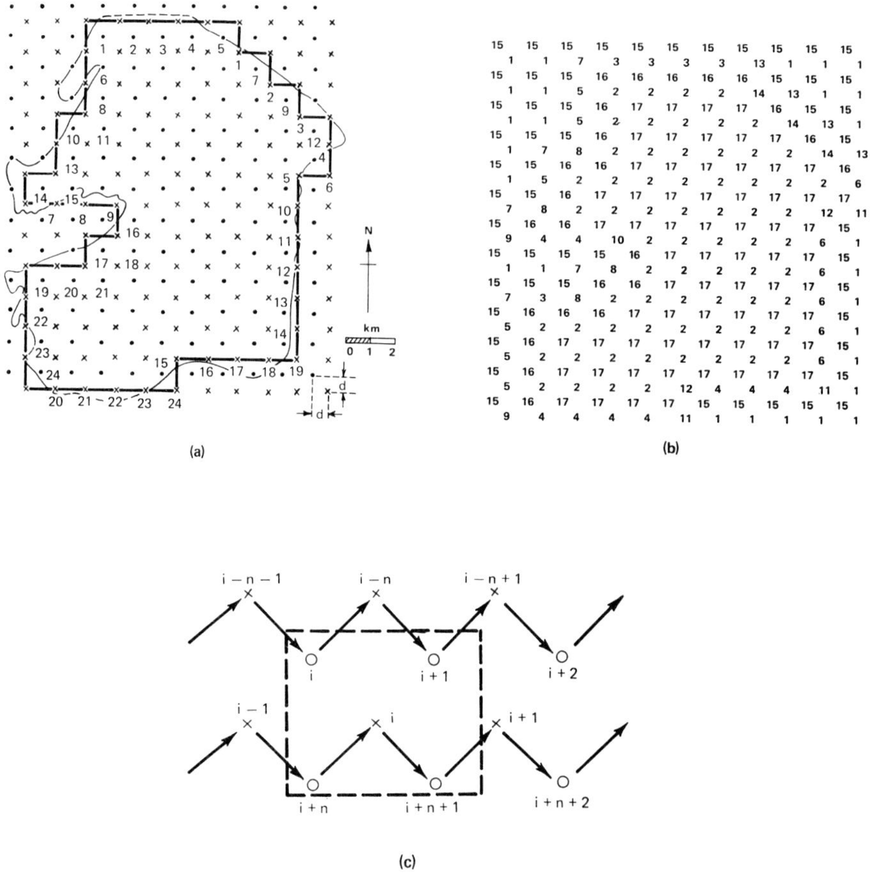

Fig. 2a. Two-dimensional array (13×11) for modelling behaviour in Lake Albert using the system of Equations (4.18). "Starting" values of Z are defined at the 24 numbered grid-points O; "end" values of P or Q occur at the numbered grid-points ×.

2b. Corresponding group number for each array point of the model.

2c. A typical grouping of four elevation points around a stream point × is shown inside the pecked lines. The order of calculation of Z (at points O) and S,D (at points ×) is shown by the arrows.

ered in increasing order of magnitude of the element number i in which they occur. For this Lake Albert model, $p = 10$ (Fig. 2a).

Since Z and Q are complex numbers, then so are the starting values s_p and the end values ε_p. Define complex k-vectors $\underset{\sim}{s}$ and $\underset{\sim}{\varepsilon}$, such that

$$\underset{\sim}{s} = \{s_p\}, \quad \underset{\sim}{\varepsilon} = \{\varepsilon_p\}.$$

Because the marching scheme is linear, each end value is linearly related to the set of k starting values. It is thus possible to write

$$\underset{\sim}{\varepsilon} = A\underset{\sim}{s} + \underset{\sim}{b} \tag{4.3}$$

where $A = (a_{\ell p})$ is a $k \times k$ complex matrix which determines the effect of the starting values on the end values and $\underset{\sim}{b}$ is column vector of k complex coefficients which describes the effect of the forcing function, through X and Y, on the end values. The correct starting values $\underset{\sim}{s}^*$ are thus determined by solving (4.3) for $\underset{\sim}{s}$ with $\underset{\sim}{\varepsilon} = \underset{\sim}{0}$, giving

$$\underset{\sim}{s}^* = -A^{-1}\underset{\sim}{b}. \tag{4.4}$$

The coefficients of A and $\underset{\sim}{b}$ must therefore be found. Firstly, $\underset{\sim}{b}$ is the end vector $\varepsilon^{(o)}$ obtained by setting $\underset{\sim}{s} = \underset{\sim}{0}$, using the particular forcing functions that act upon the lake through Equation (4.2). Then each column of A may be generated by choosing a starting vector of the Kronecker delta form, viz.

$$\underset{\sim}{s}^{(p)} = (s_\ell^{(p)}) = (\delta_{\ell p}), \tag{4.5}$$

so that, by (3.3), we have

$$\underset{\sim}{\varepsilon}^{(p)} - \underset{\sim}{\varepsilon}^{(o)} = (\varepsilon_\ell^{(p)} - \varepsilon_\ell^{(o)}) = (a_{\ell q}\delta_{qp}) = (a_{\ell p}), \tag{4.6}$$

i.e. the end values $\underset{\sim}{\varepsilon}^{(p)} - \underset{\sim}{\varepsilon}^{(o)}$ are the coefficients of the pth column of the matrix A. $(k+1)$ marches through the grid therefore provides the required coefficients of A and $\underset{\sim}{b}$.

Having determined the correct starting values $\underset{\sim}{s}^*$ from (4.4) a final run through the scheme determines the correct values of Q, Z and P at each elevation or stream point over the model lake. By this procedure we have replaced the necessity to invert a very large matrix, of the order (250×250) for this Lake Albert model, by the inversion of a much smaller matrix, of the order (10×10) in this case.

The size of the computer to be used therefore decides the number of columns which can be used in our model array, as it is this which determines the number of gridpoints where starting values of Z must be defined and hence the size of the matrix to be inverted.

Also, as mentioned in Section 3, a further limitation on the magnitude of the grid spacing in a finite-difference scheme of this kind is the manner in which errors introduced at any step of the computations, such as by rounding off processes in the computer, propagate through the grid scheme.

Suppose an error ΔZ_i is introduced in the computation of Z_i in a model lake of constant depth H, so that we obtain instead

$$Z_i^* = Z_i + \Delta Z_i. \tag{4.7}$$

From Equation (4.2c) it is seen that this error affects the calculation of both P_i and P_{i+1}, producing instead

$$P_i{}^* = P_i - \{gH/2\beta d\}\ \Delta Z_i \tag{4.8}$$

and

$$P_{i+1}^* = P_{i+1} + \{gH/2\beta d\}\ \Delta Z_i\ , \tag{4.9}$$

where $\beta = j\omega + rH^{-1}$. The errors introduced affect the computation of Q_{i+n}, using (4.2a), producing

$$Q_{i+n}{}^* = Q_{i+n} - \{2jd\omega + gH/\beta d\}\Delta Z_i\ , \tag{4.10}$$

which in turn produces an error in Z_{i+n}, calculated using (4.2b), giving

$$Z_{i+n} = Z_{i+n}{}^* + \{3 + 4jd^2\omega\beta/gH\}\Delta Z_i \tag{4.11}$$

The error in Z_{i+n} is, therefore,

$$\Delta Z_{i+n} = 3\{(1-4d^2\omega^2/3gH) + j(4rd^2\omega/3gH^2)\}\Delta Z_i \tag{4.12}$$

so in the computation of the elevation one row immediately above the grid-point at which the error was introduced, it has increased in magnitude by the factor $F_\ell^{(1)}$ where

$$F_\ell^{(1)} = 3\left\{1 - \frac{8d^2\omega^2}{3gH} + \frac{16}{9}\frac{d^4\omega^4}{g^2H^2} + \frac{16}{9}\frac{d^4\omega^2r^2}{g^2H^4}\right\}^{\frac{1}{2}} \tag{4.13}$$

For the Lake Albert model we have $d=8.5\times10^2$m, $H=2$m, $r=6\times10^{-4}$m-s^{-1}, $g=9.8$ m-s^{-2}, with frequencies in the range 0-24cpd of interest, so that $0<\omega<2\times10^{-3}$s^{-1}. Since, in this range of ω,

$$\frac{8d^2\omega^2}{3gH} >> \frac{16}{9}\ \frac{d^4\omega^4}{g^2H^2} \quad\text{and}\quad \frac{8d^2\omega^2}{3gH} >> \frac{16}{9}\ \frac{d^4\omega^2r^2}{g^2H^4}\ ,$$

then $F_\ell^{(1)} = 3-$, so that the magnitude of the error has increased nearly threefold in the first iteration for Z.

However, in the next iteration, to obtain Z_{i+2n} from Z_{i+n}, errors in the computation of Q, Z and P in the $(i+n)^{th}$ and $(i+n+1)^{th}$ elements are involved. Due to the introduction of the error ΔZ_i in Z_i we find for P_{i+n} the value

$$P_{i+n}{}^* = P_{i+n} - \frac{gH}{2\beta d}\left\{3 + \frac{4jd^2\omega\beta}{gH}\right\}\Delta Z_i\ , \tag{4.14}$$

and for Q_{i+n+1}, Z_{i+n+1} and P_{i+n+1}, we find, respectively, the values

$$Q_{i+n+1}{}^* = Q_{i+n+1} + \frac{gH}{2\beta d}\ \Delta Z_i\ , \tag{4.15a}$$

$$Z_{i+n+1}{}^* = Z_{i+n+1} - \Delta Z_i\ , \tag{4.15b}$$

$$P_{i+n+1}{}^* = P_{i+n+1} + \frac{2gH}{\beta d}\left\{1 + j\ \frac{d^2\omega\beta}{gH}\right\}\Delta Z_i\ . \tag{4.15c}$$

It now follows that we find for Q_{i+2n} and Z_{i+2n}, respectively, the values

$$Q_{i+2n}{}^* = Q_{i+2n} + \left\{\frac{8d^3\omega^2\beta}{gH} - \frac{5gH}{\beta d} - 12jd\omega\right\}\Delta Z_i\ , \tag{4.16a}$$

$$Z_{i+2n}{}^{*} = Z_{i+2n} + \left\{13 + \frac{16\beta^2 d^4 \omega^2}{g^2 H^2} + \frac{28 j d^2 \omega\beta}{gH}\right\}\Delta Z_i \, . \qquad (4.16b)$$

The error in the calculation of Z_{i+2n} is therefore found to be, on substitution for β,

$$\Delta Z_{i+2n} = \left\{\left(13 - \frac{28d^2\omega^2}{gH} - \frac{16d^4\omega^2 r^2}{g^2 H^4} + \frac{16d^4\omega^4}{g^2 H^2}\right)\right.$$

$$\left. + j\left(\frac{28d^2\omega r}{gH^2} - \frac{32d^4\omega^3 r}{g^2 H^3}\right)\right\}\Delta Z_i \, . \qquad (4.17)$$

Consideration of the appropriate values of d, ω, H and r for the Lake Albert model, indicates that the magnitude of the error has now increased to approximately ten times the magnitude of the original error ΔZ_i, with

$$F_\ell^{(2)} = 10- .$$

Proceeding in this fashion through the whole array we find that the largest error, which occurs in the westernmost value of Z computed along the northernmost (m^{th}) row, is of the order of $10^{m/2}$ times the original error, in magnitude.

It follows that the number of rows in our array is governed by the propagation of errors through our model array. The round off error in the first-computed value of Z is magnified within the computed "end values" of Q, which in turn affects the computed "starting values" of Z found by the inversion of the matrix A, the coefficients of which consist of "end values" Q. The number of rows in our array is therefore a function of the size of the word used in the computer. For instance, in the University of Adelaide CDC 6400, computations are made to 24 figure accuracy in double precision calculations. If we require results accurate to four significant figures, our array can have the maximum number of rows in it that produce errors in the last 20 figures in the final iteration after calculating the required "starting values", i.e. we can have at least 40 rows in our array.

P,Q computed at same position

The grid scheme adopted here is similar to that described by Heaps (1969). Shown in Fig.2a is a model of Lake Albert constructed from this scheme. A brief description of the grid scheme and algorithms used is given below: a full description is given in Walsh and Noye (1974).

For a given ω, $Z(x,y)$ is evaluated only at points marked "0", while $P(x,y)$, $Q(x,y)$ are evaluated only at points marked "X". A rectangular array of consecutive rows of elevation and stream points completely covers the lake, the array dimensions being 13×11, i.e. 13 rows and 11 columns of both elevation and stream points. The horizontal and vertical array axes (x and y axes respectively) coincide with the west-east and south-north directions respectively. The array is bounded to the north and west by elevation points and to the south and east by stream points. The distance between consecutive columns and rows is denoted d. For this model $d = 630$ m.

For an $m \times n$ array, both elevation and stream points are numbered 1,2, ..., mn starting from the first row and proceeding from west to east along each row. A typical cluster of points is shown in Fig. 2c. Each stream point and each elevation point belongs to a particular <u>group</u> of points. There are 14 groups of stream points and 3 groups of elevation points, listed in Table 2. In Fig. 2b, each array point of the Lake Albert model of Fig. 2a has been designated by its group number.

For all array points interior to the lake boundary (groups 2,16,17 points) spatial derivatives are approximated by centered differences. Thus at the interior stream point i,

$$(\partial Z/\partial x)_i \sim (Z_{i+1} - Z_i + Z_{i+n+1} - Z_{i+n})/4d,$$

$$(\partial Z/\partial y)_i \sim (Z_i - Z_{i+n} + Z_{i+1} - Z_{i+n+1})/4d;$$

while at the interior <u>elevation</u> point i

$$(\partial P/\partial x)_i \sim (P_i - P_{i-1} + P_{i-n} - P_{i-n-1})/4d,$$

$$(\partial Q/\partial y)_i \sim (Q_{i-n} - Q_i + Q_{i-n-1} - Q_{i-1})/4d.$$

Using these approximations (2.4) tranforms into a system of difference equations evaluated at interior array points, viz.

$$\beta_i S_i = -\gamma_i (Z_{i+1} - Z_{i+n}) + K(X + Y) , \qquad (4.18a)$$

$$\beta_i D_i = -\gamma_i (Z_{i+n+1} - Z_i) + K(X - Y) , \qquad (4.18b)$$

$$-4dj\omega Z_i = D_i - S_{i-1} + S_{i-n} - D_{i-n-1} , \qquad (4.18c)$$

where $\beta_i = j\omega + rh_i^{-1}$, $\gamma_i = -gh_i/2d$, and unknowns S_i, D_i are given in terms of P_i, Q_i by $S_i = P_i + Q_i$, $D_i = P_i - Q_i$.

"Degenerate" forms of (4.18) are satisfied by each group of boundary stream point, and are described in the Appendix. The boundary conditions used in such cases are: $P_i = 0$ for groups 5, 6 points, $Q_i = 0$ for groups 3,4 points, $P_i = Q_i = 0$ for groups 7-14 points. Use of a "non-trivial" condition at 270° corner points resulted in a net loss or gain of water from the model, a factor previously noted in a numerical study on Grand Traverse Bay, USA by Smith (1973).

Equations (4.18) together with the boundary equations form a closed set in the complex unknowns Z,S and D evaluated at all non-exterior elevation or stream points. A solution could be obtained immediately by the inversion of a large complex square matrix. An alternative approach which requires only the inversion of a small non-sparse matrix follows.

Rewriting (4.18) as

$$S_i = \{K(X + Y) - \gamma_i (Z_{i+1} - Z_{i+n})\}/\beta_i , \qquad (4.19a)$$

$$D_i = D_{i-n-1} + S_{i-1} - S_{i-n} - 4dj\omega Z_i , \qquad (4.19b)$$

$$Z_{i+n+1} = Z_i - \{\beta_i D_i - K(X - Y)\}/\gamma_i, \qquad (4.19c)$$

then knowing values for S and D at stream points i-n-1, i-n and i-1, and also values for Z at elevation points i, i+1 and i+n, values for

S_i , D_i and Z_{i+n+1} may be calculated. In order to initiate an iterative scheme, assume a "starting value" for Z at each group 16 point. Values for S and D for all points belonging to groups 3,5,7,8,9 and 13 are computed using the degenerate difference equations described in the Appendix. Next, proceeding west to east along each row and treating each row successively, calculate values for S and D at each group 2 point and a value for Z at each group 17 point. The order of calculations is indicated by the direction of the arrows in Fig. 2c. Along each row, stream points belonging to any one of groups 4,6,10,11,12 and 14 may be encountered; values for S and D at such points are computed according to the degenerate equations.

Values for S and D at groups 3,5,7,8,9 and 13 points have been calculated so that the relevant boundary condition at each point is satisfied, regardless of the starting values. This is not the case for groups 4,5,10,11,12 and 14 points - the relevant boundary conditions will only be satisfied at each of these if the correct starting values for Z have been chosen. At each such point define a so-called "end value" e which, if zero, ensures that the relevant boundary condition is satisfied. Details are given in the Appendix. For example, for group 4 points the boundary condition is that $S_i = D_i$, so $e_i = S_i - D_i$.

For a consistent system the number of starting values and the number of end values are the same, say k. Thus vectors $\underset{\sim}{s}$ and $\underset{\sim}{\varepsilon}$ may be formed, both of dimension $(k \times 1)$, consisting of starting values and end values respectively. The vector elements are numbered according to the position in the array of the point to which they refer. For the Lake Albert model in Fig. 2a, $k = 24$. Because each end value is linearly related to the set of starting values, we may write

$$\underset{\sim}{\varepsilon} = A \underset{\sim}{s} + \underset{\sim}{b}.$$

The k-vector $\underset{\sim}{b}$ and the $(k \times k)$ complex matrix A can be found as described previously by marching through the scheme $(k+1)$ times determining the end vectors $\underset{\sim}{\varepsilon}$ corresponding to $(k+1)$ specially chosen starting vectors $\underset{\sim}{s}$. The required starting vector is given by

$$\underset{\sim}{s}^* = -A^{-1} \underset{\sim}{b}.$$

The values of Z,P,Q are then found by running through the scheme once more with the starting vector $\underset{\sim}{s}^*$.

A discrete perturbation stability analysis for this finite difference scheme yielded the same stability criterion as that which applied to the previous system.

5. CONCLUSIONS

Using the linearised depth-integrated equations of motion and continuity, several finite difference schemes have been devised to investigate wind effects on a lake caused by a spatially homogeneous wind stress oscillating sinusoidally in time. The steady state response is found without inverting a very large matrix, a process required if conventional methods of solving the system of linear algebraic equations are applied. The one-dimensional schemes were stable at all frequencies of interest for shallow waters and practical grid sizes. Even though the method is generally unstable for two-dimensional systems, it nevertheless gives good results for models with reasonably fine grid spacing. Investigation of other systems is continuing, in order to find a stable two-dimensional scheme.

6. REFERENCES

Fischer, G. (1959) Ein Numerisches Verfahren zur Errechnung von Windstau und Gezeiten in Randmeeren. Tellus 11, 60-76.

Francis, J.R.D. (1954) A Note on the Velocity Distribution and Bottom Stress in a Wind-driven Current System. J. Mar. Res. 12,93-98.

Groen, P. and Groves, G.W. (1960) Surges. fr. The Sea Vol.1 (M.N.Hill, ed.) Interscience, New York, 611-646.

Heaps, N.S. (1969) A Two-dimensional Numerical Sea Model. Phil. Trans. Roy. Soc. A265, 93-137.

Leendertse, J.J. (1967) Aspects of a Computational Model for Long-Period Water Wave Propagation. RAND Corporation, Memorandum, RM-5294-PR.

Noye, B.J. (1973) The Response of Lake Levels to an Unsteady Wind Stress. Bull. Aust. Math. Soc. 8, 423-433.

Noye, B.J. and Walsh, P.J. (1976) Wind-induced Water Level Oscillations in Shallow Lagoons. Aust. J. Marine Freshwater Research, 27, 417-30.

Smith, E.B. (1973) Wind-driven and seiche-forced Water Motion in Grand Traverse Bay, Michigan. Ph.D. Thesis (Eng. Mech.), University of Michigan.

Walsh, P.J. and Noye, B.J. (1974) A Numerical Model of Wind-induced Circulation in the Murray Mouth Lakes, South Australia. 5th Austln. Conf. on Hyd. and Fl. Mech., University of Canterbury, Christchurch, N.Z. 2, 284-293.

APPENDIX: FINITE-DIFFERENCE SPECIFICATION OF THE BOUNDARY CONDITIONS FOR THE NUMERICAL MODEL GIVEN BY EQUATION (4.19)

For groups 3,5 boundary points specification of values for S_i, D_i is made in terms of nearby "starting values" Z_i.

Group 3

The boundary condition is $Q_i = 0$ so that $S_i = D_i (= P_i)$. In terms of the starting values Z_{i+n} and Z_{i+n+1} (refer Fig. (a)), $(\partial Z/\partial x)_i \sim (Z_{i+n+1} - Z_{i+n})/2d$ which on substitution into Equation (2.4a) gives

$$S_i = D_i = \{KX - \gamma_i (Z_{i+n+1} - Z_{i+n})\}/\beta_i \,. \qquad \text{(A1)}$$

Group 5

The boundary condition $P_i = 0$ gives $S_i = -D_i (= Q_i)$. Writing $(\partial Z/\partial y)_i \sim (Z_{i+1} - Z_{i+n+1})/2d$ (refer Fig. (b)) yields, on substitution into Equation (2.4b),

$$S_i = -D_i = \{KY - \gamma_i (Z_{i+1} - Z_{i+n+1})\}/\beta_i \,. \tag{A2}$$

Groups 7,8,9,13

The boundary condition is $P_i = Q_i = 0$ which gives

$$S_i = D_i = 0 \,. \tag{A3}$$

At boundary points belonging to other groups, an "end" value is defined which, given the correct starting values, must take a zero value.

Group 4

Again $S_i = D_i$ so that $e = S_i - D_i$ is a suitable choice for an end value. Clearly, from Fig. (c), S_i may be found from an equation similar to (A1),viz.

$$S_i = \{KX - \gamma_i (Z_{i+1} - Z_i)\}/\beta_i \tag{A4}$$

while D_i may be determined from Equation (4.18).

Group 6

Again $S_i = -D_i$ so that $e = S_i + D_i$ is a suitable choice for an end value. S_i is determined from

$$S_i = \{KY - \gamma_i (Z_i - Z_{i+n})\}/\beta_i \tag{A5}$$

(refer Fig.(d)) while D_i is found from Equation (4.18).

Groups 10,11,12,14

Again $S_i = D_i = 0$. A suitable choice for an end value is thus $e = D_i$, the value for D_i being found by application of Equation (4.18).

Fig. (a)

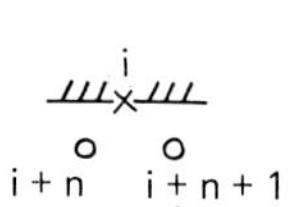

Fig. (b)

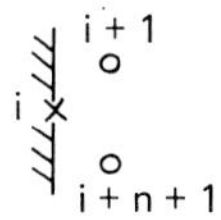

Fig. (c)

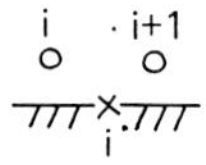

Fig. (d)

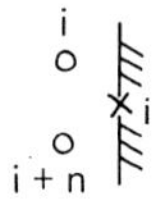

Numerical Simulation of Fluid Motion, ed. John Noye
pp. 453-465

COMPARISON OF SOME FINITE DIFFERENCE AND ANALYTICAL METHODS OF SOLVING TIDAL PROPAGATION EQUATIONS

B. Webber, M.M. Rienecker, M.D. Teubner and B.J. Noye

University of Adelaide

South Australia

SUMMARY

The propagation of tides in a bay of variable breadth and constant depth is investigated using an analytical method and three different finite-difference numerical methods. The effect of various values for the friction parameter K is investigated and a comparison made between results obtained for the usual value used, $K = .002\ ms^{-1}$.

1. INTRODUCTION

Although various finite-difference numerical schemes for modelling tidal propagation into gulfs and shallow seas have long been in use, few comparisons have been made between them. Here, two time-stepping numerical schemes based on the two-dimensional spatial grids of Lauwerier (1962) and Platzman (1958) (henceforth referred to as Model L and Model P) are compared with two frequency response methods. To do this, Models L and P are allowed to run through several tidal cycles until a steady-state is reached. The next cycle provides an output of elevations and velocities. These time-series are then Fourier analyzed and the results compared with those from a two-dimensional numerical frequency response method and a one-dimensional analytical response solution. The latter is a solution to the linearized one-dimensional approximation to channel flow. In fact, four different mathematical models are being compared - the analytical solution is an exact solution to a set of equations representing one-dimensional channel type flow as an approximation to the tidal propagation in the gulf, whilst the numerical schemes give an approximate solution to a set of differential equations which represent two-dimensional flows.

The aim of this discussion is to compare the results obtained from each of these schemes when applied to a model situation and thus to determine the validity of the use of each.

2. TIME STEPPING NUMERICAL MODEL P

The equations used for this investigation are the linearized two-dimensional depth-integrated equations described by Noye and Tronson (1978), omitting the Coriolis terms. They are:-

Continuity:
$$\frac{\partial U}{\partial x} + \frac{\partial V}{\partial y} + \frac{\partial \zeta}{\partial t} = 0 \qquad (2.1a)$$

Momentum:
$$\frac{\partial U}{\partial t} + \frac{KU}{h} + gh\frac{\partial \zeta}{\partial x} = 0 \qquad (2.1b)$$

$$\frac{\partial V}{\partial t} + \frac{KV}{h} + gh\frac{\partial \zeta}{\partial y} = 0 \qquad (2.1c)$$

where $U(x,y,t)$ and $V(x,y,t)$ are respectively the x and y components of the stream transport,

$h(x,y)$ is the depth of the basin at position (x,y),

g is the acceleration due to gravity,

K is the linear coefficient of friction,

and $\zeta(x,y,t)$ is the displacement of the surface from its mean position.

These equations are solved subject to the appropriate initial and boundary conditions.

The discretisation of Equations (2.1) is achieved by using finite-difference approximations to derivatives, based on a rectangular grid scheme essentially the same as that used by Platzman (1958). Within this grid system, elevation and the two flow components are each evaluated at distinct points (see Figure 2.1). The basic grid element consists of one point at which the elevation is calculated, a point at which a U component of flow is computed and one point at which a V component of flow is evaluated. Each individual element is classified according to its position relative to the boundary - this produces a total of 17 different element labels. The labelling system is the same as that used by Flather (1972). The label of a particular element determines which form the finite-difference algorithm will take. Basically, the scheme uses forward differences in time and centred differences in space whilst utilising the most recently computed values of each variable. A subscripted array (of size $m \times n$) is laid over the area of interest and each of the variables is referred to by its position within that array (see Figure 2.1).

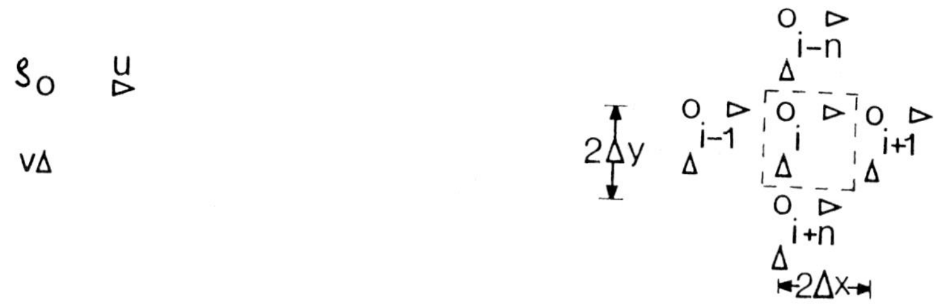

(a) (b)

Figure 2.1: *(a) Orientation of displacement (O); U flow component (▷) and V flow component (Δ) within the basic element.*

(b) The relationship of the i-th grid element to those surrounding it.

The most simple form of the finite difference equations occurs for elements not adjacent to boundaries. The equation of continuity (2.1a) yields the following algorithm for computing elevations one time-step ahead:

$$\zeta_i(t+\Delta t) = \zeta_i(t) - \frac{\Delta t}{2\Delta x}\left[U_i(t) - U_{i-1}(t)\right] - \frac{\Delta t}{2\Delta y}\left[V_{i-n}(t) - V_i(t)\right] \tag{2.2a}$$

The equations of motion (2.1b,c) yield the following algorithms for computing the U and V flow components at time $(t + \Delta t)$:

$$U_i(t+\Delta t) = U_i(t) - \frac{2K\Delta t\, U_i(t)}{h_i + h_{i+1}} - \frac{g\Delta t}{4\Delta x}(h_i + h_{i+1})\{\zeta_{i+1}(t+\Delta t) - \zeta_i(t+\Delta t)\} \tag{2.2b}$$

$$V_i(t+\Delta t) = V_i(t) - \frac{2K\Delta t\, V_i(t)}{(h_i + h_{i+n})} - \frac{g\Delta t}{4\Delta y}(h_i + h_{i+n})\{\zeta_i(t+\Delta t) - \zeta_{i+n}(t+\Delta t)\} \tag{2.2c}$$

The form of the finite difference algorithm depends on which type of element is being considered and is similar to that used by Flather (1972). The main differences between these equations and those used by Flather (1972) are due to the location of the depth point within the grid element and to the difference in the formulation of the grid array.

Analysis of the processes involved has shown that the steady-state is reached quite rapidly if the initial displacement over the whole array is set to the negative of the maximum amplitude of oscillation or, in the case of an open system, to the negative of the maximum displacement which occurs at the open boundary. With this initial condition and the appropriate boundary conditions the solution to the finite difference problem can now be obtained. A complete stability analysis shows that, provided certain relationships between grid-sizes, time-steps and depths are obeyed, the system will be stable under all conditions. The stability analysis follows procedures similar to those used by Flather (1972). The results obtained from this analysis give the restriction

$$\frac{2K\Delta t}{h^*} + \left[\left(\frac{\Delta t}{\Delta x}\right)^2 + \left(\frac{\Delta t}{\Delta y}\right)^2\right] gh^* \leqslant 4 \tag{2.3}$$

where h^* is the maximum depth.

In conjunction with the computer program of the finite difference scheme, a program is used which Fourier analyzes one cycle of the fundamental oscillation of displacements into various constituent harmonics and determines the phases and amplitudes of each. The accuracy of this analysis is largely dependent upon the number of data points used and the relative magnitudes of the fundamental and the harmonics. A series of test runs made with the Fourier scheme showed that the accuracy of the results was dependent on the number of data points used in one cycle of the fundamental oscillation. For a typical input of

$$y = 0.1 \sin t + 0.01 \sin 2t + 0.005 \sin 3t + 0.002 \sin 4t,$$

the resultant output gave, for 50 data points per cycle of the fundamental oscillation a maximum error of 6 per cent, for 100 data points a maximum error of 3 per cent, and for 1000 points a maximum error of 1 per cent in the analyzed values for the amplitudes and phases of the four constituents.

3. TIME STEPPING NUMERICAL MODEL L

The differential equations upon which this model is based are the same as those for Model P, namely Equations (2.1). Finite difference approximations to derivatives are again used in this model, but now a grid scheme similar to that used by Heaps (1969) and Noye and Tronson (1978) is applied. This is a rectangular grid scheme in which elevation is calculated at one point and the two flow components are calculated at another point as illustrated in Figure 3.1. An element of this grid scheme contains one elevation point and one flow point. According to the position of an element relative to the boundaries, it is classified into one of twenty-two possible categories as listed in Heaps (1969). Once again the form of the difference equation is determined by the type of the element being considered. The model uses forward and backward differences in time and centred differences in space. Unfortunately, this type of difference scheme tends to produce troublesome reflections at 270° corners. Smith (1973) has overcome this by modifying the difference equations when an element falls into this category. In general, the number of grid points in the system is determined by the fact that there are m rows each containing n grid points and both elevation and flow points are numbered $1,2,3,\ldots,mn$, (see Figure 3.1).

The finite difference algorithms which produce the elevation and flow components one time-step ahead are

$$\zeta_i(t+\Delta t) = \zeta_i(t)$$

$$- \frac{\Delta t}{4\Delta x}\{U_{i+n+1}(t+\Delta t) - U_{i+n-1}(t+\Delta t) + U_{i-n+1}(t+\Delta t) - U_{i-n-1}(t+\Delta t)\}$$

$$- \frac{\Delta t}{4\Delta y}\{V_{i+n+1}(t+\Delta t) - V_{i-n+1}(t+\Delta t) + V_{i+n-1}(t+\Delta t) - V_{i-n-1}(t+\Delta t)\} \quad (3.1a)$$

$$U_i(t+\Delta t) = U_i(t) - \frac{K}{h}\Delta t U_i(t)$$

$$- \frac{gh\Delta t}{4\Delta x}\{\zeta_{i+n+1}(t) - \zeta_{i+n-1}(t) + \zeta_{i-n+1}(t) - \zeta_{i-n-1}(t)\} \quad (3.1b)$$

$$V_i(t+\Delta t) = V_i(t) - \frac{K}{h}\Delta t V_i(t)$$

$$- \frac{gh\Delta t}{4\Delta y}\{\zeta_{i+n+1}(t) - \zeta_{i-n+1}(t) + \zeta_{i+n-1}(t) - \zeta_{i-n-1}(t)\} \quad (3.1c)$$

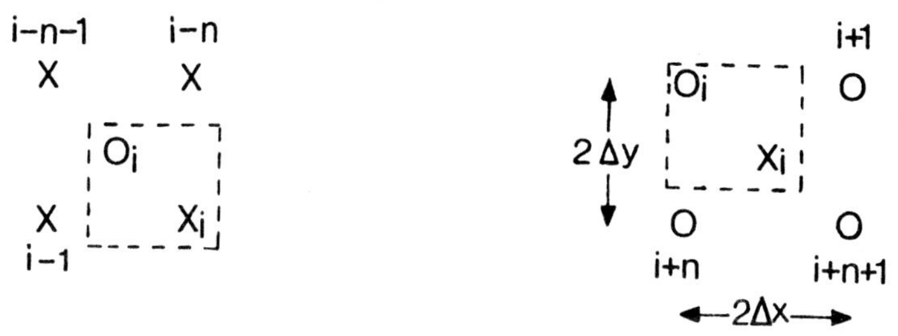

(a) 4 stream points surrounding one elevation point.

(b) 4 elevation points surrounding one stream point.

Figure 3.1: Illustration of orientation of points within the grid scheme.

A stability analysis of this scheme has shown that, provided the following restriction on relations between time and space increments are obeyed, the system will remain stable if

$$\Delta t \leqslant \frac{2\,\min(\Delta x,\ \Delta y)}{(gh^*)^{1/2}} \quad (3.2)$$

where h^* is the maximum depth in the system being considered (see Noye and Tronson, 1977) provided a certain minimum value of the friction coefficient is used. Tests used in this analysis have shown that stability is maintained for values of $K > 10^{-4}\,m.s.^{-1}$, but the minimum value may be much less than this. In fact, further analysis indicates that by modification of the finite difference equations the requirement of the presence of friction may be eliminated. However, in practice, the actual level of friction is generally greater than the value mentioned above, so this restriction is not serious.

4. NUMERICAL FREQUENCY RESPONSE METHOD

This model is also based on the Equations (2.1), modified only in that the fundamental equations involve depth-averaged velocities instead of transports. To determine the frequency response, an input

$$\zeta = Z_0 e^{j\omega t}$$

is applied to the system at the open boundary $x = L$, ω being the circular frequency of the tide. The value which ω takes in this analysis, namely 1.4×10^{-4} s^{-1}, corresponds to a tidal period of 12 hours. Since the system is linear, $\zeta(x,y,t)$, $u(x,y,t)$ and $v(x,y,t)$ are also proportional to the time factor $e^{j\omega t}$.

If $Z(x,y)$, $P(x,y)$ and $Q(x,y)$ represent the spatial factors of the elevation and x and y components, respectively, of the depth-averaged velocity, then the fundamental equations are solved subject to the conditions that $Z(L,y) = Z_0$ and that the component of the vector (P,Q) normal to the boundary is zero everywhere along the contour.

The numerical method used in this model follows that used by Noye (1978). The grid is composed of n rows of m elements, numbered $1,2,\ldots,mn$. Each element belongs to one of ten classes according to its position with respect to an open boundary or a land-sea boundary and the finite difference equations are modified accordingly. The grid is "staggered", that is Z, P and Q are evaluated at different points in each element and are positioned as shown in Figure 4.1. If Z is evaluated at the point (x,y), then P is evaluated at $(x-\Delta x,y)$ and Q at $(x,y-\Delta y)$. The values of each of these at their respective points in the i-th element are denoted Z_i, P_i and Q_i.

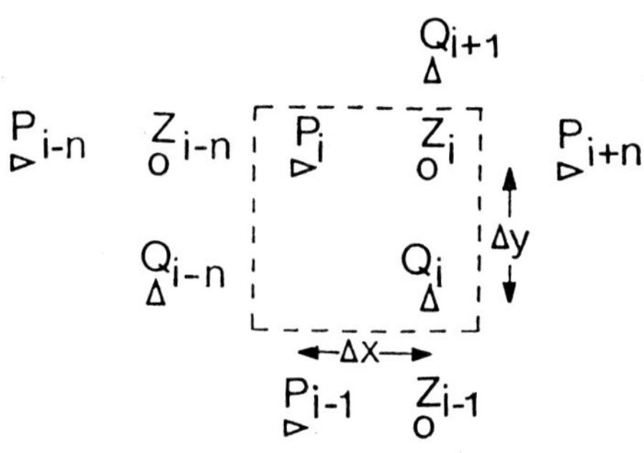

Figure 4.1: Illustration of basic element configuration and its relation to neighbouring elements.

The finite difference equations, obtained by manipulation of the centred finite difference approximations to the derivatives in Equations (2.1) with depth considered a constant, are

$$P_i = P_{i-n} + \frac{\Delta x}{\Delta y}(Q_{i-n} - Q_{i-n+1}) - \frac{2j\omega\Delta x}{h} Z_{i-n} \tag{4.1a}$$

$$Z_i = Z_{i-n} - \frac{2\Delta x}{g}\left(j\omega + \frac{K}{h}\right) P_i \tag{4.1b}$$

$$Q_i = -\left\{j\omega + \frac{K}{h}\right\}^{-1} \frac{g}{2\Delta y}(Z_i - Z_{i-1}) \tag{4.1c}$$

These equations, together with the boundary condition that $Z = Z_0$ at $x = L$, $P_i = 0$ if a vertical segment of the boundary passes through the i-th element, and $Q_i = 0$ if a horizontal segment of the boundary passes through the i-th element, form a closed set in the complex unknowns Z, P and Q evaluated for all non-exterior points.

The iterative scheme is initiated by choosing a starting value for P at the open boundary, $x = L$. The appropriate starting values which produce the required end-values, $P_i = 0$, are found using an adaptation to the iterative scheme given by Noye (1977).

If ΔZ_{i+n} is the error in the value of Z_{i+n} which results if an error ΔZ_i is introduced in Z_i, analysis shows that

$$|\Delta Z_{i+n}| = |F| \ |\Delta Z_i|$$

where the magnification factor $|F|$ is approximately 3. However the error build-up, using this finite difference scheme, is slow enough to give a good approximation to the solution if both the grid size and the number of columns of elements is restricted. This, naturally, limits the usefulness of the model, but, in cases where it is applicable, the results obtained are in good agreement with those given by the other models as shown in Table 6.1.

5. ONE-DIMENSIONAL ANALYTICAL FREQUENCY RESPONSE SOLUTION

Consider the propagation of a tide into the bay shown in Figure 5.1

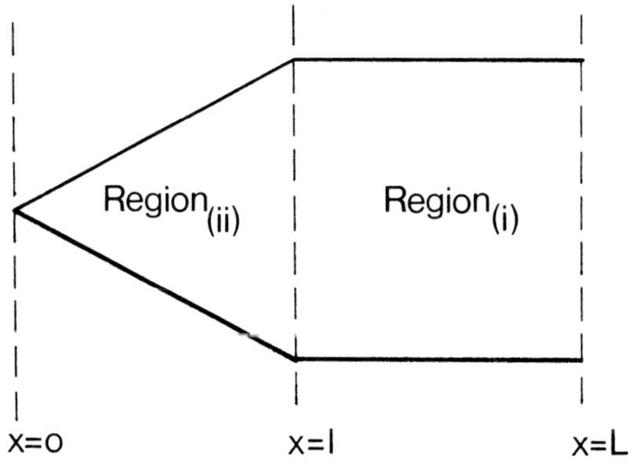

Figure 5.1: Illustration of bay geometry

In region (i), $h(x) = h_0$,
$b(x) = b_0 \ell$,

In region (ii), $h(x) = h_0$,
$b(x) = b_0 x$,

where $h(x)$ is the depth of the bay,
$b(x)$ is the breadth of the bay,
h_0, b_0 are constants determining the depth of the bay and the angle at $x = 0$,

x is the variable parameter measuring the distance from the vertex of the bay to the open end.

ℓ, L are the lengths as illustrated in Figure 5.1.

The steady-state analytical solution to this problem is found using the linearized one-dimensional channel flow approximation to Equations (2.1), viz:

$$\frac{\partial u}{\partial t} = - g \frac{\partial \zeta}{\partial x} - \gamma u , \tag{5.1a}$$

$$\frac{\partial \zeta}{\partial t} = - \frac{1}{b} \frac{\partial}{\partial x} (hbu) , \tag{5.1b}$$

where $\zeta(x,t) = Z(x) e^{j\omega t}$ is the elevation of the water surface, $u(x,t) = P(x)e^{j\omega t}$ is the mean velocity over the breadth in the x-direction, γ is the frictional parameter given by $\gamma = K/h$ where K is the constant defined previously.

Substitution of these expressions into (5.1a) and (5.1b) and subsequent rearrangement yields the following differential equation for Z in region (i)

$$\frac{d^2 Z}{dx^2} - k^2 Z = 0 , \tag{5.2}$$

where
$$k^2 = - \frac{\omega(\omega^2 + \gamma^2)}{gh_0(\omega + j\gamma)} .$$

Therefore the solution in this region has the form

$$Z(x) = Ae^{kx} + Be^{-kx} ,$$

where A and B depend upon the boundary conditions.

For region (ii) the resultant equation is

$$x \frac{d^2 Z}{dx^2} + \frac{dZ}{dx} + \lambda^2 x Z = 0 , \tag{5.3}$$

where
$$\lambda^2 = \frac{\omega(\omega^2 + \gamma^2)}{gh_0(\omega + j\gamma)}$$

and hence the solution is of the form

$$Z(x) = CJ_0(\lambda x) + DY_0(\lambda x) ,$$

where C and D are constants (see McLachlan (1955)).

Since the bay terminates at $x = 0$, where $Y_0(x)$ ceases to be a solution to Equation (5.2), D must be zero, and hence $Z(x) = CJ_0(\lambda x)$.

To evaluate A, B and C, the following conditions are used:

(a) $Z(x)$ is continuous at $x = \ell$,

(b) $\frac{dZ}{dx}$ is continuous at $x = \ell$,

(c) $Z = Z_0$ (a constant amplitude across the open boundary) at $x = L$.

Applying these conditions it is found that

$$A = \frac{\alpha Z_0}{\alpha e^{kL} + \beta e^{-kL}} ,$$

$$B = \frac{\beta Z_0}{\alpha e^{kL} + \beta e^{-kL}} ,$$

$$C = \frac{2kZ_0}{\alpha e^{kL} + \beta e^{-kL}} ,$$

where $\alpha = k\, J_0(\lambda\ell) e^{-k\ell} - \lambda J_1(\lambda\ell) e^{-k\ell}$,

and $\beta = \lambda J_1(\lambda\ell) e^{k\ell} + k J_0(\lambda\ell) e^{k\ell}$,

k and λ being as previously defined.

The solution for $\zeta(x,t)$ is thus given by

Region (i) : $0 \leqslant x \leqslant \ell, \quad \zeta(x,t) = \dfrac{2kZ_0}{\alpha e^{kL} + \beta e^{-kL}} J_0(\lambda x) e^{j\omega t}$ (5.4a)

Region (ii) : $\ell \leqslant x \leqslant L, \quad \zeta(x,t) = \dfrac{Z_0(\alpha e^{kx} + \beta e^{-kx})}{\alpha e^{kL} + \beta e^{-kL}} e^{j\omega t}$ (5.4b)

6. RESULTS

The methods described in the preceding sections have been applied to a basin with geometry shown in Figure 5.1 using the following dimensions

$$\ell = 90 \text{ km}$$

$$L = 135 \text{ km}$$

$$h_0 = 20 \text{ m}$$

$$b_0 = 2 \text{ (i.e. the angle at the vertex is } 90^\circ)$$

With these dimensions each of the numerical models used a square grid, i.e. $\Delta x = \Delta y = d$, with $d = 10$ km.

For each model, a tidal input defined by,

$$\zeta_{in}(t) = 0.1 \cos \omega t$$

was applied across the mouth, where ω (the circular frequency of oscillation) took the value $2\pi/12$ rad.hr.$^{-1}$, so that the period of oscillation was 12 hours.

The models L and P, which must pass through a transient stage before steady-state oscillations are reached, needed two complete tidal cycles to reach this stage. Results from the third cycle were stored for analysis. The subsequent results from this analysis and those obtained from the numerical and analytical frequency response methods are shown in Table 6.1 for a value of the friction coefficient of 0.002 ms^{-1}. The results given for the numerical models were obtained by taking a weighted average across a section so that they could be compared with the one-dimensional analytic results. Across any section amplitudes at the boundaries were weighted to have only half the effect of amplitudes farther from the boundaries; that is, for a section n grid points

TABLE 6.1

Comparisons of amplitude and phase-lag of surface elevation for the four methods of solution for a frictional coefficient K = .002 ms^{-1}.

Distance from Mouth (in kms.)	Amplitude of oscillation (in metres)			
	Analytic	Frequency-Response Model	Model L	Model P
0	.100	.100	.100	.100
10	.106	.105	.105	.104
20	.118	.115	.116	.113
30	.133	.127	.129	.125
40	.148	.140	.142	.138
50	.162	.154	.157	.153
60	.177	.170	.172	.170
70	.190	.185	.186	.185
80	.201	.199	.199	.200
90	.212	.212	.211	.214
100	.220	.225	.220	.226
110	.227	.231	.228	.234
120	.231	.236	.232	.240

Distance from Mouth (in kms.)	Phase-Lag Relative to Mouth (in hours)			
	Analytic	Frequency-Response Model	Model L	Model P
0	0	0	0	0
10	.59	.52	.54	.55
20	1.07	.95	.98	1.01
30	1.43	1.29	1.31	1.37
40	1.70	1.53	1.56	1.65
50	1.88	1.77	1.77	1.90
60	2.03	1.97	1.94	2.11
70	2.15	2.13	2.08	2.28
80	2.24	2.25	2.19	2.41
90	2.32	2.35	2.28	2.52
100	2.37	2.42	2.34	2.60
110	2.39	2.47	2.39	2.65
120	2.45	2.51	2.41	2.68

in width and denoting the amplitude at each grid point by a_i, then

$$\text{Average amplitude} = \{\tfrac{1}{2}(a_1 + a_n) + \sum_{i=2}^{n-1} a_i\}/(n-1)$$

Comparisons were also made between the various methods for frictional values of $K = 0.04$ and $0.0001\ ms^{-1}$. The results obtained for these values showed similar agreement.

Differences between the values obtained by the four methods vary according to the frictional value used and the position within the basin. The variation between the three numerical methods is relatively small and since the results obtained from these methods stemmed from a two-dimensional analysis, it is probable that they are more realistic values than those given by the analytic model.

The agreement between numerical and analytical results is also good and so it appears that each of the numerical models approximates well the exact solution of the two-dimensional system.

The following factors should be considered when looking at these results:

(a) the analytical model treats the flow as being one-dimensional and hence the analytical solution is a width average;

(b) each of the numerical methods is two-dimensional and hence the narrowing of the basin causes the presence of the sides to have an increasing effect on the flow and hence causes a variation in both elevation and phase-lag across the basin;

(c) the representation of the straight sides by a "stepped" rectangular grid can cause reflection and production of various harmonics of the fundamental frequency of oscillation within the time-stepping models. For example, model L (for $K = .002$) produces a first harmonic which has amplitudes ranging from .001 metres at the mouth to .005 metres at the closed end with the corresponding phase-lags ranging from 0 hours to 2.77 hours. The amplitude of the second harmonic ranges from 0 metres at the mouth to .003 metres. For different frictional values the magnitude and phase-lags of the various harmonics will differ from these values.

To emphasize the importance of (a) and (b) in evaluating these results, a comparison of examples of the two-dimensional results is given using one of the time stepping models and the frequency response model. Tables 6.2 and 6.3 give the spatial results for the amplitude and phase lag over the basin for $K = .002$ for Model P and the frequency response model. (Because the flow is symmetrical about the centre line only the upper half of the results is shown.) In each case there is a marked difference between the values of elevations and stream flow at the edge and those at the centre. The variation in the elevation, or in the phase lag, from one grid point to the next decreases as the centre line is approached indicating the lessening effect of the basin geometry upon the flow.

7. CONCLUSION

This investigation shows that each of the three numerical schemes described can be used to solve, with reasonable accuracy, the equations describing the tidal propagation into a bay or shallow sea where the depth is large enough for second order terms to be unimportant and dimensions small

TABLE 6.2

Two dimensional array of (a) elevation amplitudes and (b) elevation phase lags, from Time Stepping Numerical Model P.

(a)	.100	.105	.112	.120	.125								
	.100	.105	.112	.121	.129	.137							
	.100	.104	.112	.122	.133	.143	.152						
	.100	.104	.113	.124	.136	.148	.159	.170					
	.100	.104	.113	.125	.139	.152	.166	.178	.189				
	.100	.104	.113	.127	.141	.156	.171	.184	.196	.206			
	.100	.104	.114	.128	.143	.159	.175	.189	.202	.213	.222		
	.100	.104	.114	.128	.144	.161	.177	.192	.205	.217	.226	.234	
	.100	.104	.114	.129	.145	.162	.179	.194	.207	.219	.229	.236	.241
(b)	0	.46	.84	1.11	1.27								
	0	.48	.87	1.17	1.39	1.57							
	0	.50	.92	1.25	1.50	1.70	1.86						
	0	.53	.97	1.32	1.59	1.80	1.97	2.10					
	0	.55	1.02	1.39	1.67	1.89	2.06	2.20	2.30				
	0	.57	1.06	1.44	1.74	1.96	2.13	2.27	2.37	2.46			
	0	.59	1.09	1.49	1.78	2.00	2.18	2.32	2.42	2.51	2.57		
	0	.60	1.12	1.51	1.81	2.04	2.22	2.35	2.46	2.54	2.60	2.63	
	0	.61	1.13	1.53	1.83	2.06	2.23	2.37	2.47	2.55	2.61	2.66	2.69

TABLE 6.3

Two dimensional array of (a) elevation amplitudes and (b) elevation phase lags, from Numerical Frequency Response Model.

(a)	.100	.105	.113	.121	.127								
	.100	.105	.113	.122	.131	.138							
	.100	.105	.114	.124	.134	.144	.153						
	.100	.105	.114	.126	.138	.149	.160	.170					
	.100	.105	.115	.127	.140	.154	.166	.178	.188				
	.100	.105	.115	.128	.143	.157	.171	.183	.195	.204			
	.100	.105	.116	.129	.145	.160	.175	.188	.200	.210	.219		
	.100	.105	.116	.130	.146	.162	.177	.191	.203	.214	.223	.230	
	.100	.105	.116	.130	.147	.163	.178	.193	.205	.216	.225	.232	.236
(b)	0	.44	.80	1.05	1.20								
	0	.46	.83	1.11	1.31	1.47							
	0	.48	.87	1.17	1.41	1.59	1.73						
	0	.50	.92	1.24	1.49	1.68	1.84	1.96					
	0	.53	.96	1.30	1.56	1.76	1.92	2.05	2.15				
	0	.55	1.00	1.35	1.62	1.83	1.99	2.11	2.21	2.29			
	0	.56	1.03	1.39	1.66	1.87	2.03	2.16	2.26	2.34	2.39		
	0	.57	1.05	1.41	1.69	1.90	2.06	2.19	2.29	2.37	2.42	2.47	
	0	.58	1.06	1.43	1.70	1.92	2.08	2.20	2.30	2.38	2.44	2.48	2.51

enough for Coriolis forces to have little effect. Each of the schemes used can, in fact, be adapted to include Coriolis forces but no comparison of these extensions is made here.

The main advantage of the numerical frequency response method is the very great saving in computer time. However, it is only of practical use if the error accumulation due to the instability of the scheme is small enough. The two time-stepping models may be extended to include non-linear terms which is not true for the frequency response method.

Hence, when considering a particular problem the physical aspects must be examined to enable an appropriate choice of numerical scheme to be made. In most cases each of the models will be suitable but, for example, some geometrical configurations will be better suited to the application of one scheme in preference to another.

REFERENCES

Flather, R.A. (1972), Analytical and Numerical Studies in the Theory of Tides and Storm Surges, Ph.D. Thesis (Liverpool).

Heaps, N.S. (1969), A Two-dimensional Numerical Sea Model. Phil. Tran. Roy. Soc., Ser. A., 265, 93-137.

McLachlan, N.W. (1955), Bessel Functions for Engineers, O.W.P.

Noye, B.J. (1978), Finite Difference Schemes for the Solution of Linearised Wind Effect Equations. Numerical Simulation of Fluid Motion, ed. John Noye (North-Holland), 435-451.

Noye, B.J. and Tronson, K.C.S. (1978), Finite Difference Techniques Applied to the Simulation of Tides and Currents in Gulfs. Numerical Simulation of Fluid Motion, ed. John Noye (North-Holland), 285-355.

Platzman, G.W. (1958), A Numerical Computation of the Surge of 26th June, 1954, on Lake Michigan. Geophysica, 6, 407-438.

Proudman, J. (1953), Dynamic Oceanography. Methuen, London.

Smith, E.B. (1973), Wind-driven and Seiche Forced Water Motion in Grand Traverse Bay, Michigan, Ph.D. Dissertation (Eng. Mech.), University of Michigan.

Numerical Simulation of Fluid Motion, ed. John Noye
pp. 467-481

BOUNDARY CONDITIONS FOR THE NAVIER - STOKES EQUATIONS

M.M. Gupta

The Papua New Guinea University of Technology

Lae, Papua New Guinea

SUMMARY

This paper considers the stream function - vorticity formulation of the two-dimensional steady state Navier-Stokes equations. The natural boundary conditions provide the values of stream function and its normal derivative on the solid boundaries. However, the values of vorticity are undefined on the no-slip surfaces and these values are approximated using the boundary data and values of stream function inside the fluid flow region. This paper suggests two kinds of boundary approximations for the unknown vorticity values. The effect of these approximations is to speed up the iterative procedures for solving the Navier-Stokes equations without significantly affecting the accuracy of the final solutions.

1. INTRODUCTION

Consider the stream function - vorticity formulation of the two-dimensional Navier - Stokes equations representing steady flow of a viscous incompressible fluid:

$$\Delta\psi \equiv \frac{\partial^2\psi}{\partial x^2} + \frac{\partial^2\psi}{\partial y^2} = -\omega \tag{1.1a}$$

$$\Delta\omega + R\left(\frac{\partial\psi}{\partial x}\cdot\frac{\partial\omega}{\partial y} - \frac{\partial\omega}{\partial x}\frac{\partial\psi}{\partial y}\right) = 0 \tag{1.1b}$$

Here R is the dimensionless Reynolds number. For simplicity, we consider the prototype cavity flow problem in a square cavity $0 \le x,y \le 1$. The boundary conditions describing the flow of an incompressible fluid in this cavity are given by

$$\psi = 0,\ \frac{\partial\psi}{\partial x} = 0 \text{ when } x = 0,\ 0 < y < 1; \tag{1.2a}$$

$$\psi = 0,\ \frac{\partial\psi}{\partial y} = 0 \text{ when } y = 0,\ 0 < x < 1; \tag{1.2b}$$

$$\psi = 0,\ \frac{\partial\psi}{\partial x} = 0 \text{ when } x = 1,\ 0 < y < 1; \tag{1.2c}$$

$$\psi = 0,\ \frac{\partial\psi}{\partial y} = -1 \text{ when } y = 1,\ 0 < x < 1. \tag{1.2d}$$

Let the interior of the square cavity be denoted by D and its boundary by ∂D. In order to solve this problem using finite difference methods, we introduce a uniform mesh $\bar{D}_h = \{(x_i, y_j):\ x_i = ih,\ y_j = jh,\ 0 \le i, j \le n,\ nh = 1;\ h > 0\}$. The set of mesh points lying inside the cavity D is called D_h and the set of mesh points on the boundary ∂D is ∂D_h.

Let ψ_{ij} and ω_{ij} denote the approximate values of ψ and ω at a mesh point (x_i, y_j). The usual finite difference approximation of the stream function Equation (1.1a) is given by the five point formula:

$$\begin{aligned}\Delta_h\psi_{ij} &\equiv h^{-2}[\psi_{i+1,j} + \psi_{i,j+1} + \psi_{i-1,j} + \psi_{i,j-1} - 4\psi_{i,j}] \\ &= -\omega_{ij},\ i,j = 1,2,\ ---,\ (n-1).\end{aligned} \tag{1.3}$$

The approximation for the vorticity transport Equation (1.1b) is given by

$$\begin{aligned}L_h\omega_{ij} \equiv k_1\,\omega_{i+1,j} &+ k_2\,\omega_{i,j+1} + k_3\,\omega_{i-1,j} + k_4\,\omega_{i,j-1} \\ &- k_0\,\omega_{ij} = 0,\end{aligned} \tag{1.4}$$

where

$$\begin{aligned}k_1 &= 1 - \tfrac{1}{4}R(\beta_{ij} - |\beta_{ij}|), \\ k_2 &= 1 + \tfrac{1}{4}R(\alpha_{ij} + |\alpha_{ij}|), \\ k_3 &= 1 + \tfrac{1}{4}R(\beta_{ij} + |\beta_{ij}|), \\ k_4 &= 1 - \tfrac{1}{4}R(\alpha_{ij} - |\alpha_{ij}|), \\ k_0 &= 4 + \tfrac{1}{2}R(|\alpha_{ij}| + |\beta_{ij}|);\end{aligned} \tag{1.5}$$

and

$$\alpha_{ij} = (\psi_{i+1,j} - \psi_{i-1,j}),$$

$$\beta_{ij} = (\psi_{i,j+1} - \psi_{i,j-1}) \qquad (1.6)$$

for $\quad i,j = 1,2, ---, (n-1).$

The above approximation (1.4) is the so called upwind difference approximation obtained by using the non-central finite differences for discretizing ω_x and ω_y in the vorticity equation (1.1b). The set of Equations (1.4) has a diagonally dominant coefficient matrix which ensures the convergence of the iterative procedures for arbitrary Reynolds numbers.

2. THE BOUNDARY CONDITIONS

The boundary data (1.2) consists of the values of ψ and its normal derivative ψ_n. However, in order to solve the Equations (1.3) and (1.4) we require values of ψ and ω on ∂D_h. The values of ω can be approximated using the following formula

$$\omega_{ij} = -\Delta_h \psi_{ij} \quad , \quad (x_i \ , \ y_j) \text{ on the boundary } \partial D_h. \qquad (2.1)$$

Assuming that (x_i, y_j) lies on the left boundary $x = 0$, this approximation becomes

$$\omega_{0,j} = -h^{-2}[\psi_{1,j} + \psi_{0,j+1} + \psi_{0,j-1} + \psi_{-1,j} - 4\psi_{0,j}] \qquad (2.2)$$

where $\psi_{-1,j}$ is undefined as the mesh point $x = -h$, $y = jh$ lies outside the fluid flow region. However, since the value of ψ_x is known on the left boundary, a further approximation can be made

$$(\psi_x)_{0,j} \approx \frac{\psi_{1,j} - \psi_{-1,j}}{2h} \qquad (2.3)$$

or $$\psi_{-1,j} \approx \psi_{1,j} - 2h\,(\psi_x)_{0,j} \qquad (2.4)$$

Substituting (2.4) into (2.2) we get

$$\omega_{0,j} - -h^{-2}[2\psi_{1,j} + \psi_{0,j+1} + \psi_{0,j-1} - 4\psi_{0,j} - 2h(\psi_x)_{0,j}] \qquad (2.5)$$

Similar approximations can be defined on other parts of the boundary ∂D_h. The approximation (2.5) is the "conventional approximation" for boundary values of the vorticity.

Some authors divide the set D_h into two subsets $D_{h,1}$ and $D_{h,2}$ where $D_{h,1}$ consists of those mesh points of D_h which are one mesh distance away from the boundary ∂D_h and $D_{h,2} = D_h - D_{h,1}$. Greenspan (1968) solved the discrete stream function equation (1.3) only on the set $D_{h,2}$ and interpolated the ψ values on $D_{h,1}$ using approximations of the following type

$$\psi_{1,j} = \frac{1}{4}[\psi_{2,j} + 3\psi_{0,j} + 2h\,(\frac{\partial\psi}{\partial x})_{0,j}] \qquad (2.6)$$

Similar formulas are valid for other parts of $D_{h,1}$. Such formulas have been successfully used by Greenspan (1968, 1969), Gupta and Manohar (1972) and others. For further references, see Gupta (1975 b).

Since the boundary values of ω_{ij} using the formula (2.5) are computed immediately after the values of ψ have been interpolated on $D_{h,1}$ using (2.6), the effective approximation for $\omega_{0,j}$ is a combination of (2.5) and (2.6):

$$\omega_{0,j} = -h^{-2}[\frac{1}{2}\psi_{2,j} + \psi_{0,j+1} + \psi_{0,j-1} - \frac{5}{2}\psi_{0,j} - h(\frac{\partial\psi}{\partial x})_{0,j}] \tag{2.7}$$

Exactly same definition of $\omega_{0,j}$ would be obtained if in place of (2.3), the following approximation were used for eliminating $\psi_{-1,j}$ from (2.2):

$$(\psi_x)_{0,j} \approx \frac{1}{6h}[\psi_{2,j} + 3\psi_{0,j} - 4\psi_{-1,j}] \tag{2.8}$$

It may be noted that the approximations (2.7) and (2.8) involve values of $\psi_{2,j}$ as compared to the conventional approximation (2.5) which involves values of $\psi_{1,j}$. Moreover, considerable success has been reported with the introduction of the modification (2.6) of the conventional approximation (see references in Gupta, 1975 b).

3. NEW BOUNDARY APPROXIMATIONS

As noted above, the conventional boundary approximation (2.5) involves, besides the known boundary data, the unknown values of $\psi_{1,j}$ which are one mesh distance away from the boundary ∂D_h. The modified approximation (2.7) involves, besides the known boundary data, the unknown values of $\psi_{2,j}$ which are two mesh distance away from the boundary. The following question may be posed: What happens when a boundary approximation involving ψ - values further away from the boundary (still inside the fluid flow region) are used?

We propose two kinds of new boundary approximations for ω. The first formula, in the notation of Section 2, is given by:

$$\omega_{0,j} = -h^{-2}[\frac{2}{p^2}\psi_{p,j} + \psi_{0,j+1} + \psi_{0,j-1} - \frac{2h}{p}(\psi_x)_{0,j} - 2(1 + p^{-2})\psi_{0,j}] \tag{3.1}$$

where p is a positive integer.

This formula involves the unknown value of ψ at $x = ph$, $y = jh$ which is at a distance ph away from the boundary. Similar formulas can be written for other segments of the boundary. The truncation error of the above approximation is of order h. Moreover, this formula reduces to (2.5) when $p = 1$ and (2.7) when $p = 2$.

A second type of boundary approximation is defined by

$$\omega_{0,j} = -h^{-2}[-q^3\alpha\psi_{p,j} + p^3\alpha\psi_{q,j} + \psi_{0,j+1} + \psi_{0,j-1} - 2(\frac{p+q}{pq})\,h(\psi_x)_{0,j} - (2 + p^3\alpha - q^3\alpha)\,\psi_{0,j}] \tag{3.2}$$

where $\alpha = \frac{2}{p^2q^2(p-q)}$, $p \neq q$

and p,q are positive integers. This approximation involves two unknown ψ values at (ph, jh) and (qh, jh) and has a truncation error of order h^2. The approximation (3.2) is expected to provide a better solution as more information inside D_h is used to define the boundary values of ω.

The above boundary approximations have been successfully used for the biharmonic equation (Equations (1.1) with $R = 0$). It has been shown that iterative procedure used for solving the discrete equations becomes faster as the boundary approximations with large values of p and q are used. It has also been shown that the use of the approximation (3.2) yields a more accurate solution (Discretization error of h^2) as compared to the use of (3.1) which yields a discretization error of order $h^{3/2}$ (see Gupta, 1975 a; Ehrlich and Gupta, 1975).

4. IMPLEMENTATION

Since the boundary values of ψ, ψ_x, ψ_y for the prototype cavity flow problem are given in (1.2), the boundary values for ω may be approximated as follows:

Using the first order approximation (3.1) for any positive integer p, we obtain the following boundary conditions:

$$
\begin{aligned}
\omega_{0,j} &= -\frac{2}{p^2h^2}\psi_{p,j} && \text{on the left boundary, } x = 0\\
\omega_{n,j} &= -\frac{2}{p^2h^2}\psi_{n-p,j} && \text{on the right boundary } x = 1;\\
\omega_{i,0} &= -\frac{2}{p^2h^2}\psi_{i,p} && \text{on the bottom boundary } y = 0;\\
\omega_{i,n} &= -\frac{2}{h^2}\left[\frac{1}{p^2}\psi_{i,n-p} - \frac{h}{p}\right] && \text{on the top boundary } y = 1.
\end{aligned}
\tag{4.1}
$$

Using the second order approximation (3.2) for any positive integers p and q $(p \neq q)$, we obtain the following boundary conditions:

$$
\begin{aligned}
\omega_{0,j} &= -h^{-2}\alpha[-q^3\psi_{p,j} + p^3\psi_{q,j}];\\
\omega_{n,j} &= -h^{-2}\alpha[-q^3\psi_{n-p,j} + p^3\psi_{n-q,j}];\\
\omega_{i,0} &= -h^{-2}\alpha[-q^3\psi_{i,p} + p^3\psi_{i,q}]\\
\omega_{i,n} &= -h^{-2}[-q^3\alpha\psi_{i,n-p} + p^3\alpha\psi_{i,n-q} - \frac{2h(p+q)}{pq}]
\end{aligned}
\tag{4.2}
$$

where $\alpha = \dfrac{2}{p^2q^2(p-q)}$, $p \neq q$.

In order to solve the discrete stream function - vorticity equations, we define an "outer" iterative procedure as follows:

(a) Start with some initial approximations $\psi^{(0)}$ and $\omega^{(0)}$

(b) Solve the discrete Poisson equations to obtain $\psi^{(1)}$ (Inner iterations for ψ)

(c) Obtain the boundary values of ω from $\psi^{(1)}$ using (4.1) or (4.2).

Call these values $\bar{\omega}^{(1)}$. Obtain the modified boundary values of ω from

$$\omega^{(1)} = (1 - \delta)\bar{\omega}^{(1)} + \delta\omega^{(0)} , \quad 0 \le \delta \le 1. \tag{4.3}$$

(d) Solve the discrete vorticity equation to obtain $\omega^{(1)}$.
(Inner iterations for ω)

(e) Repeat the steps (b), (c), (d) with the new values of $\psi^{(m)}$ and $\omega^{(m)}$, for m = 1, 2, 3, ... until some convergence criterion is satisfied.

The number δ in equation (4.3) is a damping parameter and is chosen to make the above outer iterative procedure converge rapidly.

The above procedure obviates the need of dividing the set D_h into two sets $D_{h,1}$ and $D_{h,2}$. This also eliminates the need to redefine the values of ψ on the set $D_{h,1}$. Such redefinition was only used to give a modified value of ω on the boundary of ∂D_h. Our boundary approximations (4.1) and (4.2) achieve this in one step.

5. NUMERICAL EXAMPLES

We carried out computations using the outer iteration procedure defined in section 4 to test the effectiveness of the boundary approximations (4.1) and (4.2).

We used a non-iterative elimination subroutine to solve the discrete Poisson and vorticity equations at each outer iterative step. We require a near optimal value of the damping parameter δ in Equation (4.3) and an initial estimate is obtained by first computing the growth factors of the overall iterative scheme for various boundary approximations. We carried out a small number of iterations (usually 5) with $\delta = 0$ to obtain the rate ρ of growth of errors in stream function values at successive outer iterates. We found that these growth factors ρ remained virtually constant for all values of the Reynolds number. An initial estimate of the near optimal value of δ is obtained from

$$\delta_{opt} \approx \rho/(\rho + 2) \tag{5.1}$$

Below we give some growth factors and corresponding values of δ_{opt}.

Boundary Approximation (4.1)	Growth factor ρ	Near optimum value of δ(Equation 5.1)
p = 1	3.726	0.6507
p = 2	1.60	0.4444
p = 3	1.10	0.3548
p = 4	0.678	0.2531

Table 1: Mesh size h = 0.1

We discovered that as a large value of p was chosen, the growth factor of the iterative procedure reduced. We used the values of δ of Table 1 to compute solutions for R = 10, 50, 100 and 500. In each case, the number of iterations required for convergence reduced as p increased. In Table 2, we summarize some of these results. In this table N is the number of outer iterations required and E is the error which is the maximum difference between successive ψ iterates after N outer iterations. It can be seen that the convergence in each case is accelerated with large values of p.

Boundary Approximation (4.1)	δ	R = 10		R = 50		R = 100		R = 500	
		N	E	N	E	N	E	N	E
p = 1	0.6507	60	0.099	50	2.5(-7)	33	1.0(-7)	50	5.2(-7)
p = 2	0.4444	32	1.6(-7)*	16	9.4(-8)	19	1.6(-7)	33	1.2(-7)
p = 3	0.4558	14	1.1(-7)	13	1.0(-7)	16	1.1(-7)	24	1.9(-7)
p = 4	0.2531	11	7.8(-8)	12	6.6(-8)	16	1.1(-7)	19	8.8(-8)

*1.6(-7) = 1.6×10^{-7}

Table 2: Mesh size h = 0.1

In case of h = 0.05, we carried out 5 outer iterations with $\delta = 0$ to determine the growth factors for various values of R. As in case of Table 1, these factors remain nearly constant for all R. These results and corresponding δ are given in Table 3.

p	q*	Growth factor ρ	Nearly optimum $\delta (= \frac{\rho}{2+\rho})$
1	0	10.6	0.84
2	0	4.86	0.71
3	0	2.98	0.60
4	0	2.07	0.51
2	1	16.25	0.89
3	2	8.6	0.81
4	3	5.75	0.74
5	4	4.23	0.68
6	3	4.80	0.71
7	2	6.43	0.76

*q = 0 indicates the boundary approximation (4.1)
q $\neq$ 0 indicates the boundary approximation (4.2)

Table 3: Mesh size h = 0.05

We have found that the values of δ given in Table 3 provide a good estimate to the optimum values for most Reynolds numbers. In Tables 4, 5, 6 and 7 we give results of our computations with R = 10, 50, 100 and 500. In the following tables, Error $-\psi$ and Error $-\omega$ are maximum differences between successive ψ and ω iterates after N outer iterations.

p	q*	δ	N	Error -ψ	Error -ω
1	0	0.85	58	4.00(-8)+	8.9(-6)
2	0	0.695	51	9.0(-8)	1.0(-5)
2	1	0.905	87	4.8(-8)	1.0(-5)
3	2	0.82	54	1.5(-8)	8.34(-6)

*See Table 3
+See Table 2

Table 4: h = 0.05, R = 10

p	q	δ	N	Error -ψ	Error -ω
1	0	0.85	66	1.67(-7)	9.30(-6)
2	0	0.66	66	6.15(-8)	7.30(-6)
3	0	0.58	23	1.9(-7)	6.8(-6)
2	1	0.89	91	1.64(-7)	9.77(-6)
3	2	0.81	54	1.53(-7)	8.58(-6)
4	3	0.73	38	1.57(-7)	8.11(-6)

Table 5: h = 0.05, R = 50

p	q	δ	N	Error -ψ	Error -ω
1	0	0.85	82	2.67(-7)	9.54(-6)
2	1	0.89	99	1.04(-6)	3.23(-5)
3	2	0.83	74	2.00(-7)	9.06(-6)

Table 6: h = 0.05, R = 100

p	q	δ	N	Error -ψ	Error -ω
1	0	0.86	99	1.92(-5)	1.34(-3)
2	0	0.70	92	1.34(-7)	9.30(-6)
3	0	0.60	63	2.08(-7)	8.61(-6)
3	2	0.84	99	8.7(-6)	9.2(-4)
4	3	0.75	99	4.3(-7)	6.3(-5)

Table 7: h = 0.05, R = 500

In each of the above cases we note that the outer iteration procedures converge faster when the boundary approximations with larger values of p and q are used. As in the case of the biharmonic equation (Ehrlich and Gupta, 1975), some of the higher order boundary approximations (4.2) are slower than the approximations (4.1). However, it is possible to use a boundary approximation whichis faster and more accurate than the conventional approximation (2.5).

Since there are no quantitative methods available to judge the accuracy of our numerical solutions, we attach some stream line and equivorticity curves for $R = 50$ in the appendix. It may be noted that each boundary approximation yields a qualitatively correct flow pattern.

This research work is still continuing. We are in the process of examining the effectiveness of our boundary approximations for other Reynolds numbers and also for other boundary configurations. We expect to achieve especially good results when a smaller mesh is used near the solid boundaries. In such cases, the boundary approximation (4.2) with large values of p and q should give a rapid convergence without significantly affecting the overall accuracy.

6. REFERENCES

Ehrlich, L.W. and Gupta, M.M. (1975) SIAM J. Numer. Anal., 12, 773-790.

Greenspan, D. (1968) Lectures on the Numerical Solution of Linear, Singular and Nonlinear Differential Equations, Prentice-Hall.

Greenspan, D. (1969) Computer Journal, 12, 89-94.

Gupta, M.M. and Manohar, R. (1972) International J. Numer. Methods Engng, 4, 251-260.

Gupta, M.M. (1975a) SIAM J. Numer. Anal., 12, 364-377.

Gupta, M.M. (1975b) Numerical Methods for Viscous Flow Problems, Tech. Rep. P.N.G. University of Technology, Lae.

APPENDIX: STREAMLINES AND EQUIVORTICITY CURVES FOR VARIOUS BOUNDARY APPROXIMATIONS.

Reynolds number = 50

Mesh size = 0.05

Plate (a) Conventional approximation Equation (2.5) [Boundary approximation (4.1) with $p = 1$].

Plate (b) Modified approximation Equation (2.7) [Boundary approximation (4.1) with $p = 2$].

Plate (c) New boundary approximation (4.1) with $p = 3$.
Plate (d) New boundary approximation (4.2) with $p = 2$, $q = 1$.
Plate (e) New boundary approximation (4.2) with $p = 3$, $q = 2$.
Plate (f) New boundary approximation (4.2) with $p = 4$, $q = 3$.

Legend Stream line contours are drawn for ψ-values of 0.005, 0.01, 0.03, 0.05, 0.06, 0.07 and 0.08.

Equi-vorticity curves are drawn for ω-values of 15.0, 2.0, 0.5, 0.1, -0.1, -0.3 and -0.5.

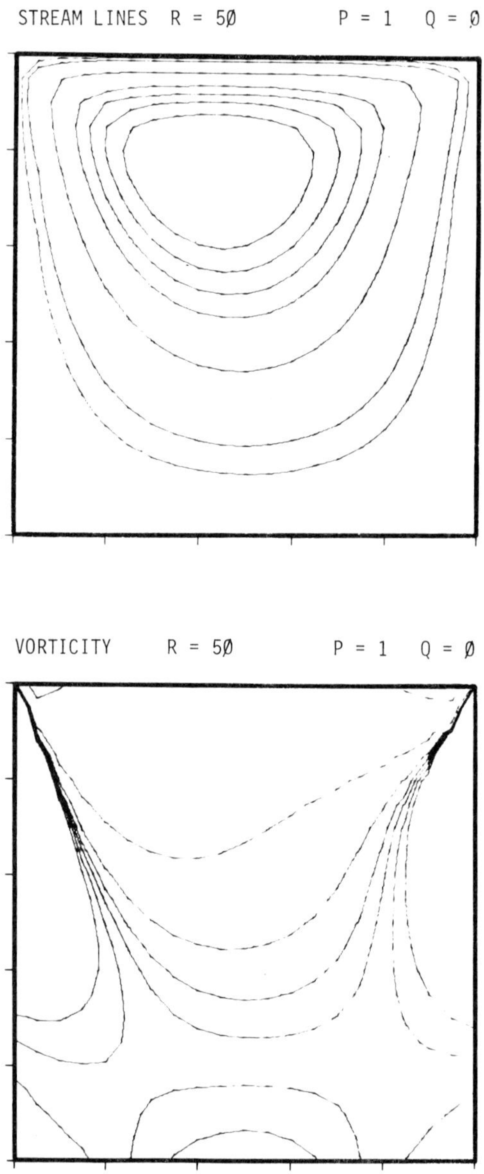

Plate (a) Conventional approximation (2.5)

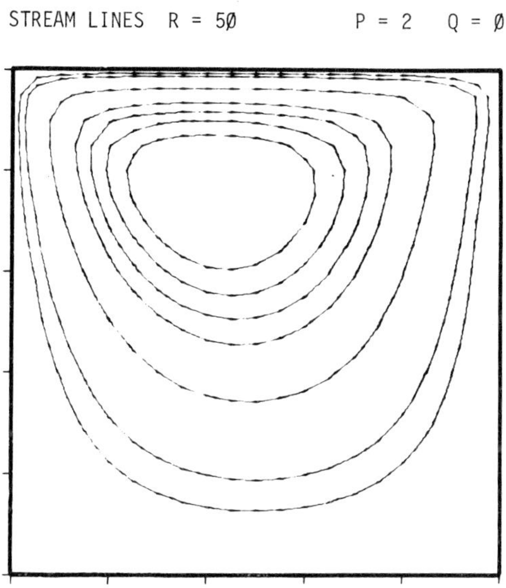

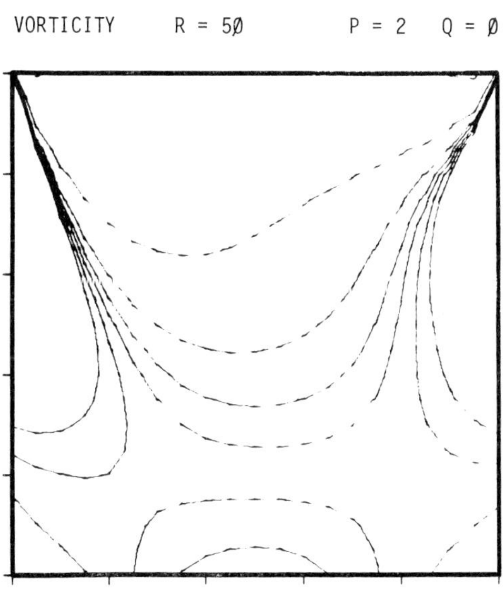

Plate (b) Modified approximation (2.7)

STREAM LINES R = 5∅ P = 3 Q = ∅

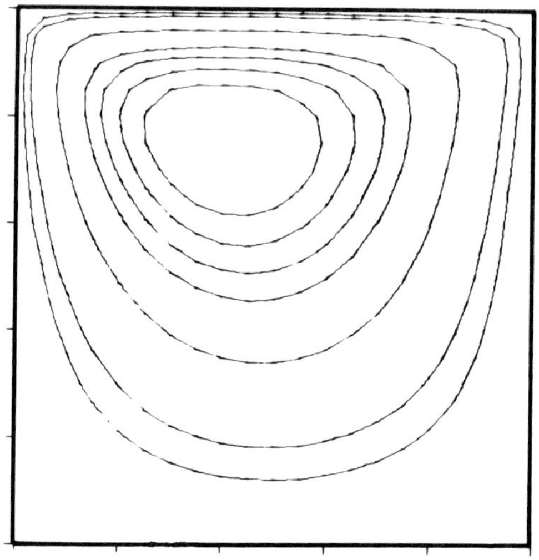

VORTICITY R = 5∅ P = 3 Q = ∅

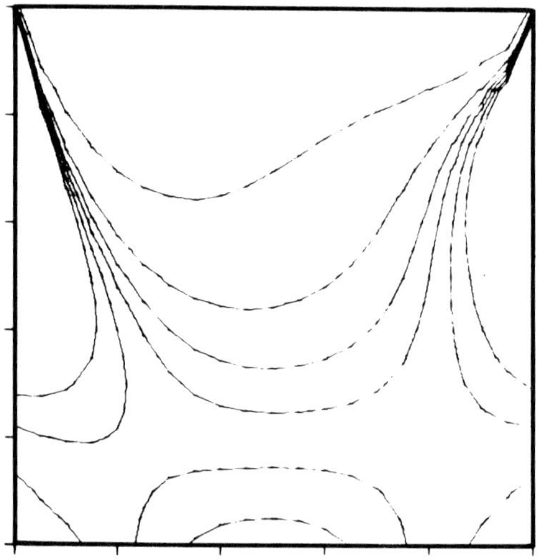

Plate (c) New Boundary approximation (4.1)

STREAM LINES R = 5∅ P = 2 Q = 1

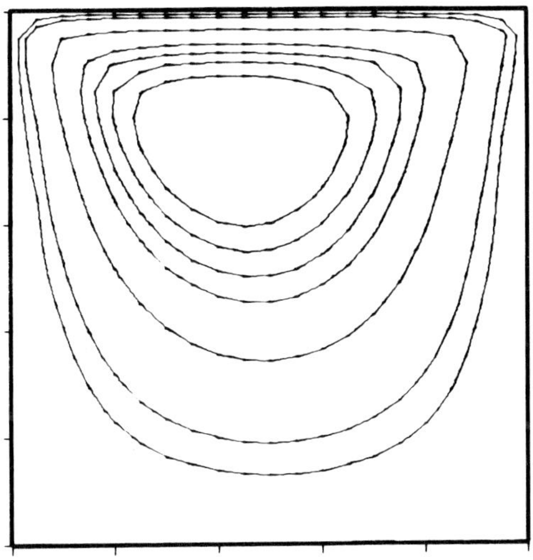

VORTICITY R = 5∅ P = 2 Q = 1

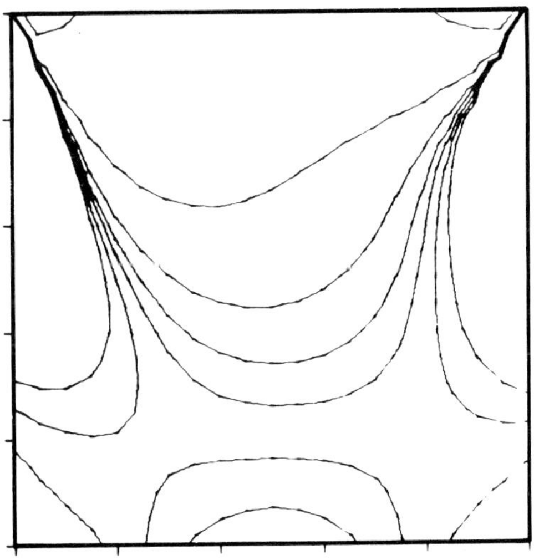

Plate (d) New Boundary Approximation (4.2)

STREAM LINES R = 5∅ P = 3 Q = 2

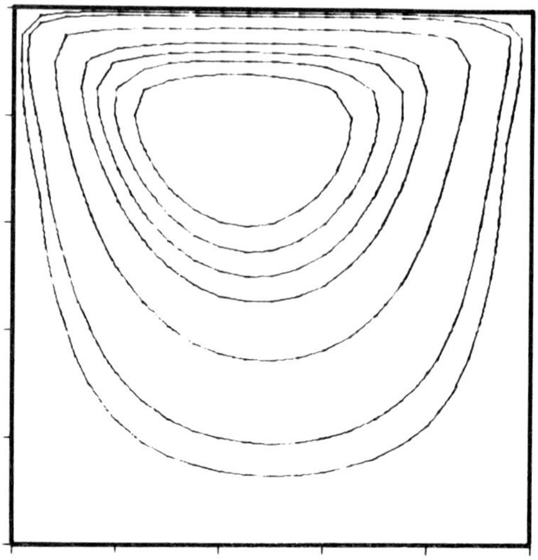

VORTICITY R = 5∅ P = 3 Q = 2

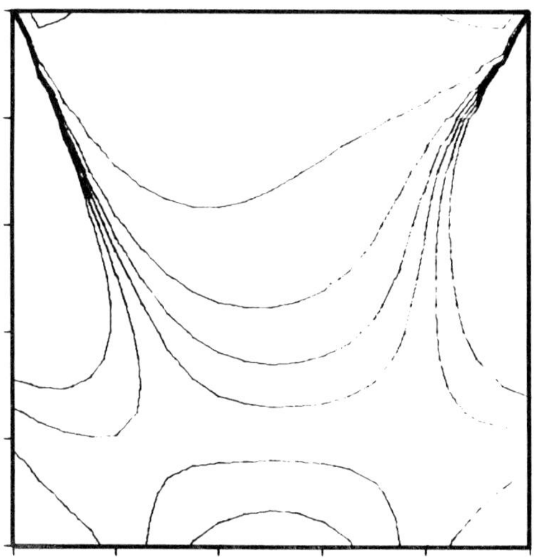

Plate (e) New Boundary Approximation (4.2)

STREAM LINES R = 5Ø P = 4 Q = 3

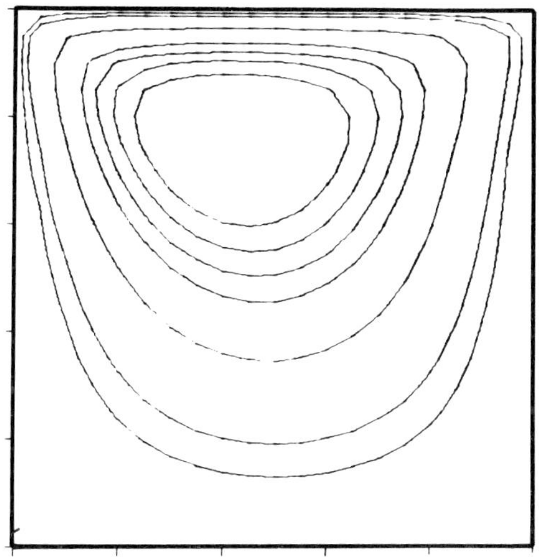

VORTICITY R = 5Ø P = 4 Q = 3

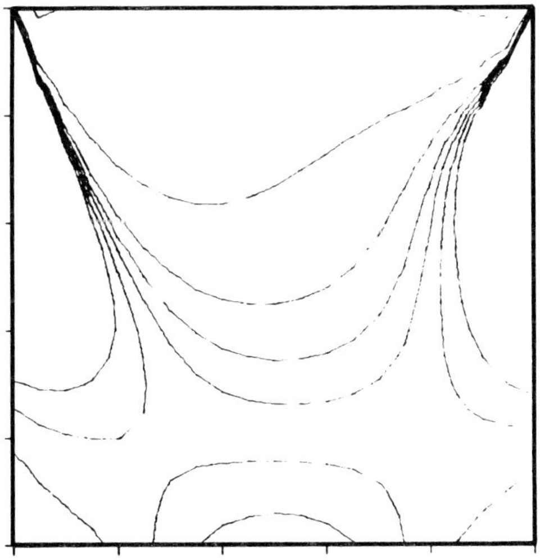

Plate (f) New Boundary Approximation (4.2)

Numerical Simulation of Fluid Motion, ed. John Noye
pp. 483-492

NATURAL CONVECTION IN A CLOSED CAVITY - A NUMERICAL EXPERIMENT

J.C. Patterson

University of Western Australia

Perth, Australia

SUMMARY

A modified version of the numerical technique due to Chorin (1968) is applied to two dimensional problems of natural convection in a closed cavity with differentially heated end walls. Two values (reasonably high and low) of Rayleigh number are used. It is envisaged that even higher values of the Rayleigh number will be possible, that tilted configurations will be easily soluble, and that the method will apply three dimensional problems with little modification.

1. INTRODUCTION

The numerical solution of problems involving natural convection in a closed cavity with differentially heated end walls has generally been confined to two dimensional problems with relatively low Rayleigh number. The introduction of a non-uniform grid which depends on the Rayleigh number and the use of the raw variables of pressure and velocity in a time dependent procedure allows the development of high Ra flows from rest to be observed in either two or three dimensions.

In the following, a modification to the method devised by Chorin (1968) is developed, and applied in two dimensions to a low Ra and, with further modification, to a high Ra problem. The extension of the technique to three dimensions should be straightforward.

2. THE EQUATIONS AND METHOD

In both cases, a square cavity of side ℓ, enclosing fluid of temperature T_0, coefficient of thermal expansion α, kinematic viscosity ν, and conductivity κ, is, at time $t = 0$, heated on the left to temperature $T_0 + \Delta T$, and cooled on the right to $T_0 - \Delta T$. The ends are maintained at these temperatures and the horizontal sides are insulated. The Rayleigh number is defined as

$$Ra = \frac{g\ \alpha\ \ell^3\ \sigma\ \Delta T}{\nu^2}$$

where g = acceleration due to gravity,

and $\sigma = \frac{\nu}{\kappa}$ = Prandtl number.

Experimental evidence (Imberger (1974)) indicates that, for high Ra, a boundary layer of thickness $0(Ra^{-\frac{1}{4}})$ is formed. For low Ra, however, the characteristic length scale is the box depth ℓ. A Rayleigh number of about $0(10^5 - 10^6)$ may be interpreted as being the upper limit for which ℓ is the characteristic length. Hence for a $Ra < 10^5$, a relatively coarse grid is applicable; for $Ra > 10^6$, a boundary layer is formed, the characteristic length is $0(Ra^{-\frac{1}{4}})$, and a much finer grid is required in the vicinity of the layer.

For the low Ra case, with distances non-dimensionalised with respect to ℓ, velocities with respect to ν/ℓ, and time with respect to $\frac{\ell^2}{\nu Ra}$, and temperature as $T^* = \frac{T - T_0}{\Delta T}$, the Boussinesq equations and boundary conditions become

$$\frac{\partial u}{\partial t} + \frac{1}{Ra}\left(u\frac{\partial u}{\partial x} + v\frac{\partial u}{\partial y}\right) = -\frac{\partial p}{\partial x} + \frac{1}{Ra}\nabla^2 u \qquad (2.1)$$

$$\frac{\partial v}{\partial t} + \frac{1}{Ra}\left(u\frac{\partial v}{\partial x} + v\frac{\partial v}{\partial y}\right) = -\frac{\partial p}{\partial y} + \frac{1}{Ra}\nabla^2 v + \frac{T}{\sigma} \qquad (2.2)$$

$$\frac{\partial u}{\partial x} + \frac{\partial v}{\partial y} = 0 \qquad (2.3)$$

$$\frac{\partial T}{\partial t} + \frac{1}{Ra}\left(u\frac{\partial T}{\partial x} + v\frac{\partial T}{\partial y}\right) = \frac{1}{Ra\sigma}\nabla^2 T \qquad (2.4)$$

with

$$\left.\begin{array}{lll} u = v = 0 & \text{on } x = 0, 1 \; ; \; y = 0, 1 \\ T = 1 & x = 0 \; , \\ T = -1 & x = 1 \; , \\ \dfrac{\partial T}{\partial y} = 0 & y = 0 \; , \; 1 \; , \end{array}\right\} t > 0$$

$$\left.\begin{array}{lll} u = v = 0 \;\forall & x, y \\ T = 0 & 0 < x < 1 \; ; \; 0 < y < 1 \end{array}\right\} t = 0$$

Chorin's method involves writing (2.1) and (2.2) as

$$\frac{\partial u_i}{\partial t} + \frac{\partial p}{\partial x_i} = F_i(\underline{u}, T) \tag{2.5}$$

Assuming that U_i^n , p^n, T^n represent the discrete approximation to the solution at time t^n, an auxiliary velocity field U_i^A may found from

$$U_i^A - U_i^n = \Delta t \; (F_i(\underline{U}, T))^{(n,n+1)}$$

where the superscript $(n,n+1)$ indicates that either implicit, explicit, or mixed methods may be used. Then using (2.5),

$$U_i^{n+1} = U_i^A - \Delta t \frac{\partial p^{n+1}}{\partial x_i} \tag{2.6}$$

and continuity (2.3) remains to be satisfied. In the centred-difference approximation, (2.3) is, for any interior point (i,j) , with equal mesh spacings

$$U_{i+1,j}^{n+1} - U_{i-1,j}^{n+1} + V_{i,j+1}^{n+1} - V_{i,j-1}^{n+1} = 0 \; ,$$

and using (2.6)

$$p_{i+2,j}^{n+1} + p_{i,j+2}^{n+1} - 4p_{i,j}^{n+1} + p_{i,j-2}^{n+1} + p_{i-2,j}^{n+1} = G(U^A, V^A)i,j \; . \tag{2.7}$$

Near the boundary, the no-slip condition is used, to yield, e.g. at $i = 1$,

$$p_{3,j}^{n+1} - 3p_{i,j}^{n+1} + p_{1,j+2}^{n+1} + p_{1,j-2}^{n+1} = G(U^A, V^A) \; 1,j \tag{2.8}$$

and on the boundary, e.g. $i = 0$, the condition $\frac{\partial U}{\partial x} = 0$ yields

$$4\left(p_{2,j}^{n+1} - p_{0,j}^{n+1}\right) - \left(p_{3,j}^{n+1} - p_{i,j}^{n+1}\right) = G(U^A, V^A) \; 0,j \tag{2.9}$$

Equations (2.7), (2.8), and (2.9) then provide a set of algebraic equations for the values of $p_{i,j}^{n+1}$ over the whole mesh, which may subsequently be used in (2.6) to provide the $U_{i,j}^{n+1}$, $V_{i,j}^{n+1}$, automatically satisfying continuity at every point (i,j). However, this set is over a group of interwined meshes, connected only at the boundary by relations of the form (2.9). Even at low Ra , extremely high accuracy is necessary in the solution of equations (2.7), (2.8) and (2.9) to prevent the appearance of an oscillation between grid points, particularly in the x direction. A mild improvement may be made by using boundary conditions from the original equations, e.g. on surfaces $x = 0$, $x = \ell$, the form $\frac{\partial p^{n+1}}{\partial x} = \frac{1}{Ra} \frac{\partial^2 U^{n+1}}{\partial x^2}$, gives a better connection between meshes. The same problem is, however, still evident at high Ra.

With the loss of some accuracy but with a considerable saving in computing time, this problem may be avoided in the following manner. If the mesh points are numbered 0 to n_x in the x direction and 0 to n_y in the y direction, where n_x and n_y are even numbers, those points for which i and j are both odd constitute one complete mesh independent of values of p^{n+1} on the boundary. The boundary condition on p has been effectively replaced by the no-slip condition on velocity. The mesh may now be thought of as the well known form of a group of boxes, with the pressures evaluated at the centres, U^{n+1} on the vertical sides, and V^{n+1} on the horizontal sides. If (2.7) and (2.8) are solved over this reduced mesh and the velocities calculated from (2.6), continuity is satisfied in each box. The number of equations to be solved has been reduced by a factor of four. The velocities at the remainder of the points may then be found by interpolation.

Hence, in the modified scheme, given at time t^n, U^n, V^n, T^n over the whole mesh, and p^n over the reduced mesh, the order of calculation is

(1) Solve for T^{n+1} over the whole mesh, using U^n, V^n

(2) Solve for U^A, V^A over the whole mesh, using T^{n+1}

(3) Solve for p^{n+1} over the reduced mesh

(4) Calculate U^{n+1}, V^{n+1} over the whole mesh using p^{n+1} and interpolation .

(5) Increment time and go to step (1)

In this scheme, T^{n+1} is calculated using the velocities from the previous step and is not updated until the next step. It is thought that this error is small, although it should be possible to calculate an auxiliary field as for velocity which is updated in the same step.

In the actual solution of the problems below, the mixed implicit explicit scheme (Gourlay (1970)) was used over four sub time intervals ($\Delta t/4$) for steps (1) and (2), and step (3) solved every time step (Δt) by a standard relaxation method.

3. RESULTS

3.1 Low Rayleigh number results

The problem for low Ra was solved with $Ra = 1.4 \times 10^4$,$\sigma = 7.0$. The flow fields in the form of velocity vectors and isotherms for various times up to steady state are shown in Figure 1. Steady state was deemed to have been reached when the Nusselt number and the velocities at a number of particular points became constant. The development of the Nusselt number is shown in Figure 2, the steady state value being 3.16 . The results here are qualitively and quantitatively similar to those obtained in previous numerical results (De Vahl Davis (1968), Cormack, Leal, and Seinfeld (1974) e.g.).

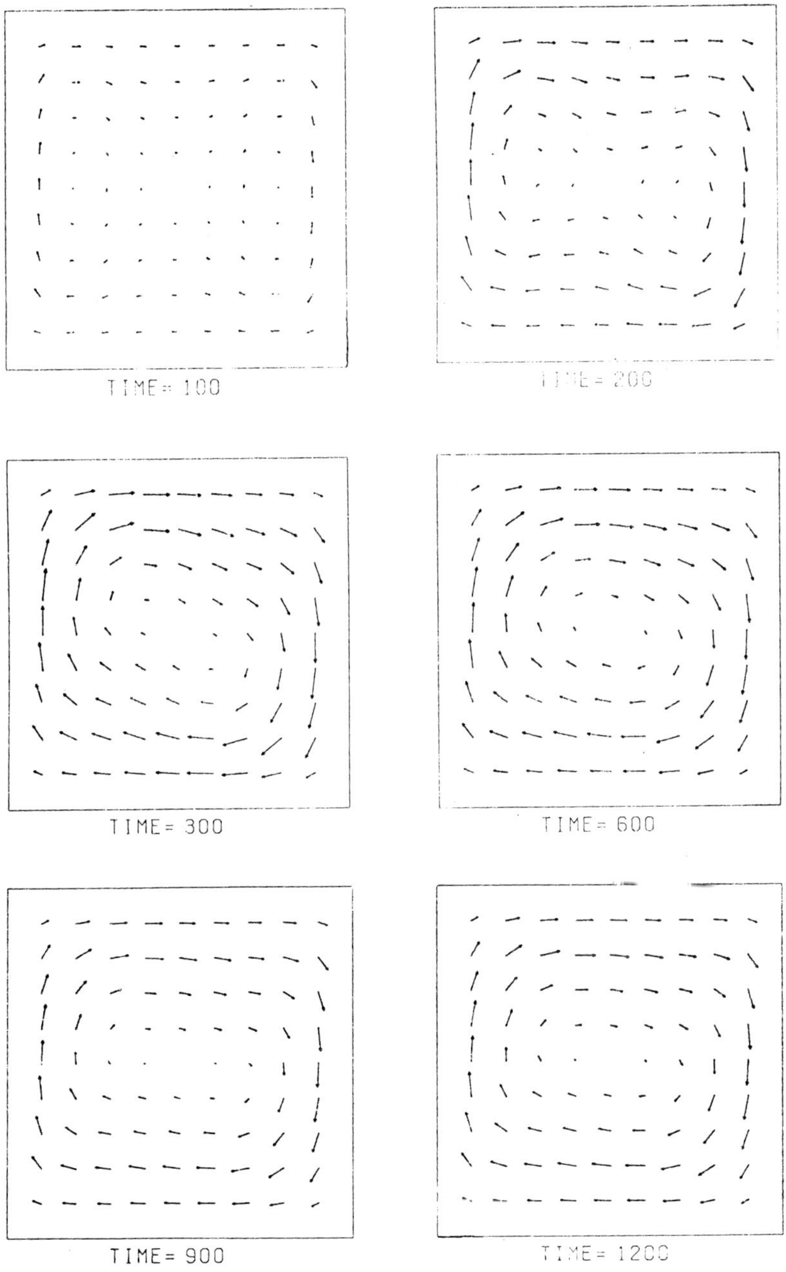

Fig. 1(a) *The development to steady state of the flow field, with* Ra = 1.4×10^4, $\sigma = 7$.

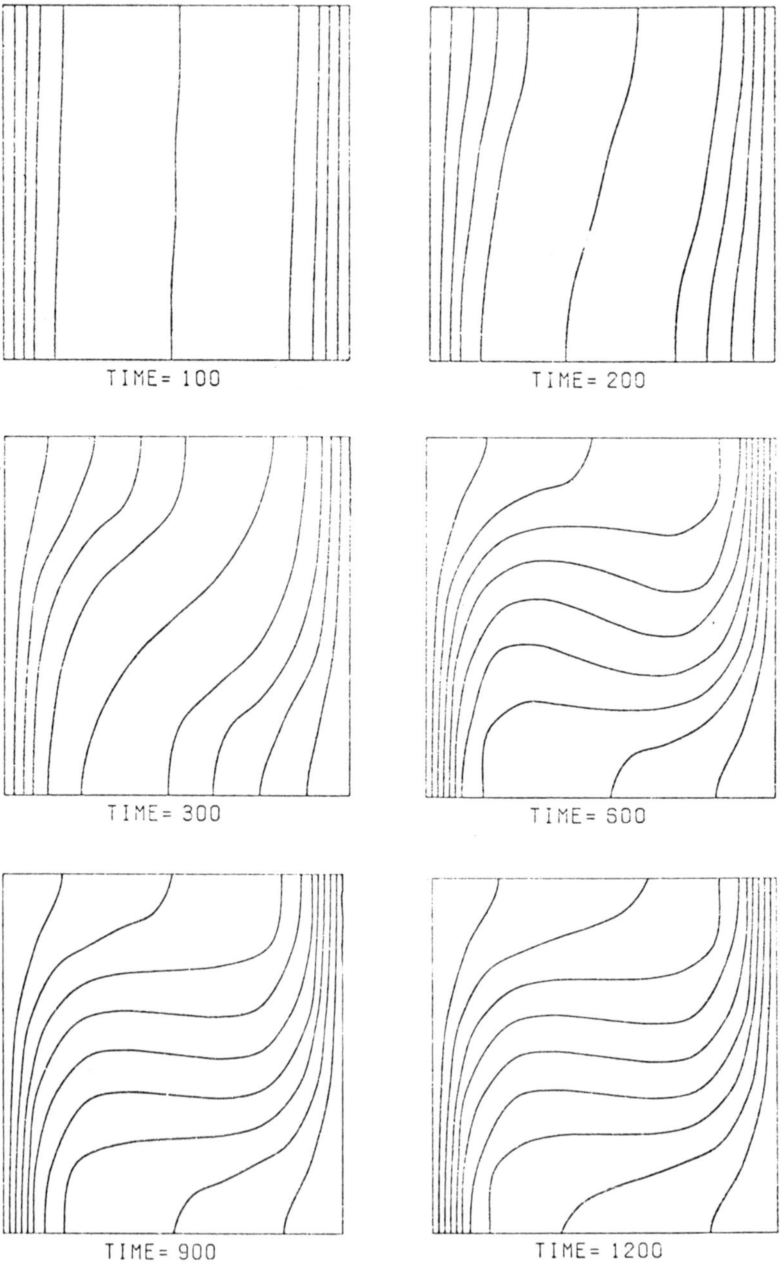

Fig. 1(b) *The development to steady state of the temperature field, with* Ra = 1.4×10^4, $\sigma = 7$

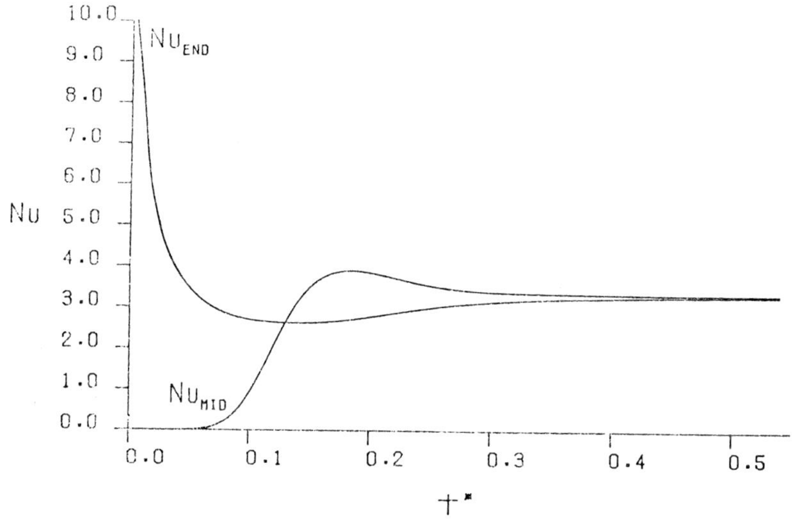

Fig.2 *The growth of the Nusselt number to its steady value, for* $Ra = 1.4 \times 10^4$, $\sigma = 7$, *measured at the heated end* (Nu_{end}) *and the centre line* (Nu_{mid}) *of the cavity.*

3.2 High Rayleigh number results

For high Ra, the existence of the boundary layer of thickness $O(Ra^{-\frac{1}{4}})$ necessitates a much finer mesh at the ends of the cavity at least. The scaling must also be modified to allow steady state to be reached without excessive computation. The most suitable scaling appeared to be

$$t^* = \frac{tRa^{\frac{1}{2}}\nu}{\ell^2 \sigma^{\frac{1}{2}}} \qquad U^* = \frac{U\ell}{\kappa Ra^{\frac{1}{4}}\sigma^{\frac{1}{4}}} \qquad V^* = \frac{V\ell}{\kappa Ra^{\frac{1}{2}}\sigma^{\frac{1}{2}}}$$

$$x^* = \frac{x}{\ell} \qquad y^* = \frac{y}{\ell} \qquad T^* = \frac{T-T_0}{\Delta T} ,$$

obtained from a balance between the convection and conductive temperature terms. To obtain the finer grid spacing at either end, a transformation depending on Ra is required. The transformation used was

$$X = \left(\frac{x^* + 1}{2}\right) \tanh \left(x^* Ra^{\frac{1}{4}}\right) + \left(\frac{2 - x^*}{2}\right) \tanh \left((x^* - 1)Ra^{\frac{1}{4}}\right) + 1 , \quad 0 \le X \le 2 \tag{3.1}$$

Dropping the asterisks, the full equations become

$$\frac{\partial U}{\partial t} + \frac{1}{\sigma}\left(Uf_1 \frac{\partial U}{\partial X} + V \frac{\partial U}{\partial y}\right) = - Ra^{\frac{1}{2}} \frac{\partial p}{\partial X} + D^2 U$$

$$\frac{\partial V}{\partial t} + \frac{1}{\sigma}\left(Uf_1 \frac{\partial V}{\partial X} + V \frac{\partial V}{\partial y}\right) = - \frac{\partial p}{\partial y} + D^2 V$$

$$f_1 \frac{\partial U}{\partial X} + \frac{\partial V}{\partial y} = 0$$

$$\frac{\partial T}{\partial t} + \frac{1}{\sigma}\left(Uf_1 \frac{\partial T}{\partial X} + V \frac{\partial T}{\partial y}\right) = - D^2 T$$

where

$$D^2 U \equiv \frac{1}{Ra^{\frac{1}{2}}} \frac{\partial^2 X}{\partial x^{*2}} \frac{\partial U}{\partial X} + f_1^{\,2} \frac{\partial^2 U}{\partial X^2} + \frac{1}{Ra^{\frac{1}{2}}} \frac{\partial^2 U}{\partial y^2}$$

$$f_1 = \frac{1}{Ra^{\frac{1}{4}}} \frac{\partial X}{\partial x^*} ,$$

with corresponding boundary conditions.

The method was applied to these equations with $Ra = 1.4 \times 10^7$, $\sigma = 7.0$. With the transformation (3.1), equal mesh spacings of 0.1 on the X scale yields points on the x^* scale of 0.003, 0.006, 0.01, 0.016, 0.029, 0.1, 0.2, 0.3, 0.4, 0.5, symmetric about $x^* = 0.5$. For this value of Ra ; the estimate of boundary layer thickness is 0.016, and three points are contained in the layer. Figure 3 shows the development of the flow from rest to time $t = 12$, at which time the Nusselt number has reached a steady value of 26.7 . It is evident that the boundary layer thickness is greater than $Ra^{-\frac{1}{4}}$, and a modification to (3.1) is required to redistribute the points between 0.0 and 0.1 on the x^* scale. A finer mesh in the vertical direction may also be desirable.

Little experimental evidence is available for comparison; however the end flow at $t = 12$ displays substantially the same features observed by Imberger (1974) in a long box heated at one end with a Rayleigh number of the same order.

At time $t = 12$, the Nusselt number has become constant, and the fluid velocities are changing very slowly; the solution is near steady state. Experiments to check further deviations and stability at higher time values are in progress.

3.3 Computational aspects

Both the low Ra and the transformed high Ra problems were calculated on uniform meshes of spacing 0.1 , on 11 x 11 mesh for the former, and 21 x 11 for the latter. The low Ra case was found to be very stable, and relatively large time steps (up to 0.1) were possible, particularly near steady state, with little effect on the results. For high Ra , a step of 0.01 was used initially, becoming 0.02 at $t = 8$. This increase in time step is introduced in an attempt to reduce CPU time. On a PDP-10 computer, the low Ra case required approximately 32 minutes of CPU time to reach steady state; in its present form, the high Ra case required 90 minutes to reach $t = 12$, apparent steady state. The CPU time depends strongly on the convergence criterion for

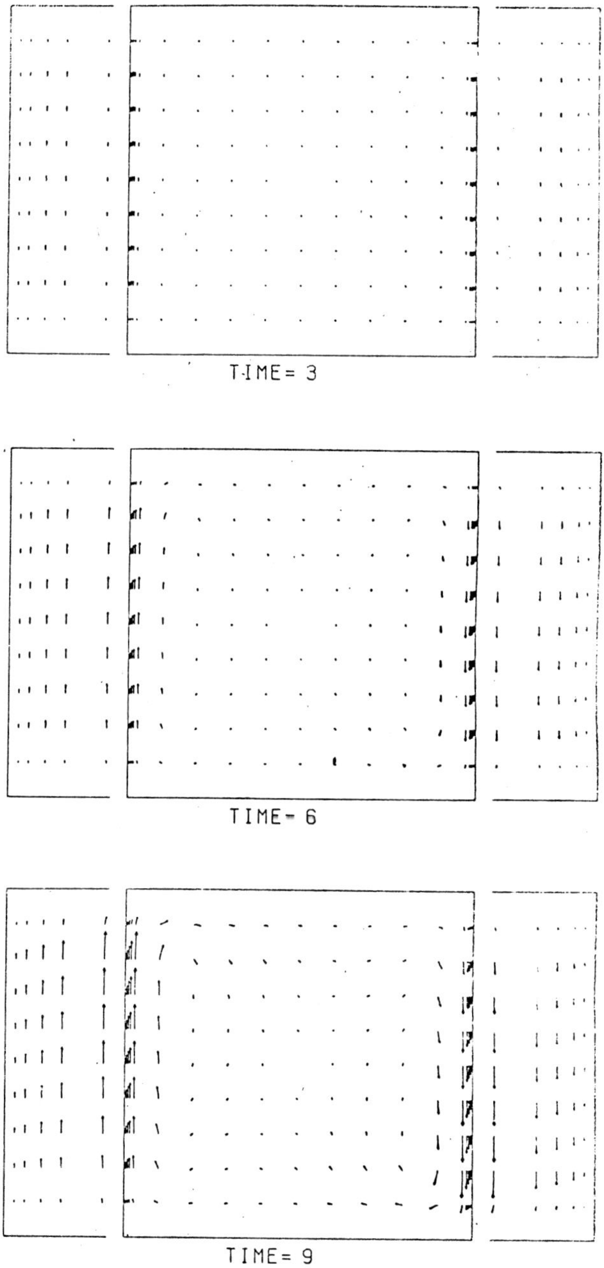

Fig. 3 For legend see following page

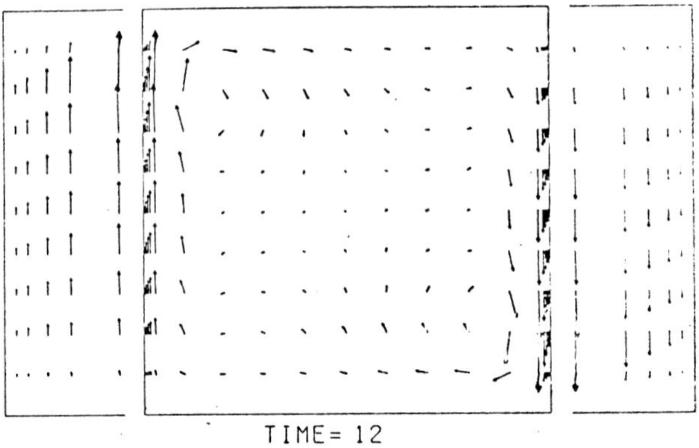

Fig. 3 *The development to apparent steady state of the flow field, with* Ra = 1.4×10^7, $\sigma = 7$.

the relaxation solution of the pressure equations. The criterion applied here for successive iterates $\nu, \nu + 1$ was

$$\frac{\max\limits_{i,j} \left| p_{i,j}^{n+1,\nu+1} - p_{i,j}^{n+1,\nu} \right|}{\max\limits_{i,j} \left| p_{i,j}^{n+1,\nu} \right|} \leq 10^{-5}$$

A less stringent condition considerably reduces CPU time.

4. Further Possibilities

With the improved resolution, it is envisaged that higher Rayleigh number problems will also be soluble with little modification. Tilted configurations should be easily solved by adjustment to the body force terms, and the transition to three dimensions, for low Ra and horizontal configurations at least, may now be easily accomplished.

5. REFERENCES

Chorin, A.J. (1968) Maths. Comp. 22, 745

Cormack, D.E., L.G. Leal and J.H. Seinfeld (1974) J. Fluid Mech. 65, 230

De Vahl Davis, G. (1968) Int. J. Heat Mass Transfer, 11, 1675

Gourlay, A.R. (1970) J. Inst. Maths Applics, 6, 375

Imberger, J. (1974) J. Fluid Mech, 65, 247

Numerical Simulation of Fluid Motion, ed. John Noye
pp. 493-502

DEPENDENCE OF NUMERICAL RESULTS ON TRUNCATION ERRORS

R.A. Pearson and P.G. Williams

Monash University

Victoria, Australia

SUMMARY

For a circularly symmetric heat island, or sea breeze, the form of the solution seems to be influenced by the island size. In a line heat island no such differences are observed. It is possible that the change in circulation pattern could be a realistic phenomena, however it could be influenced by numerical errors. The numerical errors and feedback mechanisms for the observed differences are discussed.

INTRODUCTION

Much of the impetus towards a better understanding of the processes involved in complicated fluid flows has come from simple model experiments. Some experiments use physical laboratory models, others use analytical or numerical solutions that arise from simplifications of the Navier Stokes Equations. Numerical experiments generally involve the solution of non-linear differential (usually partial differential) equations. In the mathematical sense the existence and uniqueness of such solutions has not been generally shown. For the numerical representations to have any validity it is necessary that, as the resolution increases, the solution will converge (in some sense) to the solution of the original equation. The criteria for convergence of numerical techniques, for all possible equations, is very difficult to prove. In most applications Lax's theorem (Lax & Nirenberg 1966) is assumed to hold.

Most of the work on convergence, and the related error bounds is of a theoretical nature. In general the cost effective use of manpower, computer time and space does not warrant a complete numerical study of convergence. An analytical approach has yielded valuable insight on stability and convergence. Estimates and examples of error bounds for a simple wave equation are given in Orszag & Israeli (1974), while a useful summary is found in Kreiss & Oliger (1973 and 1974).

Recently some computer solutions have demonstrated convergence in actual practical examples. In a model of convection between flat plates Denny & Clever (1974) show that the finite difference method converges from above while the Galerkin solutions converge from below. They also show that the errors in each technique are distributed in different regions of the flow. The finite difference model has small errors in the boundary layer and the core flow, with larger local errors in the intermediate region. The errors in practical examples do not necessarily follow those anticipated from the analytical results. In a model of baroclinic instability Merilees (1975) found that the reduced truncation error of the finer resolution had little effect. The advantages were due to a decrease in the viscosity required to damp numerical noise.

As well as convergence, as the resolution increases, it is useful to consider convergence as the order of the numerical method increases. It can be shown theoretically that error bounds decrease as the order of the finite difference scheme increases. In particular, for simple time dependent systems the growth of errors is much slower for higher order schemes. Analytical examination of this growth (Kreiss & Oliger 1973) shows a theoretical computational advantage in the higher order schemes. Numerical tests of this error are usually limited to simple systems where the analytical results have been shown to be valid. More complicated models do not necessarily follow these rules. In a numerical experiment with a baroclinic model (Merilees 1975) the results of increasing the order were similar to those quoted above for increasing resolution. The improvement was caused by a reduction in the damping required for realistic solution.

In numerical experiments designed to further understanding of complex geophysical systems it is impractical to use very fine resolution, or to study how the solution converges. The main results are obtained with fairly crude resolution. Often the main interest is how large fluctuations could be influenced by relatively small changes in the model parameters (e.g. in climate models). In other cases the numerical experiment examines the functional dependence of the solution's behaviour on initial conditions.

NUMERICAL RESULTS AND TRUNCATION ERRORS

It is important in such cases that modelling assumptions and numerical truncation errors do not influence the final result. It is also necessary that any instabilities inherent in the real world are present in the model These difficulties are usually overlooked and the numerical techniques and model assumptions are chosen for convenience.

An example of this convenience is the removal of numerical noise by implicit damping or some gross eddy parameterisation. The use of such models is usually justified by vague statements about the real world having enough damping to maintain a stable state. However although some gross properties are examined very few modellers attempt to show that the numerical damping bears any resemblance to the real world. For this reason it is not certain that possible instabilities have not been modified or removed by the modelling assumptions.

As a parallel consider two ordinary first order coupled equations. As time varies every solution follows a path which can be followed in a plane. The solution space, called a phase space, can be characterised by certain properties of this plane. The behaviour of solutions with different initial conditions can be grouped into regions, or a "topology" of the plane. As an example consider a stable focal point. Solutions with nearby initial conditions will eventually approach this point. A stable limit cycle is a closed curve (or surface) in phase space which solutions follow with time while in any nearby region any solution will approach the curve. There are also unstable limit cycles and unstable focal points.

Unfortunately there is a strong possibility that the inclusion of damping will significantly alter the solution space, for example, stable limit cycles could disappear into a stable focal point. Worse, the entire structure of phase space could be altered.

As a practical extension of this idea the empirical study of climate and weather data suggests that climate is "quasi-stable" in the sense that it oscillates between a number of states; such as ice ages and inter-glacial epochs (S.M.I.C. 1971). Each of these states could be imagined as semi-stable limit cycles. There is no guarantee that a numerical model with damping will not change or completely remove these quasi-stable limit cycles. This is particularly important as a small change in initial conditions near an unstable point can completely change the character of the solution. If a numerical model of climate is used to assess the impact of a pollutant completely spurious physical inferences could result because of a change in the character of the multidimensional phase space.

This paper will consider initially a model where the physical properties of a steady state solution (or stable critical point) are constant with different initial conditions. However the path in solution space is completely altered by different physical damping, or by relatively small changes in the initial conditions. These changes are probably not realistic and could be caused by the growth of a numerical error. The possible dynamics of the error and reason for the changes of the solution are discussed. This forms the basis for considering other published results of the sea breeze circulation which show a similar phenomena.

IPTION, EQUATIONS AND BOUNDARY CONDITIONS

del is that used by Williams (1974) in his study of a circular The main aims were to examine the steady state solution as a heat island size and Rayleigh Number, examine possible open nditions, and compare solutions obtained from both axisymmetric eat islands (Leslie & Morton 1973). The transient solutions in this paper were a means of obtaining the steady state.

The model considers motion in an initially neutral environment over ılar heated region. The Boussinesq form of the equations of motion ılid (Dutton & Fichil 1969). This leads to the following set of ions.

vorticity equation is

$$\frac{D\zeta}{Dt} = \frac{\partial\zeta}{\partial t} + J\left(\psi, \frac{\zeta}{r}\right) = -\frac{g}{T_o}\frac{\partial T}{\partial r} + \kappa\left[\frac{\partial}{\partial r}\left(\frac{1}{r}\frac{\partial}{\partial r}(r\zeta)\right) + \frac{\partial^2\zeta}{\partial z^2}\right] .$$

The heat equation is

$$\frac{DT}{Dt} = \frac{\partial T}{\partial t} + \frac{1}{r} J(\psi, T) = \kappa\left[\frac{1}{r}\frac{\partial}{\partial r}\left(r\frac{\partial T}{\partial r}\right) + \frac{\partial^2 T}{\partial z^2}\right] .$$

Here ψ is a stream function $\left(u = -\frac{1}{r}\frac{\partial\psi}{\partial z}, v = \frac{1}{r}\frac{\partial\psi}{\partial r}\right)$; ζ the vorticity $\left(\zeta = \frac{\partial u}{\partial z} - \frac{\partial v}{\partial r}\right)$; T is the temperature; T_o the initial ambient temperature; and κ an eddy diffusivity.

The heat source has an initial temperature $T_s = T_o + \Delta T$ at $z = 0$ inside a region $0 \leqslant r \leqslant \ell$. Flow is symmetrical about $r = 0$; at $z = 0$, $u = v = 0$, at $z = H$ a rigid lid ($v = 0$) is imposed. The lateral boundary conditions at $x = L$ were varied but do not change the features of the flow considered here.

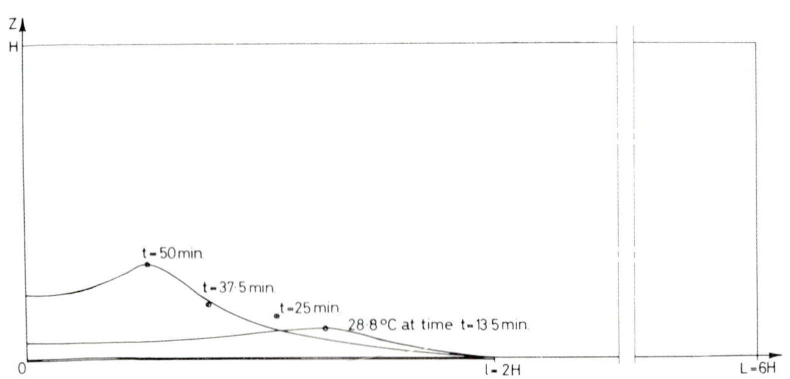

Figure 1: The large island case. The limits of the 28.8°C temperature at 13.5 minutes and 50 minutes. The position of the highest point at 25 minutes and 37.5 minutes is also shown.

All motion is induced by buoyancy forces and the relevant non-dimensional parameter is the Pseudo-Rayleigh number

$$R_a = \frac{g\ \Delta T\ H^3}{\kappa^2\ T_o} .$$

The other characteristic numbers are the two aspect ratios H/ℓ, H/L. These three non-dimensional numbers completely determine the form of the steady state solution (Williams 1974).

For the numerical simulation a simple second order, finite difference analogue was used for the vorticity generation, Arakawas Jacobian (1966) was used for the non-linear terms, while the diffusive terms were modelled using the method of Dufort & Frankel. Another numerical realisation using leap-frog time integration with lagged diffusion terms did not influence the results for the large island, nor did the use of the constituent parts of Arakawas Jacobian.

2. RESULTS

For one form of the transient solutions consider the large island ($H/\ell = 1/2$) at a large Rayleigh number (Ra = 3800). Initially no flow occurs and heat conduction from the hot source into the air above results in a flow field. The early temperature field in the air above the heated region shows a maximum just inside the outer limits of the source (Figure 1). Associated with the maximum are positive and negative regions in the vorticity field. The associated stream function field has a "two cell" structure with low level flow outwards to the end of the island from the interior and inwards from the outer flow field towards the island (Figure 2). As time progresses the maximum temperature moves inwards and upwards. Eventually all flow at low levels is towards the axis with outflow at the upper levels. This single cell structure forms the stable steady state solution (Figure 3 a) & b)).

With a lower Rayleigh number (Ra = 420) the temperature at the early stages shows an almost uniform rise over the island. The temperature gradients are always of the same sign. At any level the maximum value occurs at the axis. The stream function field for this flow has a single cell structure, with flow towards the axis at low levels and outflow at the upper levels. This single cell structure is maintained as the steady state solution is obtained.

For the small island ($H/\ell = 1/1$) only one cell forms. The path towards steady state is similar to the large island with low Rayleigh number.

This difference in path should be contrasted to that of a line heat island, where at all Rayleigh numbers and for all sizes of island, only one cell is formed.

3. DISCUSSION

One characteristic of all the solutions is that the largest heat fluxes occur just at the ends of the island. For the axisymmetric case the heat flux per grid point also increases as r increases. As the largest

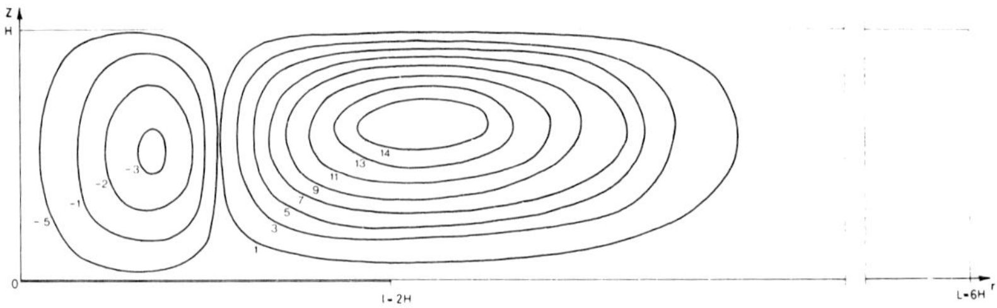

Figure 2: *The contours of stream function corresponding to 50 minutes temperature profile shown in Figure 1.*

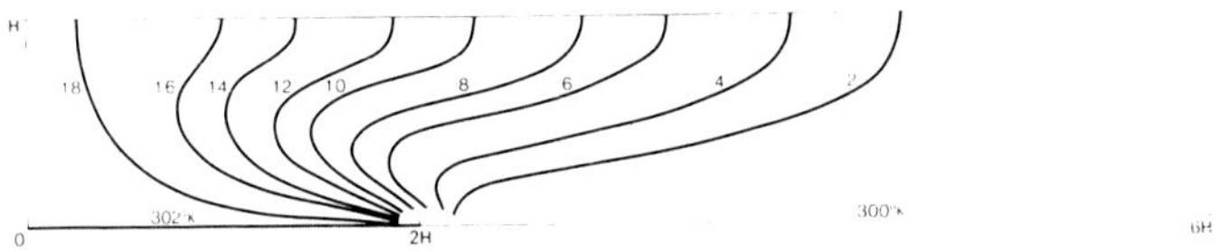

Figure 3a: *Steady state isotherms for case with $Ra = 3800$, $H/\ell = 1/2$, $H/L = 1/6$, $T = T_0$ for inflow boundary. Contours vary from 300.2 to 301.8°K.*

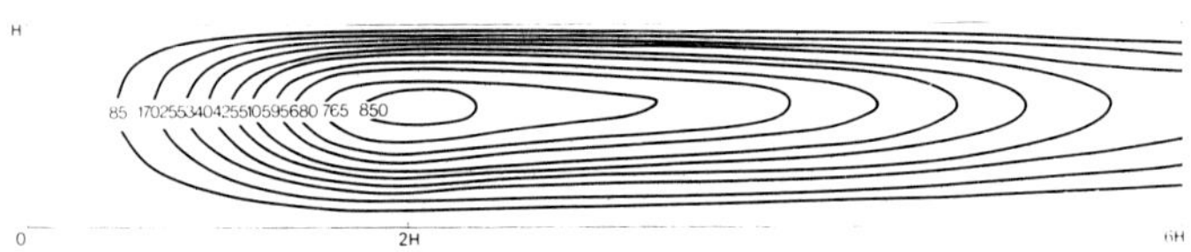

Figure 3b: *Steady state streamlines for case with $Ra = 3800$, $H/\ell = 1/2$, $H/L = 1/6$, $T = T_0$ for inflow boundary. Contours vary from 0.0 to $8.5 \times 10^5 m^3/sec$, scaled by 1.0×10^{-3}.*

gradient of heat flux is at the edge of the island this is the most likely region for large numerical errors.

No basic reason for the difference between different Rayleigh numbers and different island sizes is obvious especially as the line heat island does not show at all the "two cell" phenomena. However the growth and existence of a double cell can be rationalised. A possible mechanism is as follows:- Assume that a small perturbation, either a real physical phenomena or numerical noise, results in a small excess temperature at low levels near the island edge. The resulting temperature gradient generates a vorticity dipole. The corresponding flow brings more air from both directions towards the island edge. This results in a rise in level of the bump with a concentrated updraft of warm air at the island boundary. Because of the effects of the advection of cold air there are larger heat fluxes on either side of the island edge, especially towards the axis. These processes increase the size of the temperature perturbation, thus closing the feedback loop. Schematically the growth of the two cell flow is:- temperature excess → vorticity dipole → plume (temperature excess advected upwards) → more vorticity generation.

This loop does not occur at low Rayleigh numbers because there is sufficient viscosity for numerical noise to be effectively removed. The viscosity also acts as a physically realistic damping for the plume. The loop is also more likely for the large island as the heat flux per grid point is larger (elemental area proportional to r) thus there is more gradient in the fluxes and more likely errors. The smaller equivalent grid point heat flux in the smaller island gives less likely numerical errors, which can be easily damped by low "κ" with only a single cell resulting.

4. THE SEA BREEZE

The occurrence and rationalisation of a double cell in the sea breeze circulation is more complicated. For a straight coastline the existence of a double cell depends on the model. Estoque (1960) and McPherson (1968,1970) predict two cells, yet another version of the same model (Neuman & Mahrer 1971) shows only one cell. Differences between the axisymmetric and line versions of the same model are also apparent. In the axisymmetric case Neumann & Mahrer predict two cells for the large island (Figure 4) and show one for the small island (Figure 5).

All models quoted above fix the temperature at the ground and use basically the same form of boundary layer (Estoque 1963). This parameterisation allows the eddy viscosity (and hence heat flux) to be a function of shear and vertical temperature gradients. Its action is complex. One effect is to give larger heat fluxes for larger values of the shear, increasing the positive dynamic feedback of the form:- Numerical noise → small temperature "bump" → vortex pair → flow → increased shear over boundary → greater heat flux → much larger plume → much larger vortex pair.

Although the boundary layer parameterisation is essentially of the same form each author chooses slightly different constants. These different constants change the relationship between heat flux and damping, allowing one model to damp the two cell structure while another does not. The axisymmetric model also gives a velocity structure, and shear, above the lower boundary which will be very different from the line symmetric model. This

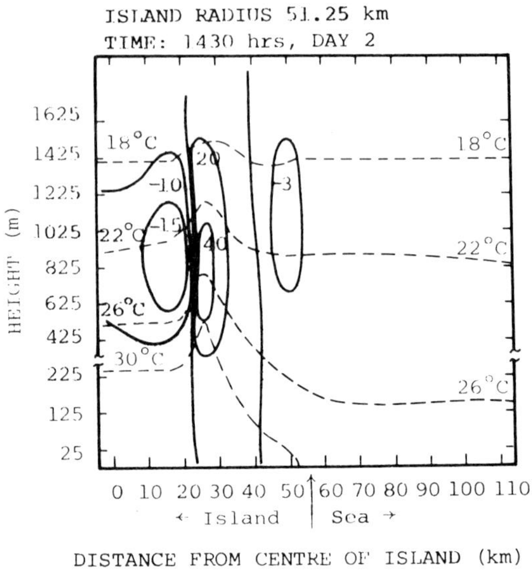

Figure 4: Contours of vertical velocity (solid lines) in metres/sec^{-1} and contours of the temperature field at 1430 hrs for the large island case (after Neumann & Mahrer 1974).

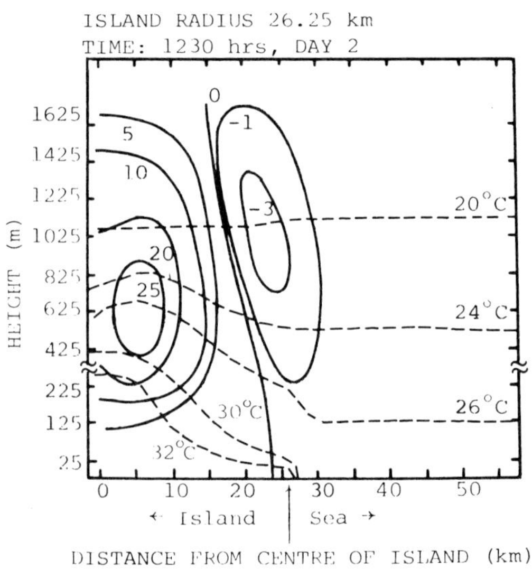

Figure 5: Contours of vertical velocity (solid lines) in metres/sec^{-1} and contours of the temperature field at 1230 hrs for the small island case (after Neumann & Mahrer 1974).

further meshes the physical realistic differences with the possible numerical model. An essential part of the feedback loop in both cases above is the increased heat flux at the edge of the warmer area. This can be removed by specifying the heat flux. Another disadvantage of the fixed temperature boundary condition is that it does not allow a temperature change to move inland with the sea breeze. A disadvantage of the fixed heat flux is that no heat storage is allowed in the ground. The completely realistic model of the sea breeze, or heat island, would have a lower boundary condition between these extremes.

To further understand the possible existence of a double cell, some experiments (Pearson 1972) have been made with a fixed heat flux model. In this case a single cell forms unless an increased heat flux is specified at the boundary. Further experiments with this model, with the additional possibility of a fixed temperature boundary condition, will allow an examination of the dynamical significance of the above mechanisms. It is also intended that an axisymmetric version of this model will be generated to study the possibility of a physically realistic double cell structure for the circular island.

REFERENCES

Arakawa, A. (1966). Computational design for long-term numerical integration of equations of fluid motion. Two-dimensional incompressible flow. Part 1. J. Comput. Phys. Vol. 1, pp 119-143.

Denny, V.E. & Clever, R.M. (1974). Comparisons of Galerkin and finite difference methods for solving highly non-linear thermally driven flows. J. Comput. Phys. Vol. 16, No. 3, pp 271-284.

Dufort, E.C. & Frankel, S.P. (1953). Stability conditions in the numerical treatment of parabolic differential equations. Math. Tables and Other Aids to Computation, Vol. 7, No. 135.

Dutton, J.A. & Fichil, G.H. (1969). Approximate equations of motion for gases and liquids. J. Atmos. Sci., Vol. 26, pp 241-254.

Estoque, M.A. (1960). A theoretical investigation of the sea breeze. Quart. Jour. Royal. Met. Soc., Vol. 87, pp 136-146.

Estoque, M.A. (1963). A numerical model of the atmospheric boundary layer. J. Geophysical Res., Vol. 68, No. 4, pp 1103-1113.

Kreiss, H.O. & Oliger, J. (1972). Comparison of accurate methods for the integration of hyperbolic equations. Tellus, Vol. 14, No. 3, pp 199-215.

Kreiss, H.O. & Oliger, J. (1973). Methods for the approximate solution of time dependent flows. G.A.R.P. Publication Ser. No. 10, World Met. Org.

Lax, P.D. & Nirenberg, L. (1966). On stability for difference schemes; a sharp form of Gardings inequality. Comm. Pure & Appl. Math., 19, 473-492.

Leslie, L.M. & Morton, B.R. (1973). Convective stagnation under an inversion - a numerical experiment. Proc. First Aust. Conf. on Heat & Mass Transfer, Monash Uni., Melbourne, May 1973.

McPherson, R.D. (1968). A three dimensional numerical study of the Texas Coast sea breeze. Ph.D. Dissertation, Uni. of Texas, Austin.

McPherson, R.D. (1970). A numerical study of the effect of a coastal irregularity on the sea breeze. Jour. Appl. Met., Vol. 19, pp 767-777.

Merilees, P.E. (1975). The effect of grid resolution on the instability of a simple baroclinic model. Mon. Wea. Rev., Vol. 103, pp 101-104.

Neumann, J. & Mahrer, Y. (1971). A theoretical study of the land and sea breeze circulation. Jour. of Atmos. Sci., Vol. 28, pp 532-542.

Neumann, J. & Mahrer, Y. (1974). A theoretical study of the land and sea breezes of circular islands. J. Atmos. Sci., Vol. 31, No. 8, pp 2027-2039.

Orszag, S.A. & Israeli, M. (1974). Numerical simulation of viscous incompressible flows. Ann. Rev. Fluid Mech., pp 281-317.

Pearson, R.A. (1972). An analytical and numerical model of the sea breeze. Ph.D. Dissertation, Uni. of Miami.

S.M.I.C. (1971). Inadvertent climate modification: Report of the study of man's impact on climate. M.I.T. Press.

Williams, P.G. (1974). A numerical study of an axisymmetric heat island. Proc. Fifth Aust. Conf. on Hydraulics & Fluid Mech., Uni. of Canterbury, Christchurch, N.Z.

Numerical Simulation of Fluid Motion, ed. John Noye
pp. 503-520

SOME EFFECTS OF THE LINE DYNAMICS ON AN HYDRAULIC SYSTEM

B.W. Barnard

Caulfield Institute of Technology

Victoria, Australia

SUMMARY

The numerical integration of a large set of non-linear differential equations has been made more feasible by the development of translator type computer packages. Such packages use a FORTRAN style language and sort the equations into a suitable sequence for solution. It is attractive to use these packages to predict the time dependent transient behaviour of dynamic systems, and in this case an oil hydraulic system is chosen for study.

The first major task is to derive the relevant equations describing the mechanical and fluid properties of the system. The power bond graph method for systems analysis has been used to obtain the equations of the system and to organize them for solution. Despite the aid provided by the bond graph technique, the analyst is obliged to make many simplifying assumptions. Some of these assumptions are discussed in this paper, particularly the effect of various approximations to the dynamics of oil in the connecting hoses. A few mathematical difficulties are also mentioned. Experimental work has been carried out to allow comparison of the theoretical results with measured values.

1. INTRODUCTION

Hydraulic systems using oil to transmit mechanical power are now in widespread use. Many early or simple analyses assumed the oil to be incompressible. But since high forces and rapid accelerations are often required, compressibility is a significant effect.

On occasions engineers have been disappointed by their designed systems which have lacked the dynamic response required. This has led to an interest in predicting the transient performance of such systems. Since the energy source (pump) is connected by a length of hose to the point at which the energy is used by the hydraulic cylinder or motor, the part played by the oil in the hoses has received some attention over the past few years.

2. HYDRAULIC SYSTEM

A fairly conventional system was chosen for study. It comprises an electrically driven pump, a control valve which connects through hoses of 1.8 m length and a hydraulic cylinder which lifts a 140 kg mass on the end of a 1.5 m beam. The system is shown schematically in Figure 1.

For each component, the physical aspects which will be included in the mathematical model are partly decided by experience and judgement. Decisions can be confirmed only by a great amount of effort, particularly as each additional aspect included may increase the order of the differential equation describing the system. The important aspects of each component are discussed below.

2.1 Pump

The fixed displacement pump is considered to run at constant speed, and deliver oil at a fixed flow rate (Q_s). (The notation is listed in Appendix 1). However leakage depends on pressure, so this loss is subtracted to obtain the output flow:

i.e. $Q_{out} = Q_s - (P_s/R_\ell)$

where P_s is the pump pressure and R_ℓ is leakage resistance

Furthermore the pressure relief valve resistance, R_r, has a sharp discontinuity; it discharges oil out of the system only when the pressure exceeds some particular value.

A further characteristic considered in the pump is the oil compressibility. A change in pressure causes an apparent loss of volume. Thus the final flow rate from the pump becomes:

$$Q'_p = Q_s - (P_s/R_\ell) - R_r^{-1}(P_s) - \dot{P}_s/C_s \qquad (2.1)$$

where R_r^{-1} is the relief valve step function, a conductance

C_s is the pump capacitance (volume/bulk modulus)

$\dot{P}_s$ is the time derivative of P_s

The simplifying assumptions involve neglecting any effects of load on pump or motor speed and neglecting pressure drops in the hoses from the pump.

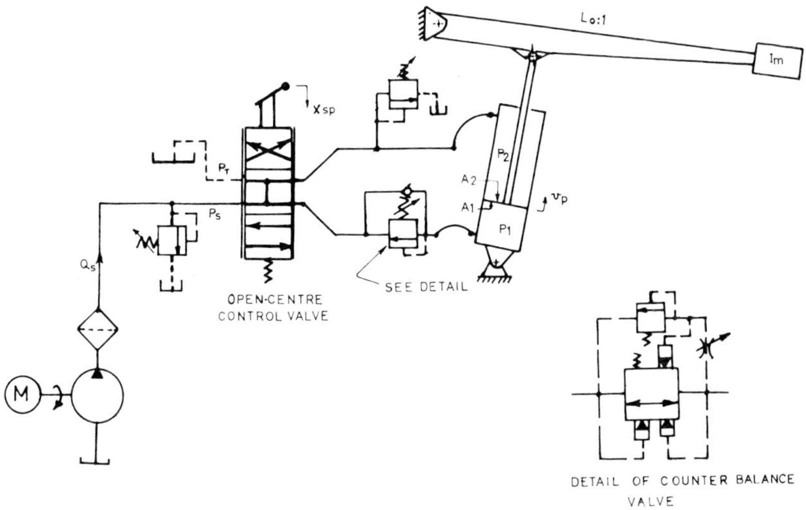

Fig. 1 Schematic Circuit of Hydraulic System showing pump, valve, cylinder and load and including various pressure relief valves and counter balance valve.

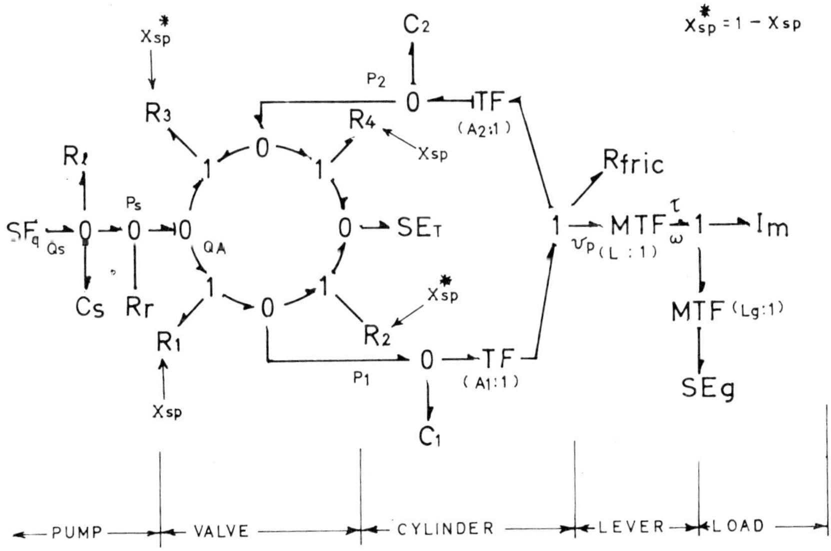

Fig. 2 Fourth Order Model of Hydraulic System in Power Bond Graph Notation

2.2 Control Valve

This is a complex network of flow orifices which are variable in area (and thus hydraulic resistance). They allow the operator to control pressure drops, and so flows, either to or from the cylinder ends. Thus each flow is a function of some external input variable X_{sp}. Whilst flows are non linear functions of pressure drops, they are assumed not to depend significantly on any time derivatives.

2.3 Hoses

Many previous studies merely consider these to offer increased pressure drop. Whereas the flow from the pump to the valve remains roughly constant, thus giving a constant pressure drop, the flow to (and from) the cylinder varies from zero to a maximum. Thus the pressure drop cannot be ignored in the hose.

The flow in the hoses accelerates. Inertia forces on this oil can be significant due to a "hydraulic ram" effect (Healey, 1966). In fact, if a mass is added to the piston to compensate for the inertia forces of oil in the hose it can be shown that the mass required is:

$$\text{Equivalent Mass} = (A/a_h)^2 . m_{oil}$$

where: A is the piston area
a_h is the hose bore cross section area
m_{oil} is mass of oil in the hose

Thus this is worth considering in any model. However, whereas the model so far has been synthesised by ideal "lumps", the fluid in the hose is a continuous system, each elemental length having mass, compressibility and causing some pressure drop through fluid friction. Representation can be quite complex and fluid line models are discussed later.

2.4 Cylinder

Each side of the cylinder contains a sizable volume of oil. Thus its compressibility cannot be ignored. Furthermore, the volume changes as the piston moves, making the differential equation non linear. The cylinder is primarily a device for transforming hydraulic pressure into a mechanical force and, since the energy is conserved, piston velocity into a fluid flow rate. The friction at the seal is both of viscous and coulomb types.

2.5 Load

The output force from the piston is converted into a torque applied to the lever, and the load then becomes the rotational inertia due to the mass on the end of the beam. An alternative model might consider the flexibility and damping of the beam to be significant, or even the mass of the piston and the oil moving with it.

3. MODELLING THE SYSTEM

Whatever method is used to develop the model a thorough understanding of the physical system is needed. A useful technique for model development is the power bond graph method. This technique yields a diagram similar to a block diagram but rather than each line denoting one signal, the line represents a pair of signals chosen to specifically denote energy transfer. The line is thus an energy coupling or bond. The method is explained by Karnopp and Rosenberg (1968, 1975) and summarized by Dransfield and Barnard (1973).

To read a power bond graph, the seven standard symbols must be known. Each has a precise mathematical definition and is shown in a little detail in Appendix 2. Figure 2 shows the power bond graph model of the system described above. Each and every feature of the system just described in sections 2.1 to 2.5 above can be identified in that diagram. Briefly the symbols mean:

I Newton's Second Law
C Capacitive relation - the integrated flow rate yields the pressure rise
R A resistive function, pressure loss related to flow rate
0 Continuity of flow equation, implying summation of flows
1 Summation of pressures or forces

Because mechanical parts and hydraulic hoses transmit power and the line on the bond graph denotes power flow, there is often a close correspondence between the power bond graph and the system schematic. This can be seen by comparing Figure 1 and Figure 2.

The logical procedure followed with this method is very convenient:

(a) Decide the features to be included and sketch out the bond graph showing all the power flows.

(b) Try out, directly on the bond graph, the sequence of substitution in the various equations implied by the bond graph by using the simple rules and symbols - the "causal bar", Appendix 3. This is similar in effect to, but simpler than, deciding the directions of arrows on a block diagram.

(c) Finally write down the equations as a set of integral equations directly from the bond graph. The equations can be written immediately in the form needed for the computing package (e.g. MIMIC). No further algebra is required, thus reducing the risk of errors. However, a valuable check on the signs of the equations is available, Rosenberg, 1971.

In particular, step (b) permits the analyst to specify that all derivative equations will be of integral form. The presence of a differentiation term can indicate an error or that a change in the model is desirable. Another source of difficulty are implicit equations in one unknown. The usual practice is to manipulate the equation to obtain a discrete expression for the unknown. However, using packages or simulation languages such as MIMIC or CSMP, the presence of the implicit equation is not apparent. Furthermore, many non-linear equations cannot be suitably solved analytically. Appendix 3 sets out a rule for detecting such equations before preparing the MIMIC coding, thus allowing either a model change (e.g. including another integration) or the inclusion of the sub-iteration subroutine of MIMIC.

For linear equations Rosenberg (Rosenberg and McPherson, 1975) has written a computer program which fully automates step (b), and works directly from the bond graph description.

4. SIMULATION OF THE SYSTEM

4.1 Numerical Integration Package

For the non linear simulation of the system a Control Data Corporation package, MIMIC, was employed on Melbourne University's Cyber 6400 computer. MIMIC uses a fourth order Runge-Kutta numerical integration routine, using a variable step length to adjust the relative error to below 10^{-6}. The package accepts equations in FORTRAN form, with a few additional functions, especially the integration function.

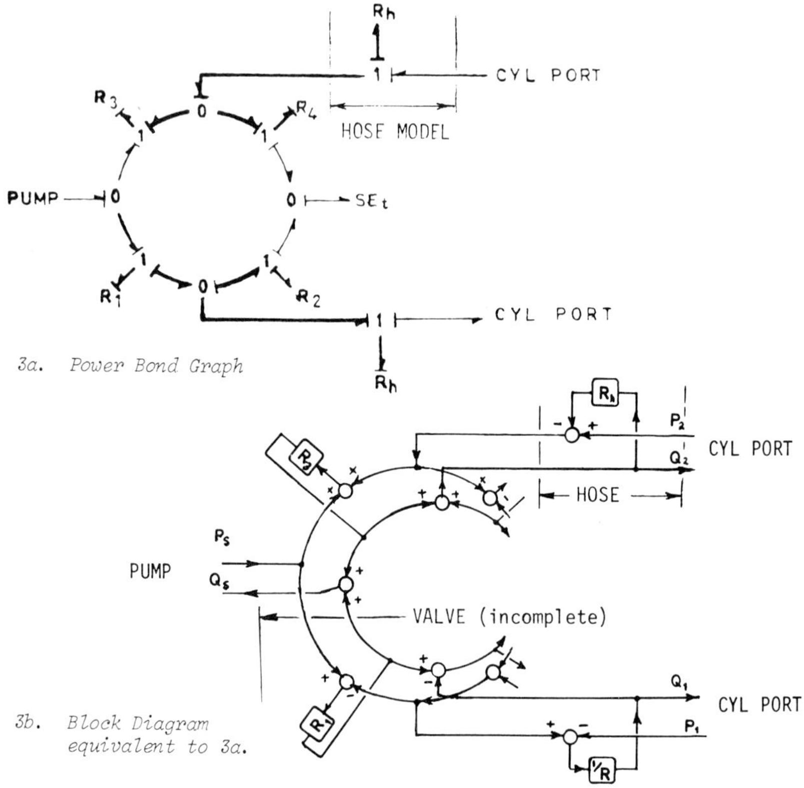

3a. *Power Bond Graph*

3b. *Block Diagram equivalent to 3a.*

Fig. 3 Segment from Figure 2 showing insertion of Hose Models

4.2 Fourth-Order Model

The simplest model incorporating the effects discussed in Section 2 is shown as a bond graph in Figure 2. The diagram completely describes the fourth order equation set which forms the model - each C and I denoting an integration. The equations are given in some detail in Appendix 4. The relations for any non linear elements such as friction (R_{fric}) or the valve resistances (R_1 to R_4) must also be defined as discussed earlier. However, this model does not include the pressure drops in the hoses from the valve to the cylinder. The model in Figure 2 may be separated where the power bonds connect valve and cylinder, and the simple hose models of Figure 3 (a) inserted. For illustration, Figure 3 (b) shows the same models in traditional block diagram form. However the model now has a group of four implicit equations embedded in it, and would fail to compute if implemented directly (since $Q_2 = fn(Q_2, P_2, P_3)$. It is thus useful to employ Appendix 3 to locate such problems.

The simplest method of avoiding the problem is to modify the magnitudes of the other fluid resistances (R_1 and R_4) to compensate approximately. Power Bond Graphs make it apparent that including the oil compressibility in the hose model would also avoid the problem. But these integrations make the model sixth order.

4.3 Eighth Order Model

If the oil compressibility in the hose is to be included, it may be argued that other similar effects should also be included. Thus several other aspects were judged to be worth including. First the beam stiffness, and to control the oscillations, the small but significant damping. It can be seen that the beam and the hydraulic cylinder are like two springs in series. Since the piston velocity is being computed, it seems worthwhile to include the piston mass to lower the oscillation frequency. These effects, when added to the model of Figure 2 give Figure 4.

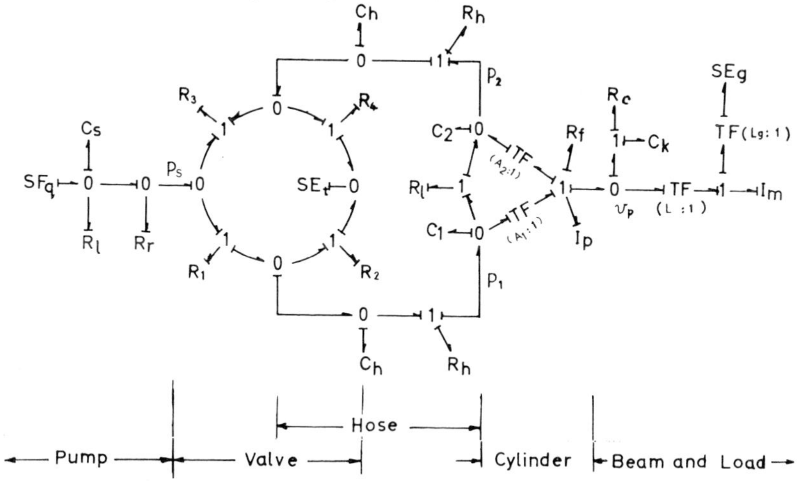

Fig. 4 Eighth order model of system including hose compliance and beam stiffness

Initially the simulation was unstable. However when the initial values of pressure at each end of each hose were altered so that there was no longer zero pressure difference across each hose, the problem vanished! Figure 5 shows one result for the flow rate along one hose. A 90 hz flow pulsation is obvious. Hand calculations indicated that very large pressure differences would be needed just to accelerate the small mass of oil flowing in the lines in the manner predicted. This was taken to indicate that the inertia of the oil in the lines could not be neglected.

4.4 Tenth Order Model

Inclusion of the oil inertia changes Figure 4 to produce Figure 6. Whilst requiring more computing time, simulation of this produces the results shown in Figure 7 (dotted line), giving piston tr-vel and pressures. Superimposed for comparison are the simulation of the fourth order model (computer print out) and experimental measurements (full line). The input command opened the valve rapidly (i.e. reduced resistances R_1 and R_4), held it open for 0.65 seconds and rapidly closed it back to the central, neutral position.

Of particular interest is the pressure trace of the return side of the cylinder, P_2 where a 37 hz pressure fluctuation is visible at the start of motion. The experimental run from which this pressure trace was taken had a rather longer valve operation time and this is why the traces are of different length. However, the pressure fluctuations correspond remarkably in number of cycles, spacing and amplitude. Both the simulation and the measured results exhibit a slight hesitation in the pressure when the relief valve operates.

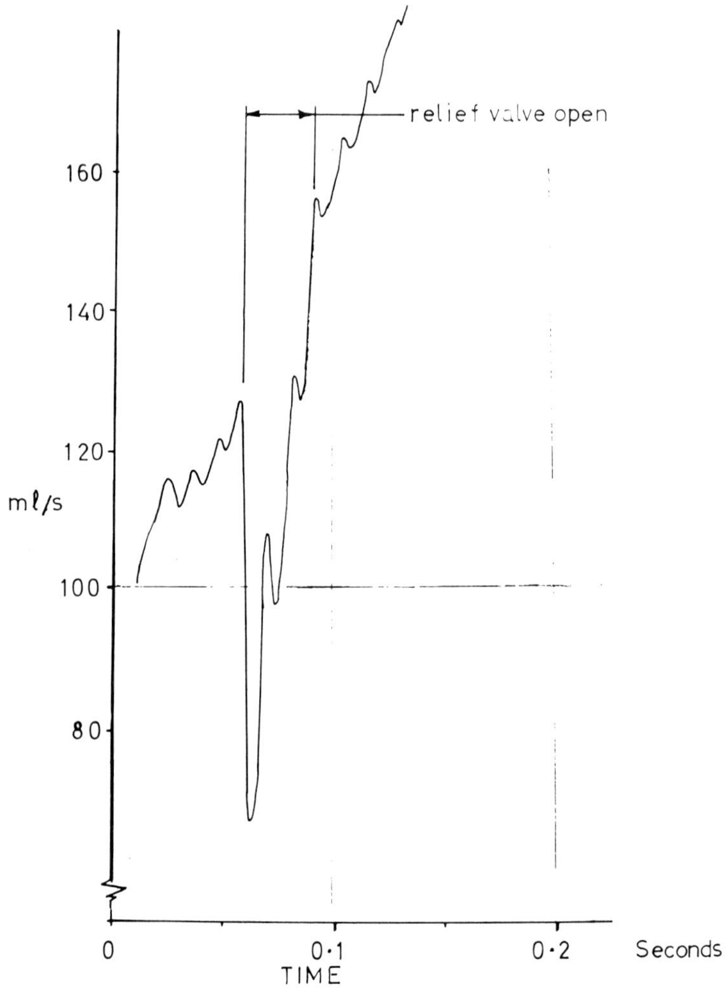

Fig. 5 Flow rate to cylinder as computed by eighth order model

4.5 Other Line Models

The single lump approach used above to approximate the line dynamics, whilst adequate for this system study, neglects the pressure waves traversing the length of the hoses.

In order to simulate the effects of these waves and their reflections greater sophistication is needed. The general rule of thumb has been to use 10 lumps per wavelength at the highest frequency of interest. Oldenberger, 1970, offers a more rational mathematical basis for selecting the lumps, using Laplace transforms and transfer functions to obtain an infinite number of factors to describe the line. He then selects the most significant of these. A simpler method is to use normal mode analysis and include one "lump" to cover the fundamental and higher modes as desired (Karnopp and Rosenberg, 1975). Effectively the method used in Figure 6 considers the first modes only.

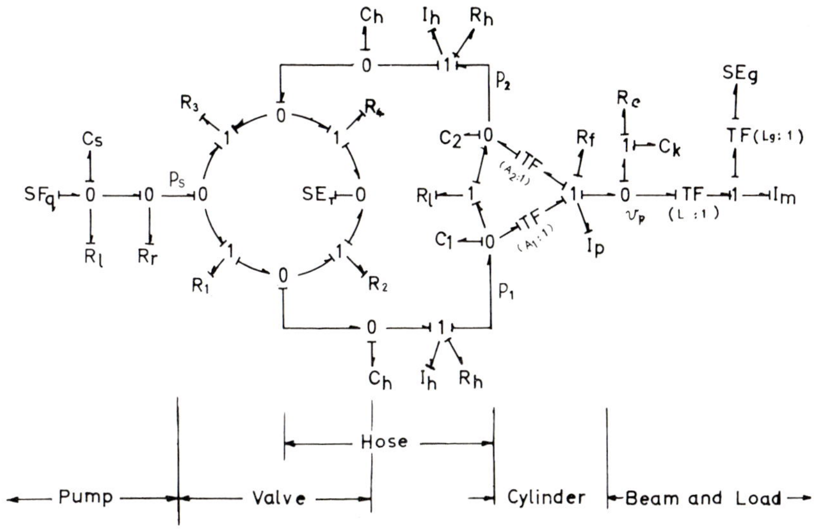

Fig. 6 Tenth order model with oil inertia included in hose models

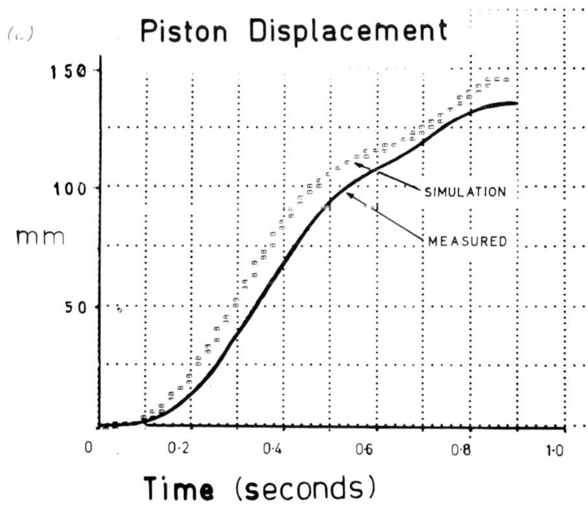

Fig. 7 Simulation results and experimental results from models 7a. Piston displacement

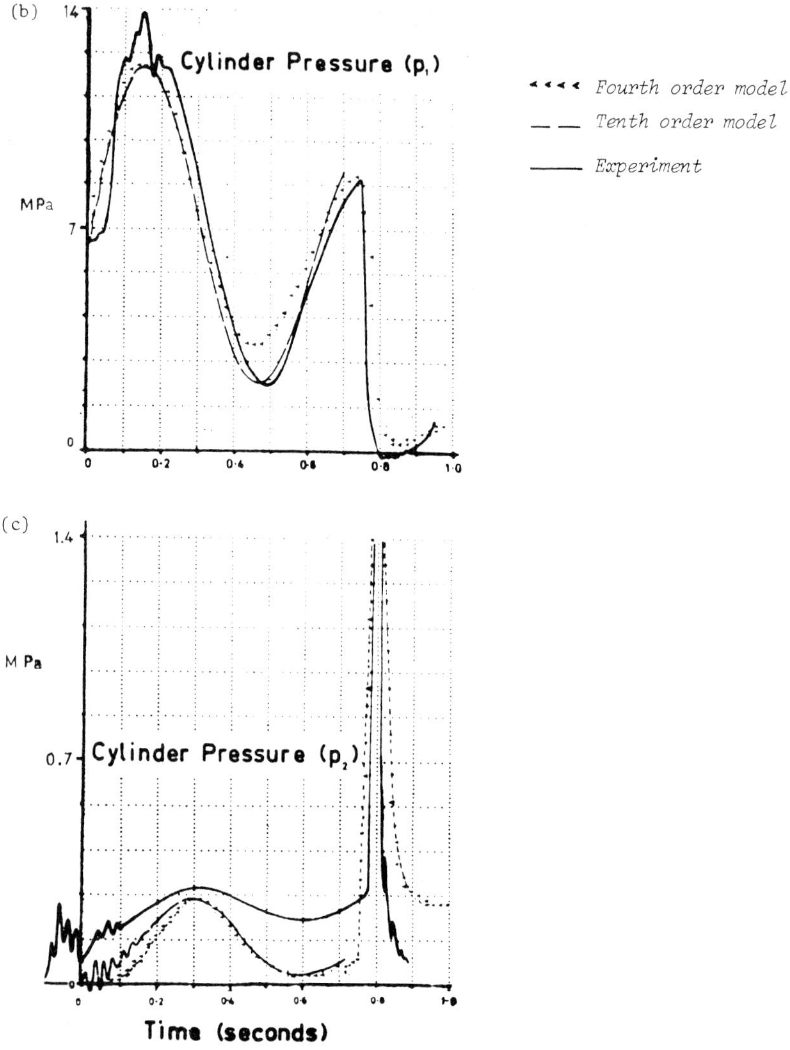

Fig. 7 Simulation and experimental results from models
7b. Supply pressure
7c. Return pressure

4.6 Linearized Models

Several standard computer programs are available for analysis of linear systems. Professor Rosenberg of Michigan State University provided a copy of ENPORT (Rosenberg and McPherson, 1975) which accepts the description of the system in bond graph terms, and computes the eigenvalues of the system (strictly, of the A-matrix as in Appendix 4). Results for four different models are given in Figure 8 as plots on the left hand half of a complex plane. The third order model lumps both sides of the cylinder into one "fluid spring". The fourth order model is as discussed previously and the sixth and eighth order models each include parts of the tenth order version. From Figure 8 it can be seen that two eigenvalues, known in control theory as the dominant poles, change very little between models (see inset) and these give the results in Figure 7 their primary frequency.

However, the third and other eigenvalues (or poles) shift considerably. An intriguing observation in all but the third order model is that the centroid of the eigenvalues lies in the same place. It is suspected that this may not merely be accidental. Consider a simple case of two values each separated a small distance, b, on the real axis from the centroid, c; then in the real time (i.e. inverse Laplace) domain, part of the response Y, will be

$$Y(t) = e^{(c-b)t} + e^{(c+b)t} \tag{4.6 a}$$

and by expanding these terms it can be shown that provided bt is very small

$$e^{(c-b)t} + e^{(c+b)t} \simeq 2.e^{ct} \tag{4.6 b}$$

However, it cannot be claimed that the various eigenvalues are close to their centroid here. It would be convenient if some rule like this did justify approximation of high order d.e.'s by one of lower order.

The eighth order model also shows reasonable agreement with the experimental mental results, indicating a lightly damped oscillation at 42 hz.

5. CONCLUSION

The simulation of oil hydraulic systems presents the engineer with a number of difficulties. In this paper, attention has been concentrated on the choice of aspects to be included in the model, and their affect on the transient response of the system. Mention might also have been made of the difficulties of simulating logical devices such as the counterbalance valve which is required to prevent reverse flow at the start. Even obtaining a physical description of the system can demand direct measurement. The effective fluid bulk modulus is greatly affected by air entrained within it, as well as by hose design and even the O-ring seal design in the cylinder.

Fortunately, while attention has been given to flow rates and pressures, these are simply the most sensitive variables to inspect. The mathematics of the physical system is such that pressure is integrated to obtain load velocity, and displacement is the integral of velocity. These two integrations substantially smooth out the high frequency components visible in flow and pressure traces, and so the displacement time trace is not greatly affected by small changes in the model. Of course, some engineering systems are very susceptible to peak pressures, and these cases have received considerable attention in water hammer studies.

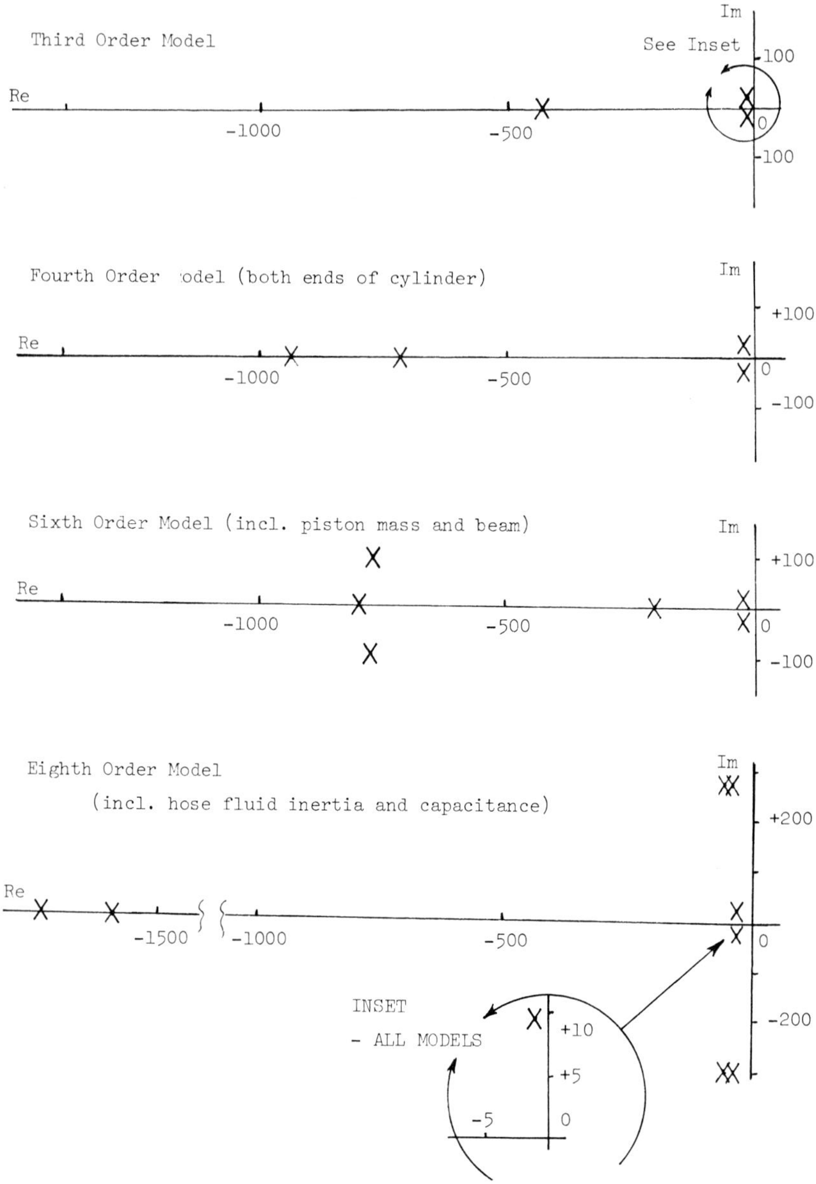

Fig. 8 Eigen-value plots for various linearized models of differing complexity

As a result of this study the author believes that it is highly desirable, when modelling dynamic systems, to include both inertia and compliance (or stiffness) effects as almost inseparable pairs. Not to do so seems to encourage spurious, high frequency oscillations and difficulties with numerical simulation. Power bond graphs highlight the difference between the inertia and compliance although both involve integration. The effects each has on the system matrix equation might be worthy of further mathematical study.

Despite the uncertainties and difficulties, numerical simulation packages coupled with power bond graphs to facilitate obtaining the equations, together make simulation a more viable engineering tool.

6. REFERENCES

Barnard, B.W. (1973) Predicting the Dynamic Response of an Hydraulic System using Power Bond Graphs. M.Eng.Sc. Thesis, Monash University.

Dransfield, P. and Barnard, B.W. (1973) Power Bond Graph Modelling of Hydraulic Control Systems. I.F.A.C. Symposium on Automatic Control in Mining and Metal Processing, Sydney.

Healey, M. (1966) Effect of Oil Inertia on Resonant Frequency of Hydraulic Drives. Control 10, No. 95.

Karnopp, D. (1968) Computer Representation of Continuous Vibratory Systems Using Normal Modes. Simulation, March 1968, p. 129.

Karnopp, D. and Rosenberg, R. (1968) Analysis and Simulation of Multiport Systems. The M.I.T. Press.

Karnopp, D. and Rosenberg, R. (1975) System Dynamics: A Unified Approach. Wiley-Interscience.

Oldenberger, R. (1970) Theory of Distributed Systems. Trans. ASME, Ser. D., 92 (1), 1.

Rosenberg, R. (1971) State-space Formulation for Bond Graph Models of Multiport Systems. Trans. ASME, Ser. G., 93 (March), 35.

Rosenberg, R. and McPherson, M. (1975) A Users Guide to ENPORT 4. Wiley Interscience.

Swisher, G. H. and Doebelin, E. O. (1970) Lumped Parameter Modelling vs. Distributed Parameter Modelling for Fluid Control Lines. J. Spacecraft and Rockets, 7 (6), 766.

APPENDIX 1

NOTATION

A	Piston area
a_h	Area of hose bore
b	A small distance on a complex plane
C	Capacitive function on a bond graph, or hydraulic capacitance $(= \frac{V}{\beta})$
c	A point on the real axis of a complex plane
d	A small-difference operator
E	A generalized effort variable (e.g. pressure)
F_p	Force output from piston
F	A generalised flow variable (e.g. flow rate)
g	Force of gravity on unit mass
I	Inertance function on a bond graph, or mass moment of inertia
L	Lever arm of piston
L_g	Lever arm of gravity force on output arm
m	Mass
P	Pressure
Q	Volume flow rate
R	Resistance function on a bond graph or the hydraulic resistance
SE	An effort source (pressure or gravity force) on a bond graph
TF	An energy transforming function on a bond graph
V	A volume
X_{sp}	Input displacement of piston
X_p	Output displacement of piston
β	Fluid bulk modulus
τ_m	Torque applied to load
ω_m	Angular velocity of load

Suffices:

1	to lifting side of cylinder	ℓ	relating to leakage
2	to return side	r	relating to relief valve
f	relating to friction		
h	relating to hose	s	relating to pump

APPENDIX 2

SUMMARY OF POWER BOND GRAPH SYMBOLS

Power bond graphs are possible because the mathematics describing different energy forms is similar. If an effort variable, E, is defined (e.g. force, torque, pressure, voltage, etc.) then the other variable needed to give power transfer is a flow rate variable, F, (e.g. velocity or volume flow rate). The mathematics for energy storage as kinetic or potential energy or for dissipation is analogous whether it be for rotatory or linear, fluid or electric energy; only the constant in the equations needs to be defined from fundamental theory. Table 1 summarizes the symbols and their meaning.

Notice that the power bond(s) on each symbol implies that one variable is known and the other will be computed using the function on that bond. Thus in the I-element, E is known and F is computed. Which is computed is decided by the causul bar ——|. Each definition is given in terms of the general symbols E and F. If these are each replaced by pressure, P for E and flow rate, Q for F (or force and velocity for E and F) some familar equations result.

Signs at the 0 or 1 junctions are obtained from a definition of the direction of power flow.

Energy Form	Bond Graph Symbol	Definition	Comments
Kinetic	I |——	$F = \frac{1}{I}\int E.dt$	e.g. Newton's Law and I = m
Potential	C ——|	$E = \frac{1}{C}\int F.dt$	e.g. Mechanical Spring with C = 1/K
Dissipation	R |——	$F = \frac{E}{R}$	e.g. Viscous Drag
	R ——|	$E = F.R.$	
Transformation	—1— TF —2— (A:1)	$E_1 = A.E_2$ and $F_2 = A.F_1$	Notice that both equations apply
Sources	SE ——|	$E = \text{constant}$	Not affected by returning signal
	SF |——	$F = \text{constant}$	Not affected by E, e.g. a constant flow pump
Flow Junction	—1—| 0 —2—| (bond 3 below, causal bar ⊥)	$F_1 = F_2 + F_3$ and $E_2 = E_1$, $E_3 = E_1$	Continuity at a tee junction
Effort Junction	—1—| 1 —2—| (bond 3 below, causal bar T)	$E_2 = E_1 - E_3$ and $F_1 = F_2$, $F_3 = F_2$	e.g. Summation of forces on a free body

Table 1 - Summary of Power Bond Graph Symbols

APPENDIX 3

IDENTIFYING IMPLICIT FUNCTIONS WITHIN R-FIELDS

A power bond may be defined as a pair of oppositely directed signals - one effort signal and one flow signal. Causality indicates the direction of the effort signal, thus ——| means p →
← q

Figure 9 shows an R-field consisting of three one-port R-elements. The causality of the mesh is dictated by the causality forced on its 'free' ends (presumably by the S, C or I elements of the remainder of the system). By noticing the causality of the bonds joining the R-elements (drawn boldly), the following rule may be deduced from the definition of the power bond.

RULE TO IDENTIFY IMPLICIT EQUATIONS

If the power-bonds of a portion of a bond graph all have a uniform causal bar direction from one R-element to one or more other R-elements the node output variables are described by equations which are implicit in the output variables concerned. The bond is unaffected by passing through a TF element, but is reversed through a GY element. For example, in Figure 9 (a) one implicit relation exists between the causal outputs of effort on bond 2 and flow on bond 8. However, in Figure 9 (b) the causal bars do not flow from one R-element to the next. E2 is then independent of the outputs from bonds 8 and 5.

In many systems the implicit nature of the relations is quite obscure without the use of this rule. The R-field is not disturbed by C or I elements attached to it unless they alter causality.

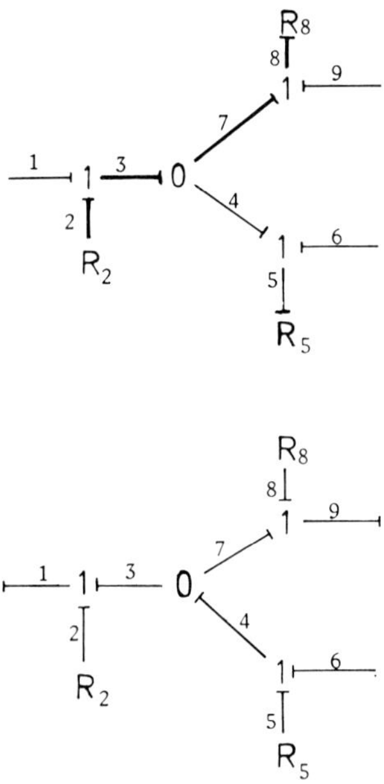

(a) Implicit Relation between E2 and F8 (another exists between E2 and E5)

(b) Explicit Relation between E2 and F8

Fig. 9 Identifying implicit functions from causality "flow"

APPENDIX 4

EQUATIONS FROM POWER BOND GRAPH OF FOURTH ORDER MODEL

The simplest method of obtaining the equations from the bond graph (Figure 2) is to write the whole list of short first order equations.

For example, for the load mass element, I_m:-

$$\omega_m = (\frac{1}{I_m})\int \tau_m.dt \qquad (A.4a)$$

where ω_m and τ_m are the angular velocity and torque applied to the lever and I_m the mass moment of inertia.

For the cylinder fluid volume:-

$$P_1 = (\frac{1}{C_1})\int dQ_1.dt \qquad (A.4b)$$

where dQ_1 is the flow rate due to fluid compressibility.

The 1- and 0- junctions give very simple equations, e.g. for the 1-junction near the mass element I_m, the equation is a sum of torques:-

$$\tau_m = (L.F_p) - Lg.(m.g) \qquad (A.4c)$$

where F_p is the nett piston force and L its lever arm (from the TF element)

and mg is the gravity force on the mass (from SE_g)

and L_g its lever arm (from the related TF element)

When all the necessary equations have been written, if desired they can be reduced to a standard matrix form,

$$[\dot{X}] = [A]\ [X] + [B]\ [U] \qquad (A.4d)$$

where $[\dot{X}] = \begin{bmatrix} dQ_s \\ dQ_1 \\ \tau_m \\ dQ_2 \end{bmatrix}$ $\quad [B] = \begin{bmatrix} 1 & 0 \\ 0 & 0 \\ 0 & -L_g \\ 0 & 0 \end{bmatrix}$ $\quad [U] = \begin{bmatrix} Q_s \\ mg \end{bmatrix}$

and

$$[A] = \begin{bmatrix} \sum_{i=1,4}\left(\frac{1}{R_i . C_s}\right) & \frac{1}{R_1 . C_1} & 0 & \frac{1}{R_2 C_2} \\ \frac{1}{R_1 . C_s} & -\left[\frac{1}{R_1 . C_1} + \frac{1}{R_3 C_1}\right] & -\frac{L.A_1}{I_m} & 0 \\ 0 & \frac{L.A_1}{C_1} & \frac{L^2 . R_f}{I_m} & -\frac{L.A._2}{C_2} \\ \frac{1}{R_2 . C_s} & 0 & \frac{L.A_2}{I_m} & -\left[\frac{1}{R_2 C_2} + \frac{1}{R_4 . C_2}\right] \end{bmatrix}$$

The output variables can be obtained from the simple matrix.

$$\begin{bmatrix} P_s \\ P_1 \\ \omega_m \\ P_2 \end{bmatrix} = \begin{bmatrix} \frac{1}{C_s} & 0 & 0 & 0 \\ 0 & \frac{1}{C_1} & 0 & 0 \\ 0 & 0 & \frac{1}{I_m} & 0 \\ 0 & 0 & 0 & \frac{1}{C_2} \end{bmatrix} .[X] \qquad \text{(A.4e)}$$

However the equations are not valid in this form except for linearized analysis as the functional R_1 is the square root of the difference of two variables from the X-matrix. Matrix representation is obviously not adequate for non linearities of this type.

The A-matrix is interesting because it still contains some of the diagonal skew symmetry which forms the method of checking proposed by Rosenberg (1971). Full details of the check procedure applied to this problem are given by Barnard (1973).

Numerical Simulation of Fluid Motion, ed. John Noye
pp. 521-535

THE TRANSMISSION OF SOUND IN NONUNIFORM DUCTS USING A MODIFIED GALERKIN METHOD

W. Eversman

University of Canterbury

Christchurch, New Zealand

SUMMARY

The method of weighted residuals in the form of a modified Galerkin Method with boundary residuals is developed for the study of the transmission of sound in nonuniform ducts carrying a steady compressible flow. In this formulation the steady flow is modelled as essentially one-dimensional but with a kinematic modification to force tangency of the flow at the duct walls.

Good agreement has been obtained with known results in hard walled ducts up to near sonic velocities. To attempt to investigate the fundamental soundness of the Galerkin Method using basis functions which don't individually satisfy the boundary conditions, the problem of transmission in a straight acoustically treated duct with a uniform flow has been formulated and the modified Galerkin Method used with basis functions derived from the case when flow is absent. A QR root finding scheme has been used to compute the axial wave numbers. It is found that at low frequencies these basis functions do not yield results in satisfactorily close agreement with exact calculations. An improved set of basis functions is introduced and computations show good agreement with exact values.

1. INTRODUCTION

The method of weighted residuals (MWR) in the form of a Galerkin technique with boundary residuals has been shown to be a useful method for the investigation of the transmission of sound in nonuniform ducts with no flow (Eversman et.al.,1974). In this paper the method is extended to treat the case when the duct carries a steady nonuniform flow.

The introduction of steady mean flow into the study of nonuniform duct propagation complicates the problem in two ways. The acoustic field equations no longer reduce to either the simple wave equation of the no flow case or the convective wave equation of the uniform flow case. In addition, the field equations represent acoustic perturbations on a steady nonuniform flow field which in itself is difficult to describe. As a consequence of these complications, no simple theories of propagation have been developed and investigations appearing in the literature have been relatively few.

The majority of studies appearing have been based on a one-dimensional theory which treats the propagation as plane waves moving in a one-dimensional nozzle flow. Powell (1960), Eisenberg and Kao (1969), King and Karamcheti (1973) and Davis and Johnson (1974) have studied one-dimensional models by a variety of techniques.

It is the purpose of this investigation to study multi-modal propagation in nonuniform ducts with flow. Published work dealing with this problem has been limited to approximate methods for nonuniformities in cross sections, lining and flow properties which are slowly varying. Tam (1971) considered hard walled ducts with slowly varying cross section by using a Born type of approximation and Fourier Transform methods. Nayfeh et.al.,(1973) have used the method of multiple scales and include slowly varying area, lining and boundary layer. While Tam's work allows modal coupling, Nayfeh's does not.

The method described here treats the complete acoustic field equations and is approximate only in the sense that the solutions are represented in terms of a superposition of a limited number of specified functions. We will be concerned primarily with the method of solving the field equations and relatively little with the difficulty presented by describing the nonuniform steady flow field. Instead, we will use an approximation for the flow field based on one-dimensional compressible flow with a kinematic modification to allow for flow tangency at the duct walls.

2. CO-ORDINATE SYSTEM AND GEOMETRY

In the following development we will consider only two-dimensional ducts. The method employed generalizes to the three-dimensional case, but is then, of course, more computationally demanding. We will also consider explicitly only ducts of infinite length, although as has been shown (Eversman et.al., 1974) the extension to ducts which are of finite length is not difficult provided that the reflection properties of the termination are known.

Figure 1 shows the type of configuration under consideration in which two semi-infinite uniform duct sections of height b_0 and b_ℓ are joined by a transition section of variable height $b(x)$. The duct wall at $y = 0$ is hard and the wall at $y = b(x)$ in the transition section has variable impedance $Z(x)$. The uniform sections have impedance Z_0 and Z_ℓ at $y = b_0$ and $y = b_\ell$, respectively, and have a hard wall at $y = 0$. This representation can be considered as a model of a duct of this configuration or of one symmetric with respect to the x axis with symmetric propagation. We will consider only continuous variations in duct height; however discontinuous changes in impedance will be permitted at the ends of the nonuniformity.

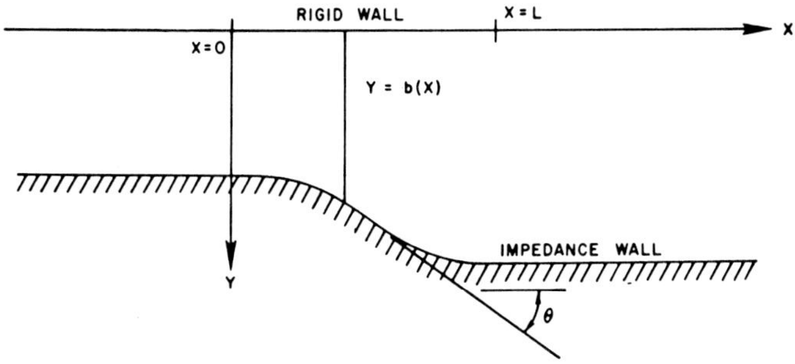

FIGURE 1
DUCT GEOMETRY FOR EXAMPLE CASE

3. GOVERNING EQUATIONS

The equations of motion for acoustic propagation in a steady compressible flow field are obtained from the momentum and energy equations and the isentropic equation of state (continuity and energy become equivalent, but the standard form of the energy equation is conveniently in terms of pressure instead of density). The assumption of small perturbations, with harmonic time dependence $e^{i\omega t}$, is made to obtain

$$ik_r\vec{V} + \vec{V}\cdot\text{grad}\ \vec{V}_o + \vec{V}_o\cdot\text{grad}\ \vec{V} + \frac{1}{\rho_o}\ \text{grad}\ p - \frac{1}{\gamma}\ \frac{p}{p_o\rho_o}\ \text{grad}\ p_o = 0 \tag{3.1}$$

$$ik_r p + \vec{V}_o\cdot\text{grad}\ p + \gamma p_o\ \text{div}\ \vec{V} + \vec{V}\cdot\text{grad}\ p_o + \gamma p\ \text{div}\ \vec{V}_o = 0 \tag{3.2}$$

The acoustic pressure and particle velocity p, $\vec{V}$ are nondimensionalized with respect to $\rho_r c_r^2$ and c_r respectively, where ρ_r is a reference density and c_r is a reference speed of sound. The steady flow pressure, velocity and density is denoted by p_o, $\vec{V}_o$, ρ_o and is similarly nondimensional, with ρ_o being nondimensionalized by ρ_r. The reference conditions are taken at a point where the flow is uniform, generally in the uniform duct segment on the sound source side. The wave number $k_r = \omega/c_r = 2\pi/\lambda_r$ where λ_r is the wave number for plane waves in the reference state. The boundary condition at the rigid wall is simply that the normal velocity component must vanish. At the acoustically treated wall the boundary condition is characteristic of a normally reacting acoustic lining, namely

$$V_\nu = Ap$$

where V_ν is the nondimensional velocity in the lining of the wall, in the direction of $\vec{\nu}$ the unit outward normal, p is the nondimensional acoustic pressure at the wall, and A is the specific acoustic admittance of the wall based on the reference density and speed of sound, i.e. $A = \rho_r c_r / Z$ where Z is the impedance of the wall. The acoustic particle velocity in the fluid is related to the wall velocity in harmonic motion by

$$\vec{V}\cdot\vec{\nu} = V_\nu + \frac{U_\tau}{ik_r}\frac{\partial V_\nu}{\partial \tau}$$

where τ is distance measured tangential to the duct wall. The soft wall boundary condition is thus

$$\vec{V}\cdot\vec{\nu} = Ap - \frac{iU_\tau}{k_r}\frac{\partial}{\partial \tau}(Ap)$$

The steady flow field is given by

$$U_o(x,y) = \bar{U}_o(x)$$

$$V_o(x,y) = \bar{U}_o(x)\,\frac{y}{b}\frac{db}{dx}$$

$$p_o(x,y) = \bar{p}_o(x)$$

$$c_o(x,y) = \bar{c}_o(x)$$

where the one-dimensional gas dynamic equation of motion is

$$\frac{dM_o}{dx} = -\frac{M_o}{1-M_o^2}\left[1 + \frac{\gamma-1}{2}M_o^2\right]\frac{1}{b}\frac{db}{dx}$$

and related equations are

$$\bar{\rho}_o = \frac{\left[1 + \frac{\gamma-1}{2}M_r^2\right]^{\frac{1}{\gamma-1}}}{\left[1 + \frac{\gamma-1}{2}M_o^2\right]^{\frac{1}{\gamma-1}}} \qquad \bar{p} = \frac{1}{\gamma}\frac{\left[1 + \frac{\gamma-1}{2}M_r^2\right]^{\frac{\gamma}{\gamma-1}}}{\left[1 + \frac{\gamma-1}{2}M_o^2\right]^{\frac{\gamma}{\gamma-1}}}$$

$$\bar{c}_o^{\,2} = \frac{1 + \frac{\gamma-1}{2}M_r^2}{1 + \frac{\gamma-1}{2}M_o^2} \qquad \bar{U}_o = M_o\frac{\left[1 + \frac{\gamma-1}{2}M_o^2\right]^{\frac{1}{2}}}{\left[1 + \frac{\gamma-1}{2}M_r^2\right]^{\frac{1}{2}}}$$

As noted this approximation is basically the one-dimensional result on the straight wall (or duct centreline) modified to force tangency at the duct walls. It is considered a satisfactory approximation for the present study.

With these results we can write the field equations and boundary conditions in component form

$$U_o\frac{\partial u}{\partial x} + \frac{1}{\rho_o}\frac{\partial p}{\partial x} + V_o\frac{\partial u}{\partial y} + \left(ik_r + \frac{\partial U_o}{\partial x}\right)u - \frac{1}{\gamma p_o \rho_o}\frac{\partial p_o}{\partial x}\,p = 0 \tag{3.3}$$

$$U_o\frac{\partial v}{\partial x} + V_o\frac{\partial v}{\partial y} + \frac{1}{\rho_o}\frac{\partial p}{\partial y} + \frac{\partial V_o}{\partial x}u + \left(ik_r + \frac{\partial V_o}{\partial y}\right)v = 0 \tag{3.4}$$

$$\gamma p_o\frac{\partial u}{\partial x} + U_o\frac{\partial p}{\partial x} + \gamma p_o\frac{\partial v}{\partial y} + V_o\frac{\partial p}{\partial y} + \frac{\partial p_o}{\partial x}u + \left[ik_r + \gamma\left(\frac{\partial U_o}{\partial x} + \frac{\partial V_o}{\partial y}\right)\right]p = 0 \tag{3.5}$$

$$v \cos\theta - u \sin\theta = Ap - \frac{iU_o}{k_r}\left[\frac{\partial}{\partial x}(Ap) + \tan\theta\, A\frac{\partial p}{\partial y}\right] \text{at } y = b(x) \tag{3.6}$$

$$v = 0 \text{ at } y = 0$$

where $\tan\theta = {}^{db}/dx$, the wall slope.

4. APPLICATION OF THE GALERKIN METHOD

In Eversman et.al.(1974) the method of weighted residuals (MWR) in the form of a modified Galerkin Method with boundary residuals was employed to study acoustic propagation in nonuniform ducts without flow. We will use this method with minor changes in the present study.

We seek solutions to the field equations, Equations (3.3) - (3.5) and the boundary conditions, Equations (3.6) in the form

$$p_N(x,y) = \sum_{n=1}^{N} p_n(x)\psi_n(x,y)$$

$$u_N(x,y) = \sum_{n=1}^{N} u_n(x)\psi_n(x,y)$$

$$v_N(x,y) = \sum_{n=1}^{N} v_n(x)\phi_n(x,y)$$

The basis functions ψ_n (x,y) and ϕ_n (x,y) are chosen to satisfy the boundary condition v = 0 at y = 0 but are not required to individually satisfy the boundary condition at y = b(x).

The success of the MWR is dependent on an appropriate choice of basis functions. In the no flow case of Eversman et.al.(1974), the basis functions were chosen to be the transverse modal functions which would exist in a uniform duct with the properties existing locally in the nonuniform duct. This philosophy can be carried over to the duct with a steady mean flow. In this case the appropriate choice of basis functions would be the transverse modal functions for a uniform duct with flow and geometry properties existing locally in the nonuniform duct.

The major disadvantage in using this type of basis function is that there are different modes for upstream and downstream propagation so that to obtain the same resolution as in the no flow case we would use twice as many modes, half representing upstream propagation and half representing downstream propagation.

A second possibility is the use of the no flow modes as was done in Eversman et.al.(1974). These modes are not generally near solutions so that we would expect the process of convergence to be slower than if the full set of flow modes were used. The principal advantage of using the no flow modes lies in the great savings in dimensionality. Numerical experimentation has indicated that even though rapidity of convergence suffers somewhat, the number of basis functions required when the no flow modes are used is considerably less than when the flow modes are used. In the present study we have used the no flow duct modes as basis functions.

The no flow basis functions are

$$\psi_n(x,y) = \cos\,[\kappa_n(x)y]$$

$$\phi_n(x,y) = \sin\,[\kappa_n(x)y]$$

where the sequence of eigenvalues of $\kappa_n(x)$ are locally determined from the eigenvalue equation

$$\kappa_n b \tan \kappa_n b = ikbA$$

Eversman (1975) discusses a method of obtaining the sequence of eigenvalues required by a method of numerical integration, while Cremer (1953) discusses the physical problem which generates these basis functions.

The modified Galerkin Method employing boundary residuals follows the same basic approach as the standard Galerkin Method in that residuals of the field equations (errors which remain when the assumed solution is used) are forced to be orthogonal on the duct cross section to a complete set of functions (conveniently taken as the basis functions themselves), and in this way forced to vanish (since a function which is orthogonal to every member of a complete set of functions must vanish). The modification in the approach is the use of a set of basis functions which do not individually satisfy the boundary condition but which will be forced to on the aggregate. Following the argument of Finlayson (1972) and Zinn and Powell (1968), this can be achieved by appending to the energy equation a boundary residual, or equivalently, using the boundary residual to eliminate boundary terms which arise in the energy equation (the energy equation is the appropriate one to which to append the boundary residual since the boundary condition is directly related to terms which appear in the residual of the energy equation when the orthogonality requirement is enforced). In essence, the argument is as follows: If the assumed solution is chosen to identically satisfy the boundary condition (which can occur even if the individual basis functions don't satisfy the boundary conditions, if the summation of basis functions is chosen to identically satisfy the boundary condition), then in applying the Galerkin orthogonalization procedure certain boundary terms which arise either vanish or are replaced as a result of the boundary condition. The unknown coefficients are then determined from the set of equations which results (ordinary differential equations in the present application). We can't generally pick the summation of basis functions in an a priori way to identically satisfy the boundary condition. However, we can form a boundary residual and apply the Galerkin orthogonalization process to it. This residual can then be used to eliminate or replace terms which arise in the orthogonalization process of the field equations. The resulting equations (differential equations) will be identical to those which would occur if we started out assuming the assumed solution identically satisfies the boundary condition. The solution of the resulting equations for the coefficients should then generate a summation of the basis functions which approximately satisfies the field equations and the boundary condition.

In the present problem the field equation and boundary condition residuals are

$$R_1 = U_o \frac{\partial u_N}{\partial x} + \frac{1}{\rho_o} \frac{\partial p_N}{\partial x} + V_o \frac{\partial u_N}{\partial y} + \left(ik_r + \frac{\partial U_o}{\partial x}\right) u_N - \frac{1}{\gamma p_o \rho_o} \frac{\partial p_o}{\partial x} p_N$$

$$R_2 = U_o \frac{\partial v_N}{\partial x} + V_o \frac{\partial v_N}{\partial y} + \frac{1}{\rho_o} \frac{\partial p_N}{\partial y} + \frac{\partial v_o}{\partial x} u_N + \left(ik_r + \frac{\partial v_o}{\partial y}\right) v_N$$

$$R_3 = \gamma p_o \frac{\partial u_N}{\partial x} + U_o \frac{\partial p_N}{\partial x} + \gamma p_o \frac{\partial v_N}{\partial y} + V_o \frac{\partial p_N}{\partial y} + \frac{\partial p_o}{\partial x} u_N + \left[ik_r + \gamma\left(\frac{\partial U_o}{\partial x} + \frac{\partial V_o}{\partial y}\right)\right] p_N$$

$$R_B = v_N \cos\theta - u_N \sin\theta - A \left\{ p_N - \frac{iU_o}{k_r}\left[\frac{\partial p_N}{\partial x} + \tan\theta \frac{\partial p_N}{\partial y} + \frac{1}{A}\frac{dA}{dx} p_N\right]\right\}\Bigg|_{y=b}$$

The orthogonalization process is

$$\int_0^b \psi_n R_1 \, dy = 0 \qquad \int_0^b \phi_n R_2 \, dy = 0$$

$$\int_0^b \psi_n R_3 \, dy = 0 \qquad \psi_n(x,b) R_B = 0$$

$$n = 1,2,3 \ldots N$$

The orthogonality is enforced on the cross section for the field equations (a unit depth is assumed) and on the curve enclosing the cross section for the boundary residual (reduces to y = b in this case). When the operations indicated are carried out and we use rules for differentiation of integrals, integration by parts, the condition of flow tangency at the walls, and eliminate boundary terms in the energy equation residual using the boundary residual, the following set of linear ordinary differential equations is obtained

$$\begin{bmatrix} N_{nm} & \frac{U_o}{\gamma p_o}\bar{N}_{nm} & 0 \\ \rho_o U_o N_{nm} & N_{nm} & 0 \\ 0 & 0 & M_{nm} \end{bmatrix} \begin{Bmatrix} \frac{dU_m}{dx} \\ \frac{dp_m}{dx} \\ \frac{dv_m}{dx} \end{Bmatrix} = \begin{bmatrix} -\bar{U}^u_{nm} & -\bar{U}^p_{nm} & -\bar{U}^v_{nm} \\ -\bar{P}^u_{nm} & -\bar{P}^p_{nm} & -\bar{P}^v_{nm} \\ -\bar{V}^u_{nm} & -\bar{V}^p_{nm} & -\bar{V}^v_{nm} \end{bmatrix} \begin{Bmatrix} u_m \\ p_m \\ v_m \end{Bmatrix} \tag{4.1}$$

where

$$\bar{U}^u_{nm} = \frac{dN_{nm}}{dx} + \frac{d\kappa_n}{dx} L_{nm} + \frac{1}{\gamma p_o}\frac{dp_o}{dx} N_{nm}$$

$$\bar{U}^p_{nm} = \frac{U_o}{\gamma p_o}\left[\frac{dN_{nm}}{dx} - \frac{1}{b}\frac{db}{dx} N_{nm}\right] + \frac{1}{\gamma p_o}\left[ik_r + \gamma\left(\frac{dU_o}{dx} + \frac{\partial V_o}{\partial y}\right)\right] N_{nm}$$

$$+ \frac{U_o}{\gamma p_o}\frac{1}{b}\frac{d\kappa_n b}{dx} L_{nm} + \frac{A}{\cos\theta}\left\{1 + \frac{iU_o}{k_r}\left[\frac{d\kappa_m b}{dx}\tan\kappa_m b - \frac{1}{A}\frac{dA}{dx}\right]\right\}\cos\kappa_n b \, \cos\kappa_m b$$

$$\bar{U}^v_{nm} = \kappa_n M_{nm}$$

$$\bar{P}^u_{nm} = \rho_o U_o\left[\frac{dN_{nm}}{dx} - \frac{1}{b}\frac{db}{dx}N_{nm}\right] + \rho_o\left(ik_r + \frac{dU_o}{dx}\right)N_{nm} + \rho_o U_o \frac{1}{b}\frac{d\kappa_n b}{dx} L_{nm}$$

$$\bar{P}^p_{nm} = \frac{dN_{nm}}{dx} + \frac{d\kappa_n}{dx} L_{nm} - \frac{1}{\gamma p_o}\frac{dp_o}{dx} N_{nm} - \frac{db}{dx}\cos\kappa_n b \, \cos\kappa_m b$$

$$\bar{P}^v_{nm} = 0$$

$$\bar{V}^u_{nm} = \left[\frac{1}{U_o}\frac{dU_o}{dx}\frac{1}{b}\frac{db}{dx} + \frac{1}{b}\frac{d^2 b}{dx^2} - \left(\frac{1}{b}\frac{db}{dx}\right)^2\right] L_{nm}$$

$$\bar{V}^{p}_{nm} = \frac{1}{\rho_o U_o}\left[\sin\kappa_n b \;\; \cos\kappa_m b \; - \; \kappa_n N_{nm}\right]$$

$$\bar{V}^{v}_{nm} = \frac{dM_{nm}}{dx} - \frac{1}{b}\frac{db}{dx}M_{nm} + \frac{1}{U_o}\left(ik_r + \frac{\partial V_o}{\partial y}\right)M_{nm} - \frac{1}{b}\frac{d\kappa_n b}{dx}J_{nm}$$

$$N_{nm} = \int_0^b \psi_n\psi_m dy \; , \quad M_{nm} = \int_0^b \phi_n\phi_m \, dy$$

$$L_{nm} = \int_0^b y\phi_n\psi_m dy \; , \quad J_{nm} = \int_0^b y\psi_n\phi_m \, dy$$

$$\bar{N}_{nm} = N_{nm} - \frac{i}{k_r}\frac{\gamma p_o}{\cos\theta} A \cos\kappa_n b \;\; \cos\kappa_m b$$

5. MATCHING OF THE SOLUTION IN THE NONUNIFORM SECTION TO THE CLASSICAL SOLUTION

As shown in Figure 1, the nonuniform duct segment is considered to be a transition section between two uniform infinite ducts, although as indicated in Eversman et.al.(1974) this restriction can be relaxed to allow terminated uniform ducts. In our computations to date the uniform ducts have been taken as having hard walls so that the classical solution (Eversman et.al.(1972)) applies

$$u = c_o\sum\left[a_n^+\alpha_n^+ e^{-ik^+_{x_n}x} + a_n^-\alpha_n^- e^{-ik^-_{x_n}x}\right]\cos\frac{n\pi}{b}y$$

$$p = \rho_o c_o^2\sum\left[a_n^+ e^{-ik^+_{x_n}x} + a_n^- e^{-ik^-_{x_n}x}\right]\cos\frac{n\pi}{b}y$$

$$v = c_o\sum\left[a_n^+\beta_n^+ e^{-ik^+_{x_n}x} + a_n^-\beta_n^- e^{-ik^-_{x_n}x}\right]\sin\frac{n\pi}{b}y$$

$$\alpha_n^+ = \frac{k^+_{x_n}/k}{1-M\frac{k^+_{x_n}}{k}} \qquad \alpha_n^- = \frac{k^-_{x_n}/k}{1-M\frac{k^-_{x_n}}{k}}$$

$$\beta_n^+ = -\,i\,\frac{n\pi/kb}{1-M\frac{k^+_{x_n}}{k}} \qquad \beta_n^- = -\,i\,\frac{n\pi/kb}{1-M\frac{k^-_{x_n}}{k}}$$

$$\frac{k_{x_n}}{k} = \frac{1}{1-M^2}\left[-M \pm \sqrt{1-(1-M^2)\left(\frac{n\pi}{kb}\right)^2}\right] \tag{5.1}$$

The axial propagation constant $k_{x_n}^{+}$ is for waves propagating in the positive x direction, while $k_{x_n}^{-}$ corresponds to the opposite direction. Positive propagation corresponds to the positive sign in Equation (5.1) as long as $(1-M^2)(n\pi/kb)^2 < 1$. The opposite prevails when the radical becomes imaginary (Eversman el.al.(1972)). The form of u,p,v shown is nondimensionalized with respect to ρ_r and c_r. In the uniform duct where ρ_r, c_r are defined, ρ_o and c_o are unity. However, both uniform ducts might not have the same flow conditions due to a difference in area.

In addition to these acoustic modes it will be shown in Section 7 that hydrodynamic modes also exist. These hydrodynamic modes are swept along at the local flow velocity and have a propagation constant.

$$\frac{k_x}{k} = \frac{1}{M}$$

There are N such modes in an N basis function expansion and these must be properly accounted for in matching the uniform duct solutions to the nonuniform section solution. Several methods of achieving the matching have been considered but are all straightforward modifications of the method used in the no flow case (Eversman et.al.(1974)). The matching procedure is used to generate a transition matrix between the modal amplitudes on the incident and transmitted sides of the form

$$\begin{Bmatrix} b^+ \\ b^- \end{Bmatrix} \quad \begin{bmatrix} TA_{11} & TA_{12} \\ TA_{21} & TA_{22} \end{bmatrix} \begin{Bmatrix} a^+ \\ a^- \end{Bmatrix}$$

from which is obtained

$$\{a^-\} = -[TA_{22}]^{-1}[TA_{21}]\{a^+\} = [REF]\{a^+\}$$

$$\{b^+\} = \left([TA_{11}] + [TA_{12}]\ [REF]\right)\{a^+\} = [TRAN]\{a^+\}$$

[REF] and [TRAN] are the matrices of reflection and transmission coefficients.

6. COMPUTER IMPLEMENTATION

The mathematical formulation discussed above has been implemented in a digital computer program and extensive numerical experimentation has been carried out on the IBM Models 44 and 65 and Burroughs B6718 facilities. The major goal has been to investigate the accuracy with which we are able to compute reflection and transmission coefficients for duct nonuniformities.

The major problem in this program has been the almost total lack of any results against which to make comparisons and evaluation. As noted in Eversman et.al.(1974) we were forced to develop several alternative computational schemes to make an evaluation of the results of the no flow computational scheme. None of the alternative schemes of the no flow case are available to us in the flow case. The only approach we have been able to take is the reduction of the general scheme to certain special cases. Of particular interest in this regard is the computational scheme of Davis and Johnson (1974) in which they treat plane wave propagation through a one-dimensional compressible flow in a variable cross section duct. Their computational scheme should allow us to generate results which serve as low frequency comparisons for the computational schemes developed here. We have adapted this formulation to treat problems of the type encountered in the present investigation. The approach is to substitute an integration of their governing differential equation for the transfer matrix integration scheme in the program we have developed. The matching of the nonuniform segment to infinite uniform ducts is done in essentially the same way as explained in Eversman et.al.(1974). The Davis and Johnson (DJ) formulation can be considered a low frequency approximation, since only at low frequencies can we be assured of primarily plane wave propagation, as assumed in their analysis.

To generate comparison with the general program we have considered inlet (flow opposite to propagation) and exhaust (flow and propagation directions coincide) cases for a duct with a cosine shaped converging taper. In the DJ formulation the walls are hard and in the general formulation we take $A = 0.0001 + i\ 0.0001$. The reduced frequency is $kb_0 = 1.0$, with $b_0 = 1.0$, $b_\ell = 0.75$ and $\ell = 2.0$. Mach numbers at $x = 0$ in the range $-0.5 \leqslant M_0 \leqslant 0.5$ are considered. These correspond to Mach numbers at $x = \ell$ in the range $-0.93 \leqslant M \leqslant 0.93$. Figures 2 and 3 show comparisons of the transmission and reflection coefficients in the plane wave mode of propagation. Both figures have two Mach number axes, one calibrated in terms of M at x = 0 and the other at $x = \ell$. The figures show that agreement is very good throughout the entire Mach number range. We consider this correlation to be of major significance in supporting the validity of the MWR.

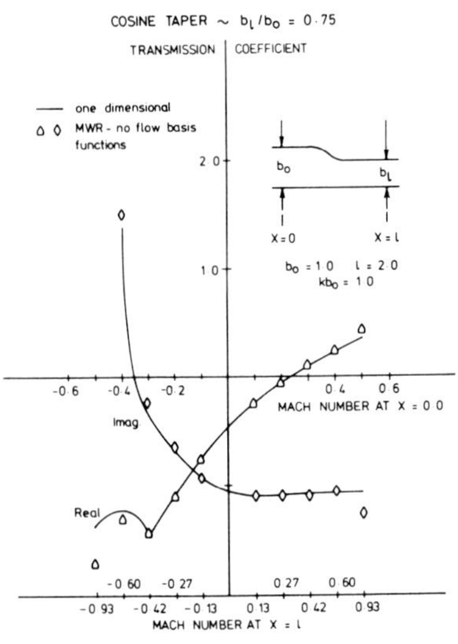

Fig.2. Transmission coefficient in the plane wave for a converging duct. Comparison of MWR and plane wave theory.

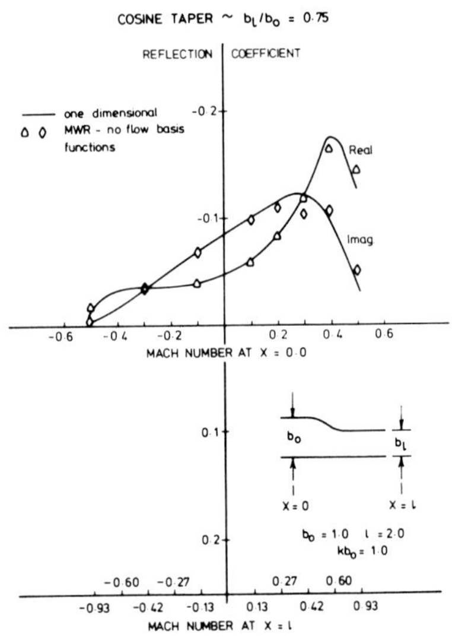

Fig.3. Reflection coefficient in the plane wave for a converging duct. Comparison of MWR and plane wave theory.

In order to investigate the extent to which the Galerkin solution converges to the correct results for lined ducts, we have considered the case of a uniform lined duct carrying a uniform flow. This problem has been formulated as noted above in the form of a series of basis functions (no flow basis functions and thus not individually satisfying the boundary condition) and the differential equations for the undetermined coefficients have been obtained. They are

$$\begin{Bmatrix} \dfrac{du_n}{dx} \\ \dfrac{dp_n}{dx} \\ \dfrac{dv_n}{dx} \end{Bmatrix}_{3N\times 1} = \underset{3N\times 3N}{[L]} \begin{Bmatrix} u_n \\ p_n \\ v_n \end{Bmatrix}_{3N\times 1}$$

In the uniform duct case the coefficient matrix [L] is constant and solutions in the form

$$\begin{Bmatrix} u_n \\ p_n \\ v_n \end{Bmatrix} = \{q\}\, e^{-ik_x x} \qquad (6.2)$$

can be assumed. The determination of the axial wave numbers k_x then becomes an eigenvalue problem

$$i[L]\{q\} = k_x\{q\}$$

and the k_x can be rapidly computed by using the general QR algorithm written to compute complex eigenvalues and eigenvectors of complex matrices. The eigenvalues can then be compared against the exact eigenvalues computed as explained in Eversman (1975). Table 1 shows the results of the case of a uniform lined duct with $M = -0.5$, $\ell = 1.0$, $b = 1.0$, $kb = 1.0$ and $A = 0.720 + 0.420i$. The first column shows the exact computations for the first six pairs of positive and negative propagating and cutoff modes (positive x direction and negative x direction). The second, third, fourth and fifth columns show results from the Galerkin Method using two, three, five and six basis functions. It is noted that convergence to the exact result is definite but more rapid convergence would be desirable.

In addition to the 2N modes which are converging toward the exact result, there are at any stage N eigenvalues

$$\frac{k_x}{k} = \frac{1}{M}$$

These extra eigenvalues represent a spatial velocity variation which is convected with the flow and which produces no pressure variation. This can be shown by using the assumed exponential solution, with k_x defined as in Equation (6.2) and substituting in the differential equations. They are hydrodynamic disturbances, carried with the flow, and in the uniform duct case are the rotational part of the solution, the acoustic modes being irrotational. It is assumed that a similar hydrodynamic character exists in the nonuniform duct and that Equation (4.1) has in its total solution elementary solutions which correspond with these hydrodynamic modes in the uniform ducts. Appropriate care must be taken in matching to account for these.

Since it has been observed that convergence is slow at low frequency it is appropriate to re-examine the basis functions. The basis functions which we use are the no flow modes of propagation, $\psi_n = \cos\kappa_n y$, $\phi_n = \sin\kappa_n y$. The eigenvalues κ_n are defined by

$$\kappa b \tan\kappa b = ikbA$$

For large κb (high order modes) the eigenvalue equation approaches

$$\tan\kappa b = 0$$

which also defines the hardwall ($A = 0$) eigenvalues. The boundary residual in the general case is

$$R_B = v_N \cos\theta - u_N \sin\theta - A\left\{p_N - \frac{iU_o}{k_r}\left[\frac{\partial p_N}{\partial x} + \tan\theta\,\frac{\partial p_N}{\partial y} + \frac{1}{A}\frac{dA}{dx}\,p_N\right]\right\}\Bigg|_{y=b}$$

Specialized to the uniform duct, this becomes

$$R_B = v_N - A\left(p_N - i\frac{M}{k_r}\,\frac{\partial p_N}{\partial x}\right)\Big/_{y=b}$$

At $y = b$

$$p_N = \sum p_N \cos\kappa_n b$$

$$v_N = \sum v_N \sin\kappa_n b$$

and the eigenvalue equation for large κb yields

$$\lim_{n\to\infty} \sin\kappa_n b = 0$$

$$\lim_{n\to\infty} \cos\kappa_n b = 1$$

It is seen that for increasing n, terms are added to v_N which themselves tend to vanish and in the limit can't provide any help in forcing the boundary residual to vanish unless the corresponding coefficients become large. This problem would be most acute at low frequency since in this case the $n\to\infty$ limit is approached very rapidly.

This problem doesn't occur if the duct has hard walls or if the flow vanishes since the basis functions for the uniform duct case are then exact solutions. The problem is likely to be less severe if the duct is not of uniform cross section, since the contribution to the axial velocity, u_N, doesn't vanish for the higher modes and leverage is still available to drive the residual toward zero. Support for this proposition can be found in the results of the no flow case of Eversman et.al.(1974) where it was found possible to generate good results for nearly hardwalled ducts of nonuniform cross section even though the nearly hard-walled modes yield $v_N \simeq 0$ and most of the adjustment dictated by the boundary residual must be absorbed in u_N.

The choice of basis functions would possibly be better if the eigenvalues approached the soft wall flow eigenvalues as $n \to \infty$. The eigenvalue equation for uniform ducts when flow is present is

$$\kappa b \tan \kappa b = ikbA\left(1 - M \frac{k_x}{k}\right)^2$$

$$\frac{k_x}{k} = \frac{1}{1-M^2}\left[-M \pm \sqrt{1-(1-M^2)\left(\frac{\kappa b}{kb}\right)^2}\right]$$

which gives rise to pairs of eigenvalues for positive and negative directed mode pairs (Eversman (1975)). It can be shown that for large κb (high order eigenvalues) the eigenvalues approach those defined by

$$\kappa b \tan \kappa b = -ikbA \frac{M^2}{1-M^2}\left(\frac{\kappa b}{kb}\right)^2 \tag{6.3}$$

In order to attempt to introduce a proper high κb limit we have formulated the Galerkin-QR problem for the uniform duct segment using basis functions $\psi_n = \cos \kappa_n y$, $\phi_n = \sin \kappa_n y$ with κ_n defined by

$$\kappa b \tan \kappa b = ikbA\left(1 - \frac{M^2}{1-M^2}\left(\frac{\kappa b}{kb}\right)^2\right)$$

The eigenvalues will have the small κb limit of the no flow case and the high κb limit of the flow case. In addition the limiting cases of $M \to 0$, $kb \to 0$ and $A \to 0$ are still preserved. We have no a priori evidence of the completeness of the resulting set of basis functions and must rely on computational success as an indication. We note that in this case not even the ψ_n functions are orthogonal in the classical sense.

As an indication of the potential usefulness of this type of basis function and as a motivation for further investigation of the choice of basis functions we have shown in Table 2 a comparison of the exact results for the axial propagation constant, k_x/k, as compared with analyses using six basis functions of the original type and of this new type. This is a low frequency analysis of $kb = 1.0$ at which point the new functions should be particularly useful because of the rapid onset of the high κb limit. We note much improved accuracy in all modes.

7. CONCLUSIONS

The development of a Method of Weighted Residuals computational scheme for the investigation of the transmission of sound in nonuniform ducts with a high subsonic mean flow has been described. Extensive numerical experimentation has shown that good results are obtained for flow Mach numbers up to near sonic for ducts with rigid walls. The case when the walls of the duct are acoustically soft has required more detailed study and good results have been verified at low frequencies for Mach numbers up to $M = 0.5$. Numerical experimentation is still in progress to fully establish the range over which good results are obtained and to build further improvements into the basic numerical scheme.

ACKNOWLEDGEMENTS

This work has been supported by the National Aeronautics and Space Administration Lewis Research Centre under NGR 17-003-018 and the Nacelles and Noise Abatement Technology Group of the Boeing Company, Wichita Division.

REFERENCES

Cremer, L. (1953) Theorie der Luffschall - Dampfung in Rechtekkanal mit Schluckender Wand und Das Sich Dabei Ergebende Hochste Dampfungsmass. Acoustica 3, 249-263.

Davis, S.S. and Johnson, M.L. (1974) Propagation of Plane Waves in a Variable Area Duct Carrying a Compressible Subsonic Flow. Paper KK-2, 87th. Meeting of the Acoustical Soc. America, N.Y. April 23-26.

Eisenberg, N.A. and Kao, T.W. (1969) Propagation of Sound Through a Variable Area Duct with a Steady Compressible Flow. J.Acous.Soc.Am. 49, 169-175.

Eversman, W. (1975) Computation of Axial and Transverse Wave Numbers for Uniform Two-dimensional Ducts with Flow Using a Numerical Integration Scheme. J.Sound and Vibration 41(2), 252-255.

Eversman, W., Cook, E.L. and Beckemeyer, R.J. (1974) A Method of Weighted Residuals for the Investigation of Sound Transmission in Nonuniform Ducts without Flow. J.Sound and Vibration 37 (3), 105-123.

Eversman, W., Nelsen, M.D., Armstrong, D. and Hall,Jr., O.J. (1972) Design of Acoustic Linings for Ducts with Flow. J.Aircraft 9 (8), 548-556.

Eversman, W., Astley, R.J., and Vo Phu Thanh (1977) Transmission in Nonuniform Ducts - A Comparative Evaluation of Finite Element and Weighted Residuals Computational Schemes. American Institute of Aeronautics and Astronautics Paper 77-1299.

Finlayson, B.A. (1972) The Method of Weighted Residuals and Variational Principles. Academic Press.

Huerre, P. and Karamcheti, K. (1973) Propagation of Sound Through a Fluid Moving in a Duct of Varying Area. Proc. Interagency Symp. Uni.Res. Transportation Noise, Stanford U., March, 28-30.

King, L.S. and Karamcheti, K. (1973) Propagation of Plane Waves in the Flow Through a Variable Area Duct. Proc. Am.Inst. Aero and Astro. Aero-Acoustics Conf., Seattle, Washington, Oct. 15-17. Paper 73 - 1009.

Nayfeh, A.H., Telionis, D.P. and Lekoudis, S.G. (1973) Acoustic Propagation in Ducts with Varying Cross-sections and Sheared Mean Flow. Proc. Am.Inst.Aero. and Astro. Aero-Acoustics Conf., Seattle, Wash. Oct. 15-17. Paper 73-1008.

Powell, A. (1960) Theory of Sound Propagation Through Ducts Carrying High Speed Flow, J. Acous. Soc. Am. 32. 1640-1646.

Tam, C.K.W. (1971) Transmission of Spinning Acoustic Modes in a Slightly Nonuniform Duct. J.Sound and Vibration 18, 339-351.

Zinn, B.T. and Powell, E.A. (1968) Application of the Galerkin Method in the Solution of Combustion Instability Problems. Proc.Int. Astronautical Cong. 3, 59-73.

APPENDIX

TABULAR DATA

TABLE 1 CONVERGENCE OF AXIAL WAVE NUMBER TO EXACT RESULTS FOR A UNIFORM LINED DUCT USING THE MODIFIED GALERKIN METHOD

M = -0.5, ℓ = 1.0, b = 1.0, kb = 1.0, A = 0.720 + 0.420i

	EXACT	2 MODES	3 MODES	5 MODES	6 MODES
1+	1.52 - i1.01	1.51 - i1.15	1.52 - i1.09	1.52 - i1.05	1.52 - i1.05
1-	-0.772 + i0.126	-0.783 + i0.112	-0.778 + i0.117	-0.775 + i0.121	-0.774 + i0.122
2+	0.679 - i4.54	-1.27 - i5.05	0.517 - i5.00	0.607 - i4.74	0.622 - i4.70
2-	0.897 + i2.23	0.771 + i2.40	0.822 + i2.34	0.856 + i2.29	0.863 + i2.28
3+	0.465 - i8.501		-3.36 - i6.00	0.398 - i9.02	0.403 - i8.90
3-	1.29 + i5.64		1.02 + i6.12	1.16 + i5.91	1.18 + i5.86
4+	0.455 - i12.32			0.486 - i13.15	0.453 - i12.93
4-	1.18 + i9.03			1.04 + i9.54	1.08 + i9.44
5+	0.477 - i16.05			-4.98 - i5.51	0.536 - i16.91
5-	1.04 + i12.59			0.882 + i13.36	0.942 + i13.18
6+	-5.92 - i 4.08				-5.26 - i5.28
6-	0.955 + i16.21				0.823 + i17.03

TABLE 2 COMPARISON OF SIX BASIS FUNCTION EXPANSION WITH EXACT RESULTS FOR k_x/k ORIGINAL BASIS FUNCTIONS AND NEW BASIS FUNCTIONS

M = -0.5, ℓ = 1.0, b = 1.0, kb = 1.0, A = 0.720 + i0.420

	EXACT	6 MODES - ORIGINAL BASIS FUNCTIONS	6 MODES - NEW BASIS FUNCTIONS
1+	1.52 - i1.01	1.52 - i1.05	1.52 - i1.01
1-	-0.772 + i0.126	-0.774 + i0.122	-0.772 + i0.126
2+	0.679 - i4.54	0.622 - i4.70	0.678 - i4.54
2-	0.897 + i2.23	0.863 + i2.28	0.896 + i2.23
3+	0.465 - i8.501	0.403 - i8.90	0.463 - i8.50
3-	1.29 + i5.64	1.18 + i5.86	1.29 + i5.64
4+	0.455 - i12.32	0.453 - i12.93	0.453 - i12.32
4-	1.18 + i9.03	1.08 + i9.44	1.18 + i9.03
5+	0.477 - i16.05	0.536 - i16.91	0.474 - i16.05
5-	1.04 + i12.59	0.942 + i13.18	1.04 + i12.60
6+	-5.92 - i 4.08	-5.26 - i5.28	-5.92 - i4.11
6-	0.955 + i16.21	0.823 + i17.03	0.957 + i16.22

Numerical Simulation of Fluid Motion, ed. John Noye
pp. 537-550

APPLICATION OF AN IMPROVED METHOD OF INTEGRAL RELATIONS TO THE SUPERSONIC BOUNDARY LAYER FLOW ABOUT CONES AT LARGE ANGLES OF ATTACK

C.A.J. Fletcher

Weapons Research Establishment

Salisbury, South Australia

SUMMARY

An improvement to the Method of Integral Relations has been made by the introduction of orthonormal weighting functions. The use of the same orthonormal functions as trial functions (i.e., a Galerkin procedure) and the numerical evaluation of certain key integrals make the present procedure suitable for the numerical solution of boundary layer-type flows. Comparisons have been made with an exact solution for the supersonic boundary layer flow on a cone at zero angle of attack. Skin friction is predicted to within 0.1% and heat transfer to within 1% of the exact results. The same technique has been applied to the supersonic flow about cones at large angles of attack. Comparisons have been made with experimental data and with other numerical techniques and the agreement for skin friction and heat transfer is very good. Results have been obtained for Reynolds numbers up to 10^6 and for angles of attack up to 50° which is roughly twice the maximum angle of attack for which previous numerical solutions have been obtained. This paper will describe the underlying method and briefly highlight the subsequent results.

INTRODUCTION

The Method of Integral Relations was first applied to the solution of the boundary layer equations by Dorodnitsyn(ref.1) in 1960. More recently, the Method of Integral Relations has come to be included in the wider category of "Methods of Weighted Residuals"(ref.2).

In the weighted residual methods it is customary to introduce a trial solution with prescribed approximating functions, U_k, which presume the functional behaviour in one or more coordinate directions.

e.g.
$$U_T(x,y) = \sum_{k=1}^{N} C_k(x) \cdot U_k(y). \tag{1}$$

If the governing equation is :

$$L(U) = 0 \tag{2}$$

then the substitution of equation (1) into (2) would produce some residual, γ, i.e.,

$$L(U_T) = \gamma. \tag{3}$$

The weighted residual methods take their name from the imposed condition :

$$\int f_k(y) \cdot \gamma \cdot dy = 0 \tag{4}$$

where $f_k(y)$ is a prescribed weighting function. In practice, sufficient independent weighting functions are introduced successively into equation (4) to evaluate the unknown coefficients C_k in equation (1). With increasing values of N (equation (1)), it is to be expected that smaller residuals will result (equation (3)) and that the trial solution will converge to the exact solution as N approaches infinity.

At this stage nothing has been said about the choice of $f_k(y)$ in equation (4). If $f_k(y)$ is chosen to equal $U_k(y)$ in equation (1) then the Galerkin method is obtained. Traditionally, the weighting and approximating functions have been made continuous throughout the region. However, if these functions are piecewise polynomials, i.e., the approximating and weighting function are zero except adjacent to a particular node, then we can derive some familiar results. Namely, the finite difference equations if a rectangular grid is used and the Galerkin formulation of the finite element method if an arbitrary grid is used.

In both the traditional formulation of the weighted residual methods and in finite difference and finite element methods, the results are obtained from the solution of a matrix equation.

$$[A] \cdot [x] = [B]. \tag{5}$$

In the weighted residual methods, x represents the unknown coefficients $C_k(x)$. In the finite difference and finite element methods, x represents the nodal values $\bar{U}(x, y)$.

The character of [A] is strongly dependent on the method. In the traditional weighted residual methods using continuous approximating and weighting functions the matrix [A] is dense and increasing difficulty is found in inverting it as N (in equation (1)) is increased. On the other hand, with finite difference and finite element methods, the matrix [A] is sparse with the non-zero terms clustered close to the main diagnal. No great difficulty is found in inverting [A] even for large systems of equations.

Thus the traditional weighted residual methods are limited in accuracy unless special procedures (which may require additional algebraic manipulation) are introduced. However, because of the higher order of the approximating function used, such a method will give more accurate results than a finite difference or finite element method with the same number of unknowns. Traditional weighted residual methods have tended to be used where "engineering" orders of accuracy (1% - 5%) are required with modest demands on computer execution time and storage.

Boundary layer flows are described by a parabolic governing equation with the streamwise (x) direction playing a time-like role. Thus, it is then natural to specify analytic representation across the boundary layer (y) and hence to reduce the problem to the solution of a set of simultaneous ordinary differential equations in the x direction. Because weighted residual methods are less sensitive to disturbances than are finite difference methods, they have been particularly successful in dealing with problems involving boundary layer separation.

The rest of this paper will concentrate on one example of the weighted residual methods, namely the Method of Integral Relations. Techniques will be described that make this method more efficient from the computational standpoint.

2. THE TRADITIONAL METHOD OF INTEGRAL RELATIONS

The Method of Integral Relations (MIR) can be most easily described with reference to a simple example, namely two dimensional incompressible boundary layer flow. For this example, the governing equations are :

$$\frac{\partial u}{\partial x} + \frac{\partial v}{\partial y} = 0 \tag{6a}$$

$$\frac{u\partial u}{\partial x} + \frac{v\partial u}{\partial y} = -\frac{1}{\rho} \cdot \frac{dp}{dx} + \nu \frac{\partial^2 u}{\partial y^2} . \tag{6b}$$

After non-dimensionalisation the equations are combined into a single equation by forming the expression f_k x (6a) + df_k/du x (6b) where f_k is the weighting function. The equations are integrated across the boundary layer with respect to y and then the change of independent variables, from (x,y) to (x,u), is introduced. The particular combination of equations (6a), (6b) and the weighting function f_k is made so that the resulting equation, which is :

$$\frac{d}{dx} \int_0^1 f_k \frac{u}{\tau} du = \frac{Ue'}{Ue} \int_0^1 \frac{f_k}{\tau} (1 - u^2) du - \left[f_k' \tau \right]_{wall} - \int_0^1 f_k \cdot \tau \cdot du \tag{7}$$

does not depend explicitly on v.

In equation (7) τ is the shear stress (proportional to $\partial u/\partial y$) and Ue is the velocity at the outer edge of the boundary layer (prescribed). Equation (7) is essentially equivalent to equation (4) after substitution of equation (3).

In Dorodnitsyn's formulation of the Method of Integral Relations the dependent variable is taken to be the inverse of the velocity gradient, i.e., $\theta = 1/\tau$; and the independent variables are taken to be the velocity component, u and the coordinate, x. There are two reasons for working with independent variables u, x rather than y, x. Firstly specifying equal increments of u (i.e., in order to evaluate the coefficients a_j) gives more weight to the region close to the wall (see figure 1) than specifying equal increments of y. Thus a more accurate prescription, for the same number of unknown coefficients, is achieved. Secondly, an integration over an infinite region (of y) is replaced by a finite region (of u).

The following dependences of θ on u is assumed,

$$\theta = \frac{1}{1-u}\left[a_o + \sum_{j=1}^{N-1} a_j . u^j .\right] \tag{8}$$

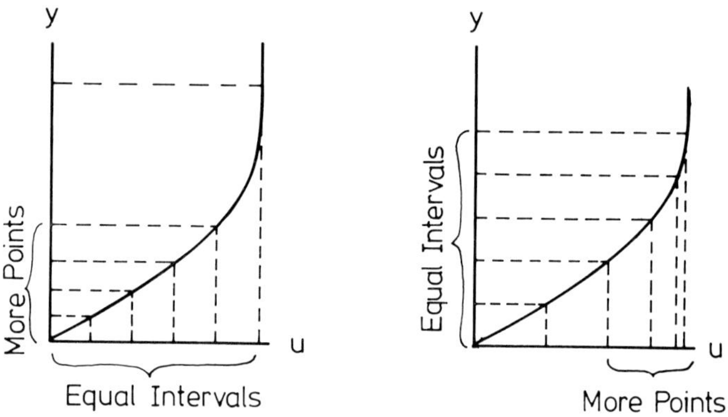

FIGURE 1. TYPICAL VELOCITY DISTRIBUTION IN BOUNDARY LAYER

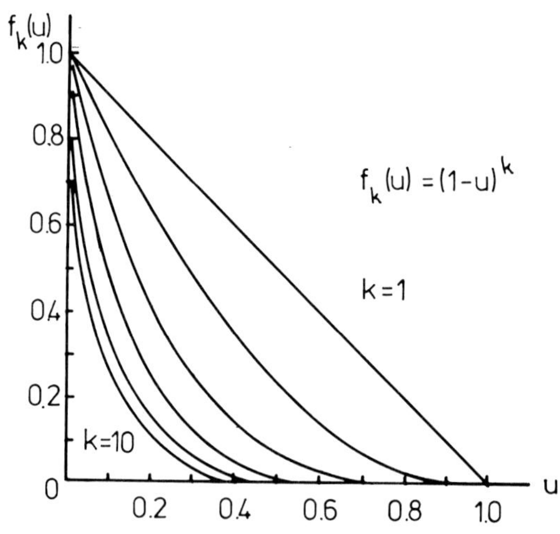

FIGURE 2. DORODNITSYN WEIGHTING FUNCTION

The factor $1/(1-u)$ is required in order to ensure that θ rather than τ or some other variable is determined by the form of the left hand side of equation (7).

N equations are required in order to evaluate the N unknown coefficients a_j in equation (8). These are obtained from equation (7) with N different weighting functions f_k. For f_k Dorodnitsyn chose the linearly independent set :

$$f_k = (1-u)^k. \tag{9}$$

It may be seen from figure 2 that as k increases more weight is given to the region adjacent to the wall. Examination of figure 2 shows up a fundamental difficulty if more accurate solutions are sought by increasing N. The equations that result from application of f_N and f_{N-1} in equation (7) lead to almost linear dependence of the coefficients a_j.

One way of overcoming this problem would be to divide the u domain into N equal intervals and to specify piecewise polynomials across each in turn (in a manner similar to the finite element method). This technique will not be pursued here.

The result of evaluating equation (7) with N different values of f_k may be expressed in matrix notation as :

$$[B] \cdot [\partial aj/\partial x] = [C] \tag{10a}$$

$$[\partial a_j/\partial x] = [B]^{-1} \cdot [C]. \tag{10b}$$

Thus at each step of the integration for the a_j's, [B] and [C] must be evaluated from the current values of τ, Ue etc. and [B] must be inverted. For increasing N it follows from equation (9) that [B] will be progressively more ill-conditioned and hence [B] will be difficult to invert accurately.

A second difficulty associated with increasing N is the shear algebraic labour of evaluating [B] and [C] analytically. For the present example the effort is bearable for N<4 or 5 but for more complicated problems in three dimensional, compressible or separated flow, the algebraic effort is horrendous for N>2 or 3.

How the two problems, of almost dependent equations and overwhelming algebraic effort for large N, are overcome is described in the next section.

3. IMPROVEMENTS TO THE METHOD OF INTEGRAL RELATIONS

The problem of excessive algebraic effort can be eliminated by numerically evaluating the integrals of equation (7) that contribute to the elements of matrix [C] in equation (10). Clearly the numerical evaluation of the integrals introduces an additional error. But this error is more than compensated for by the increase in the order of approximation (N) that is possible when numerical evaluation of the integrals is permitted.

The other problem, that of the unsuitability of the weighting function when N is large and the resultant ill-conditioned matrix [B], is more fundamental. The basis of the present improvement is to take Dorodnitsyn's weighting functions, f_k, and to generate from them a set of orthonormal functions, g_j. The functions g_j then replace f_k in equation (7); consequently the linear independence of the equations that form (10) is ensured. The functions g_j are also used to replace the powers of u, i.e., u^j, in the analytic representation for θ (equation (8)). This has the advantage that the inversion of [B] is made trivial.

The way in which this "Galerkinised" Method of Integral Relations is set up is described as follows. Firstly g_j is defined as follows :

$$g_j = \sum_{k=1}^{j} C_{kj} \cdot f_k. \tag{11}$$

f_k are the Dorodnitsyn weighting functions defined by equation (9) and the coefficients C_{kj} are evaluated using the Gram-Schmidt orthonormalisation process(ref.3).

The form of the left hand side of equation (7), after substitution of an appropriate analytic representation similar to equation (8), requires that the functions g_j should be orthonormal with respect to a given weighting function, w, that is to be determined. Thus the following inner product is defined,

$$(g_i, g_j) = \int_a^b g_j (x) . g_j (x) . w (x) . dx . \tag{12a}$$

$$\{g_i (x)\} \text{ is a set of orthonormal functions if } (g_i, g_j) = 0 \text{ when } i \neq j \\ = 1 \text{ when } i = j. \tag{12b}$$

To make use of this property it is necessary to adopt a suitable analytic representation to replace equation (9). This is :

$$\theta = \frac{1}{1-u}\left[b_o + \sum_{j=1}^{N-1} b_j\, g_j\right]. \tag{13}$$

The non-orthogonal leading coefficient, b_o, is retained in order to satisfy conditions on θ at the outer edge of the boundary layer.

Substitution of equation (13) into equation (7), and the replacement of $f_k(u)$ with $g_k(u)$ produces the following governing equation,

$$\frac{d}{dx}\int_o^1 \left[b_o + \sum_{j=1}^{N-1} b_j\, g_j\, (u)\right] . g_k\, (u)\, \frac{u}{1-u} . du = C'(k),\ k = 1 \ldots N \tag{14}$$

where

$$C'(k) = \frac{Ue'}{Ue}\int_o^1 \theta . g'_k(u) . (1-u^2)\, du - \left[g'_k(o) . \tau\right]_{wall} - \int_o^1 g''_k\, (u) . \tau . du. \tag{15}$$

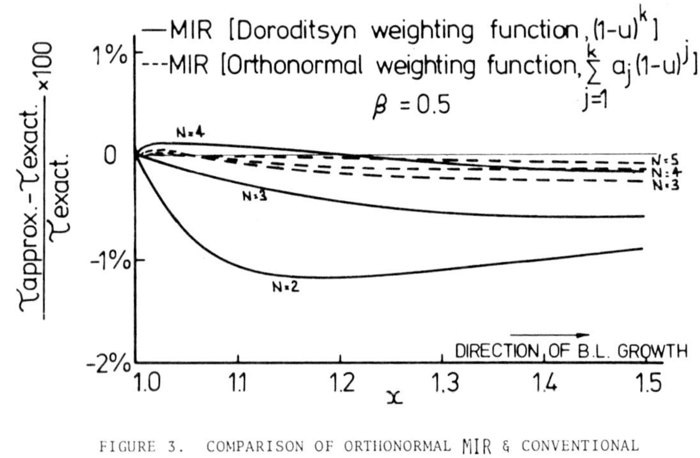

FIGURE 3. COMPARISON OF ORTHONORMAL MIR & CONVENTIONAL MIR FOR FALKNER-SKAN SOLUTIONS

C'(k) is evaluated numerically using a local Simpson's rule and m evaluations of the integrand at equal intervals in the range u = 0 to 1. Typically for $m \geqslant 30$ C'(k) is accurate to six decimal places.

Comparison of equations (12a) and (14) indicates that the orthonormal property (12b) can be exploited if the orthonormal weighting function is chosen to be w(u) = u/(1-u). Equation (14) then becomes :

$$\frac{db_0}{dx}\int_0^1 g_k \,.\, w(u) \,.\, du + \frac{db_k}{dx} = C'(k), \; k = 1 \ldots N\text{-}1. \tag{16}$$

When k = N equation (14) reduces to :

$$\frac{dbo}{dx}\int_0^1 g_N \,.\, w(u) \,.\, du = C'(N). \tag{17}$$

From (16) and (17) it follows that :

$$\frac{db_k}{dx} = C'(k) - C'(N)\left[\frac{\int_0^1 g_k \; w(u) \; du}{\int_0^1 g_N \; w(u) \; du}\right]. \tag{18}$$

Integrals of the form $\int_0^1 g_k \; w(u) \; du$ can be integrated once and for all. It is apparent that equation (18) gives an explicit expression for db_k/dx and the inversion of the ill-conditioned matrix [B] (equation (10) has been avoided.

To compare with the conventional application of MIR equations (18) and (15) have been integrated numerically using a fourth-order Ruge-Kutta scheme. Starting data for the integration are provided by the Falkner-Skan exact solutions at x = 1. The external conditions are chosen to be :

$$\beta = \sqrt{2x} \,.\, Ue'/Ue = 0.5.$$

Figure 3 shows the percentage difference between the approximate and exact wall shear stress plotted against x. The purpose of this comparison is to demonstrate that the improved MIR can be easily extended to larger values of N and that these solutions approach the exact solution more closely as N is increased.

4. APPLICATION OF THE IMPROVED METHOD OF INTEGRAL RELATIONS TO AXISYMMETRIC, SUPERSONIC FLOW ABOUT A CONE

The compressible, boundary layer flow over a cone at zero angle of attack is more complex than the example considered in Section 3 because density and temperature variations are now involved. However, this example is still simple enough to possess an exact solution with which the present solution can be compared.

The purpose of this section is to compare the present solution with the exact solution for various values of N and m. N is the number of unknown coefficients in the analytic representation and m is the number of integral evaluations in the numerical evaluation of C'(k). An attempt is made to obtain a limiting value of the present solution as N and m tend to infinity.

For this problem the governing equations, equivalent to equation (7) in Section 2, are :

$$\int_0^1 g_k(u) \cdot \frac{u}{\tau}\, du = e_1 \left[g_k'(o)\tau(o) + \int_0^1 g''_k(u) \cdot \tau\, du \right] \qquad (19)$$

and

$$\int_0^1 g_k(u) \cdot \frac{s.u.}{\tau}\, du = e_1 \cdot g_k'(o) \cdot s(o) \cdot \tau(o) + e_3 \cdot g_k(o) \cdot \frac{\partial s}{\partial u}(o) \cdot \tau_o +$$

$$(e_1+e_2)\int_0^1 g_k'(u) \cdot \tau \cdot \frac{\partial s}{\partial u} \cdot du + e_1\int_0^1 g''_k(u) \cdot \tau \cdot s \cdot du \qquad (20)$$

$$e_3\int_0^1 g''_k(u) \cdot \tau \cdot u \cdot du.$$

Here τ is the modified normal gradient of the velocity along the cone generator, s is the non-dimensional total enthalpy, g_k is the orthonormal weighting function, and e_1, e_2 and e_3 are parameters that depend on the boundary conditions.

Analytic representations, like equation (13), are introduced for the groups on the left hand side of equations (19) and (20). These are :

$$\frac{u}{\tau} = \frac{1}{(1-u)} \left[b_{o1} + \sum_{j=1}^{N-1} b_{j1} \cdot g_j(u) \right] \qquad (21)$$

and

$$\frac{s.u.}{\tau} = \frac{1}{(1-u)} \left[b_{02} + \sum_{j=1}^{N-1} b_{j2} \cdot g_j(u) \right]. \qquad (22)$$

Certain of the coefficients b_{ji} are determined by satisfying the boundary conditions. At the outer edge of the boundary layer s = 0 and $g_j(1) = 0$ for all j. Thus in equation (22) $b_{02} = 0$. At u = 0, s = s_w, which has to be given a priori. This leads to the following relations amongst the coefficients :

$$b_{01} + \sum_{j=1}^{N-1} b_{j1} \cdot g_j(o) = 0 \qquad (23)$$

$$\sum_{j=1}^{N-1} b_{j2} \cdot g_j(o) = 0 \tag{24}$$

and

$$\sum_{j=1}^{N-1} b_{j2}\, g_j'(o) = s_w \cdot \left[\sum_{j=1}^{N-1} b_{j1} \cdot g_j'(o) \right]. \tag{25}$$

To obtain the other 2(N-2) coefficients, equation (19) and (20) are evaluated (N-2) times, with different weighting functions $g_k(u)$, $k = 1, 2 \ldots N-2$. As in the previous example, the right hand sides of equations (19) and (20) are evaluated numerically using a local Simpson's rule and m evaluations of the integrand.

It can be seen that introduction of equations (21) and (22) reduce equations (19) and (20) to algebraic relations that require a particular choice of the coefficients b_{ji} for their solution. To do this an iterative technique based on a quasi-least squares function minimisation method(ref.4) has been employed. The technique, which is peripheral to the present problem is described elsewhere (ref.5).

In order to compare the present solution with the exact solution, we require a suitable characteristic parameter of the flow. A suitable parameter is the non-dimensional skin friction parameter, $Re_x^{\frac{1}{2}} \cdot C_f$. Re_x is the local Reynolds number and C_f is the skin friction coefficient. $Re_x^{\frac{1}{2}} \cdot C_f$ is proportional to the value of τ at the wall. In the limit of N and m tending to infinity it is expected that the approximate solution represented by equations (21) and (22) will approach the exact solution to equations (19) and (20). To check this, solutions have been obtained for values of N and m up to N = 14 and m = 128. The values of the skin friction parameter for various values of N and m are shown in Table 1.

TABLE 1

Variation of $Re_x^{\frac{1}{2}} C_f$ with N and m

N	m = 24	m = 40	m = 64	m = 96	m = 128	m = ∞
3	1.28452	1.28767	1.28818	1.28955	1.28996	
4	1.15341	1.15570	1.15698	1.15768	1.15803	
5	1.14788	1.14966	1.15064	1.15118	1.15145	
6	1.14768	1.14934	1.15025	1.15074	1.15099	
7	1.14724	1.14879	1.14961	1.15006	1.15028	1.15050
8	1.14716	1.14868	1.14947	1.14989	1.15010	1.15031
9	1.14711	1.14862	1.14939	1.14979	1.14999	1.15019
10	1.14708	1.14859	1.14935	1.14975	1.14994	1.15013
11	1.14705	1.14857	1.14933	1.14972	1.14990	1.15008
12		1.14856	1.14932	1.14970	1.14988	1.15006
13		1.14855	1.14931	1.14969	1.14987	1.15005
14		1.14859	1.14930	1.14969	1.14987	1.15005
∞			1.14929	1.14969	1.14987	1.15005

It can be seen that the tabulated values of $Re_x^{\frac{1}{2}} C_f$ vary smoothly enough to permit extrapolation to $N = \infty$ and $m = \infty$, as long as entries corresponding to small m coupled with large N are discounted. Clearly this is not an efficient combination in any case.

The value of $Re_x^{\frac{1}{2}} . C_f$ given by the exact solution is 1.15028. The extrapolated value in Table 1, i.e., $Re_x^{\frac{1}{2}} . C_f = 1.15005$, compares favourably with the exact result.

A more demanding comparison is obtained if the non-dimensional heat transfer parameter, $Pr.Re_x^{\frac{1}{2}}.St$, is considered. Pr is the Prandtl number, St is the Stanton number and the parameter. $Pr.Re_x^{\frac{1}{2}}.St$ is proportional to the gradient $\partial s/\partial u$ at the wall. The variation of $Pr.Re_x^{\frac{1}{2}}.St$ with N and m is shown in Table 2.

TABLE 2

Variation of $Pr.Re_x^{\frac{1}{2}}.St$ with N and m

N	m = 24	m = 40	m = 64	m = 96	m = 128
3	0.29225	0.29038	0.29131	0.28992	0.28970
4	0.47669	0.47652	0.47663	0.47660	0.47657
5	0.48676	0.48659	0.48652	0.48643	0.48646
6	0.48855	0.48825	0.48813	0.48807	0.48804
7	0.49553	0.49472	0.49454	0.49448	0.49446
8	0.49832	0.49705	0.49675	0.49667	0.49664
9	0.50546	0.49947	0.49865	0.49848	0.49843
10	0.49455	0.50058	0.49994	0.49975	0.49970
11	0.61802	0.50629	0.50147	0.50080	0.50067
12		0.49778	0.50165	0.50146	0.50137
13		0.55076	0.50552	0.50258	0.50212
14		0.24730	0.50039	0.50238	0.50240

It is apparent that an unrealistic oscillation with increasing N develops when m is small. From these results, and others not shown, it was found necessary to increase m as N is increased. For a Prandtl number, Pr = 0.7, the exact solution is $Pr.Re_x^{\frac{1}{2}}.St = 0.50644$. The tabulated value at N = 14 and m = 128 is within 1% of the exact result. It is noticeable that a fairly large value of N is required before the heat transfer parameter becomes close to the exact result. This comparison is quite reasonable when it is recalled that the test parameter, $Pr.Re_x^{\frac{1}{2}}.St$, depends on differentiating equations (21) and (22) twice.

5. APPLICATION TO CONES AT LARGE ANGLE OF ATTACK

This is the research problem for which the improvements to the Method of Integral Relations have been developed. The overall task is to obtain flow solutions about cones inclined at large angles of attack (up to 50°) in a supersonic freestream. The approach used is to obtain a solution of the outer inviscid region and a simultaneous solution of the boundary layer region adjacent to the cone surface.

The governing equations in the outer region are mixed elliptic/hyperbolic depending on the local cross-flow Mach number. In the present example the elliptic inviscid region is solved by the GTT Method and the hyperbolic inviscid region is solved by the method of characteristics. Since the solution of the inviscid region is not central to an understanding of the present problem, reference 7 may be consulted for further details.

The purpose of this section is to compare the solution obtained by the present method with a finite difference solution of the full Navier-Stokes equations and with some experimental results.

For this problem the governing equations of motion (corresponding to equation 7 in Section 2) are of the form :

$$\frac{\partial}{\partial \zeta} \int_0^1 \frac{w}{\tau} \cdot g_k(u) \cdot du = C_1(k,u) \tag{26}$$

$$\frac{\partial}{\partial \zeta} \int_0^1 \frac{w^2}{\tau} \cdot g_k(u) \cdot du = C_2(k,u) \tag{27}$$

$$\frac{\partial}{\partial \zeta} \int_0^1 \frac{s.w.}{\tau} \cdot g_k(u) \cdot du = C_3(k,u). \tag{28}$$

In equations (26) to (28) ζ is a non-dimensional circumferential coordinate, w is the circumferential velocity component, τ is proportional to the normal gradient of the radial velocity component and s is a non-dimensional enthalpy. $g_k(u)$ is the orthonormal weighting function and $C_1(k,u)$ to $C_3(k,u)$ are complicated integral functions of the independent variables (similar to equation 15) and are evaluated numerically as before. All the results shown in this section have used 32 evaluations of the integrand.

Because of the structure of equations (26) to (28) the following analytic representation are postulated for w/τ, w^2/τ and sw/τ :

$$\frac{w}{\tau} = \frac{1}{1-u}\left[b_{01} + \sum_{j=1}^{N-1} b_{j1} \cdot g_j(u)\right] \tag{29}$$

$$\frac{w^2}{\tau} = \frac{1}{1-u}\left[b_{02} + \sum_{j=1}^{N-1} b_{j2} \cdot g_j(u)\right] \tag{30}$$

$$\frac{s.w}{\tau} = \frac{1}{1-u}\left[b_{03} + \sum_{j=1}^{N-1} b_{j3} \cdot g_j(u)\right]. \tag{31}$$

These equations are equivalent to equation (13) in Section 3. As before certain of the coefficients b_{ji} are determined by the boundary conditions. The other coefficients are determined by successive application of equations (26) to (28) with different values of the weighting function g_k. Clearly equations (26) to (28) become ordinary differential equations for b_{ji} as functions of ζ when equations (30) and (31) are introduced. Integration of these equations starts at the windward line of symmetry, where equations (26) to (28) reduce to an algebraic form, and proceeds until boundary layer separation occurs. The details are given in reference 8.

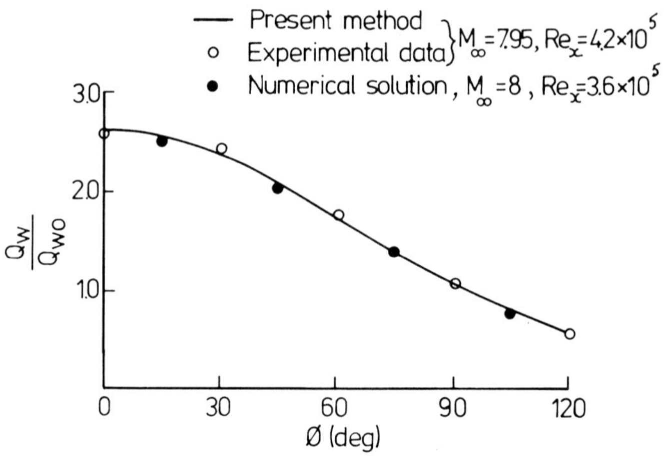

FIGURE 4. CIRCUMFERENTIAL VARIATION OF HEAT TRANSFER

All the results shown in this section feature seventh order analytic representations (N = 7 in equations (30) and (31)). The accuracy of the present method may be gauged by the results for heat transfer at the cone surface that are presented in figure 4.

Comparisons of heat transfer are considered a particularly stringent test. These results were obtained at the same freestream conditions as in Tracy's experiments(ref.9) which are also shown. For comparison the results of Lubard and Helliwell(ref.10) have also been included. Lubard and Helliwell solved a modified form of the full Navier-Stokes equations by a finite different method. We may note, in passing, that solutions by the present method require typically 40-60secs on a CDC 7600 whereas the results of Lubard and Helliwell require 1 hour on a CDC 7600. Thus the present method achieves adequate accuracy for practical problems and is considerably more economical and simpler to program than other numerical techniques.

Up to the present time, the largest angles of attack for which solutions to the flow about a cone had been obtained were 24° for the boundary layer flow and 30° for the inviscid flow. The results for heat transfer shown in figure 5 are included to demonstrate the extremes of freestream conditions that the present method will cope with. These results were obtained from solutions to both the inviscid and boundary layer region up to angles of attack of 50°, i.e. roughly twice what was previously possible.

In conclusion, improvements to the Method of Integral Relations have been described that make the method a powerful technique for obtaining accurate and economical solutions to boundary layer flows.

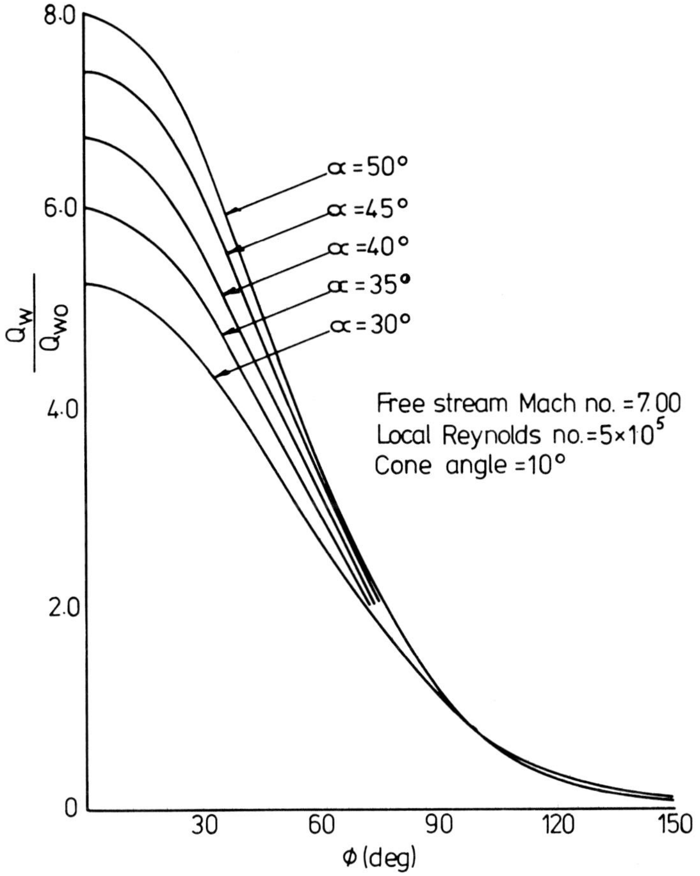

FIGURE 5. CIRCUMFERENTIAL VARIATION OF HEAT TRANSFER FOR VARIOUS ANGLES OF ATTACK

REFERENCES

No.	Author	Title
1	Dorodnitsyn, A.A.	in "Advances in Aeronautical Sciences" pp 207-219, Pergamon Press, New York, 1962.
2	Ames, W.F.	"Non-linear Partial Differential Equations in Engineering, Vol. II", pp 155-166, Academic Press, New York, 1972.
3	Isaacson, E. and Keller, H.B.	"Analysis of Numerical Methods", pp 199, 1st Edn., Wiley, New York, 1966.

4	Fletcher, C.A.J.	"Powit : A Program for Solving Non-linear Algebraic Equations" WRE-TN-1431 (WR&D), June 1975.
5	Fletcher, C.A.J. and Holt, M.	"An Improvement to the Method of Integral Relations" Journal of Computational Physics, Vol. 18, No. 2, pp 154-164, 1975.
6	Young, A.D.	in "Modern Developments in Fluid Dynamics, High Speed Flow", Vol. I, p 426, Oxford University Press, London, 1953.
7	Fletcher, C.A.J.	"The Vortical Singularity Behind a Highly Yawed Circular Cone" AIAA Journal, Vol. 13, No. 8, 1975.
8	Fletcher, C.A.J. and Holt, M.	"Supersonic Flow Over Cones at Large Angles of Attack" (To be published)
9	Tracy, R.R.	"Hypersonic Flow Over a Yawed Circular Cone", Cal. Inst. of Techn, Memo No. 69.
10	Lubard, S.C. and Helliwell, W.S.	"Calculation of Flow on a Cone at High Angle of Attack", AIAA Journal, Vol. 12, No. 7, 1974.

Numerical Simulation of Fluid Motion, ed. John Noye
pp. 551-566

FINITE ELEMENT SOLUTION FOR A FLUID SUBJECT TO THERMAL AND BUOYANCY EFFECTS

G.P. Steven

University of Sydney

New South Wales, Australia

SUMMARY

The equations modelled in this work are the two dimensional Navier-Stokes equations with a buoyancy term added, together with the heat conduction equation. A least squares criterion is applied to a finite element formulation for an approximate solution to these equations. Physical situations investigated to date include two dimensional models of a katabatic wind generated by nocturnal ground cooling and fire winds generated by a localised ground hot spot. Solutions obtained show some resemblance to previous ones produced by finite difference techniques.

1. INTRODUCTION

The work reported herein arose from a desire to show that the finite element method could be a very useful tool for the solution of a large range of problems in fluid mechanics relating to geophysical situations. The whole programme of work is still at an embryonic stage and consequently the solutions reported in this paper are not as detailed or extensive as could be desired. However, they should serve to illustrate the potential of the finite element approach to the problems described and if the advantages of this approach over others for the approximate solution to field problems are considered, then it is clear that further effort is worthwhile. These advantages of the finite element method mentioned, include short computational time, simple coding, arbitrary geometry of solution domain, simplicity of applying and changing boundary conditions, ease of incorporating inhomogeneous physical properties and many more.

The two dimensional equations governing the unsteady motion of a viscous incompressible fluid subject to buoyancy effects and the equation governing heat transport within the solution domain are adopted. These are taken in their non dimensional stream function (ψ) and vorticity (ω) format. The cartesian gradients of the three independent variables are introduced as auxiliary functions thereby altering the three second order differential equations to nine first order equations. Upon application of the least squares criterion for an approximate solution these auxiliary functions can be regarded as independent variables in their own right.

A triangular finite element is adopted. The auxiliary functions are interpolated linearly within each element and the base variables (ψ,ω,T) are interpolated quadratically. This leads to a thirty six degree of freedom element. To illustrate this finite element approach two sample problems are considered. In one a katabatic wind produced by the nocturnal cooling of an isolated hill is simulated and in the other the wind generated by a localised ground hot spot are predicted. Both these problems have been investigated previously by Leslie and Smith (1974) and Smith, Morton and Leslie (1975) using a finite difference solution procedure and comparison of the two sets of solutions indicate that, despite the poor resolution of the finite element grid that has been used, the elemental approach does appear to predict similar results. The emphasis in this paper is on the solution procedure rather than on the physical nature of the solution.

2. THE EQUATIONS

In stream function and vorticity format the momentum and heat transport equations together with the continuity equation are, where the subscript denotes differention with respect to the indicated coordinate,

$$K(\omega_{xx} + \omega_{yy}) - \psi_y\omega_x + \psi_x\omega_y + gT_x/T_0 - \omega_t = 0 \tag{2.1}$$

$$(\psi_{xx} + \psi_{yy}) + \omega = 0 \tag{2.2}$$

$$H(T_{xx} + T_{yy}) - \psi_y T_x + \psi_x T_y - T_t = 0 \tag{2.3}$$

where t is time, T_0 is some reference temperature, the velocity components are $u = \psi_y$, $v = -\psi_x$ and the vorticity is $\omega = v_x - u_y$. In the derivation of these equations the perfect gas law holds and thus the density changes in the

the buoyancy term can be directly related to temperature changes. The coefficient K in (2.1) is a turbulent eddy diffusivity and H in (2.3) is the heat diffusivity. For detailed derivation of these equations see Smith et al. (1975).

Starting from some set of initial conditions, equations (2.1) to (2.3) may be solved at any point of time whereupon the pressure (p) field can be determined using the following Poisson type equation,

$$(P_{xx} + P_{yy}) = 2\psi_{xx}\psi_{yy} - 2\psi_{xy}\psi_{xy} \ . \tag{2.4}$$

We now introduce auxiliary functions $(\alpha,\beta,\lambda,\delta,\varepsilon,\eta)$ to represent the cartesian gradients of the independent variables (ψ,ω,T),

$$\text{i.e.} \quad \alpha,\lambda,\varepsilon = \chi_x \text{ and } \beta,\delta,\eta = \chi_y \text{ where } \chi = \psi,\omega,T. \tag{2.5}$$

Equations (2.1) and (2.3) are non linear and it is the author's experience that, when the form of such equations are reduced to a series of non linear algebraic equations for solution, of which there may be several thousand, then direct solution using Newton-Raphson, conjugate gradient or search techniques does not work. Consequently the equations for solution are linearised á prioré and an iterative scheme is adopted. Also it has been found convenient to incorporate the time stepping algorithm into the iterative process. The time stepping scheme used is such that,

$$\omega_t = (\omega^k - \omega^{k-1})/\Delta t, \ \Delta t = \text{time increment}$$

and although very simple this scheme has been found to be as efficient as some of the more complex ones (see Wood and Lewis (1975), Desai and Abel (1972)).

With the auxiliary functions, the linearisation and time stepping algorithm introduced, the equations describing the physical process at the kth time step are, in matrix form, as follows,

$$\begin{bmatrix} 1 & 0 & 0 & 0 & 0 & 0 & -\frac{\delta}{\delta x} & 0 & 0 \\ 0 & 1 & 0 & 0 & 0 & 0 & -\frac{\delta}{\delta y} & 0 & 0 \\ 0 & 0 & 1 & 0 & 0 & 0 & 0 & -\frac{\delta}{\delta x} & 0 \\ 0 & 0 & 0 & 1 & 0 & 0 & 0 & -\frac{\delta}{\delta y} & 0 \\ 0 & 0 & 0 & 0 & 1 & 0 & 0 & 0 & -\frac{\delta}{\delta x} \\ 0 & 0 & 0 & 0 & 0 & 1 & 0 & 0 & -\frac{\delta}{\delta y} \\ \frac{\delta}{\delta x} & \frac{\delta}{\delta y} & 0 & 0 & 0 & 0 & 0 & 1 & 0 \\ 0 & 0 & K\frac{\delta}{\delta x} & K\frac{\delta}{\delta y} & \frac{g}{T_0} & 0 & 0 & -\frac{1}{\Delta t} & 0 \\ 0 & 0 & 0 & 0 & H\frac{\delta}{\delta x} & H\frac{\delta}{\delta y} & 0 & 0 & -\frac{1}{\Delta t} \end{bmatrix} \begin{bmatrix} \alpha \\ \beta \\ \lambda \\ \delta \\ \varepsilon \\ \eta \\ \psi \\ \omega \\ T \end{bmatrix}^{k} = \begin{bmatrix} 0 \\ 0 \\ 0 \\ 0 \\ 0 \\ 0 \\ 0 \\ -\alpha\delta+\beta\lambda-\omega/\Delta t \\ -\alpha\eta+\beta\varepsilon-T/\Delta t \end{bmatrix}^{k-1} \tag{2.6}$$

i.e.

$$L \;\{\chi\} = \{f\}, \tag{2.7}$$

where L is a linear matrix differential operator, $\{\chi\}$ is a vector of independent variables and $\{f\}$ is a vector which is a function of the solution vector at the previous iteration.

Note that the auxiliary functions are now classed as independent variables, thus in general $\alpha \neq \psi_x$, etc. This is due to the nature of the least squares criterion which will be applied. The technique will ensure that the deviation of α away from ψ_x, for example, will be made to vanish in a least squares sense element wise throughout the solution domain. The purpose of introducing the auxiliary functions is primarily to reduce the order of differentiation involved in the governing equations. This reduction is an inherent feature of variational and Galerkin type solution criteria. This reduction thereby reduces the continuity requirements on the variables across element boundaries. As a secondary feature it greatly facilitates the application of boundary conditions since there is control over both the base functions (ψ,ω,T) and their gradients at most boundaries.

3. THE LEAST SQUARES CRITERION

The least squares criterion approach to the solution of problems by finite elements has been used previously by Akin (1973) for the solution of some non linear differential equations, by Lynn (1973), (1974) for the solution of various field problems and the non linear equations governing laminar boundary layers and by Milthorpe (1975) for the solution of the Navier-Stokes equations.

If an approximate solution for $\{\chi\}$ is $\{\chi^*\}$ then the residual error for the solution domain D is, using equation (2.7),

$$R = L \;\{\chi^*\} - \{f^*\} \tag{3.1}$$

The least squares criterion tells us that,

$$\int_D R^2 dD, \tag{3.2}$$

shall be a minimum. However the residual is a scalar quantity and expression (3.2) can be regarded as the sum of the contributions over a finite number of sub divisions of D called finite elements and denoted by the superscript e,

i.e.

$$\int_D R^2 dD = \sum_{\text{ELEMENTS}} \int_{A^e} R^{e^2} dA^e \;. \tag{3.3}$$

If now we express the field distributions of the independent variable solution vector $\{\chi^*\}$ as a series of discrete values at various points (nodes) within D, these nodes lying on the boundaries of the finite elements. Denoting this vector of nodal values as $\{\overline{\chi}^*\}$ of which that subset pertaining to a single element is $\{\overline{\chi}^*\}^e$ then the minimisation of (3.3) is achieved by simultaneously differentiating with respect to each value in $\{\overline{\chi}^*\}$.

i.e.

$$\frac{\delta}{\delta\{\bar{\chi}^*\}} \sum_{\text{ELEMENTS}} \int R^{e^2} dA^e = 0 \ .$$

i.e.

$$\sum \int \frac{\delta}{\delta\{\bar{\chi}^*\}} R^{e^2} dA^e = 0$$

i.e.

$$\sum \int \frac{\delta}{\delta\{\bar{\chi}^*\}^e} R^{e^2} dA^e = 0 \qquad (3.4)$$

Since R^e is a function of $\{\bar{\chi}^*\}^e$ only then the last step towards (3.4) should be clear.

Attention is now focused on the form of the element residual in terms of $\{\bar{\chi}^*\}^e$. To do this we must represent the variation of the nodal values on the perimeter of the element. The auxiliary functions are represented as linear area functions (Steven (1978), Zienkiewicz (1971)) of the values at the corners of a triangular element (see Figure 1).

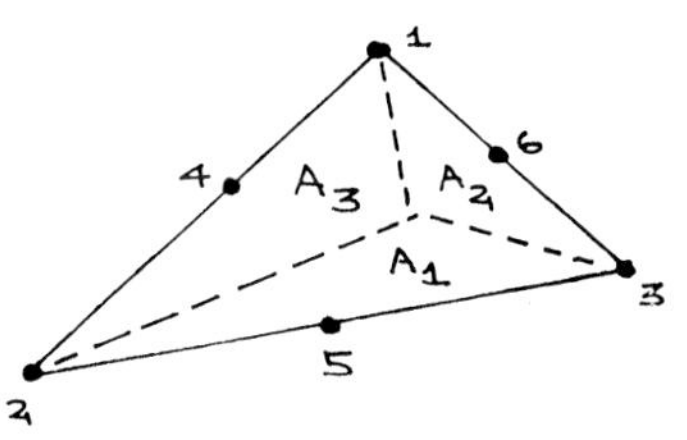

Figure 1.

$$\theta = \xi_1\theta_1 + \xi_2\theta_2 + \xi\theta_3 \qquad (3.5)$$

where

$$\theta = (\alpha\beta\lambda\delta\varepsilon\eta)$$

$$\xi_i = \frac{A_i}{\Sigma A_i}$$

The base functions are interpolated as quadratic area functions of the corner nodes and the mid point nodes,

$$\theta = \xi_1(2\xi_1-1)\theta_1+\xi_2(2\xi_2-1)\theta_2+\xi_3(2\xi_3-1)+4\xi_1\xi_2\theta_4+4\xi_2\xi_3\theta_5+4\xi_3\xi_1\theta_6 \tag{3.6}$$

where

$$\theta = \psi,\omega,T.$$

Thus the vector of independent variables within an element can be represented by an interpolation matrix N operating on a vector of nodal values of the variables.

$$\{\chi^*\}^e = N\,\{\bar{\chi}^*\}^e\ . \tag{3.7}$$

On an element basis the residual in Equation (3.3) can now be expressed as follows.

$$\begin{aligned} R^e &= L\ \{\chi^*\}^e - \{f^*\}^e \\ &= L\ \ N\ \{\bar{\chi}^*\}^e - \{f^*\}^e \\ &= K\ \{\bar{\chi}^*\}^e - \{f^*\}^e. \end{aligned} \tag{3.8}$$

If this is substituted into equation (3.4) and the differentiation carried out we are left with the final form for the equations to be represented in the computer.

$$\sum_{\text{ELEMENTS}} \int\left(K^T K\ \{\bar{\chi}^*\}^e - K^T\{f^*\}^e\right) dA^e = 0 \tag{3.9}$$

The quantity $\int K^T K\ dA^e$ is the element characteristic or stiffness matrix and $\int K^T\{f^*\}^e dA^e$ is the right hand side or load vector. The stiffness matrix will be 36 x 36 and is symmetric, an advantage over the Galerkin approach to the governing differential equations; its evaluation together with that of the load vector is a lengthy manual process which will not be detailed herein.

4. NUMERICAL SOLUTION

A FORTRAN programme of about six hundred cards has been written to model the equations described by (3.9) above. This has five major segments;

- i Point input and boundary condition details,
- ii Input of element details,
- iii Generation and assembly of element stiffness matrices and triangulation of assembled global stiffness matrix,
- iv Generation of right hand side vector within iterative loop on time,
- v Solution of equations and printing.

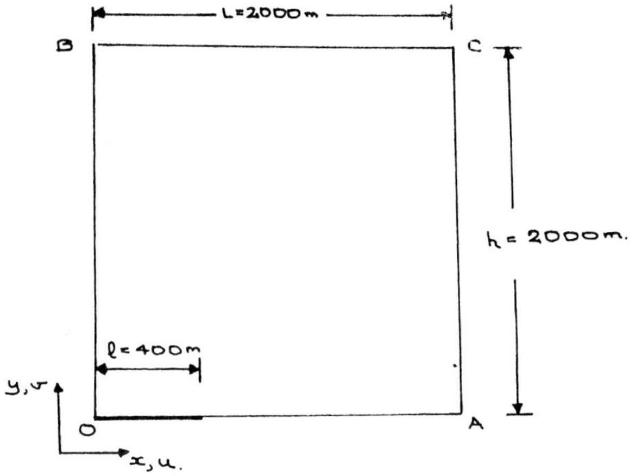

Figure 2.

The equation solver is a related form of the compact elimination methods for the Cholesky scheme by Crout and Doolittle as described in Fox (1964). No details of the programme are given here but may be obtained on request to the author.

We shall consider initially the fire wind generation model of Smith et al. (1975). The region is a square OACB Figure 2, of side 2000m; the boundary conditions are given in Table 1, where $T_g = 330^{o}K$, $0 < x \leq \ell$, $T_g = 300^{o}K$, $\ell < x \leq L$.

TABLE 1

SIDE	α	β	λ	δ	ε	η	ψ	ω	T
$y_o = 0$	0	0	?	?	0	?	0	?	T_g
y = h	0	?	0	?	?	0	0	0	?
x = 0	?	0	0	0	0	?	0	0	?
x = L	0	?	?	?	0	?	?	?	?

The ground boundary conditions (y = 0) are no slip ones with the stream function arbitrarily set zero, vorticity is shed from the ground and the temperature conditions reflect the localised hot spot with the remainder at an ambient temperature of $300^{o}K$. These on the upper boundary (y = h) may be considered as due to the presence of a thermal inversion higher up through which the plume of hot air rising from the ground is unable to penetrate; however, the present solution technique allows for many other possible conditions. On x = 0 we have the symmetry conditions with no horizontal velocity component and the stream function remains zero on this surface. Finally, on x = L is assumed to be sufficiently downstream that parallel flow conditions can be taken.

The finite element grid adopted has 100 nodes and 41 elements and is shown in Figure 3 from which it is noticed the higher intensity of elements near the hot spot to allow for higher gradients in that region.

The thermal and eddy diffusivities (H and K, Equations (2.3) and (2.1)) are taken to be the same and results are presented for two values of this, namely K = 100 and K = 200 m^2/sec. Since the flow is driven by buoyancy forces and thermal gradients there is no natural velocity scale and the only dynamical parameter is the pseudo-Rayleigh number,

$$R_a = g\left(\frac{330-300}{300}\right)\frac{\rho^3}{K^2} ,$$

which for the two values of K above is 6400 and 1600 respectively.

In Figures 4 and 5 are shown the steady state temperature and stream function fields for R_a = 6400 and in Figures 6 and 7 are the same for R_a = 1600. For the temperature fields, the higher diffusion coefficient in Figure 5 reflects in the slightly higher far field temperatures away from the hot spot; there is not a great difference in the temperature gradients near the core of the plume. These temperature fields although similar to those of Smith et al. (1975) are somewhat wider. This could be due to slight differences in ground temperature conditions and the fairly poor resolution of the finite element grid.

Since there is no velocity scale the stream function values are arbitrary and streamlines have been scaled to give suitable representation. The scaling is the same in both Figures 5 and 7. The obvious feature of both these figures is that what comes in near the ground on x = L goes out on the same boundary, but higher up and that the incoming velocities are much lower than the outgoing ones. At the higher pseudo-Rayleigh number the temperature gradients are slightly higher and this reflects in somewhat higher velocities throughout the field. However the most interesting feature of both plots is the significant dip in the lower value streamlines as they come under the influence of the thermal plume. Although no streamline actually splits thus creating stagnation points with some rotating flow, it appears that, at higher pseudo-Rayleigh number, this may happen.

The programme as yet does not feature solution of the Poisson type Equation (2.4) for pressure, however, work on this is under way. It should be noted that this problem filled 64K of a Cyber 72 computer and therefore to increase the resolution or scope of problem it is necessary either to have more core or more sophisticated use of disc units. Note also, that only the steady state solution was obtained for the above problem. This can be achieved by eliminating all terms involving Δt from the programme. Convergence was achieved in three iterations and the whole process took three minutes on the Cyber computer.

The second case to be modelled is the katabatic wind generation by the nocturnal cooling of an isolated hill. The solution domain is a region 3000 m long and 1500 m high with a simple hill in the lower left hand corner being

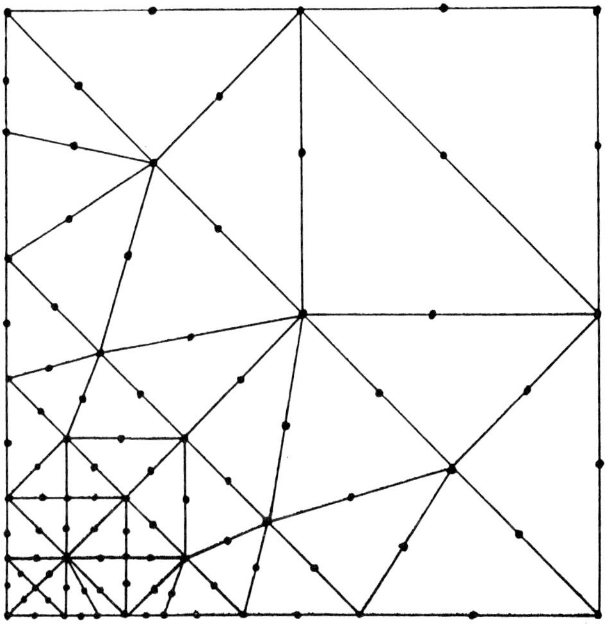

Figure 3 - Grid for Fire Wind

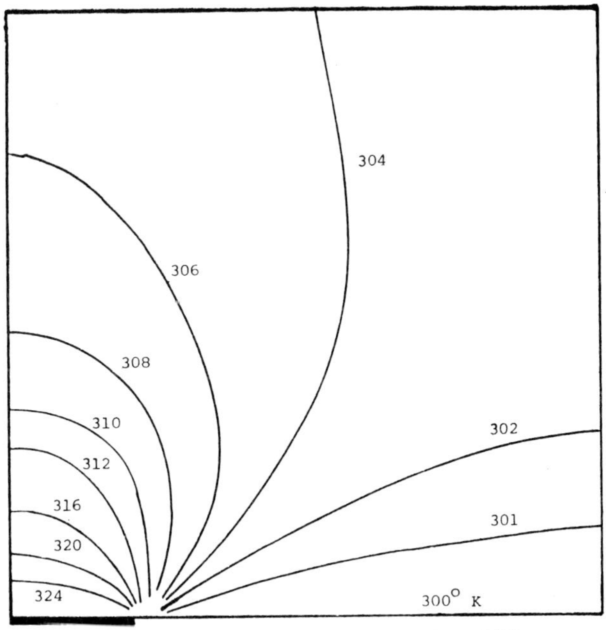

Figure 4 - ISOTHERMS Ra = 6400

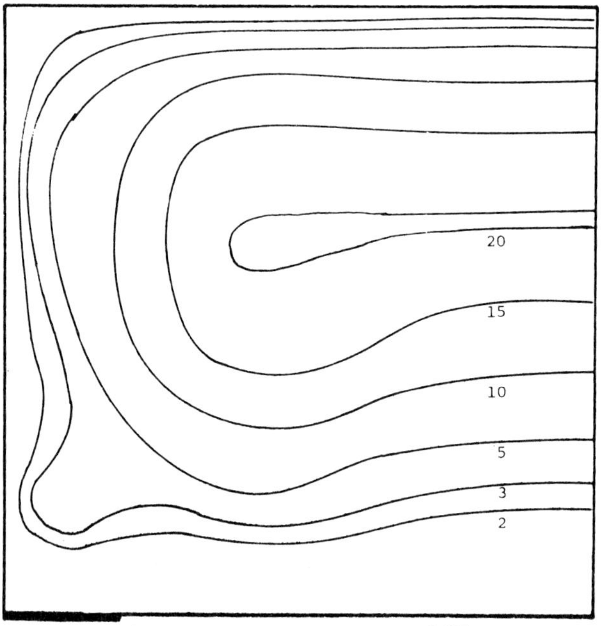

Figure 5 - STREAMLINES Ra = 6400

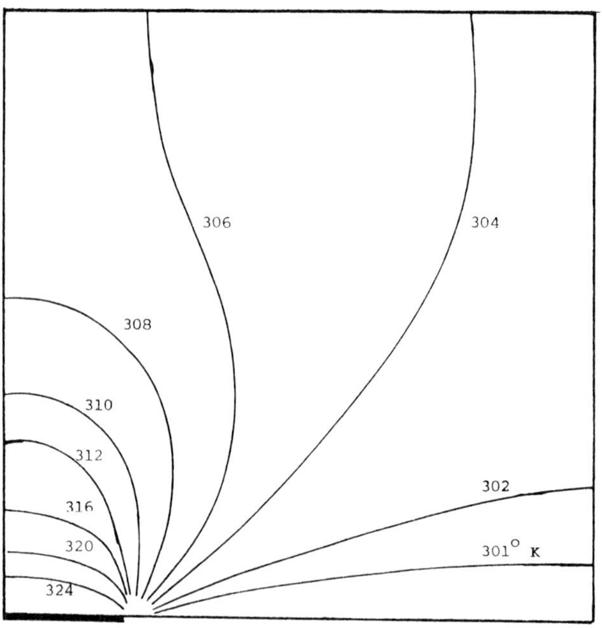

Figure 6 - ISOTHERMS Ra = 1600

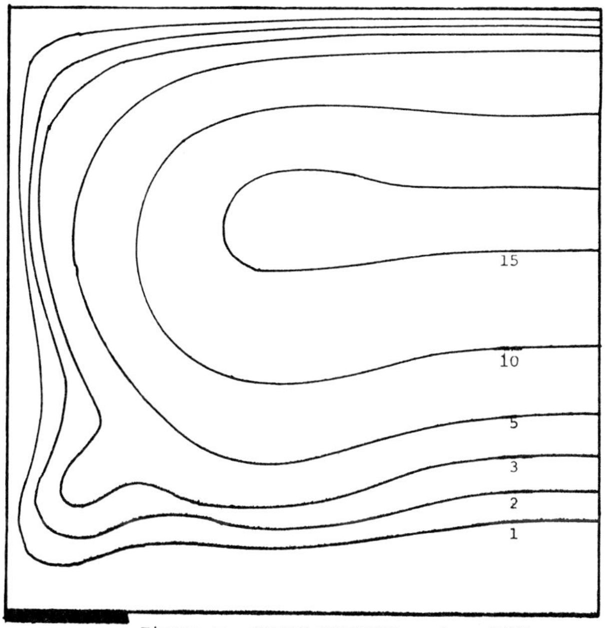

Figure 7 - STREAM FUNCTION Ra = 1600

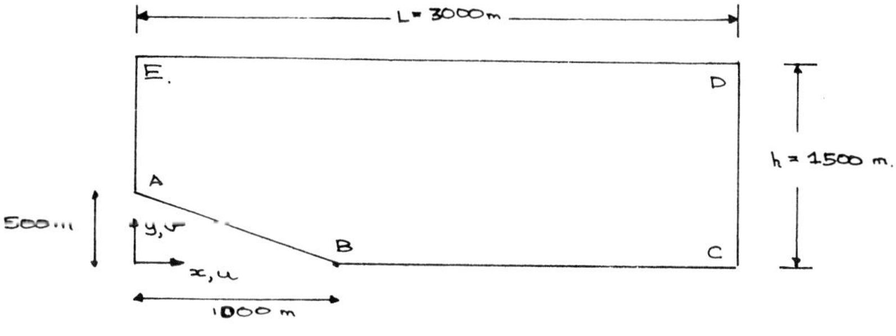

Figure 8.

1000 m long and 500 m high (see Figure 8). This domain portrays one symmetric half of the hill and surrounding region. The boundary conditions on the base variables (ψ,ω,T) and their cartesian gradients, the auxiliary functions, are given in Table 2.

TABLE 2

SIDE	α	β	λ	δ	ε	η	ψ	ω	T
AB | y=500-x/2	0	0	?	?	?	?	0	?	T_g
BC | y=0	0	0	?	?	0	?	0	?	T_g
ED | y=h	0	?	0	?	?	0	0	0	?
AE | x=0	?	0	0	0	0	?	0	0	?
CD | x=L	0	?	0	?	0	?	?	?	?

$$\text{Where } T_g = 300 - 20\sin(\pi t/18 \times 3600)\ ^oK.$$

On the ground the boundary conditions are again no slip $(u = v = 0)$ with vorticity being shed and the temperature varies with time according to the expression for Tg; note that on the hill $T_x = \varepsilon \neq 0$. For the upper surface we again model a thermal inversion in the upper atmosphere. The line AE has symmetry conditions while the remote downstream vertical boundary is assumed to be sufficiently remote that parallel flow conditions may be taken.

The finite element grid used is shown in Figure 9. For reasons of space only one set of results are presented with the diffusivities of heat and eddies (K) set at $20m^2/sec$ and Δt set at 240 seconds. At times of 2, 8 and 12 hours, plots of temperature and stream function are given in Figures 8 to 13.

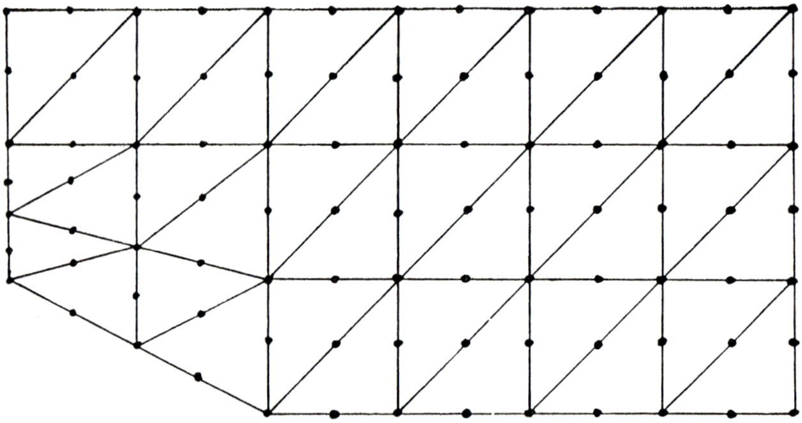

Figure 9 - GRID FOR KATABATIC WIND MODEL

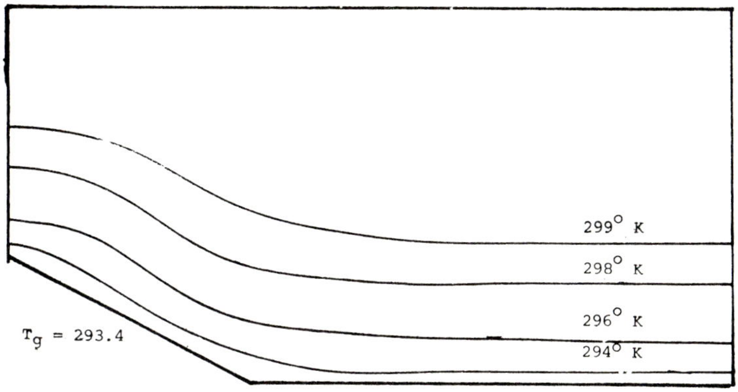

Figure 10- Time = 2 hours - K = 20 M^2/SEC.

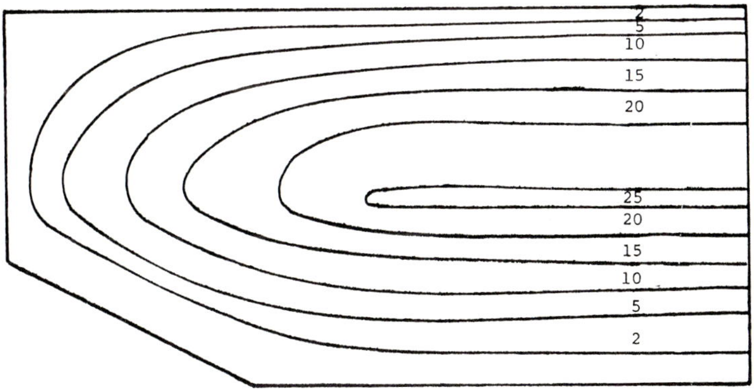

Figure 11- Time = 2 hours - K = 20 M^2/SEC.

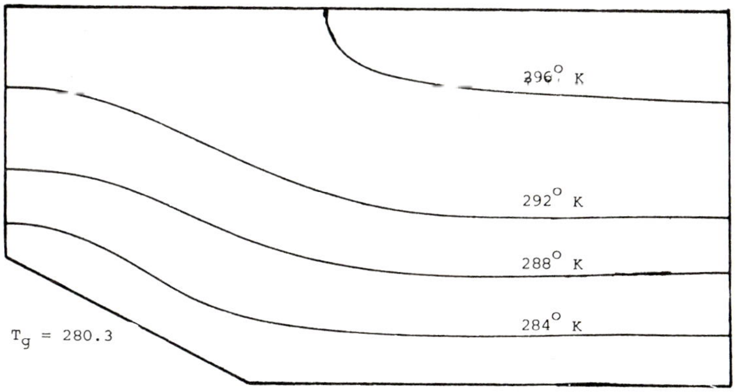

Figure 12 - Time = 8 hours - K = 20 M^2/SEC.

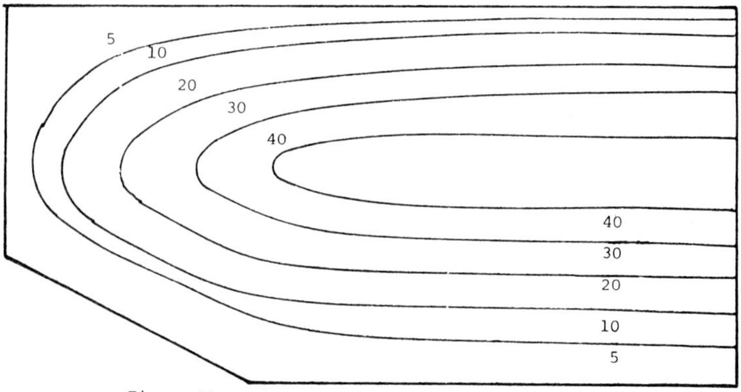

Figure 13 - Time = 8 hours - K = 20 M^2/SEC.
STREAM FUNCTION

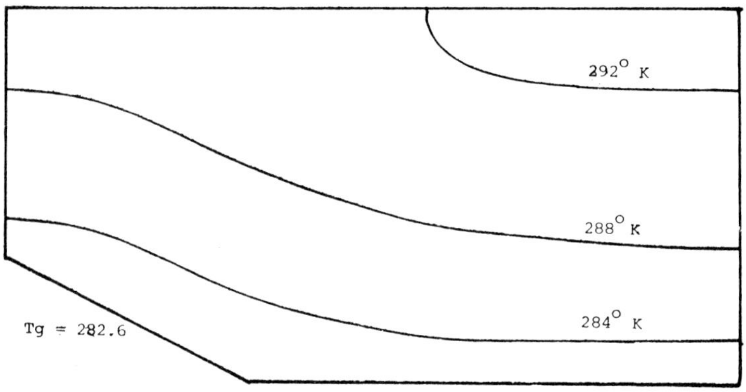

Figure 14 - Time = 12 hours - K = 20 M^2/SEC.
ISOTHERMS

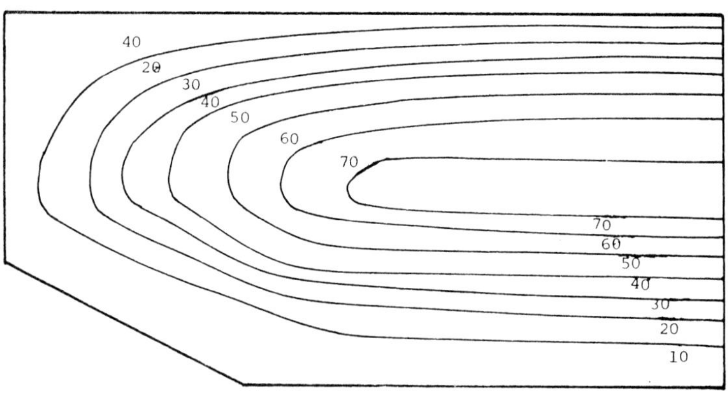

Figure 15 - Time = 12 hours - K = 20 M^2/SEC.
STREAM FUNCTION

From the three temperature profiles, Figures 10, 12 and 14, we see that at 2 hours the ground temperature is just beginning to diffuse outwards and at 8 hours has reached almost its minimum values while at 12 hours the temperatures have passed through their minimum values and are beginning to rise again. This process is much as would be expected. It is to be noticed that all three temperature profiles exhibit a slightly lower vertical temperature gradient near the bottom of the hill due to the convective effect of the katabatic wind carrying warmer air from higher altitudes.

Looking at the three stream function profiles the overall characteristics are clear. Air enters across the downstream boundary CD at the higher level, moves towards the symmetry plane, is deflected downwards and around until parallel with the hill. It then proceeds downhill and along the ground until it leaves the solution domain on the lower portion of CD parallel to the ground. Due to the no slip boundary conditions on the ground and the full slip conditions on the upper surface ED the entry velocities are higher than the exit velocities. From observation of the complete set of computed results there is a noticeable lag in the shear function profiles as the temperature profiles pass through their minimum values. This lag on stream function is somewhere between two and three hours.

5. CONCLUSION

Despite their brevity and lack of detail it is hoped that the examples presented in the last section suffice to demonstrate the power of the finite element method for the solution of such problems. Given more computing power it would not be too difficult a task to generate three dimensional models, the only additional component to the foregoing analysis would be the incorporation of the third dimension terms in the differential matrix operator [L] and into the interpolation matrix [N] , otherwise the algebraic process and computing is identical to this current two dimensional solution. It has also to be appreciated that, although simple solution domains were adopted in this work for reasons of comparison with existing finite difference solutions, it is a simple matter to have any sort of ground profile.

It has not been possible to pay attention to many of the finer points associated with the finite element process towards the given solutions, the reader is referred to the literature for this. In the future it is hoped to incorporate pressure field evaluation and automatic plotting of contours into the computer programme, also to study other forms that the equations governing fluid flow may have and eventually model the three dimensional case.

REFERENCES

Akin, S.E. (1973). A Least Squares Finite Element Solution of Non-Linear Operators. The Mathematics of Finite Elements and Applications, ed. Whiteman, Academic Press.

Desai, C.S., and Abel, J.F. (1972). Introduction to the Finite Element Method. Van-Nostrand-Reinhold.

Fox, L. (1964). An introduction to Numerical Linear Algebra, O.U.P.

Leslie, L.M., and Smith, R.K. (1974). A Numerical Study of Katabatic Winds and their Effect on Pollutant Dispersal in Urban Areas. Fifth A/Asian Conf. on Fluid Mechanics, Christchurch, 2, 553.

Lynn, P.P. (1974). Least Squares Finite Element Analysis of Laminar Boundary Layer Flows. Int. J. Num. Math. Engng. 8, 865.

Lynn, P.P., and Arya, S.K. (1973). Use of the Least Squares Criterion in the Finite Element Formulation. Int. J. Num. Math. Engng. 6, 75.

Milthorpe, J.F. (1975). The use of the Finite Element Method in Numerical Integration of the Navier-Stokes Equations. M. Eng. Sci. Thesis, The University of Sydney.

Smith, R.K., Morton, B.R., and Leslie, L.M. (1975). The Role of Dynamic Pressure in Generating Fire Winds. J. Fluid Mech. 68, 1.

Steven, G.P., (1978). On the Application of the Finite Element Method to Problems in Fluid Flow. Numerical Simulation of Fluid Motion, ed. John Noye, North-Holland, 171.

Wood, W.L., and Lewis, R.W. (1975). A Comparison of Time Marching Schemes for the Transient Heat Conduction Equation. Int. J. Num. Math. Engng. 9, 679.

Zienkiewicz, O.C. (1971). The Finite Element Method in Engineering Science, McGraw Hill.

Numerical Simulation of Fluid Motion, ed. John Noye
pp. 567-573

OBJECTIVE ANALYSIS OF GEOPHYSICAL FIELD DATA

C.E. Wallington

Canberra College of Advanced Education

Australian Capital Territory

SUMMARY

The techniques of converting a set of observations of a fluid system into best fitting orthogonal functions on which subsequent deductions and calculations can be based is desribed. These techniques give greater accuracy in fitting high degree polynomials to data than the commonly used method of fitting polynomial surfaces by a least square error procedure. They also enable constraints, physical assumptions and subjective modifications to be built into a surface fitting scheme.

1. INTRODUCTION

A problem common to many oceanographic, meteorological and hydrological studies is that of analysing observations made with varying and often uncertain percision of divers elements at various times over networks of irregularly spaced observing points. The usual objective is to convert the set of observations into coherent patterns or best fitting functions on which to base subsequent deductions or calculations. When an oceanographic cruise, for example, provides temperature, salinity and miscellaneous flow data along a zig-zag course during a period of several weeks, the physical oceanographer will want to know whether the variegated data reveals new flow patterns or confirms certain forms of the hydrodynamical equations believed to govern the flow. Dedicated analysts have uncovered new and interesting current systems by carefully drawn patterns to fit the data as well as the eye can judge. But there are severe practical limitations on such methods of analysis. They are too time consuming to apply to much of the data now available, and even when hand drawn patterns are constructed with considerable care these patterns are not generally accurate enough for differential analysis.

A numerical analysis technique commonly applied to meteorological fields is to fit polynomial surfaces by a least square error procedure to groups of observations. The procedure usually entails solving a set of linear simultaneous equations to find the required coefficients of terms in the polynomial. For geophysical fields, however, these simultaneous equations are usually ill-conditioned for high power polynomials and attempts to extend the polynomials to degrees higher than about the fourth power is usually unrewarding because of the loss of numerical accuracy.

The problem of ill-conditioning can be avoided by using orthogonal function fitting techniques. These techniques allow not only greater accuracy in fitting higher degree polynomials, they also enable the analyst to build into a surface fitting scheme restraints, such as boundaries to a flow, physical assumptions, such as geostrophic relationships, and subjective modifications, such as assessments of the quality of the data.

This paper, describing the principles of the techniques, follows an approach made by Dixon (1969 and 1970).

2. THE BASIC TECHNIQUE

Suppose we have a set of observations, h_i, for a set of space and time co-ordinates x_i, y_i, z_i, t_i, where the subscript i denotes the i th observation of the series, and we wish to fit an expression

$$h = \sum_{r=1}^{r=m} a_r f_r(x,y,z,t) \qquad (2.1)$$

to the set of n observations, a_r being a set of coefficients of the m functions $f_r(x,y,z,t)$. For many practical purposes it may be convenient to choose the functions, f_r, as a set of increasing powers of x,y,z,t, e.g. $1,x,y,z,t,x^2, xy,\ldots$ But these functions need not be restricted to such a set; they can be any finite functions of the independent variables.

Instead of using traditional least square error methods to calculate the coefficients a_r, let us form another series of the form

$$h = \Sigma A_r F_r \qquad (2.2)$$

where F_r is a set of orthogonal functions of $f(x,y,z,t)$, i.e. $(F_p.F_q) = 0$ for $p \neq q$, where the notation $(F_p.F_q)$ denotes the summation of the product of F_p and F_q over the set of n observations.

The coefficients A_r can then be determined by noting that

$$(h.F_r) = A_r(F_r.F_r) \tag{2.3}$$

Thus the coefficients can be calculated from a series of independent equations, and the problem of ill-conditioning does not arise. It can also be shown that the coefficients so determined are those that produce a least square error fit of the observations.

The crux of the problem is that of forming the functions F_r from the set of functions $f(x,y,z,t)$. The most convenient general purpose method is the Gram-Schmidt orthogonalisation process described by Synge (1957). This process is best illustrated by the following equations.

$$\left.\begin{aligned}
F_1 &= f_1 \\
F_2 &= f_2 - \frac{(F_1.f_2)}{(F_1.F_1)} F_1 \\
F_3 &= f_3 - \frac{(F_1.f_3)}{(F_1.F_1)} F_1 - \frac{(F_2.f_3)}{(F_2.F_2)} F_2 \\
&\cdots\cdots\cdots\cdots \\
F_p &= f_p - \sum_{q=1}^{q=p-1} \frac{(F_q.f_p)}{(F_q.F_q)} F_q \\
&\cdots\cdots\cdots\cdots
\end{aligned}\right\} \tag{2.4}$$

This is a simple method of generating the orthogonal functions required, but if a long sequence of functions, F, are generated the round-off errors will tend to accumulate. Thus the functions may not be strictly orthogonal. This problem is the counterpart of the ill-conditioning difficulty in the method of determining coefficients from the simultaneous equations arising from the traditional least square error method, but it is not so pronounced as the ill-conditioning problem.

The build-up of round-off errors can be suppressed by repeated normalisation of each orthogonal function as it is formed. For example, using E to denote functions and superscripts for successive formulations of orthogonal functions, we can form

$$E_1^1 = f_1$$

$$F_1^1 = E_1^1 \, (E_1^1.E_1^1)^{-\frac{1}{2}}$$

for the first orthogonal function. The second function will be formed in the four stages:

$$\left.\begin{aligned} E_2^1 &= f_2 \\ F_2^1 &= E_2^1\,(E_2^1.E_2^1)^{-\frac{1}{2}} \\ E_2^2 &= F_2^1 - (F_2^1.F_1^1)\,F_1^1 \\ F_2^2 &= E_2^2\,(E_2^2.E_2^2)^{-\frac{1}{2}} \end{aligned}\right)\qquad (2.5)$$

Successive functions will be formed in similar, but extended stages, e.g.

$$\left.\begin{aligned} E_3^1 &= f_3 \\ F_3^1 &= E_3^1\,(E_3^1.E_3^1)^{-\frac{1}{2}} \\ E_3^2 &= F_3^1 - (F_3^1.F_1^1)\,F_1^1 \\ F_3^2 &= E_3^2\,(E_3^2.E_3^2)^{-\frac{1}{2}} \\ E_3^3 &= F_3^2 - (F_3^2.F_2^2)\,F_2^2 \\ F_3^3 &= E_3^3\,(E_3^3.E_3^3)^{-\frac{1}{2}} \\ &\ldots\ldots\ldots\ldots\ldots\ldots \\ &\ldots\ldots\ldots\ldots\ldots\ldots \\ &\ldots\ldots\ldots\ldots\ldots\ldots \end{aligned}\right)$$

It can be shown that using F_r^r in place of F_r will suppress the round-off error. Because the orthogonal functions are normalised $((F_r^r.F_r^r) = 1)$ in this process

$$A_r = (h.F_r^r) \qquad (2.6)$$

3. WEIGHTING THE DATA

If we wish to weight the data by assigning a set of weights, W_i, to the set of data, we can do so by including these weights in the sums of products, i.e. by substituting $(F_r.F_r.W_r)$ for $(F_r.F_r)$, etc., where $(F_r.F_r.W_r)$ denotes the triple product of F_r, F_r and W_r summed over the set of n observations.

Specification of the weights will normally be a matter of subjective assessment, but objective guidelines may be determined if sufficient data has been accrued for statistical analysis of probable errors.

4. FITTING FLOW PATTERNS

Flow pattern data often comprises a mixture of pressure or pressure level data and current data. If there is an acceptable linear relationship between the two types of data they can both be accommodated in a single fitting process. Suppose, for example, the flow is geostrophic. For approximately horizontal flow over the Earth's surface, the appropriate relationship can be written in the form

$$\left.\begin{aligned} -lu/g &= \frac{\partial h}{\partial y} \\ lv/g &= \frac{\partial h}{\partial x} \end{aligned}\right)\qquad (4.1)$$

where

u) denotes the (eastward) component of the flow in the direction
v) (northward) towards which the flow is coming

g denotes the vertical acceleration due to gravity

$l = 2\omega \sin \phi$ where ω = the angular rate of rotation of the earth and ϕ denotes latitude North.

If Equation (2.1) is being used to fit given values of h, given values of u and v can be included in the fitting process by substituting $-lu/g$ or lv/g for h, and $(\partial h/\partial y)$ and $(\partial h/\partial x)$ respectively for $f_r(x,y,z,t)$. In effect the mixed data is used to form lists of given values of $h, -lu/g, lv/g$ and functions of x,y,z,t - such as those shown below:

Data	$f_1(x,y,z,t)$	$f_2(x,y,z,t)$	....
h_1	$f_1(x_1,y_1,z_1,t_1)$	$f_2(x_1,y_1,z_1,t_1)$	....
h_2	$f_2(x_2,y_2,z_2,t_2)$	$f_2(x_2,y_2,z_2,t_2)$	....
....	....	....	....
$-(\frac{lu}{g})_m$	$\frac{\partial f_1}{\partial y}(x_m,y_m,z_m,t_m)$	$\frac{\partial f_2}{\partial y}(x_m,y_m,z_m,t_m)$	....

A given flow component can be fitted by noting that, if s_m is a given speed of the component towards a given direction θ_m, the item of data $(ls/g)_m$ corresponds to the set of functions

$$\cos \theta_m \frac{\partial f_r(x_m,y_m,z_m,t_m)}{\partial x} - \sin \theta_m \frac{\partial f_r(x_m,y_m,z_m,t_m)}{\partial y}$$

5. HYDRODYNAMIC CONSTRAINTS

5.1 Continuity

A vector field of the type

$$\underset{\sim}{V} = \sum_{r=1}^{r=m} a_r \underset{\sim}{f}_r(x,y,z,t) \qquad (5.1.1)$$

can be fitted to a set of n given vectors, $\underset{\sim}{V}_i$, in the same way as that already described for the scalar fitting process.

If this vector field is intended to describe a flow pattern in an incompressible fluid, the fitted functions should be compatible with the requirement that $\nabla \underset{\sim}{V} = 0$. It is not normally difficult to form suitable functions. For example, we could choose

$$\begin{aligned} \underset{\sim}{V} = \underset{\sim}{a}_o &+ b_1(\underset{\sim}{i}x-\underset{\sim}{j}y) + b_2(\underset{\sim}{j}y-\underset{\sim}{k}z) + b_3(\underset{\sim}{k}z-\underset{\sim}{i}x) \\ &+ c_1(\underset{\sim}{i}x^2-2\underset{\sim}{j}xy) + c_2(\underset{\sim}{j}y^2-2\underset{\sim}{k}yz) + c_3(\underset{\sim}{k}z^2-2\underset{\sim}{i}xz) + \ldots \qquad (5.1.2) \\ &+ \underset{\sim}{i}\, f_i(y,z,t) + \underset{\sim}{j} f_j(x,z,t) + \underset{\sim}{k} f_k(x,y,t) \end{aligned}$$

where a_o is a constant vector, i, j, k are unit vectors in the x,y,z directions, and the functions f_i, f_j, f_k are independent of x,y,z, respectively.

Equation (5.1.2) is a particular form of a more general expression

$$\underset{\sim}{V} = \sum_{r=1}^{r=m} a_r \nabla \wedge \underset{\sim}{f}_r(x,y,z,t) \qquad (5.1.3)$$

that satisfies the continuity condition, $\nabla \underset{\sim}{V} = 0$.

5.2 Vorticity

As the vorticity, $\nabla \wedge \underset{\sim}{V}$, is also a linear function it can also be included in a linear constraint in the function fitting process. If the analyst wishes to set the vorticity in the fitted flow to zero, or to a constant, or to a linear function of x,y,z,t, he can easily do so. Whether such a constraint is hydrodynamically realistic is a matter for the analyst to decide.

5.3 Boundary conditions

A typical boundary condition is that there should be zero or some other specified flow across a boundary. One method of setting such a condition for a velocity vector field, $\underset{\sim}{V}$, is to fit an expression of the type

$$\underset{\sim}{V} = \underset{\sim}{B} + \sum_{r=1}^{r=m} a_r \hat{\underset{\sim}{B}} \wedge f_r(x,y,z,t) \qquad (5.3.1)$$

where $\underset{\sim}{B}$ is the specified vector flow field normal to the boundary, and $\hat{\underset{\sim}{B}}$ is a unit vector normal to the boundary.

For streamline flow patterns a typical condition is that rigid boundaries must be streamlines. For a contour pattern, h, with the geostrophic relationship, for example, this condition can be set by fitting expressions of the type

$$h = A_1B_1 + A_2B_2 + A_3B_3 + \ldots. + \sum_{r=1}^{r=m} a_r(B_1-1)(B_2-1)(B_3-1)\ldots.f_r(x,y,z,t) \qquad (5.3.2)$$

where B_q is a function of the independent variables, equal to unity on boundary q and zero on all other boundaries, and the A's are coefficients to be determined in the fitting process.

To fit a boundary condition approximately rather than exactly, the condition of zero or specified flow across a boundary can be set approximately by adding to the data a set of zero or specified flow components at a series of points along the boundary. This set of boundary data can be weighted according to the degree of approximation required.

5.4 Non-linear constraints

Constraints imposed by non-linear terms such as $(\underset{\sim}{V}\nabla)\underset{\sim}{V}$ cannot be applied exactly except for a few special and often unrealistic flow patterns. But such a constraint can be introduced approximately by an iterative process in which $(\underset{\sim}{V}\nabla)\underset{\sim}{V}$ is replaced by $(\underset{\sim}{V}_{p-1}\nabla)\underset{\sim}{V}_p$ in the pth fitting process. Thus it is possible, though cumbersome, to constrain the fitted function to conform approximately to the Navier-Stokes equation of motion. The constraint can be applied over a specified region either by integration or by specifying it to apply at a set of points in the region. Again this form of data can be weighed according to the required balance between the hydrodynamic constraints and the actual data.

6. DEDUCTION OF ADVECTIVE FLOW PATTERNS

A problem that often arises is that of deducing flow patterns from successive fields of elements carried by the flow. It is possible, for example to obtain successive mesoscale patterns of sea surface temperatures by radiometer measurements from low flying aircraft. Can such measurements be used to deduce a probable surface flow pattern? If we assume that the mesoscale changes are due mainly to surface advection and that the surface flow satisfies a stream function relationship, $u = -\partial\psi/\partial y$ $v = \partial\psi/\partial x$, we can determine a best fitting function, ψ, over the region of interest by imposing the linear constraint

$$\frac{\partial T}{\partial t} - \frac{\partial \psi}{\partial y}\frac{\partial T}{\partial x} + \frac{\partial \psi}{\partial x}\frac{\partial T}{\partial y} = 0 \qquad (6.1)$$

where T denotes a best fitting function of the temperature data. In general it cannot be proved that the ψ field is the true advective pattern, but it is likely that the deduced flow will contain a useful element of truth.

7. CONCLUSIONS

In applying a function fitting technique an experimenter is simply trying to fit a coherent pattern as best he can to data and to his views on the data. The patterns so derived cannot be said to be correct or incorrect; they are best described in terms of plausibility, realism, and usefulness to the experimenter. Subjectivity is still present in that the experimenter chooses functional forms, but it is restricted to basic premises rather than applied in an irregular, uneven manner over a set of data and hypotheses. Fitted functions that appear plausible and realistic are useful not merely for deriving artistic field patterns; they provide an intrinsic quality control over the derivation of differential fields often used in hydrodynamic studies. Vorticity and vorticity advection fields derived from fitted functions are generally smoother and almost certainly more realistic than those derived by more direct, but coarser, methods from basic data. The inclusion of time variations in a set of data automatically introduces an element of history into a fitted function and may, therefore, enable the function to be used for prediction, especially if it is also constrained to fit what are believed to be relevant equations of change.

REFERENCES

Dixon, R. (1969) Orthogonal Polynomials as a Basis for Objective Analysis. Met. Office, Sci Paper No. 30, H.M.S.O. London.

Dixon, R. (1970) The Three-dimensional Analysis of Meteorological Data, Met. Office, Sci.Paper No. 31, H.M.S.O. London.

Synge, J.C. (1957) The Hypercircle in Mathematical Physics. Cambridge University Press. London.

CONFERENCE DELEGATES

A.J. ALLAN, *Department of Public Works, Hydraulics Laboratory, King Street, Manly Vale, New South Wales, 2093.*

D.J. ALLEN, *45 Burrendong Street, Duffy, Australian Capital Territory, 2611.*

R.A. ANTONIA, *Department of Mechanical Engineering, University of Newcastle, New South Wales, 2308.*

P. ARNOT-SMITH, *Oceanics Australia Pty. Ltd., P.O. Box 203, North Brisbane, Queensland, 4000.*

B.W. BARNARD, *Department of Mechanical Engineering, Caulfield Institute of Technology, Caulfield East, Victoria, 3145.*

I.W. BARNES, *77 Myrtle Road, Seacliff, South Australia, 5049.*

A.G. BARNETT, *Power Division, Ministry of Works and Development Head Office, P.O. Box 12-041, Wellington, New Zealand.*

R.C. BELL, *Commonwealth Scientific and Industrial Research Organisation, Division of Atmospheric Physics, P.O. Box 77, Mordialloc, Victoria, 3195.*

A.A. BHARMAL, *School of Mathematical Science, Flinders University, Bedford Park, South Australia, 5042.*

A.F. BENNETT, *Geophysical Fluid Dynamics Laboratory, Monash University, Clayton, Victoria, 3168.*

P.J. BILLS, *Department of Applied Mathematics, University of Adelaide, South Australia, 5000.*

W.E. BODLEY, *Department of Civil Engineering, University of Adelaide, South Australia, 5000.*

W. BOURKE, *Australian National Meteorological Research Centre, 150 Lonsdale Street, Melbourne, Victoria, 3000.*

B.C. BOWERS, *Weapons Research Establishment, Box 2151 G.P.O., Adelaide, South Australia, 5000.*

L.W.B. BROWNE, *Department of Mechanical Engineering, University of Newcastle, New South Wales, 2308.*

G. BYRNE, *Department of Housing & Construction, 17 Yarra Street, Hawthorn, Victoria 3122.*

E. BRADFORD, *Applied Mathematics Division, Department of Scientific and Industrial Research, P.O. Box 8030, Wellington, New Zealand.*

R.A.D. BYRON-SCOTT, *School of Earth Sciences, Flinders University, Bedford Park, South Australia, 5042.*

A. CABELLI, *Commonwealth Scientific and Industrial Research Organisation, Division of Mechanical Engineering, Highett, Victoria, 3190.*

J.P. CALEGARI, *Department of Mathematics, Royal Melbourne Institute of Technology, 124 LaTrobe Street, Melbourne, Victoria, 3000.*

J. CHAMBERS, *Department of Mechanical Engineering, University of Newcastle, New South Wales, 2308.*

G.R. CLARK, *8 Jordan Grove, Glen Waverley, Victoria, 3150.*

H.A. CLARK, *Electricity Trust of South Australia, 220 Green-hill Road, Eastwood, South Australia, 5063.*

H.J. CONNELL, *33 Roberts Street, Essendon, Victoria, 3040.*

E. CONWAY, *10 Gilmore Road, Doncaster, Victoria, 3108.*

J.R. CRAWFORD, *Commonwealth Scientific and Industrial Research Organisation, Division of Building Research, P.O. Box 66, Highett, Victoria, 3190.*

D.J. CRAZE, *Mechanical Development Section, Australian Atomic Energy Commission Research Establishment, Private Mail Bag, Sutherland, New South Wales, 2232.*

M.R. DAVIDSON, *Australian Atomic Energy Commission Research Establishment, Lucas Heights, New South Wales, 2232.*

G. DE VAHL DAVIS, *Department of Mechanical Engineering, University of New South Wales, P.O. Box 1, Kensington, New South Wales, 2033.*

P.E. DEXTER, *Commonwealth Bureau of Meteorology, P.O. Box 1289K, Melbourne, Victoria, 3001.*

A.K. EASTON, *Mathematics Department, Caulfield Institute of Technology, Caulfield East, Victoria, 3145.*

W. EVERSMAN, *Department of Mechanical Engineering, University of Canterbury, Christchurch, New Zealand.*

R.L. FALCONER, *Commonwealth Bureau of Meteorology, P.O. Box 1289K, Melbourne, Victoria, 3001.*

P. FAIRLIE, *Aeronautical Research Laboratory, Box 4331 G.P.O., Melbourne, Victoria, 3001.*

G. FIX, *Department of Mathematics, Carnegie-Mellon University, Pittsburgh, U.S.A.*

C.A.J. FLETCHER, *Dynamic Group, Aerospace Division, Weapons Research Establishment, G.P.O. Box 2151, Adelaide, South Australia, 5001.*

D.A. FRITH, *Aeronautical Research Laboratory, Box 4331 G.P.O., Melbourne, Victoria, 3001.*

M.T. FOX, *1094 Heatherton Road, Noble Park, Victoria, 3174.*

D. GAUNTLETT, *Australian National Meteorological Research Centre, G.P.O. Box 5089AA, Melbourne, Victoria, 3001.*

J.A. GEYER, *Department of Mathematics, Mitchell College of Advanced Education, Bathurst, New South Wales, 2795.*

N. GILBERT, *Aeronautical Research Laboratory, Box 4331 G.P.O., Melbourne, Victoria, 3001.*

P.M. GILL, *Department of Applied Mathematics, University of Adelaide, Adelaide, South Australia, 5000.*

M.J. GOODSPEED, *Commonwealth Scientific and Industrial Research Organisation, Division of Land Use Research, P.O. Box 1666, Canberra City, Australian Capital Territory, 2601.*

D.J. GORHAM, *18/51 Pacific Parade, Dee Why, New South Wales, 2099.*

M.M. GUPTA, *Department of Mathematics, University of Technology, P.O. Box 798, Lae, Papua-New Guinea.*

R.P. HALE, *Mathematics Department, Gordon Institute of Technology, Geelong, Victoria, 3220.*

A.L. HALMOS, *Department of Chemical Engineering, Monash University, Clayton, Victoria, 3168.*

N. HARITOS, *Department of Civil Engineering, University of Melbourne, Parkville, Victoria, 3052.*

B.A. HARPER, *Unit 6, Dulkara, Merrima Avenue, Caloundra, Queensland, 4551.*

J.B. HINWOOD, *Department of Mechanical Engineering, Monash University, Clayton, Victoria, 3168.*

R.R. HUILGOL, *School of Mathematical Sciences, Flinders University, Bedford Park, South Australia, 5042.*

S.G. HUTTON, *Department of Civil Engineering, University of Adelaide, South Australia, 5000.*

J.O. JENSEN, *Department of Mathematics, Monash University, Clayton, Victoria, 3168.*

J.C.E. JONES, *Ports and Harbours Authority, 168 Exhibition Street, Melbourne, Victoria, 3000.*

R.A. JONES, *Broken Hill Proprietary Company Limited, P.O. Box 1968, Newcastle, New South Wales, 2300.*

N. JOYCE, *Aeronautical Research Laboratory, Box 4331, G.P.O., Melbourne, Victoria, 3001.*

R.C. JOYNT, *Herman Central Scientific Laboratory, State Electricity Commission of Victoria, Howard Street, Richmond, Victoria, 3121.*

L.T. KHOA, *27 Murray Square, Mayfield, New South Wales, 2304.*

C. KING, *Aeronautical Research Laboratory, Box 4331, G.P.O., Melbourne, Victoria, 3001.*

J.H. KNIGHT, *Commonwealth Scientific and Industrial Research Organisation, Division of Mathematics and Statistics, P.O. Box 1965, Canberra City, Australian Capital Territory, 2601.*

W. KORNIENKO, *Aeronautical Research Laboratory, Box 4331, G.P.O., Melbourne, Victoria, 3001.*

N. KOSHNITSKY, *Department of Applied Mathematics, University of Adelaide, Adelaide, South Australia, 5000.*

K.H. LLOYD, *Weapons Research Establishment, Box 2151, G.P.O., Adelaide, South Australia, 5001.*

N.A.E. MACH, *Department of Aeronautics, University of Sydney, Sydney, New South Wales, 2006.*

P. MANINS, *Commonwealth Scientific and Industrial Research Organisation, Division of Atmospheric Physics, P.O. Box 33, Mordialloc, Victoria, 3195.*

J.V. MANSBRIDGE, *Department of Mathematics, Monash University, Clayton, Victoria, 3168.*

M.J. MANTON, *Commonwealth Scientific and Industrial Research Organisation, Division of Cloud Physics, P.O. Box 134, Epping, New South Wales, 2121.*

G. McDONOUGH, *Department of Mechanical Engineering, University of Adelaide, Adelaide, South Australia, 5000.*

B.J. McAVANEY, *Australian National Meteorological Research Centre, Box 5089AA, G.P.O., Melbourne, Victoria, 3001.*

J. McGREGOR, *Australian National Meteorological Research Centre, Box 5089AA, G.P.O., Melbourne, Victoria, 3001.*

G. MERRINGTON, *Aeronautical Research Laboratory, Box 4331, G.P.O., Melbourne, Victoria, 3001.*

D.J. MILLS, *Department of Mechanical Engineering, Monash University, Clayton, Victoria, 3168.*

B.R. MORTON, *Department of Mathematics, Monash University, Clayton, Victoria, 3168.*

B.J. NOYE, *Department of Applied Mathematics, University of Adelaide, Adelaide, South Australia, 5000.*

J.J. O'BRIEN, *Department of Meteorology and Oceanography, Florida State University, Tallahassee, Florida, 32306, U.S.A.*

S. O'CONNOR, *Mathematics Department, Monash University, Clayton, Victoria, 3168.*

I.C. O'NEILL, *School of Engineering, University of Melbourne, Parkville, Victoria, 3052.*

M.J. O'NEILL, *Department of Mathematics, Mitchell College of Advanced Education, Bathurst, New South Wales, 2795.*

J.C. PATTERSON, *Department of Civil Engineering, University of Western Australia, Nedlands, Western Australia 6009.*

R.A. PEARSON, *Department of Mathematics, Monash University, Clayton, Victoria, 3168.*

M.C.E. PETERSEN, *Health Physics Research Section, Australian Atomic Energy Commission Research Establishment, Private Mail Bag, Sutherland, New South Wales, 2232.*

J.R. PHILIP, *Commonwealth Scientific and Industrial Research Organisation, Division of Environmental Mechanics, Canberra City, Australian Capital Territory, 2601.*

R.L. POPE, *Weapons Research Establishment, Box 2151, G.P.O., Adelaide, South Australia, 5001.*

J. PRICE, *Mechanical Development Section, Australian Atomic Energy Commission Research Establishment, Private Mail Bag, Sutherland, New South Wales, 2232.*

K. PURI, *Australian National Meteorological Research Centre, Box 5089AA, G.P.O., Melbourne, Victoria, 3001.*

J.M. REILLY, *17 Hatter Street, Oakleigh, Victoria, 3168.*

M. RIENECKER, *Department of Applied Mathematics, University of Adelaide, Adelaide, South Australia, 5000.*

T. RYALL, *Aeronautical Research Laboratory, Box 4331, G.P.O., Melbourne, Victoria, 3001.*

R. RYAN, *Commonwealth Scientific and Industrial Research Organisation, Department of Cloud Physics, P.O. Box 134, Epping, New South Wales, 2121.*

T.W. SAG, *School of Mathematical Sciences, Flinders University, Bedford Park, South Australia, 5042.*

B.B. SHARP, *Department of Civil Engineering, University of Melbourne, Parkville, Victoria, 3052.*

R.J. SOBEY, *Department of Engineering, James Cook University, Townsville, Queensland, 4811.*

I.B. STARK, *77 Myrtle Road, Seacliff, South Australia, 5049.*

K.P. STARK, *Department of Engineering, James Cook University, Townsville, Queensland, 4811.*

G.P. STEVEN, *Department of Aeronautics, University of Sydney, Sydney, New South Wales, 2006.*

P.B. STONE, *Public Works Department Hydraulics Laboratory, King Street, Manly Vale, New South Wales, 2093.*

W.C. SUMMERFIELD, *Department of Mathematics, University of Newcastle, Newcastle, New South Wales, 2308.*

R.I. TANNER, *Department of Mechanical Engineering, Sydney University, Sydney, New South Wales, 2006.*

M.D. TEUBNER, *Department of Applied Mathematics, University of Adelaide, Adelaide, South Australia, 5000.*

G.D. TONG, *International Course in Hydraulic Engineering, 95 Oude Delft, Delft, The Netherlands.*

F.L. TRAJER, *Physical Research Section, Commonwealth Bureau of Meteorology, 150 Lonsdale Street, Melbourne, Victoria, 3000.*

G.J. WALKER, *Civil and Mechanical Engineering Department, University of Tasmania, Box 252C, G.P.O., Hobart, Tasmania, 7001.*

C.E. WALLINGTON, *School of Applied Science, Canberra College of Advanced Education, P.O. Box 381, Canberra, Australian Capital Territory, 2601.*

I.G. WALLIS, *338 Highett Street, Richmond, Victoria, 3121.*

P.J. WALSH, *Commonwealth Scientific and Industrial Research Organisation, Division of Building Research, Graham Road, Highett, Victoria, 3190.*

B. WEBBER, *Department of Applied Mathematics, University of Adelaide, Adelaide, South Australia, 5000.*

I.M. WHYTE, *Broken Hill Proprietary Company Limited, P.O. Box 1968, Newcastle, New South Wales, 2300.*

D.L. WILKINSON, *Water Research Laboratory, University of New South Wales, King Street, Manly Vale, New South Wales, 2093.*

B.J. WILLIAMS, *Centre for Environmental Studies, University of Melbourne, Parkville, Victoria, 3052.*

P.G. WILLIAMS, *Central Scientific Laboratory, State Electricity Commission of Victoria, Howard Street, Richmond, Victoria, 3121.*

D.H. WOOD, *Department of Aeronautics, Imperial College of London, London, England.*

M.F. YEO, *Department of Civil Engineering, University of Adelaide, Adelaide, South Australia, 5000.*

N.W. YOUNG, *Glaciology Section, Department of Meteorology, University of Melbourne, Parkville, Victoria, 3052.*